DYNAMICS OF STRUCTURES AND MACHINERY

Problems and Solutions

Gregory Szuladzinski

Analytical Service Co.

A WILEY-INTERSCIENCE PUBLICATION
JOHN WILEY & SONS
New York · Chichester · Brisbane · Toronto · Singapore

We hope that this book will be used in making decisions
about the design and operation of machines and structures. We have tried to state the facts and our opinions
correctly, clearly, and with their limitations. But because
of uncertainties inherent in the subject and the possibility
of errors we cannot assume any liability. We urge the
readers to spend time in verification commensurate with
the risk they will assume.

Copyright © 1982 by John Wiley & Sons, Inc.

All rights reserved. Published simultaneously in Canada.

Library of Congress Cataloging in Publication Data:

Szuladzinski, Gregory, 1940–
   Dynamics of structures and machinery.

  "A Wiley-Interscience publication."
  Includes Index.
   1. Structural dynamics—Problems, exercises,
etc. 2. Machinery, Dynamics of—Problems,
exercises, etc. I. Title.

TA654.S98   624.1′71′076   80-10487
ISBN 0-471-09027-1   AACR2

Printed in the United States of America

10 9 8 7 6 5 4 3 2 1

DYNAMICS OF STRUCTURES
AND MACHINERY

To Conrad and Michael

Preface

This book is essentially a collection of solved problems in structural dynamics. The theoretical basis, which is presented in a concise, yet fairly extensive manner, is intended to make this text a self-contained learning tool as well as an aid in formal studies. Although the text is written primarily for beginners, it is hoped that also an experienced dynamicist may find some new thoughts and approaches. Simplicity is the main purpose and the keynote of the book.

The idea of creating the book occurred to me as I realized how many practicing engineers as well as engineering students were frustrated by their first encounter with the subject. There was an apparent need for a text that would satisfy all the following requirements:

1. Address a broad range of topics in the field of structural dynamics.
2. Present the reader with a large number of solved problems.
3. Fulfill these tasks without burying the subject matter under many complicated mathematical expressions.
4. By a proper selection of problems and a comprehensive presentation of theory, provide a meaningful reference in the modern, computer-oriented environment.

The last item requires some elaboration. Owing to the development and availability of the general-purpose structural programs, any large analytical task of that nature is performed by computer nowadays. When the calculation must have some features that

such a program cannot provide, a special-purpose code is usually created. That tends to substantially limit the importance of manual calculations in general and of certain analytical methods in particular. Some of the situations in which manual work is and will remain useful in the foreseeable future are:

1. The preparatory phase of large-scale calculations where a dynamic model is being generated. At this stage prudent analysts make a number of checks to ensure that they neither miss something important nor incorporate too many unnecessary details, which always makes waste.
2. An indirect verification of computer-generated results by using a greatly simplified model or a calculation method. (This not only may explain some unbelievable results, but also helps to guard against hidden, gross errors.)
3. Getting some preliminary figures on the anticipated dynamic properties and response of a system that is in an early design stage and for which a full-scale computation is not very practical.

Thus it is quite clear that the only sensible options for today's engineer embarking on a hand calculation are the methods and approaches that are brief and relatively easy. Anything more mundane might as well be done by computer. This was the major criterion in the selection of problems and methods.

Although the book emphasizes manual calculations, it attempts, as much as possible, to follow the pattern of work of structural computer programs. The majority of problems are formulated in terms of discrete (or finite) elements, and matrix notation is used extensively in the description of multiple degree-of-freedom systems. The last chapter provides an extensive discussion of how to generate models for computer analysis depending on the nature of the problem.

Too much reliance on sophisticated mathematics in the presentation of the subject has been making structural dynamics unnecessarily difficult for its students. The meaning of some very complete, very general, and otherwise very correct formulas is, unfortunately, often difficult to grasp. This text avoids, as much as possible, confusing the reader with involved expressions. Also, explanations and physical reasoning are normally employed in place of formal derivations. It is believed that once students gain a sound understanding of the subject, they should have little difficulty in progressing to more intricate notation, if desired.

The book is written with the needs of two distinct groups of readers in mind: college students and engineers in industry. In the latter category those with a good knowledge of static stress/structural analysis, who want to broaden their horizons, will undoubtedly predominate. Since there is no emphasis on any particular type of structure, those involved in mechanical aspects of aeronautical, automotive, nuclear, and civil engineering, as well as those in general machine design, may equally benefit. From the point of view of traditional division of the subject matter into graduate and undergraduate levels, the text is a fair mix of both.

To make successful use of this book, the reader should be familiar with calculus, engineering mechanics, and the strength of materials. Some knowledge of matrix algebra can be helpful, but is not necessary, since one of the appendices summarizes the subject.

A number of people have contributed their time and effort to help shaping this book. Robert B. McCalley, Jr., Sukumar Ghosh, Robert Holman, and Richard W. Winslow have reviewed portions of the text. Verl A. Stanford has carefully checked much of the text and has made many valuable suggestions. And last, but not least, my wife Ela, as a student of the subject, went over the entire text in detail. I am deeply indebted to all these persons.

All comments and criticism regarding the contents of the book should be sent to me, P.O. Box 30, Bankstown 2200, Australia. They will be most gratefully received and answered, if possible.

Gregory Szuladzinski

Bankstown, Australia
October 1981

Contents

DYNAMICS OF STRUCTURES
AND MACHINERY

Introduction

This three-part book is organized into chapters, each consisting of a theoretical summary and problems. About a half of the problems are solved; the rest are given as exercises, with the answers provided near the end of the book. It is recommended that the reader tackle both problem groups concurrently, since they often complement each other.

The theory is presented briefly and the derivations are outlined only when they are deemed essential to an understanding of a particular formula or when a simple derivation, not available in the well-established literature, can be given. The reader who desires a deeper insight into theory of structural dynamics should review Refs. 2, 6, 13, and 19.

Because of the nature of the subject matter and the manner in which it is presented, there are many cross-references throughout the text. When we refer to a particular problem, we may be addressing either the problem statement itself or something in its solution or, as in the case of exercises, the answer to the problem.

The illustrations in the theoretical part are numbered sequentially within each chapter, (e.g., Fig. 6.1, 6.2, etc., in Chapter 6). The figures related to individual problems appear in conjunction with each problem itself and do not bear numbers; the cross-references to these items, however, are given by hyphenated numbers (e.g., "Fig. 6-19" indicates the illustration accompanying Prob. 6-19).

The English system of units is consistently used throughout. Length is given in inches (in.), time in seconds (s), and force or weight in pounds (Lb). Only on rare occasions is a unit different from these three employed.

Symbols and Abbreviations

a	Acceleration (in./s^2)		(Lb-s^2-in.)
a_{ij}	Flexibility coefficient (in./Lb)	k	Stiffness of translational spring (Lb/in.)
A	Area (in.^{2})	k_{gij}	Geometric stiffness coefficient
A	Amplitude	k_{ij}	Elastic stiffness coefficient
b	Constant in exponential fatigue law	$\tilde{k}_{ij}$	Apparent stiffness coefficient, $k_{ij} + k_{gij}$
c	Damping coefficient (Lb-s/in.)	K	Stiffness of angular spring (Lb-in./rad)
c	Speed of propagation (in./s)	L	Length (in.)
c_c	Critical damping coefficient, $2M\omega$	L	Sound pressure level
C	Torsional constant of cross section (in.^{4})	m	Distributed mass (Lb-s^2/in.^{2})
$\mathbf{C}$	Transformation matrix	m_t	Distributed twisting moment (Lb-in./in.)
d	Diameter (in.)	M	Concentrated mass (Lb-s^2/in.)
D	Energy dissipated in damping (Lb-in.)	M_t	Twisting moment (Lb-in.)
D	Plate bending stiffness parameter (Lb-in.)	n	Overload factor
D	Damage ratio in fatigue	$\tilde{n}$	Number of stress cycles in fatigue
e	Eccentricity, static shaft unbalance (in.)	N	Axial force (Lb)
E	Young's modulus of elasticity (Lb/in.^{2} = psi)	N_e	Euler buckling force, $\pi^2 EI/L^2$
$f(t)$	Function of time	p	Beam frequency parameter, $p^4 = \omega^2 m/(EI)$ (1/in.)
f	Frequency of forcing (Hz = cycle/s)		
f_n	Natural frequency (Hz)	P	Load
F_y, F_u	Yield strength, ultimate strength (psi)	P	Probability
F_f	Endurance limit (psi)	q	Distributed load (Lb/in.)
g	Acceleration of gravity (386 in./s^2)	Q_r	Generalized force associated with the rth direction
G	Modulus of elasticity in shear (Lb/in.^{2})		
G	Sound spectral density (psi^2/Hz)	r	Constant in exponential fatigue law
h	Height or thickness (in.)	r	Radius of curvature (in.)
$H(t)$	Unit step function	R	Force of resistance
i	Radius of inertia (in.)	s	Normal or principal coordinate
i	$\sqrt{-1}$, imaginary unity	s	Laplace transform variable
I	Second area moment (area moment of inertia) (in.^{4})	S	Area under a curve
		S	Impulse (Lb-s or Lb-s-in.)
I_0	Cross-sectional polar moment of inertia (in.^{4})	S	Spectral density
j	Directional coefficient	t	Time (s)
J	Mass moment of inertia, for a disc about its axis of revolution (Lb-s^2-in.)	T	Kinetic energy (Lb-in.)
		T	Period of forcing function (s)
J'	Mass moment of inertia of a disc about its diametral axis through the center of gravity	u	Displacement (in. or rad)
		U	Work performed by applied forces (Lb-in.)

v	Velocity (in./s)		τ	Natural period of vibrations (s)
V	Velocity following rebound (in./s)		ϕ	Angle of dynamic unbalance (rad)
V	Shear force (Lb)		$\mathbf{\Phi}$	Modal matrix
W	Weight (Lb)		$\vec{\mathbf{\Phi}}_r$	r th modal vector
x, y, z	Coordinates (in.)		ψ	Coefficient of friction
X	Function of spatial variable x		ψ	Shear deflection coefficient
Z_a	Spectral response acceleration (in./s^2)		ω	Natural frequency (rad/s)
α	Angle of rotation (rad)		ω_d	Damped natural frequency (rad/s)
γ	Specific weight or weight density (Lb/in.^{3})		Ω	Forcing frequency (rad/s)
Γ_r	Participation factor of r th mode		Ω	Angular shaft velocity (rad/s)
δ	Displacement (in.)		$\mathcal{M}$	Bending moment (Lb-in.)
ε	Angular acceleration (rad/s^2)		$\mathfrak{p}$	Probability density
ζ	Damping ratio			
ϑ	Shear angle			
θ	Phase angle			
κ	Coefficient of restitution			
λ	Angular velocity (rad/s)			
Λ	Half-wave length (in.)			
μ	Dynamic magnification factor			
ν	Poisson's ratio			
Π	Strain energy (Lb-in.)			
ρ	Specific mass or mass density (Lb-s^2/in.^{4})			
σ	Direct stress (psi)			
σ	Standard deviation			
τ	Shear stress (psi)			

Abbreviations

DOF	degree of freedom
SDOF	single degree of freedom
MDOF	multiple degrees of freedom
CG	center of gravity
RMS	root-mean-square
RSS	root-sum-square (square root of sum of squares)
ub	upper bound
log	logarithm to the base e

PART

I

SINGLE DEGREE-OF-FREEDOM SYSTEMS

Calculation of Natural Frequency

Basic concepts relating to periodic motion are summarized. The natural frequencies of various simple systems are calculated using the direct method or the energy method. Translational and rotational systems are handled without much distinction between the two, since the analogy is clearly outlined. Some basic cases of geometric stiffening (the effect of prior loading on stiffness of a slender element) and its effect on vibratory frequency are illustrated. The latter topic is treated more fully in Chapter 16.

1A. Stiffness and Natural Frequency. Figure 1.1 presents two simple systems capable of vibratory motion. In Fig. 1.1a mass M is attached to the ground by a spring of stiffness k (Lb/in.). In Fig. 1.1b a disc with mass moment of inertia J is connected to the ground by a shaft of stiffness K (Lb-in./rad). The unit for M is Lb-s^2/in. and for J it is Lb-s^2-in. The terms *stiffness*, *rigidity*, and *spring constant* can be used interchangeably when describing a resistance capability of an elastic member. The reciprocal of stiffness is called *flexibility*, although the term *compliance* is also used by some authors. For translational motion, stiffness is defined as a force that causes a unit (1 in.) translation. Flexibility is a displacement caused by a unit (1 Lb) force. In angular motion the definitions are analogous, with "moment" replacing "force" and "rotation" instead of "translation."

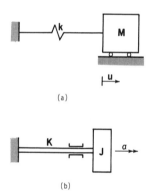

(a)

(b)

Figure 1.1 (*a*) Translational oscillator; (*b*) angular oscillator.

In this chapter we are concerned only with free vibrations, which are characterized by the absence of external forces, except perhaps that of gravity. According to the theory presented in Chapter 2, the equation of motion of an undamped translational system in Fig. 1.1a is

$$M\ddot{u} + ku = 0$$

The position of a vibrating mass with respect to some reference point is given by a function of time,

$$u = A\cos(\omega t - \beta) \qquad (1.1)$$

where A and β are constants that depend on the

initial conditions and the constant

$$\omega = \sqrt{\frac{k}{M}} \qquad (1.1a)$$

depends only on system properties. According to this equation mass M returns to the same point after every interval of time τ called the *period* of motion:

$$\tau = \frac{2\pi}{\omega} = 2\pi\sqrt{\frac{M}{k}} \qquad (1.2a)$$

The number of periods, or cycles of vibrations per second, is called the *frequency* or *natural frequency* and is designated by f:

$$f = \frac{1}{\tau} = \frac{\omega}{2\pi} \qquad s^{-1} \qquad (1.2b)$$

The unit of this frequency is one cycle per second, or hertz (Hz). The unit of ω is radian per second; ω is also referred to as the natural frequency of the system, but sometimes the term *circular frequency* is used to distinguish it from f. In translational vibration ω has little physical significance; it is used because it is convenient in the mathematical description of motion.

In Fig. 1.2a point B moves along a circular path with a constant speed, so that the radius OB has an angular velocity ω. Projecting point B on the vertical axis, we find its position determined by Eq. 1.1. Figure 1.2b, which is the plot of Eq. 1.1 as a function of time, shows the amplitude A and the natural period τ. One revolution of radius OB corresponds in this way to one vibratory cycle of motion of our actual elastic system.

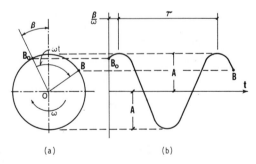

(a)　　　　　　　　　(b)

Figure 1.2　Harmonic vibration as a projection of circular motion.

The equation of motion of the rotational system in Fig. 1.1b is

$$J\ddot{\alpha} + K\alpha = 0 \qquad (1.3)$$

in which α is the angle of rotation measured from some reference position. This angle is shown as a straight-line vector throughout this book. (It is distinguished from translation by a double arrow. A curved arrow is used to represent rotation only if the plane of rotation is parallel to the plane of a drawing.) The solution is

$$\alpha = A\cos(\omega t - \beta) \qquad (1.4)$$

The circular frequency ω is not any more meaningful than it is for translation, but is used often for convenience.

$$\omega = \sqrt{\frac{K}{J}} = 2\pi f \qquad (1.5)$$

1B.　Analogy Between Translational and Rotational Motion is complete in every respect, as shown in Table 1.1.

TABLE 1.1
Analogy Between Translational Vibration (1) and Rotational Vibration (2)

Quantity	Symbol	Unit
(1) Mass	M	Lb-s^2/in.
(2) Mass moment of inertia	J	Lb-s^2-in.
(1) Force	P	Lb
(2) Torque	M_t	Lb-in.
(1) Displacement	u	in.
(2) Rotation	α	rad
(1) Velocity	$v = \dot{u}$	in./s
(2) Angular velocity	$\lambda = \dot{\alpha}$	rad/s
(1) Acceleration	$a = \dot{v} = \ddot{u}$	in./s^2
(2) Angular acceleration	$\varepsilon = \dot{\lambda} = \ddot{\alpha}$	rad/s^2
(1) Stiffness	k	Lb/in.
(2) Rotational stiffness	K	Lb-in./rad
(1) Momentum	Mv	Lb-s
(2) Angular momentum	$J\lambda$	Lb-s-in.
(1) Kinetic energy	$\frac{1}{2}Mv^2$	Lb-in.
(2) Kinetic energy	$\frac{1}{2}J\lambda^2$	Lb-in.
(1) Strain energy	$\frac{1}{2}ku^2$	Lb-in.
(2) Strain energy	$\frac{1}{2}K\alpha^2$	Lb-in.

1C. Determination of Natural Frequency by Direct Method. Stiffness k is defined as a force that is needed to move the mass by one inch in the direction of motion. This parameter is used, together with the value for a mass, in Eq. 1.1a. Quite often the weight $W = Mg$, with $g = 386$ in./s², is used instead of mass, because W is measured in pounds (Lb), a more popular unit than that of mass. The use of the symbol W is not intended to imply that the system is subjected to the force of gravity. (If gravity is to be considered, a statement is made to that effect.) For a rotational system, Eq. 1.5 is used, with the angular quantities replacing the translational ones.

1D. Determination of Natural Frequency by Energy Method. The principle of energy conservation says that if there are no damping forces and no external forces involved (beginning at some time point) we have

$$\Pi + T = C \qquad (1.6a)$$

in which C is a constant and Π is the energy of deformation or strain energy. This is the basis for the determination of natural frequency by the energy method. For a spring in Fig. 1.1a the energy of deformation is $\Pi = \frac{1}{2}ku^2$, where k is the spring rigidity and u is the deflection from the unstrained position. The kinetic energy T is $\frac{1}{2}mv^2$. The incremental form of Eq. 1.6a is

$$\Delta\Pi + \Delta T = 0 \qquad (1.6b)$$

which means that there is no net increase of mechanical energy. If gravity forces are involved, we have in place of Eq. 1.6b:

$$\Delta\Pi + \Delta T = U_g \qquad (1.7)$$

That is, if we consider the system moving from position 1 to position 2, the sum of increases of strain energy and of kinetic energy is equal to the work performed by forces of gravity U_g:

$$U_g = W(z_1 - z_2)$$

in which W is the weight of a body under consideration. The terms z_1 and z_2 denote the locations of center of gravity. The value of U_g is positive if $z_1 > z_2$ (i.e., the CG moves from a higher to a lower position). If no gravity force is involved, we have a vibratory motion that takes place between two stationary (zero velocity) positions, for which the kinetic energy is

zero. Equation 1.6a has two positive functions on the left side and if one becomes zero, the other must attain its maximum. Thus for any stationary point we have $\Pi_{max} = C$. Midway between those extreme points, there is a location for which no strain exists (neutral position) in the elastic elements, and then $T_{max} = C$. Comparing the last two equalities gives

$$\Pi_{max} = T_{max} \qquad (1.8)$$

which is a restatement of Eq. 1.6a. Once a particular form of Eq. 1.8 is known, it is sufficient to use the relation $v_{max}^2 = \omega^2 u_{max}^2$ between maximum velocity and maximum displacement to determine ω. (Refer to Chapter 2.)

1E. Types of Structural Arrangement. There are a variety of structural arrangements, that is, ways in which the elements may be connected to each other. The two cases that are found most often are called *in series* and *in parallel*. In the first one the load is the same in all members and therefore the deflections (flexibilities) are additive. The resultant flexibility is thus

$$\frac{1}{k^*} = \frac{1}{k_1} + \frac{1}{k_2} + \cdots + \frac{1}{k_n} \qquad (1.9)$$

with the summation extended to all n members of the system. In a parallel connection the displacement of all members is the same and their loads (rigidities) are additive. The resultant stiffness of a parallel connection is

$$k^* = k_1 + k_2 + \cdots + k_n \qquad (1.10)$$

As the solved problems show, in a series connection the structural elements are often placed along one continuous line, while in a parallel connection they are located side by side. The appearance, however, may sometimes be misleading and the definitions given above must always be kept in mind.

1F. Geometric Stiffening. There is an interesting effect associated with slender structural elements, namely, elements that have one dimension considerably smaller than the other two. If loading of a member of this type is carried out in two steps, the magnitude of deflections in the second step may depend on the magnitude of loads in the first. This effect is called *geometric stiffening* and is best illustrated by the behavior of a rod that has one end pinned and the other end free. As long as the element has no axial loading, it can be rotated freely. If, however, a stretching load is applied first and an attempt is then made to deflect the free end in the direction perpendicular to the axis, a resistance is encountered. This axial force that has an effect on lateral deflections is called *preload* or *prestress*.

SOLVED PROBLEMS

1-1 Find the frequency of vibration when a body of weight $W = 100$ Lb is attached to a base with two springs in series (a) and two springs in parallel (b). Assume that $k_1 = k_2 = 1000$ Lb/in., and only motion along the x-axis is possible.

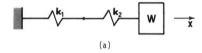

(a)

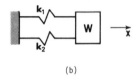

(b)

1. When load P is applied to the weight in (a), each spring is subject to the same force. This is a series connection and the individual deflections are additive.

$$u_1 = \frac{P}{k_1} \qquad u_2 = \frac{P}{k_2}$$

$$u = u_1 + u_2 = P\left(\frac{1}{k_1} + \frac{1}{k_2}\right) \equiv P\frac{1}{k^*}$$

flexibility = sum of component flexibilities

$$\frac{1}{k^*} = \frac{1}{1000} + \frac{1}{1000} \qquad \therefore k^* = 500$$

$$\omega^2 = \frac{gk^*}{W} = \frac{386 \times 500}{100} \qquad \therefore \omega = 43.93 \text{ rad/s}$$

2. Displace the weight in (b) by the distance u, which is at the same time the deflection of each spring. This is a parallel connection, in which the forces are additive.

$$N_1 = k_1 u \qquad N_2 = k_2 u$$

$$N = N_1 + N_2 = (k_1 + k_2)u \equiv k^* u$$

resultant stiffness k^* = sum of component stiffnesses

$$k^* = k_1 + k_2 = 2000 \text{ Lb/in.}$$

$$\omega^2 = \frac{gk}{W} = \frac{386 \times 2000}{100} \qquad \therefore \omega = 87.86 \text{ rad/s}$$

1-2 One of the blocks has a weight W, the other may be considered weightless. What is the frequency for the given arrangement of springs?

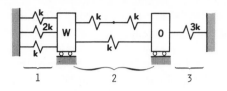

Divide the springs into three groups as shown.

1. $k_1^* = k + 2k + k = 4k$
2. For two springs in series $\dfrac{1}{k_s^*} = \dfrac{1}{k} + \dfrac{1}{k}$

$$\therefore k_s^* = k/2. \text{ Altogether:}$$

$$k_2^* = k_s^* + k = 1.5k$$

3. $k_3^* = 3k$

Since the block on the right has no mass, it also has no inertia forces during free vibrations and serves merely as a connector of the springs k_2^* and k_3^*. The latter assemblies are in effect joined in series, and their effective stiffness k_4^* is:

$$\frac{1}{k_4^*} = \frac{1}{k_2^*} + \frac{1}{k_3^*} = \frac{1}{1.5k} + \frac{1}{3k} = \frac{1}{k}$$

This assembly works in parallel with k_1^*. The resultant for the whole system is

$$k^* = k_1^* + k_4^* = 4k + k = 5k$$

$$\therefore \omega = \left(\frac{5kg}{W}\right)^{1/2}$$

1-3 Two beams, each with a rectangular cross section of width $b = 3$ in. and height $h = 2$ in. are clamped at one end and pin-joined to a rigid body with weight $W = 100$ Lb, $E = 30 \times 10^6$ psi, $L = 50$ in. Calculate the frequency of motion in the vertical direction and also the change in frequency if the pin joints are replaced by rigid, clamped connections.

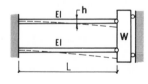

For each beam $I = bh^3/12 = 3 \times 2^3/12 = 2$ in.[4] The relation between the tip displacement u and tip load P is

$$u = \frac{PL^3}{3EI}$$

therefore

$$k = \frac{P}{u} = \frac{3EI}{L^3} = \frac{3 \times 30 \times 10^6 \times 2}{50^3} = 1440 \text{ Lb/in.}$$

The beams are connected in parallel,

$$k^* = 2k = 2880 \text{ Lb/in.}$$

$$\text{frequency } \omega_1 = \left(\frac{gk^*}{W} \right)^{1/2} = \left(\frac{386 \times 2880}{100} \right)^{1/2} = 105.4 \text{ rad/s}$$

If the connection at the right end does not permit rotation of that end of the beam, we can deduce from symmetry that each beam consists of two cantilevers $L/2$ long, connected in series. This increases the stiffness four times and the frequency twice:

$$\omega_2 = 2\omega_1 = 210.9 \text{ rad/s}$$

This example illustrates a general principle that stiffening of a system increases its natural frequency.

1-4 The beam shown in the figure is an aluminum tube with 2 in. outer diameter and wall thickness 0.058 in. The weight at the end is 50 Lb and the tube may be treated as weightless. Find the natural frequency when $E = 10.4 \times 10^6$ psi, $L = 20$ in., and $k = 2000$ Lb/in.

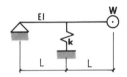

The effective stiffness of the system will be found by applying the force of gravity to calculate deflections and adding flexibilities of the components. First assume the beam to be rigid. The spring force is $2W$ and the tip deflection is

$$u_1 = \frac{2W}{k} \frac{2L}{L} = \frac{4W}{k}$$

Spring flexibility is

$$\frac{u_1}{W} = \frac{1}{k_1} = \frac{4}{k} = \frac{4}{2000} = 0.002 \text{ in./Lb}$$

Now treat the spring as rigid. Using the unit load method (or any other one, for that matter), one can find the tip deflection as $u_2 = 2WL^3/(3EI)$. For a thin-wall tube of given dimensions:

$$I = \pi r^3 t = \pi \left(\frac{2.0 - 0.058}{2} \right)^3 0.058 = 0.1668 \text{ in.}^4$$

and flexibility

$$\frac{1}{k_2} = \frac{2L^3}{3EI} = \frac{2 \times 20^3}{3 \times 10.4 \times 10^6 \times 0.1668} = 0.0031 \text{ in./Lb}$$

the effective stiffness is found from

$$\frac{1}{k^*} = \frac{1}{k_1} + \frac{1}{k_2} = 0.0051 \text{ in./Lb}$$

while frequency is

$$\omega^2 = \frac{k^* g}{W} = \frac{386}{0.0051 \times 50}; \qquad \therefore f = \frac{\omega}{2\pi} = 6.19 \text{ Hz}$$

1-5 Two masses M_1 and M_2 rest on a frictionless surface and are connected by an elastic rod of length L. This is a two degree-of-freedom system, since the movements of the masses are independent of each other to the extent permitted by the rod. Since the system is unconstrained (i.e., free to slide), one may think of its motion as the sum of two components:

1. Both masses move together in the same direction without changing their relative distance. This is called a "rigid-body mode" of motion, because this would be the only component of motion in case of a rigid rod.
2. Both masses move toward or away from each other, but their total momentum is zero, which is the vibratory mode.

This indicates that in spite of two separate masses, there is only one frequency of vibration of such a system. What is that frequency? (Stiffness constant EA is given.)

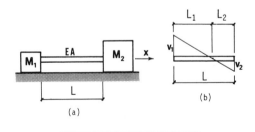

The key to solution is the statement that total momentum of the system is zero when in vibratory motion:

$$M_1 v_1 + M_2 v_2 = 0 \qquad \text{or} \qquad v_2 = -\frac{M_1 v_1}{M_2}$$

This means that if v_1 is positive, v_2 is negative. The plot of velocity distribution in (b) was made for $v_1 = 1.0$ and $M_2/M_1 = 2$. It shows there is a point of zero velocity, called a nodal point at a distance L_1 from mass M_1. If some other value for v_1 is assumed, the plot ordinates will change, but the nodal point will remain at the same location. From similar triangles we have

$$\frac{L_1}{L} = \frac{|v_1|}{|v_1| + |v_2|} = \frac{1}{1 + (M_1/M_2)}$$

If a point remains stationary during vibration, the situation is the same as if mass M_1 were attached to a rod of length L_1 fixed at the other end.

$$\omega^2 = \frac{k_1}{M_1} = \frac{EA}{L_1 M_1} = \frac{EA}{L} \left(\frac{1}{M_1} + \frac{1}{M_2} \right)$$

It may easily be checked that mass M_2 attached to the rod of length L_2 has the same frequency. (This is the expected result; otherwise vibratory motion would not take place.)

When the equation of momentum given at the outset is differentiated, the velocities are replaced by accelerations:

$$M_1 \dot{v}_1 + M_2 \dot{v}_2 = 0$$

A product of acceleration and mass is the inertia force with an opposite sign. This result indicates another property of the system, namely, that the sum of inertia forces is zero at any time.

1-6 Weight W, which can slide along the x-axis, is restrained by six symmetrically placed, pin-ended rods, each with axial stiffness EA/L. What is the frequency of vibration?

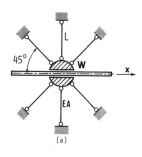

(a)

Each of six rod ends attached to the weight is made to move parallel to the x-axis. The first task is to find the resistance of one such rod shown in (b). End displacement u can be resolved into a component aligned with the rod axis, $u \cos \alpha$, and perpendicular to it, $u \sin \alpha$. The first component causes stretching with force

$$R = \frac{EA}{L} u \cos \alpha$$

while the second has no resistance associated with it, at least from the small deflection viewpoint. The resistance force in the x-direction is obtained when the axial force is projected onto x:

$$R_x = \frac{EA}{L} u \cos^2 \alpha$$

The stiffness associated with the x-axis is therefore $R_x/u = (EA/L)\cos^2 \alpha$ for a single rod. When the rigidities of all rods are added, the effective stiffness is the result. Because of symmetry, we have the contribution of one side multiplied by 2:

$$k^* = 2\frac{EA}{L} \left[\cos^2 45° + \cos^2 90° + \cos^2(-45°) \right] = 2\frac{EA}{L}$$

and then $\omega^2 = 2gEA/WL$. The vertical bars to not contribute to the overall stiffness and can be removed from the structure.

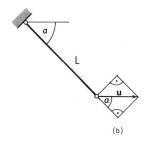

(b)

1-7 Weight W is restrained by a spring with stiffness k. Using the energy method, find the natural frequency of vibrations if (1) the weight is sliding on a frictionless, horizontal surface, and (2) if the weight is suspended vertically and subjected to the force of gravity.

For brevity we shall use the index "m" instead of "max."

1. In either extreme position the strain energy is $\Pi_m = \frac{1}{2}ku_m^2$. At the instant when the neutral point is passed by the weight, the kinetic energy is $T_m = \frac{1}{2}(W/g)v_m^2$. Using the relation between the amplitudes of displacement and velocity, $v_m^2 = \omega^2 u_m^2$, we get

$$\omega^2 = \frac{kg}{W}$$

2. The figure shows three positions of the vibrating weight: the upper extreme, the lower extreme, and the midpoint $u_2 = (u_1 + u_3)/2$. Assuming that in all three positions the spring is in tension, the spring force N is plotted as a function of u. Velocity is zero at both extreme locations, and consequently the kinetic energy is zero. The increase of strain energy when going from the top to the bottom position is due to work of gravity forces, as expressed by Eq. 1.7,

$$\frac{1}{2}ku_3^2 - \frac{1}{2}ku_1^2 = W(u_3 - u_1)$$

Rearranging terms:

$$\frac{1}{2}k(u_3 - u_1)(u_3 + u_1) = W(u_3 - u_1)$$

$$\therefore ku_2 = W, \quad \text{as} \quad u_2 = \frac{1}{2}(u_3 + u_1)$$

This tells us that the midpoint is that of static equilibrium. A significant simplification results if this position is treated as a reference level for the motion. Instead of considering the spring force as depicted by DE and IK at the extremum positions in the figure, we simply concern ourselves with the relative changes, segments EF and JK. As far as the energy balance is concerned, this is equivalent to removing the gravity forces from the picture. Conclusion: the forces of gravity may be ignored in calculating the natural frequency. This is true only when the effect of weight on an elastic component remains unchanged during motion. In the case of a pendulum, where the effective component of weight increases with the increasing deflection, the weight must be included in the analysis. These conclusions indicate that the frequency in Case (2) may be found in exactly the same manner as in (1), except that the strain energy is not a true total, but is calculated with respect to the equilibrium position. It is left to the student as an exercise to show that the reasoning holds true even if there is some resultant spring compression in the upper extreme position.

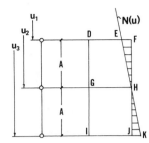

1-8 The aluminum wire with cross section $A=0.01$ in.^{2}, length $L=10$ in., and modulus $E=10\times10^6$ psi has a weight $W=1$ Lb attached to its end. The weight can move laterally in a channel and owing to a special design, the friction is so small that it can be ignored. Before the upper end was fixed, the wire was stretched so that a prestress of $\sigma_0=30,000$ psi was induced. Assuming that the wire is weightless, calculate the natural frequency. Compare the result with the frequency of axial vibrations if such were allowed.

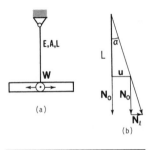

The wire displacement from the vertical position is shown in (b). The axial stretching force, which is initially denoted by N_0, grows in proportion to length. In the displaced position the force in the wire must be aligned with the direction of the wire and may be treated as a vector sum of N_0 and some additional force N_l. From the similarity of triangles we have

$$\frac{N_l}{N_0}=\frac{u}{L}\qquad\therefore\ N_l=\frac{N_0}{L}u$$

This means that to move the end of a prestressed wire perpendicular to the direction of the wire, a force of magnitude N_l is needed. This is an important difference in comparison with an unloaded axial member, which offers no resistance in that direction. We say that as a result of prestressing, the wire has geometric stiffness in addition to its elastic stiffness. The magnitude of that geometric or lateral stiffness is $k_l=N_0/L$, and it is treated in exactly the same way as the axial stiffness. Consequently, we calculate the frequency of the lateral vibrations as

$$\omega^2=\frac{gk_l}{W}$$

$$k_l=\frac{N_0}{L}=\frac{A\sigma_0}{L}=\frac{0.01\times30,000}{10}=30\ \text{Lb/in.}$$

$$\omega^2=\frac{386\times30}{1.0}\qquad\therefore\ \omega=107.6\ \text{rad/s}$$

The axial stiffness is $k_a=EA/L=10\times10^6\times0.01/10=10,000$ Lb/in., and if axial vibrations were allowed, the frequency would be

$$\omega^2=\frac{386\times10,000}{1.0}\qquad\omega=1965\ \text{rad/s}$$

In spite of a sizable initial stress, the lateral stiffness is considerably less than the axial value. This conclusion is valid for simple members like rods and wires, and need not be true for others (e.g., coil springs).

1-9 A spring has axial stiffness k_a and a lateral stiffness k_l, which was acquired as a result of preloading. One end of the spring is attached and the other forced to move along a line that makes an angle β with the spring axis. What is the effective spring stiffness along the line of movement? (*Hint.* Review Probs. 1-6 and 1-8).

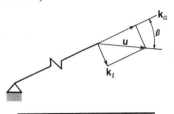

A displacement by u along the specified line is equivalent to moving by $u\cos\beta$ along the axis of the spring and by $u\sin\beta$ perpendicular to it. To enforce these deflection components, we must apply forces

$$k_a u\cos\beta\qquad\text{and}\qquad k_l u\sin\beta$$

along the respective directions. The resultant that is applied in the direction of u is found by projecting the force components above on u:

$$P=k_a u\cos^2\beta+k_l u\sin^2\beta$$

Our stiffness along u is defined as P/u, therefore

$$k_u=k_a\cos^2\beta+k_l\sin^2\beta$$

1-10 A rigid block of weight W is held by a cable, whose length is h in the position of equilibrium and axial stiffness is k. What is the natural frequency of angular motion about point A? Assume the weight of block to be uniformly distributed along its length. (H and L are also given.)

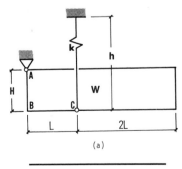

Notice that the end of cable at point C has a vertical as well as a horizontal component of movement. This means that the geometrical stiffness will also be involved. The resolution of displacement due to rotation α and spring reactions exerted on the block is shown in (b). The components of spring reaction shown are those due only to an incremental rotation by angle α. Gravity is ignored, except that it preloads the cable with force N_0. The latter can be found from the balance of moments about point A in the equilibrium position:

$$1.5LW = N_0 L \quad \text{or} \quad N_0 = 1.5W$$

The moment of resistance due to rotation α is found from (b):

$$M_r = (kL\alpha)L + \left(\frac{N_0}{h}H\alpha\right)H = K^*\alpha$$

Denoting by J the moment of inertia about point A,

$$\omega^2 = \frac{K^*}{J} = \frac{\left(kL^2 + 1.5(W/h)H^2\right)}{J}$$

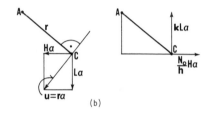

(b)

1-11 In (a) a disc is attached to the end of a shaft consisting of two segments. In (b) the same disc is placed at the transition point and the shaft is fixed at both ends. The data are as follows: $d_1 = 6$ in., $d_2 = 4$ in., $L_1 = 20$ in., $L_2 = 8$ in., $G = 11 \times 10^6$ psi, and $J = 20$ Lb-s^2-in. (mass moment of inertia). Find the natural frequencies of twisting vibration. The shafts are to be treated as weightless.

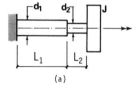

(a)

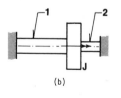

(b)

The problem is analogous to Prob.1-1.

(a) When a twisting moment M_t is applied to the disc in (a) each shaft segment is subjected to the same moment. This is a series connection in which the angles of twist are additive.

$$\alpha_1 = \frac{M_t}{K_1}; \qquad \alpha_2 = \frac{M_t}{K_2};$$

$$\alpha = \alpha_1 + \alpha_2 = M_t\left(\frac{1}{K_1} + \frac{1}{K_2}\right) = M_t\frac{1}{K^*}$$

flexibility = sum of component flexibilities

$$K = \frac{M_t}{\alpha} = \frac{GC}{L} = \frac{\pi G}{32}\frac{d^4}{L}$$

where $C = \pi d^4/32$ is the torsional constant

$$K_1 = \frac{\pi \times 11 \times 10^6}{32} \times \frac{6^4}{20} = 69.98 \times 10^6;$$

$$K_2 = 34.56 \times 10^6 \text{ Lb-in./rad}$$

$$\frac{1}{K^*} = \left(\frac{1}{69.98} + \frac{1}{34.56}\right)\frac{1}{10^6} \qquad \therefore K^* = 23.13 \times 10^6$$

$$\omega^2 = \frac{K^*}{J} = \frac{23.13 \times 10^6}{20} \qquad \therefore \omega = 1075 \text{ rad/s}$$

(b) Rotate the disc in (b) by α, which is also the angle of twist of each shaft. This is a parallel connection in which the resistance moments are additive.

$$M_{t1} = K_1\alpha; \qquad M_{t2} = K_2\alpha$$

$$M_t = M_{t1} + M_{t2} = (K_1 + K_2)\alpha = K^*\alpha$$

stiffness = sum of component stiffnesses

$$K^* = (69.98 + 34.56)10^6 = 104.54 \times 10^6 \text{ Lb-in./rad}$$

$$\omega^2 = \frac{K^*}{J} = \frac{104.54 \times 10^6}{20} \qquad \therefore \omega = 2286.3 \text{ rad/s}$$

1-12 In (a) a material point with weight W is suspended at the end of a rigid, weightless rod of length L and subject to the force of gravity. When displaced from the equilibrium position, the system will oscillate with frequency ω. In (b) there is a similar system placed on a horizontal, frictionless table, viewed from the top. Determine what the stiffness K of the angular spring should be so that the natural frequency of both systems is the same.

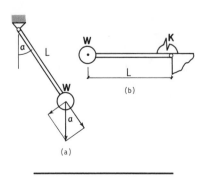

(a)

When the rod in (a) is deflected, the force of gravity W can be resolved into a component along the rod and one perpendicular to it. The latter gives a moment about the support point

$$WL \sin \alpha \approx WL\alpha$$

for small angles. In case (b) the restraining moment is $K\alpha$. A comparison of both moments leads to the conclusion that the presence of gravity is equivalent to the angular stiffness

$$K = WL$$

If a spring of stiffness WL is used in case (b), the natural frequencies will be the same because the moments of inertia are identical, $J = (W/g)L^2$. The frequency is

$$\omega^2 = \frac{K}{J} = \frac{WL}{WL^2}g = \frac{g}{L} \quad \text{or} \quad \omega = \sqrt{g/L}$$

which is the familiar result for the mathematical pendulum.

1-13 Point mass M is placed on top of a weightless rod of length L and restrained by two springs of stiffness k each (a). The pendulum in (b) is the same, except the point of support is above, rather than below the mass. At neutral position the springs are unstretched. Using the energy method, calculate the natural frequency in both cases including the effect of gravity.

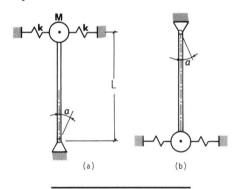

(a)   (b)

At the neutral position the vibrating pendulum in (a) has maximum kinetic energy. Using m instead of "max" for brevity, we have

$$T_m = \tfrac{1}{2}J\lambda_m^2 = \tfrac{1}{2}ML^2\lambda_m^2$$

while the strain energy is zero. At the lowest position the strain energy is

$$\Pi_m = 2\left(\tfrac{1}{2}ku_m^2\right) = k\left(\alpha_m L\right)^2$$

The work of gravity when going from the upper to the lower level is given by

$$U_g = Mg(L - L\cos\alpha)$$

Approximating $\cos\alpha_m$ by $1 - \alpha_m^2/2$ gives

$$U_g = \tfrac{1}{2}MgL\alpha_m^2$$

At the lower level the kinetic energy is zero. Equation 1.7 is now employed with respect to an energy increase when going from the upper to the lower level:

$$\left(k\alpha_m^2 L^2 - 0\right) + \left(0 - \tfrac{1}{2}ML^2\lambda_m^2\right) = \tfrac{1}{2}MgL\alpha_m^2$$

Using $\lambda_m^2 = \omega^2\alpha_m^2$ and rearranging terms:

$$\omega^2 = \frac{2k}{M} - \frac{g}{L}$$

In case (b) the derivation is analogous, with one important difference: when going from the neutral position to the extreme deflected position, the left side of the equation remains the same, but the right side changes sign. This is because the second position is above the first and the gravity performs negative work. Consequently:

$$\omega^2 = \frac{2k}{M} + \frac{g}{L}$$

The effect of turning the system upside down is therefore to increase the natural frequency. This is because in (a) the force of gravity helps to increase any deflection from the natural position, while in (b) it has the opposite effect.

1-14 A pulley of mass M and moment of inertia J is suspended on an elastic cable, which will not slacken during motion owing to the gravity force exerted on mass M. Calculate natural frequencies of the two types of motion: (1) vertical translation and (2) rotation about center. Treat the deformability of the cable in a simplified way by including the projected length of the portion that is in contact with the pulley. (The total effective length per side is L for both cases.) Assume that the pulley will not slip with respect to the cable. (A, E, and r are given.)

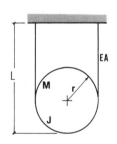

The assumptions of the problem permit one to think of the pulley as suspended by two springs, each with $k = EA/L$.

1. The effective rigidity is

$$k^* = \frac{2EA}{L}$$

and

$$\omega^2 = \frac{k^*}{M} = \frac{2EA}{ML}$$

2. A pure rotation by angle α about the center gives the moment of resistance $\mathfrak{M}=2(kr\alpha)r$. (One side of the cable is in apparent compression, which actually is only a decrease of the initial tension. This tension keeps the cable from slackening.) By definition, this moment is $\mathfrak{M}=K^*\alpha$, which gives $K^*=2kr^2$ and

$$\omega^2 = \frac{2kr^2}{J} = \frac{2EAr^2}{JL}$$

1-15 A rigid beam with mass M at one end can pivot about the other end. It is supported by three equally spaced springs. What is the natural frequency? (k and L are given.)

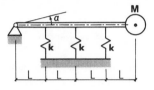

When the beam is rotated by α, the forces induced in the springs are $kL\alpha$, $2kL\alpha$, and $3kL\alpha$, respectively. This gives the total resistance moment

$$\mathfrak{M}_r = (kL\alpha)L + (2kL\alpha)(2L) + (3kL\alpha)(3L) = 14kL^2\alpha.$$

Also, $\mathfrak{M}_r = K^*\alpha$, therefore $K^* = 14kL^2$.

$$J = (4L)^2 M = 16L^2 M$$

$$\therefore \omega^2 = \frac{K^*}{J} = \frac{14k}{16M}.$$

1-16 A disc rotating with angular velocity λ has a small pendulum attached to it at point B. The entire mass M of the pendulum is lumped at its free end. Treating R, r, λ, and M as known values, calculate the natural frequency ω of the pendulum.

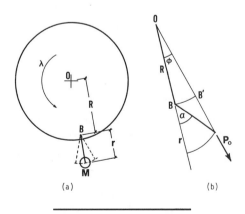

(a)   (b)

From (b) we see that the centrifugal force P_0 acting on M is

$$P_0 \approx \lambda^2 M(R+r)$$

as long as the angle α remains small. The moment of this force about point B is

$$P_0 R\phi = \lambda^2 M(R+r)R\phi$$

Noting that $(R+r)\phi = r\alpha$, we get

$$P_0 R\phi = \lambda^2 MRr\alpha$$

Let us replace the effect of the centrifugal force by a fictitious angular spring K at point B. Equating the applied moment with the moment of resistance $K\alpha$ gives us

$$K = \lambda^2 MRr$$

Moment of inertia about point B is $J = Mr^2$, therefore

$$\omega^2 = \frac{K}{J} = \lambda^2 \frac{R}{r}$$

$$\therefore \omega = \lambda \left(\frac{R}{r}\right)^{1/2}$$

EXERCISES

1-17 A body of mass $M=5$ Lb-s^2/in. is restrained in the x-direction on both sides by rods of a light material with Young's modulus $E=10\times10^6$ psi. The section areas are as indicated; $A=0.5$ in.^{2}; $L=4$ in. What is the period of vibration?

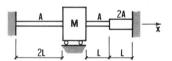

1-18 An assembly of weightless rods with section areas and lengths as shown is used to restrain the weight W. The modulus of elasticity is E. What is the natural frequency?

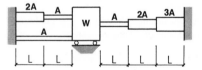

1-19 In (a) the weight W is suspended on a spring attached to the tip of the cantilever beam. In (b) the same weight is attached to the tip of the beam and supported by two springs. Given the data in the figure, what is the frequency in both cases?

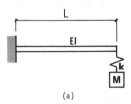

(a)

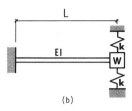

(b)

1-20 A frame consists of two elastic columns and a rigid horizontal beam. Calculate the natural frequency for the following: $h = 300$ in., $W = 50,000$ Lb, $E = 29 \times 10^6$ psi, and $I = 10,000$ in.[4]

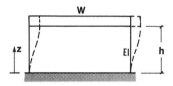

1-21 The cantilever beam is attached to the base with an angular spring of stiffness K and has mass M at the free end. (EI and L are given.) Assuming the beam to be weightless, calculate the natural frequency.

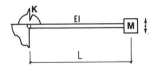

1-22 Find the frequency of vibratory motion if two masses of Prob. 1-5 are connected by three springs in series. $M_1 = 1$, $M_2 = 5$ Lb-s^2/in., $k_1 = 100$, $k_2 = 200$, $k_3 = 500$ Lb/in.

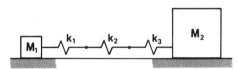

1-23 A rod with two end masses M_1 and M_2 can slide horizontally in a pair of guides. A vertical beam having moment of inertia J about its center is attached to the rod by a pin and an angular spring of stiffness K. The vertical beam is supported in a manner permitting its rotation and axial translation. Compute the natural frequency of the system.

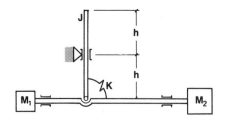

1-24 A rigid, symmetrical frame with sides of length a has mass M placed at the center and restrained by four springs. Each spring has an axial stiffness k_1 and is prestressed to an initial load N_0. Calculate the natural frequency if mass M is forced to vibrate (1) only along a horizontal line, (2) only along a diagonal, and (3) only in the direction normal to the plane of frame.

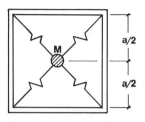

1-25 A rigid, hexagonal frame with length of side a has mass M placed at center and restrained by six springs attached to the corners. Each spring has axial stiffness k_1 and is prestressed to an initial load N_0. Calculate the natural frequency if mass M is forced to vibrate (1) only along a line formed by two springs, and (2) only normal to the plane of frame.

1-26 A gear with pitch radius $R_2 = 10$ in. and thickness $h = 4$ in. is made of steel with $\gamma = 0.284$ Lb-s^2/in.[4] The mating gears have $R_1 = 4$ in. Also, $d_1 = 4$ in., $d_2 = 6$ in., $L = 30$ in., and $G = 11 \times 10^6$ psi. Ignoring the weight of shafts and small gears, calculate the natural frequency.

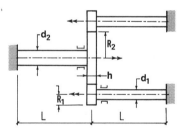

1-27 Two discs are connected with a shaft, which has a constant cross section in case (a) and consists of two segments in case (b). The system can perform only the twisting motion. Given the data shown in the figure, calculate the natural frequency of vibration as defined in Prob. 1-5 for each of the two cases.

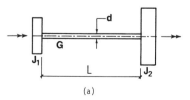

(a)

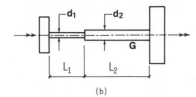

(b)

1-28 Two wheels, each with radius r and moment of inertia J about its center, have their axles connected by a rigid element. The system is on a slope with angle α and is attached to a fixed base by a spring k. If the mass of the system is M, what is its natural frequency? The arrangement is illustrated in (a), and (b) shows the virtual movement of the wheel.

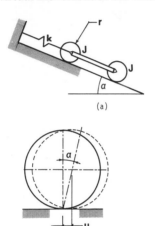

(a)

(b)

1-29 Two rigid links are pin joined in the middle of the span and pin supported at the ends. Each link has a spring attached at the center of its length. There also is an angular spring at the left support and a lumped mass at midpoint of the right link. Using the data shown in the figure, determine the frequency using the energy method.

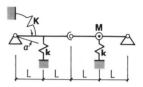

1-30 One end of a rod is attached to a wheel, which is free to rotate, and the other to a spring k. The moment of inertia of the rod about its CG is J_2, while that of the wheel is J_1. The rod is free to rotate and slide with respect to its support. Using all the data marked by symbols in the figure, derive the expression for the natural frequency of the system.

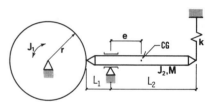

CHAPTER

2

Dynamic Response of Basic Oscillator

A simple system consisting of a mass restrained by an elastic element and a viscous element is the subject of investigation in this chapter. Since any structure with a single dynamic degree of freedom (SDOF) is reducible to such an oscillator, the devotion of an entire chapter to this topic is justified. Both the translational and the rotational versions are considered, either with damping or without. The solutions to the equation of motion are presented for periodic forcing (either harmonic or nonharmonic) and nonperiodic forcing. A number of problems are related to the use of Laplace transforms. More problems with response of SDOF systems other than oscillators are given in Chapter 7.

2A. Equations of Motion.

Any of two arrangements in Fig. 2.1 will be referred to as the basic oscillator. In Fig. 2.1a mass M is restrained by a spring of stiffness k Lb/in. and a viscous damper, which offers a resisting force $-c\dot{u}$, proportional to the magnitude of velocity $\dot{u}$, but opposite to its direction. Displacement u is measured from a position in which the spring is unstretched. Figure 2.1b illustrates the basic oscillator for angular motion. A disc with mass moment of inertia J is restrained by a shaft with rotational stiffness K Lb-in./rad. The blades attached to the shaft are in contact with a viscous medium, which provides a resisting moment $-c'\dot{\alpha}$.

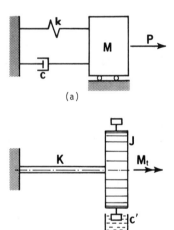

Figure 2.1 Basic oscillators with damping: (a) translational; (b) angular.

Newton's second law applied to the translational oscillator gives

$$M\ddot{u} + c\dot{u} + ku = P(t) \qquad (2.1)$$

Not only the displacement u, but also force $P(t)$ are, in general, functions of time. When $P(t)=0$, we have the so-called free motion and Eq. 2.1 may now be written as

$$\ddot{u} + 2\omega\zeta\dot{u} + \omega^2 u = 0 \qquad (2.2)$$

$$\omega^2 = \frac{k}{M}; \qquad \zeta = \frac{c}{c_c}; \qquad c_c = 2\sqrt{kM} = 2M\omega$$

19

For reasons explained later, that special value of the damping coefficient denoted by c_c is called the *critical damping*. The *damping ratio* ζ is another practical and convenient measure of damping. When $\zeta = 0$, we obtain the equation of free undamped motion:

$$\ddot{u} + \omega^2 u = 0 \qquad (2.3)$$

2B. Free, Undamped Motion. To solve either Eq. 2.2 or 2.3 for u, the initial values of velocity $\dot{u}(0) = v_0$ and displacement $u(0) = u_0$ must be prescribed. The solution of Eq. 2.3 is

$$u = u_0 \cos \omega t + \frac{v_0}{\omega} \sin \omega t \qquad (2.4)$$

or, after using some trigonometric identities,

$$u = A \cos(\omega t - \alpha) \qquad (2.5)$$

in which

$$A = \left[u_0^2 + \left(\frac{v_0}{\omega} \right)^2 \right]^{1/2} \quad \text{and} \quad \tan \alpha = \frac{v_0}{\omega u_0}$$

(Notice that there are two values of α differing by 180° and both giving the same $\tan \alpha$ in a 360° angular segment. Only one of those, however, will satisfy the initial conditions.) The motion represented by Eqs. 2.4 and 2.5 is periodic with the period $\tau = 2\pi/\omega$, which means that every τ seconds the mass begins to repeat its path. Successive differentiation of Eq. 2.5 yields velocity $v = \dot{u}$ and acceleration $a = \ddot{u}$:

$$\dot{u} = -A\omega \sin(\omega t - \alpha) \qquad (2.5a)$$

$$\ddot{u} = -A\omega^2 \cos(\omega t - \alpha) \qquad (2.5b)$$

The relationships between the maximum values are:

$$v_{max}^2 = \omega^2 u_{max}^2 \qquad (2.6a)$$

$$a_{max}^2 = \omega^4 u_{max}^2 \qquad (2.6b)$$

The graphical representation of this motion is given in Prob. 2-1. When the displacement is a sine or cosine function of time, as in Eq. 2.5, it is called *simple harmonic motion*.

A few words on how the initial displacement and/or velocity are imposed may be appropriate. Suppose we move the mass M in Fig. 2.1a from the neutral, unstrained position by a distance u_0 and release it to vibrate freely, setting $t = 0$ at the instant just before the release. This is what it means to impose an initial displacement. The initial velocity may be induced by impacting the mass and letting it vibrate afterward. The time $t = 0$ is chosen at the end of impact. If no initial displacement is to be introduced, the impact must last for such a short interval of time that theoretically no displacement of mass will occur during that impact.

2C. Forced, Undamped Motion. When there is a load P applied to an undamped translational system, Eq. 2.1 may be written as

$$\ddot{u} + \omega^2 u = \frac{P}{M} \qquad (2.7)$$

which is Eq. 2.3 with the forcing term on the right side. The solution has the form

$$u = B_1 \cos \omega t + B_2 \sin \omega t + u_p(t) \qquad (2.8)$$

while velocity and acceleration, obtained by differentiation, are

$$\dot{u} = -B_1 \omega \sin \omega t + B_2 \omega \cos \omega t + \dot{u}_p(t) \qquad (2.8a)$$

$$\ddot{u} = -B_1 \omega^2 \cos \omega t - B_2 \omega^2 \sin \omega t + \ddot{u}_p(t) \qquad (2.8b)$$

When the term *response* is used, it may mean any of the four variables: displacement, velocity, acceleration, or a spring force resulting from some external action. When an analytical expression for displacement is available, the spring force is obtained after multiplying that displacement by a spring constant. The two remaining variables result from differentiation of displacement.

B_1 and B_2 are the integration constants to be determined from the initial conditions. The first two terms in Eq. 2.8 are the same as a general solution of Eq. 2.3, while the additional function $u_p(t)$ is called the particular solution of Eq. 2.7. There are several methods of finding $u_p(t)$. The most common one is to assume that $u_p(t)$ has the same form as P/M, substitute the assumed expression in place of u in Eq. 2.7, and use the resulting identity to find the unknown coefficients.

Once the calculation is complete and u is known for any instant of time, one can divide the extreme deflection u_{max} by its value obtained from static application of the same force P, u_{st}. The ratio of the absolute values of the two is called the dynamic magnification factor:

$$\mu = \frac{|u|_{max}}{|u_{st}|}$$

If P changes with time, the peak value of P should be used for calculating u_{st}. For the mass-spring system, which we are considering at this moment, μ is the same for displacement as well as for the force in the spring.

2D. Free, Damped Motion. The form of solution of Eq. 2.2 depends on the magnitude of the damping ratio ζ. The most important case is that of small damping, that is, $\zeta < 1.0$ or $c < c_c$:

$$u = \left(u_0 \cos \omega_d t + \frac{v_0 + \omega \zeta u_0}{\omega_d} \sin \omega_d t \right) e^{-\omega \zeta t} \quad (2.9)$$

where $\omega = \left(\dfrac{k}{M} \right)^{1/2}$ and $\omega_d = \omega \left(1 - \zeta^2 \right)^{1/2}$

ω_d is called *damped circular frequency*.

The main difference between Eqs. 2.4 and 2.9 is the presence of the exponential term in the latter. Consequently, the amplitude of damped vibrations diminishes with time until it becomes insignificantly small. Equation 2.9 may also be written in a more compact form:

$$u = A e^{-\omega \zeta t} \cos(\omega_d t - \alpha_d) \quad (2.10)$$

in which

$$A^2 = u_0^2 + \frac{(v_0 + \omega \zeta u_0)^2}{\omega_d^2} \quad (2.10a)$$

$$\tan \alpha_d = \frac{v_0 + \omega \zeta u_0}{\omega_d u_0} \quad (2.10b)$$

As noted in Sect. 2B, some caution is needed in determining which of the two possible angles α will fit the initial conditions.

The cosine function is ± 1 at the extreme points. This means that the deflection described by Eq. 2.10 has the absolute value not larger than $A e^{-\omega \zeta t}$ for any given t. (See Prob. 2.2 for an illustration of this type of vibration.) The motion is not exactly periodic, because it does not repeat itself after each damped period $\tau_d = 2\pi / \omega_d$. Yet, it may be shown that not only zero points, but also the extreme points of $u(t)$ are attained after every τ_d. For this reason the term *pseudoperiodic* may be used to describe this type of oscillation.

Suppose that for a certain time point t_1 a maximum displacement u_1 is obtained. After another cycle of motion, at time $t_1 + \tau_d$, another maximum u_2 is

reached. The ratio of these maxima is, by Eq. 2.10:

$$\frac{u_1}{u_2} = \frac{e^{-\omega \zeta t_1}}{e^{-\omega \zeta (t_1 + \tau_d)}} = e^{\omega \zeta \tau_d} \quad (2.11a)$$

This means that the ratio of any two adjacent maximum deflections is always the same. One can easily show that after n cycles we have

$$\frac{u_1}{u_{1+n}} = e^{n \omega \zeta \tau_d} \equiv \exp(n \omega \zeta \tau_d) \quad (2.11b)$$

The term "ratio of amplitudes" may be used instead of "ratio of maximum displacements" for the sake of brevity only, since the amplitude is not a parameter of motion as it is in case of simple harmonic vibrations.

The logarithm of the displacement ratio in Eq. 2.11a is called *logarithmic decrement* Δ:

$$\Delta = \log \left(\frac{u_1}{u_2} \right) = \omega \zeta \tau_d \approx 2\pi \zeta \quad (2.12)$$

The near-equality sign was used in the expression above because in most practical cases the damping ratio ζ is much less than unity, which makes τ_d only very slightly larger than τ.

The case with $\zeta < 1.0$ just discussed is the only situation where vibratory motion is possible.

2E. Forced, Damped Motion. This type of motion is described by Eq. 2.1 with all the terms present, or by an alternative form

$$\ddot{u} + 2\omega \zeta \dot{u} + \omega^2 u = \frac{P}{M} \quad (2.13)$$

When damping is less than critical (i.e., $\zeta < 1.0$), the solution of this equation is

$$u = (B_1 \cos \omega_d t + B_2 \sin \omega_d t) e^{-\omega \zeta t} + u_p(t) \quad (2.14)$$

The first term of this expression is the same as a solution of Eq. 2.2, which is also referred to as the general solution. The second term, $u_p(t)$, is the particular solution of Eq. 2.13. The method of finding $u_p(t)$ is the same as that described for Eq. 2.8.

Because of the presence of the exponential expression, which is responsible for the gradual diminishing of the amplitude, the first term is also called the *transient component*. After some time it becomes insignificantly small, and then we have $u \approx u_p$. This is why u_p is referred to as the *steady-state component*.

Since u_p does not contain the integration constants, it is independent of the initial conditions and therefore it is a function of forcing and system properties only. When the phrase *transient response* is used in connection with dynamic analysis, it usually refers to the response just after the application of load, when the transient component is to be included along with the steady-state component. From a physical point of view, the transient vibratory component is a disturbance that takes place because of the transition from rest to motion and decays as time goes on, leaving only the steady-state component.

Equation 2.14 was developed and discussed only for the sake of completeness. It is rarely used in its full form for practical reasons. Once we have determined the formula for $u_p(t)$ and we proceed to find the integration constants B_1 and B_2, the resulting response expression is quite lengthy. Fortunately, there is little need to perform all those operations. If only the steady-state component, prevailing after a sufficiently long time, is of interest, we need not be concerned with the transient component at all. If we want to know what is happening during the first few cycles of motion, an alternate procedure is recommended:

1. Calculate the steady-state component $u_p(t)$ using Eq. 2.13.
2. Instead of using Eq. 2.14 to find B_1 and B_2 for the given initial conditions, employ Eq. 2.7.

The second step is equivalent to ignoring damping in the transient component and is justified only when ζ does not exceed a few percent. Of all system properties, damping is usually known with the least accuracy, and ignoring it may be thought of as an extra precaution.

2F. Superposition of Results. The right-hand term of Eq. 2.1 (or Eq. 2.13) was referred to above as a force or a load. A more general term applied to this part of the equation is *forcing function* or *excitation*. The response, be it u, $\dot{u}$, or $\ddot{u}$, is said to be linearly dependent on the forcing. This means that if $P_1(t)$ applied to the system gives us $u_1(t)$ and $P_2(t)$ acting alone and separately gives $u_2(t)$, the effect of combined forcing $q_1 P_1(t) + q_2 P_2(t)$ is $u(t) = q_1 u_1(t) + q_2 u_2(t)$, where q_1 and q_2 are some constant multipliers.

A brief restatement of the foregoing would be that a linear combination of forcing functions produces the linear combination of responses. One of the practical applications of this property of superposition

occurs when the excitation has a form that is unsuitable for direct handling. It is often possible to replace it by a series of simple components and then add the responses of each term of the series. Such superposition must be carried out with regard to the sign and time dependence of the individual responses.

Caution should be used when calculating transient response, since the initial conditions are involved. As implied before when discussing Eqs. 2.8 or 2.14, the complete solution must be known before the integration constants can be determined from the initial conditions. The superposition principle is valid only with regard to complete solutions. Of course, when only the steady-state component is of interest, this problem does not exist.

When evaluating the responses to forcing functions, which begin or end at some time point, so-called time shifting is frequently used. Suppose that we have calculated the response $u(t)$ resulting from the application of $P_1(t)$ shown in Fig. 2.2. If there is a force $P_2(t)$, which has the same shape as $P_1(t)$, but begins at $t = t_1$ rather than $t = 0$, we say that this force is shifted by t_1 with respect to the first one. The equation for P_2 is the same as for P_1, except that $t - t_1$ replaces t. The response due to the second force is obtained by putting $t - t_1$ in place of t in the formula for $u(t)$.

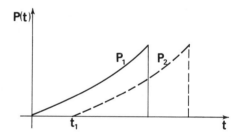

Figure 2.2 Original force P_1 and shifted force P_2.

2G. Steady-State Response to Harmonic Forcing. Resonance. When a harmonic force $P = P_0 \sin \Omega t$ is applied to the basic oscillator, the steady-state component of the displacement response in Eq. 2.14 is

$$u_p = \mu \frac{P_0}{k} \sin(\Omega t - \theta) \qquad (2.15)$$

where μ is the magnification factor,

$$\mu = \frac{1}{\left[\left(1 - \Omega^2/\omega^2\right)^2 + \left(2\zeta\Omega/\omega\right)^2\right]^{1/2}} \qquad (2.16)$$

and P_0/k is the static deflection of the mass under force P_0. The *phase angle* θ defined by

$$\tan\theta = \frac{2\zeta\Omega/\omega}{1-\Omega^2/\omega^2} \qquad (2.17)$$

shows the delay of response with respect to forcing.

It may be demonstrated that when we vary the forcing frequency Ω, the magnification factor and therefore u_p reaches maximum for some $\Omega=\Omega_c$ called the resonant frequency. When the damping ratio ζ does not exceed a few percent, which is typical for most technical applications, it is reasonably accurate to say that the resonant frequency is the same as the natural frequency of the structure, $\Omega_c=\omega$. The exact formula is

$$\Omega_c = \omega\left(1-2\zeta^2\right)^{1/2} \qquad (2.18)$$

The approximate value of the magnification factor at *resonance* (i.e., when the resonant frequency is applied) is

$$\mu_{max} = \frac{1}{2\zeta} \qquad (2.19)$$

As Eq. 2.17 shows, for small damping the phase angle is slightly larger than zero when $\Omega\ll\omega$ and somewhat smaller than 180° when $\Omega\gg\omega$. At $\Omega=\omega$, the phase angle is 90° regardless of the amount of damping. (The influence of the latter is important only in the vicinity of resonance.) Note that from Eq. 2.17 we obtain two values of angle θ in the 360° segment. It can be shown that only one, which is between 0° and 180°, gives the correct solution to the problem.

The velocity and the acceleration response are obtained by differentiation of Eq. 2.15:

$$\dot{u}_p = \mu\left(\frac{\Omega}{\omega}\right)\frac{\omega P_0}{k}\cos(\Omega t-\theta) \qquad (2.20)$$

$$\ddot{u}_p = -\mu\left(\frac{\Omega}{\omega}\right)^2\frac{P_0}{M}\sin(\Omega t-\theta) \qquad (2.21)$$

The value for which the velocity response is maximum is

$$\Omega_c' = \omega \qquad (2.22)$$

while the maximum acceleration response takes place at

$$\Omega_c'' = \frac{\omega}{\left(1-2\zeta^2\right)^{1/2}} \qquad (2.23)$$

The harmonic force $P=P_0\sin\Omega t$ applied to the basic, undamped oscillator ($\zeta=0$) gives the displacement response

$$u_p = \mu\frac{P_0}{k}\sin\Omega t \qquad (2.24)$$

which is the particular solution of Eq. 2.7. The magnification factor is

$$\mu = \frac{1}{1-\Omega^2/\omega^2} \qquad (2.25)$$

(Note that μ can be positive or negative according to this formula. This form is used to ensure that the true sign of the deflection is not lost.) The velocity and the acceleration responses are

$$\dot{u}_p = \mu\left(\frac{\Omega}{\omega}\right)\frac{\omega P_0}{k}\cos\Omega t \qquad (2.26)$$

$$\ddot{u}_p = -\mu\left(\frac{\Omega}{\omega}\right)^2\frac{P_0}{M}\sin\Omega t \qquad (2.27)$$

Note that for $\Omega=\omega$ the response is infinite according to these last four equations. Since some form of damping is always present, it is obvious that the undamped system is not a realistic model at or near $\Omega=\omega$.

2H. Periodic, Nonharmonic Forcing Function can be resolved into harmonic components by a method described in Appendix I. From the point of view of the steady-state response, the a_0 term is the constant component of the applied force, which results in the constant deflection a_0/k. The response to an nth sinusoidal component is calculated from Eqs. 2.15–2.17, except that $n\Omega$ replaces Ω. The same holds for the cosine components after replacing sin by cos. The component responses are formed into an infinite series, which is the solution to our problem. In practical applications only the first few terms of the series are significant enough to be included.

Unless the plot of $P(t)$ drastically differs from a sinusoid of the same frequency, a one-term approximation

$$P(t)\approx a_0+a_1\cos\Omega t \qquad \text{or} \qquad P(t)\approx a_0+b_1\sin\Omega t$$

may be quite sufficient. When doing so, one should of course make certain that $2\Omega, 3\Omega,\ldots$, do not come very close to the natural frequency ω.

If $P(t)$ does not have a very simple analytical expression, the integrals for a_0, a_n, and b_n from Appendix I can be found by numerical means.

2J. Application of Laplace Transforms. Quite frequently the task of solving a linear differential equation may be simplified by employing the Laplace transformation technique. The procedure transforms a differential expression (e.g., Eq. 2.1) into an algebraic equation, which is much easier to solve. The basic operations involved are summarized in Appendix III. We begin with the equation of motion given by Eq. 2.13:

$$\ddot{u} + 2\omega\zeta\dot{u} + \omega^2 u = \frac{1}{M}P(t)$$

The Laplace transformation is performed on both sides of this expression giving (see Prob. 2-16)

$$\bar{u}\left[(s + \omega\zeta)^2 + \omega_d^2\right] = (s + \omega\zeta)u_0 + v_0 + \omega\zeta u_0 + \frac{\bar{P}}{M}$$

where $\bar{u} = \bar{u}(s)$ and $\bar{P} = \bar{P}(s)$ are the transformed displacement and the forcing function, respectively. The expression is then solved for $\bar{u}$ and in the last step it is transformed back into the time variable. We again wind up with Eq. 2.14, except that $u_p(t)$ is now a specific function of t. The general part of the solution is associated with the terms containing u_0 and v_0, while the particular component results from the action of $P(t)$. In the remainder of this section we assume the homogeneous initial condition $u_0 = v_0 = 0$, which yields

$$\bar{u} = \frac{\bar{P}}{M\left[(s + \omega\zeta)^2 + \omega_d^2\right]} \qquad (2.28)$$

When $P(t)$ is defined, we can transform it into the domain of s and the inverse transformation of Eq. 2.28 gives us the displacement $u(t)$. When damping is absent,

$$\bar{u} = \frac{\bar{P}}{M(s^2 + \omega^2)} \qquad (2.29)$$

Many problems may be solved using the transform table in Appendix III. A more extensive tabulation may be found, for example, in Ref. 4. The advantage of this method is that it offers a somewhat automated solution procedure; thus we need not make any prejudgment with regard to a general and a particular solution in each case.

SOLVED PROBLEMS

2-1 Construct a diagram representing the displacement u and velocity $\dot{u}$ of harmonic motion given by Eq. 2.4 in which the values are to be shown as components of a vector rotating with an angular velocity ω.

The figure shows that when u_0 is drawn at an angle ωt with the u-axis, the first component of Eq. 2.4 is the projection of u_0 on that axis. Also, when v_0/ω is inclined at an angle $\omega t - 90°$ to the u-axis, its projection is identical with the second component of Eq. 2.4. Instead of projecting each of the two vectors separately, we can construct their sum A and project it on u-axis. This is the same as writing Eq. 2.5 instead of Eq. 2.4. The term u is the projection of its amplitude A rotating with angular speed ω.

When Eq. 2.4 is differentiated, the velocity is obtained as

$$\frac{\dot{u}}{\omega} = -u_0\sin\omega t + \frac{v_0}{\omega}\cos\omega t$$

Again, the terms of this equation correspond to projection on the $\dot{u}/\omega$-axis of the component vectors in the figure. The velocity may therefore be written as

$$\dot{u} = -A\omega\sin(\omega t - \alpha)$$

which is the same as projecting the rotating vector with length A on the $\dot{u}/\omega$ axis.

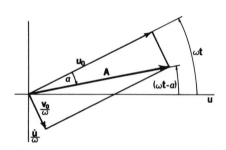

2-2 The system in Fig. 2.1a, with the following properties: $M = 10$ Lb-s^2/in., $k = 1000$ Lb/in., and $\zeta = c/c_c = 0.1$ was impacted so that its initial velocity $v_0 = 100$ in./s. Analyze and plot the displacement as a function of time.

The undamped and damped frequencies are, respectively:

$$\omega = \left(\frac{k}{M}\right)^{1/2} = \left(\frac{1000}{10}\right)^{1/2} = 10 \text{ rad/s},$$

$$\omega_d = \omega(1 - \zeta^2)^{1/2} = 10(1 - 0.1^2)^{1/2} = 9.95 \text{ rad/s}.$$

With $\omega\zeta = 1.0$, $v_0/\omega_d = 100/9.95 = 10.05$ in., and $u_0 = 0$, Eq. 2.9 becomes

$$u = 10.05\,e^{-t}\sin 9.95t$$

(u in inches if t in seconds). The period of motion

$$\tau_d = \frac{2\pi}{\omega_d} = \frac{2\pi}{9.95} = 0.6315s.$$

The u is zero whenever $\sin \omega_d t$ is zero, which is at $t = n\pi/\omega_d$:

$$t = 0, 0.3157, 0.6315, 0.9472, 1.2630 \, s, \ldots$$

By Eq. 2.11a, the ratio of two adjacent maximum displacements is

$$\exp(\omega \zeta \tau_d) = \exp(10 \times 0.1 \times 0.6315) = 1.88$$

The function $10.05e^{-t}$, shown with a dashed line, is the envelope of the plot.

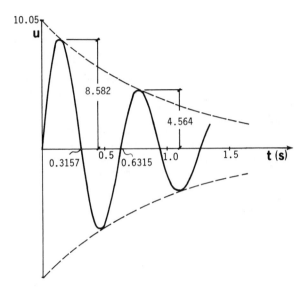

2-3 A disc mounted on a shaft was twisted and released. It was determined by measurement that the initial amplitude decreased by 41% after 30 cycles of motion. What is the damping ratio ζ for the system?

If u_0 and u_n are the starting amplitude and the amplitude after n cycles, respectively, we have, following Eq. 2.11b,

$$\frac{u_0}{u_n} = e^{n\omega \zeta \tau_d} \approx e^{2\pi n \zeta}$$

$$\log \frac{u_0}{u_n} = 2\pi n \zeta$$

$$\log \frac{u_0}{0.59 u_0} = 2\pi \times 30 \zeta \quad \therefore \zeta = 0.0028$$

2-4 A mass-spring system as in Fig. 2.1a, but without a damper, is acted on by a harmonic force with amplitude P_0, $P = P_0 \sin \Omega t$. Find the displacement response $u(t)$ and the magnification factor for $\Omega = 0.8\omega$, 0.9ω, 1.1ω, and 1.2ω, where ω is the natural frequency of the system. For the steady-state part of

deflection construct a plot showing $u(t)$ as a function of Ω/ω.

The solution of Eq. 2.7 with $P = P_0 \sin \omega t$ is given by Eq. 2.8. In the case of a sine or cosine forcing, the assumed form of $u_p(t)$ is $u_p = B_3 \cos \Omega t + B_4 \sin \Omega t$. When this is substituted in Eq. 2.7 in place of u, we have

$$\left(-B_3 \Omega^2 + B_3 \omega^2 \right) \cos \Omega t + \left(-B_4 \Omega^2 + B_4 \omega^2 \right) \sin \Omega t = \frac{P_0}{M} \sin \Omega t$$

Comparing coefficients of $\cos \Omega t$ and then $\sin \Omega t$ on both sides of the equation gives

$$B_3 = 0 \quad \text{and} \quad B_4 (\omega^2 - \Omega^2) = \frac{P_0}{M}$$

Then

$$u_p(t) = \frac{P_0}{k} \frac{1}{1 - \Omega^2/\omega^2} \sin \Omega t$$

Displacement u and velocity $\dot{u}$ are given by Eqs. 2.8 and 2.8a, respectively. In this problem motion starts from rest, therefore $u(0) = B_1 = 0$,

$$\dot{u}(0) = B_2 \omega + \frac{P_0}{k} \frac{\Omega}{1 - \Omega^2/\omega^2} = 0$$

With B_1 and B_2 known, Eq. 2.8 yields

$$\frac{u}{u_{st}} = \frac{1}{1 - \Omega^2/\omega^2} \left(\sin \Omega t - \frac{\Omega}{\omega} \sin \omega t \right)$$

where $u_{st} = P_0/k$ is the static deflection under load P_0. The ratio $|u/u_{st}|$ is the dynamic magnification factor μ. It attains the largest absolute value when $\sin \Omega t = 1$ and $\sin \omega t = -1$, or vice versa. Then

$$\mu = \left| \frac{u}{u_{st}} \right| = \frac{1 + \Omega/\omega}{|1 - \Omega^2/\omega^2|} \frac{1}{|1 - \Omega/\omega|}$$

$\dfrac{\Omega}{\omega}$	0.8	0.9	1.1	1.2
μ	5.0	10.0	10.0	5.0

The term containing $\sin \omega t$ gradually diminishes because of damping (not considered here), and in the limit we have

$$\frac{u}{u_{st}} = \frac{1}{1 - \Omega^2/\omega^2} \sin \Omega t$$

which is the steady-state part of solution. The values of u/u_{st} for $\sin \Omega t = 1$ are plotted in the figure as a function of Ω/ω in the equation above. The plot shows that when the applied force oscillates slowly (i.e., Ω/ω is near zero), the maximum deflection is close to what is obtained under statically applied force. When Ω approaches ω, the deflection tends to infinity. (This result is due to ignoring damping.) When Ω becomes larger and larger, the deflection gradually diminishes. For $\Omega < \omega$ the signs of deflection and exciting force are the same, while for $\Omega > \omega$ they are opposite. This means that in the first case the force is pulling the weight toward an extreme position, while in the second it is pushing the weight away.

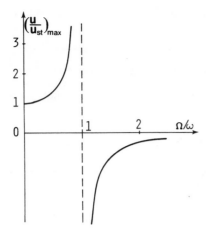

2-5 The basic oscillator in Fig. 2.1a is acted on by a harmonic force $P = P_0 \sin \Omega t$. Find the expression for displacement u and the magnification factor for $\Omega = 0.8\omega$, 0.9ω, 1.1ω, and 1.2ω. Damping ratio is $\zeta = 0.1$. Consider only the steady-state component.

The form of solution of Eq. 2.13 is given by Eq. 2.14 and the expression for $u_p(t)$ is assumed as

$$u_p = B_3 \cos \Omega t + B_4 \sin \Omega t$$

The procedure used in Prob. 2-4 is followed. Substituting u_p instead of u in Eq. 2.13 with $P = P_0 \sin \Omega t$ gives

$$\left(B_3 \omega^2 + 2 B_4 \Omega \omega \zeta - B_3 \Omega^2 \right) \cos \Omega t$$

$$+ \left(B_4 \omega^2 - 2 B_3 \Omega \omega \zeta - B_4 \Omega^2 \right) \sin \Omega t = \frac{P_0}{M} \sin \Omega t$$

Comparing the coefficients of $\cos \Omega t$ and $\sin \Omega t$ on both sides gives a system of equations from which B_3 and B_4 may be found. Using a trigonometric identity, the alternative form of u_p is

$$u_p = \frac{P_0}{k} \mu \sin(\Omega t - \theta)$$

where μ is the magnification factor dependent on the ratio of frequencies,

$$\frac{1}{\mu} = \left[\left(1 - \frac{\Omega^2}{\omega^2} \right)^2 + \left(2\zeta \frac{\Omega}{\omega} \right)^2 \right]^{1/2}$$

and the angle θ is called the *phase angle*,

$$\tan \theta = \frac{2\zeta \Omega / \omega}{1 - \Omega^2 / \omega^2}$$

The term P_0/k is the static deflection of spring under P_0. The magnification factor can now be calculated for the prescribed set of frequencies.

$\dfrac{\Omega}{\omega}$	0.8	0.9	1.0	1.1	1.2
μ	2.54	3.82	5.0	3.29	2.00

2-6 Plot the magnification factor μ and the phase angle θ obtained in Prob. 2-5 as a function of the nondimensional forcing frequency Ω/ω for three values of damping ratio: 0.1, 0.25, 1.0.

Putting $x = \Omega/\omega$ for brevity, we have

$$\frac{1}{\mu} = \left[(1 - x^2)^2 + (2\zeta x)^2 \right]^{1/2}$$

$$\tan \theta = \frac{2\zeta x}{1 - x^2}$$

As (a) shows, the magnification factor approaches unity when the forcing frequency Ω is small in comparison with the natural frequency ω. On the other hand, when Ω becomes very large, μ tends to zero. The first case is a near-static condition, where the system behaves as if a static load of magnitude $P_0 \sin \Omega t$ had been applied to it. In the second extreme case the external force changes its direction so fast that the mass is nearly in the state of rest and virtually no displacement occurs.

We know from Prob. 2-5 that if the forcing is $P = P_0 \sin \Omega t$, the displacement is expressed by $u = A \sin(\Omega t - \theta)$. When the force reaches its peak value at $\Omega t = \pi/2$, displacement does the same after Δt, calculated from $\Omega(t + \Delta t) - \theta = \pi/2$. This means that the displacement lags behind the force by $\Delta t = \theta/\Omega$.

The fact that for $\Omega = \omega$ the phase angle is 90° regardless of the amount of damping is clearly illustrated by (b).

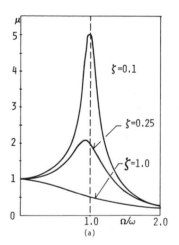

(a)

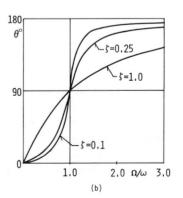

(b)

2-7 The particular solution of the equation of motion under a harmonic force has the form

$$u_p = B_3 \cos \Omega t + B_4 \sin \Omega t$$

according to Prob. 2-5. Calculate the coefficients B_3 and B_4 as a function of Ω/ω and show that the phase angle θ must be in the range

$$0 \leqslant \theta \leqslant 180°$$

when Eqs. 2.15 to 2.17 are used to define $u_p(t)$. This is to verify the statements regarding the phase angle made in Sect. 2G.

The set of equations that allows us to find the constants B_3 and B_4 can be deduced from Prob. 2-5,

$$(\omega^2 - \Omega^2)B_3 + 2\Omega\omega\zeta B_4 = 0$$

$$-2\Omega\omega\zeta B_3 + (\omega^2 - \Omega^2)B_4 = \frac{P_0}{M}$$

Dividing by ω^2 and putting $\Omega/\omega = x$,

$$(1 - x^2)B_3 + 2x\zeta B_4 = 0$$

$$-2x\zeta B_3 + (1 - x^2)B_4 = \frac{P_0}{k}$$

$$\therefore B_3 = \frac{-2x\zeta}{(1-x^2)^2 + 4x^2\zeta^2}\frac{P_0}{k} \qquad B_4 = \frac{1-x^2}{(1-x^2)^2 + 4x^2\zeta^2}\frac{P_0}{k}$$

Write the expression for u_p as

$$u_p = \mu\frac{P_0}{k}\sin(\Omega t - \theta) = \mu\frac{P_0}{k}(\sin\Omega t\cos\theta - \cos\Omega t\sin\theta)$$

By comparing this with the equation for u_p in the problem statement, we get

$$\sin\theta = 2x\zeta\mu \qquad \text{and} \qquad \cos\theta = (1-x^2)\mu$$

For $x < 1$ or $\Omega < \omega$ we have

$$\sin\theta > 0 \qquad \text{and} \qquad \cos\theta > 0$$

which means that θ is between 0° and 90°. For $x > 1$ or $\Omega > \omega$, $\sin\theta > 0$ and $\cos\theta < 0$, which shows that θ is between 90° and 180°. This verifies the appropriate statements of Sect. 2G.

2-8 The formula for the exciting force shown in the figure is $P = P_0\sin\Omega t$ for the first half of each period $T = 2\pi/\Omega$ and $P = 0$ for the second half of each period. Present this force as an infinite sum of its harmonic components.

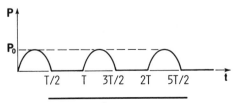

Note that we need to integrate only over the first half of the period, since the function is zero over the second half. After Appendix I:

$$a_0 = \frac{1}{T}\int_0^T P(t)\,dt = \frac{1}{T}\int_0^{T/2}P_0\sin\Omega t\,dt = \frac{1}{T}P_0\frac{2}{\pi}\frac{T}{2} = \frac{P_0}{\pi}$$

$$a_n = \frac{2}{T}\int_0^{T/2}P_0\sin\Omega t\cos n\Omega t\,dt$$

$$= \frac{2P_0}{T}\frac{T}{2\pi}\left(\frac{1}{1+n} + \frac{1}{1-n}\right) \qquad n = 2,4,6,\ldots$$

$$a_n = 0 \qquad \text{for} \qquad n = 1,3,5,\ldots$$

$$b_n = \frac{2}{T}\int_0^{T/2}P_0\sin\Omega t\sin n\Omega t\,dt$$

This is equal to zero for all n except $n = 1$.

$$b_1 = \frac{2}{T}P_0\left(\frac{T}{4}\right) = \frac{P_0}{2}$$

Calculating the values of a_n have

$$\frac{P}{P_0} = \frac{1}{\pi} + \frac{1}{2}\sin\Omega t - \frac{1}{\pi}\left(\frac{2}{3}\cos 2\Omega t\right.$$

$$\left. + \frac{2}{15}\cos 4\Omega t + \frac{2}{35}\cos 6\Omega t + \cdots\right)$$

2-9 The exciting force $P(t)$ has a rectangular, repeating pattern. If a steady state is considered, it makes little difference whether we regard point O, point A, or another arbitrary point as the origin. Yet, the form of expansion is different in each case. Show how to conclude from the appearance of the plot which harmonic components will be missing.

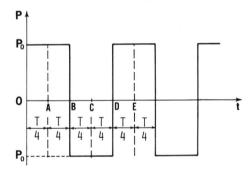

1. Point O is chosen as the origin. The period is measured from O to D. From Appendix I we have

$$a_0 = 0 \quad (\text{average value})$$

Regarding the coefficient a_n note that the function $\cos\Omega t$ is symmetric with respect to O, while $P(t)$ is antisymmetric, which makes the integral equal to zero. The term b_n can be written as

$$b_n = \frac{2}{T}\int_0^{T/2}P_0\sin n\Omega t\,dt + \frac{2}{T}\int_{T/2}^T(-P_0)\sin n\Omega t\,dt$$

By sketching $\sin n\Omega t$ for several values of n, we notice that

$$\int_0^{T/2} \sin n\Omega t\, dt = -\int_{T/2}^T \sin n\Omega t\, dt =$$

$$\int_0^{T_n/2} \sin n\Omega t\, dt = \frac{2}{\pi}\frac{T_n}{2}$$

This is true for n odd, otherwise the integrals become zero.

$$\therefore b_n = \frac{4}{T}P_0\frac{T_n}{\pi} = \frac{4P_0}{n\pi} \quad \text{since} \quad T_n = \frac{2\pi}{n\Omega}$$

$$P(t) = \frac{4P_0}{\pi}\left(\sin\Omega t + \frac{1}{3}\sin 3\Omega t + \frac{1}{5}\sin 5\Omega t + \cdots\right)$$

2. Point A chosen as the origin. The period is measured from A to E. A similar procedure gives

$$P(t) = \frac{4P_0}{\pi}\left(\cos\Omega t - \frac{1}{3}\cos 3\Omega t + \frac{1}{5}\cos 5\Omega t - \cdots\right)$$

3. If an arbitrary point is chosen as the origin, both sine and cosine terms appear.

The following criterion allows us to determine the form of the expansion. We look at the values of the function $P(t)$ on both sides of the point $t_1 = \tau$ on the time axis. If the forcing function is symmetric with respect to this point [i.e., $P(\tau+t) = P(\tau-t)$], we have a cosine expansion. If it is antisymmetric [i.e., $P(\tau+t) = -P(\tau-t)$], a sine expansion results. When neither of these two features is present, both sine and cosine terms may appear.

2-10 The forcing function in Fig. 2-9 can be presented as a series of sinusoidal components. Using Eqs. 2.24 and 2.25 develop a series representing the total, steady-state response. Discuss the problem of how many terms of the latter series have to be included in a practical calculation of response.

A harmonic force defined by $P_n = b_n \sin n\Omega t$ acting on the basic system gives

$$u_n = \frac{b_n}{k}\frac{1}{1-(n\Omega/\omega)^2}\sin n\Omega t \qquad (a)$$

on the basis of Eqs. 2.24 and 2.25. A resolution of the forcing function in Prob. 2-9 gave

$$P(t) = \frac{4P_0}{\pi}\sin\Omega t + \frac{4P_0}{3\pi}\sin 3\Omega t + \frac{4P_0}{5\pi}\sin 5\Omega t + \cdots$$

The response to the successive harmonic components of this forcing is

$$u_1 = \frac{4P_0}{\pi k}\frac{\sin\Omega t}{1-\Omega^2/\omega^2}$$

$$u_3 = \frac{4P_0}{3\pi k}\frac{\sin 3\Omega t}{1-(3\Omega/\omega)^2}$$

$$u_5 = \frac{4P_0}{5\pi k}\frac{\sin 5\Omega t}{1-(5\Omega/\omega)^2}$$

The total response is the infinite series:

$$u = u_1 + u_3 + u_5 + \cdots$$

The relative importance of these terms depends on the magnitude of the multiplier of $\sin\Omega t$ in the equation above for u_n. One tendency is the decrease of amplitude with increasing n, because of the presence of n in the denominator. The other factor is the proximity of a particular term to resonance, which happens when $n\Omega = \omega$. (The infinite response is implied here because damping was ignored.) If we had $\Omega/\omega = 2$, for example, all forcing frequencies except the first would be above resonance and writing $u \approx u_1 + u_3$ would be a very close approximation. When one of the ratios $n\Omega/\omega$ is near unity, more terms may have to be included.

2-11 The periodic function in the figure is defined numerically, by means of the following table:

t/T	P/P_0
0.05	0.45
0.15	0.82
0.25	0.98
0.35	0.87
0.45	0.50
0.55	0.62
0.65	1.95
0.75	2.73
0.85	2.71
0.95	1.20

Calculate the first five terms of Fourier expansion.

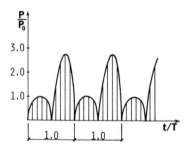

The function $P(t)$ will be treated as a constant within each segment $\Delta t = 0.1T$. The same approximation will also be applied to the harmonic functions in Eqs. A1.2. We can therefore present each integral as a finite sum extended over the 10 segments, into which the period was subdivided,

$$a_0 = \frac{1}{T}\sum_{r=1}^{10} P_r\Delta t = 0.1\sum P_r$$

$$a_n = 0.2\sum_r P_r\cos\left(2\pi r\frac{t}{T}\right)$$

$$b_n = 0.2\sum_r P_r\sin\left(2\pi r\frac{t}{T}\right)$$

Substituting the numbers, we obtain

$$a_0 = 0.1(0.45 + 0.82 + \cdots + 1.2) = 1.283$$

$$a_1 = 0.2[0.45\cos(2\pi \times 0.05) + 0.82\cos(2\pi \times 0.15) + \cdots] = 0.1843$$

$$b_1 = 0.2[0.45\sin(2\pi \times 0.05) + 0.82\sin(2\pi \times 0.15) + \cdots] = -0.8843$$

Similarly:

$$a_2 = -0.6863 \quad \text{and} \quad b_2 = -0.2281$$

Our function can therefore be represented as

$$P(t) \approx P_0\left[1.283 + 0.1843\cos\left(2\pi\frac{t}{T}\right) - 0.8843\sin\left(2\pi\frac{t}{T}\right)\right.$$
$$\left. - 0.6863\cos\left(4\pi\frac{t}{T}\right) - 0.2281\sin\left(4\pi\frac{t}{T}\right)\right]$$

2-12 The basic oscillator in Fig. 2.1a had been subjected to a step loading shown in (a). Calculate displacement $u(t)$ when damping is ignored.

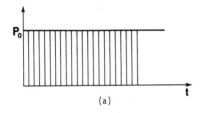

(a)

Equation 2.7 becomes

$$\ddot{u} + \omega^2 u = \frac{P_0}{M}$$

Assume the particular solution as a constant:

$$u = C, \quad \therefore \ \dot{u} = \ddot{u} = 0$$

Then $\omega^2 C = \dfrac{P_0}{M}$

$$\therefore \ C = \frac{P_0}{\omega^2 M} = \frac{P_0}{k} = u_{st}$$

The initial conditions are zero, $u(0) = \dot{u}(0) = 0$. From Eqs. 2.8 and 2.8a,

$$0 = B_1 + u_{st} \quad \text{or} \quad B_1 = -u_{st}$$

$$0 = B_2 \quad \text{and} \quad u = -u_{st}\cos\omega t + u_{st} = u_{st}(1 - \cos\omega t)$$

The system oscillates with amplitude u_{st} about the position of static deflection; see (b).

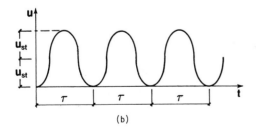

(b)

2-13 The basic oscillator is subjected to a step load, which is removed after t_0 as shown in (a). How does the spring force change with time? At what instant should the load be removed to give the largest possible spring force after removal?

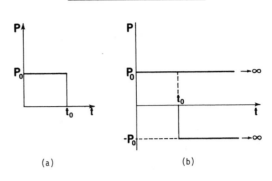

(a) (b)

The load pattern in (a) may be treated as a superposition of the step loads in (b). If the one originating at $t = 0$ gives a deflection u_1 (see Prob. 2-12),

$$u_1 = u_{st}(1 - \cos\omega t)$$

the second step beginning at $t = t_0$ will result in

$$u_2 = -u_{st}[1 - \cos\omega(t - t_0)]$$

For $t \leqslant t_0$, u_1 is the deflection response. For $t > t_0$ both terms are superposed and the total response is

$$u = u_1 + u_2 = 2u_{st}\sin\frac{\omega t_0}{2}\sin\omega\left(t - \frac{t_0}{2}\right)$$

Since u_{st} is the static deflection of spring under P_0, $u_{st} = P_0/k$, the spring force is

$$N = ku = 2P_0\sin\frac{\pi t_0}{\tau}\sin\omega\left(t - \frac{t_0}{2}\right)$$

Only the second sine term is time dependent and oscillates according to the natural frequency. The first term is a function of t_0 and the natural period τ. Its absolute value reaches unity for $t_0 = \tau/2$; $3\tau/2, 5\tau/2, \ldots$. This means that if the load P_0 is removed after any of those values of t_0, the amplitude of spring force during the free vibrations that follow will be $2P_0$. An inspection of Fig. 2-12b reveals why this happens. Notice that the amplitude has doubled in comparison with what is obtained from the step loading in Fig. 2-12a.

If t_0 is different from an odd multiple of $\tau/2$, the amplitude of spring force will be less than $2P_0$. In fact, when t_0 is equal to a multiple of τ, no vibration takes place for $t > t_0$.

2-14 An undamped torsional oscillator is subjected to a series of pulses of torque with magnitude M_{t0}. Each pulse lasts for t_1 and as shown in (a), the period of the forcing function is T. The spring rotates by angle α_{st} when M_{t0} is applied quasi-statically and the natural period of the system is τ. How should t_1 and

T be related to τ to maximize the response? What is the maximum deflection after n pulses?

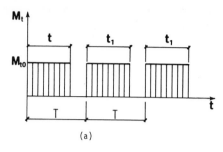

(a)

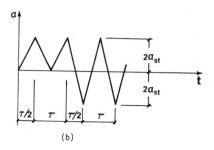

(b)

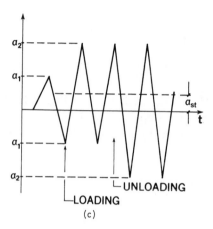

(c)

On the basis of Prob. 2-13 we know that as far as the first pulse is concerned, the removal of load must take place immediately after an odd number of half-periods $\tau/2$. For an example of what happens when this condition is fulfilled, see (b). (The sine curve is drawn with straight-line segments for simplicity.) After the load M_{t0} is removed at $t=3\tau/2$, the amplitude of vibration changes from α_{st} to $2\alpha_{st}$. Suppose now that the oscillating disc reaches its extreme negative position and begins to return in the positive direction. At this time we apply a positive torque M_{t0} again, which is the beginning of a second pulse at $t=T$. If we use the angles α_1 and α_2 as designated in (c), the following equation of work-energy can be written:

$$\tfrac{1}{2}K\alpha_2^2 - \tfrac{1}{2}K\alpha_1^2 = M_{t0}(\alpha_1 + \alpha_2)$$

which means that the change of strain energy in the spring is due to the work of torque M_{t0} on the path ($\alpha_1 + \alpha_2$). The unknown α_2 may now be calculated.

$$\tfrac{1}{2}K(\alpha_2 - \alpha_1)(\alpha_2 + \alpha_1) = M_{t0}(\alpha_1 + \alpha_2)$$

$$\therefore \alpha_2 = \alpha_1 + 2\alpha_{st}$$

where

$$\alpha_{st} = M_{t0}/K$$

Notice, however, that α_2 is not the amplitude, but merely an extreme deflection. The amplitude becomes $3\alpha_{st}$ and remains so until the load is removed again. If the removal takes place at the extreme positive position, the amplitude changes to $4\alpha_{st}$. Since t_1 is an odd multiple of $\tau/2$ and so is the time during which the system is unloaded, the sum of both quantities, which is T, must be an even multiple of $\tau/2$, or, as we may prefer to say, a multiple of τ. (This result could be expected because we are to maximize the response; consequently forcing must be tuned to natural period of vibration.)

To summarize the analysis of (b) and (c), we may say that forcing the oscillator in the manner described magnifies the amplitude by α_{st} at each load application and by α_{st} again after removal. The application shifts the neutral point of vibrations by α_{st} and the removal of the load cancels the shift. Each load pulse increases the amplitude by $2\alpha_{st}$; thus after n cycles the amplitude becomes $2n\alpha_{st}$.

2-15   A torsional oscillator is subjected to rectangular pulses as in Prob. 2-14. The period of the forcing function is equal to n natural periods of the oscillator. The damping ratio is ζ, and the angle of rotation due to quasi-static application of load is α_{st}. After a number of load pulses, a steady-state condition will be approached in which the effect of forcing will be offset by damping. What is the amplitude corresponding to that condition? Assume that within each forcing cycle the load is applied during m natural half-periods and removed during the remaining p half-periods, so that $m+p=2n$.

The figure shows a forcing cycle, from one to the next application of torque. From Prob. 2-14 we know that

$$A_1 = A_4 + \alpha_{st} \qquad \text{and} \qquad A_3 = A_2 + \alpha_{st}$$

Assuming that the system is loaded during $m\tau_d/2$ and unloaded during $p\tau_d/2$, where τ_d is the natural period of damped oscillations, we make use of Eq. 2.11b:

$$A_1 = A_2 \exp\left(\frac{m\omega\zeta\tau_d}{2}\right) \qquad \text{and} \qquad A_3 = A_4 \exp\left(\frac{p\omega\zeta\tau_d}{2}\right)$$

The values of the exponential expressions are known from the problem statement, so we can put

$$A_1 = A_2 e' \qquad \text{and} \qquad A_3 = A_4 e''$$

and solve the set of four equations for the unknown amplitudes, of which A_1 and A_3 are of particular importance,

$$A_1 = \frac{e'(e''+1)}{e'e''-1}\alpha_{st} \quad \text{and} \quad A_3 = \frac{e''(e'+1)}{e'e''-1}\alpha_{st}$$

The amplitude will be defined as one-half of the entire range between maximum and minimum deflection:

$$A = \frac{1}{2}(A_1 + A_3 + \alpha_{st}) \quad \text{or} \quad A = \frac{\alpha_{st}}{2}\left(\frac{e'+e''+2e'e''}{e'e''-1}+1\right)$$

For small damping we can replace $\omega\zeta\tau_d$ in the exponential expressions by $2\pi\zeta$ (refer to Sect. 2D). Consequently,

$$e' = \exp\left(\frac{m\omega\zeta\tau_d}{2}\right) = \exp(m\pi\zeta)$$

$$e'' = \exp\left(\frac{p\omega\zeta\tau_d}{2}\right) = \exp(p\pi\zeta)$$

$$e'e'' = \exp(2n\pi\zeta)$$

since $(m+p)/2 = n$ as given by the problem statement.

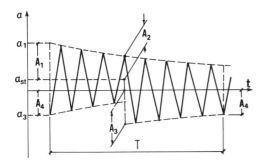

2-16 The equation of free vibration of the basic oscillator can be written as

$$\ddot{u} + 2\omega\zeta\dot{u} + \omega^2 u = 0$$

Using Laplace transformation, develop the equation for the deflection $u(t)$ when damping is small ($\zeta < 1$) and when the initial conditions are $u(0) = u_0$ and $\dot{u}(0) = v_0$.

We put $u(t) \to \bar{u}(s)$, or briefly $u \to \bar{u}$. With the aid of the table in Appendix III one can write

$$\dot{u} \to s\bar{u} - u_0 \quad \text{and} \quad \ddot{u} \to s^2\bar{u} - su_0 - v_0$$

The transformed equation of motion is

$$s^2\bar{u} - su_0 - v_0 + 2\omega\zeta s\bar{u} - 2\omega\zeta u_0 + \omega^2\bar{u} = 0$$

or

$$\bar{u}(s^2 + 2\omega\zeta s + \omega^2) = su_0 + v_0 + 2\omega\zeta u_0$$

Introducing the damped frequency ω_d according to $\omega_d^2 = \omega^2(1-\zeta^2)$

we can write

$$\bar{u} = \frac{(s+\omega\zeta)u_0 + v_0 + \omega\zeta u_0}{(s+\omega\zeta)^2 + \omega_d^2}$$

The table of transforms indicates that

$$\frac{su_0}{s^2 + \omega_d^2} \to u_0\cos\omega_d t$$

Replacing s by $(s+\omega\zeta)$ is equivalent to multiplying by $e^{-\omega\zeta t}$ in the t variable:

$$\frac{(s+\omega\zeta)u_0}{(s+\omega\zeta)^2 + \omega_d^2} \to u_0 e^{-\omega\zeta t}\cos\omega_d t$$

The remaining part of the expression for $u(s)$ transforms as

$$\frac{v_0 + \omega\zeta u_0}{(s+\omega\zeta)^2 + \omega_d^2} \to \frac{v_0 + \omega\zeta u_0}{\omega_d}e^{-\omega\zeta t}\sin\omega_d t$$

Finally:

$$u(t) = \left(u_0\cos\omega_d t + \frac{v_0 + \omega\zeta u_0}{\omega_d}\sin\omega_d t\right)e^{-\omega\zeta t}$$

2-17 The load applied to the basic oscillator is a cutoff step function (Fig. 2-13a),

$$P = \big[H(t) - H(t-t_0)\big]P_0$$

How does the spring force change with time? Assume zero initial conditions and no damping.

Using the transform table in Appendix III we get

$$H(t) \to \frac{1}{s}$$

$$H(t-t_0) \to \frac{1}{s}e^{-t_0 s}$$

$$P \to (1 - e^{-t_0 s})\frac{1}{s}P_0 = \bar{P}(s)$$

From Eq. 2.29,

$$\bar{u} = \frac{P_0}{M}\frac{1}{s(s^2 + \omega^2)}(1 - e^{-t_0 s})$$

The same table is used again to find the inverse transform. Since

$$\frac{1}{s^2 + \omega^2} = \frac{1}{\omega}\frac{\omega}{s^2 + \omega^2} \to \frac{1}{\omega}\sin\omega t$$

then

$$\frac{1}{s}\frac{1}{s^2 + \omega^2} \to \int_0^t \frac{1}{\omega}\sin\omega t\, dt = \frac{1}{\omega^2}(1 - \cos\omega t)$$

Multiplication by $e^{-t_0 s}$ in the domain of s is equivalent to translation and cutoff in the time variable:

$$\frac{1}{s}\frac{1}{s^2 + \omega^2}e^{-t_0 s} \to \frac{1}{\omega^2}H(t-t_0)\big[1 - \cos\omega(t-t_0)\big]$$

Finally

$$u(t) = \frac{P_0}{M} \frac{1}{\omega^2} \{1 - \cos \omega t - H(t - t_0)[1 - \cos \omega (t - t_0)]\}$$

This single expression gives us the displacement both prior to unloading at $t = t_0$ as well as after $t = t_0$. The function $H(t - t_0)$ is zero for $t \leqslant t_0$, so we have the spring force

$$N = ku = P_0 (1 - \cos \omega t)$$

for this period of time. For $t > t_0$

$$N = P_0 [\cos \omega (t - t_0) - \cos \omega t]$$

This can be rewritten as

$$N = 2 P_0 \sin \frac{\pi t_0}{\tau} \sin \omega \left(t - \frac{t_0}{2}\right)$$

in which $\tau = 2\pi/\omega$. Our results agree with the solution of Prob. 2-13 achieved by more conventional means. Although there are a few manipulations needed in the transformation method, one does not have to solve the differential equation of motion.

2-18 An undamped oscillator is subjected to harmonic forcing,

$$P = P_0 \cos \Omega t \quad \text{or} \quad P = P_0 \sin \Omega t$$

Determine the steady-state response using Laplace transforms.

––––––––––––––

Using the identity

$$e^{i\Omega t} = \cos \Omega t + i \sin \Omega t$$

allows us to handle both functions in a single operation, since the right side of the equation represents a set of two independent functions of time. The equation of motion is

$$\ddot{u} + \omega^2 u = \frac{P_0}{M} e^{i\Omega t}$$

Upon transforming both sides and using homogeneous initial conditions we obtain

$$\bar{u} = \frac{P_0}{M} \frac{s + i\Omega}{s^2 + \Omega^2} \frac{1}{s^2 + \omega^2}$$

From items 18 and 19 in the table of transforms (Appendix III) we get

$$\frac{s}{(s^2 + \Omega^2)(s^2 + \omega^2)} \rightarrow \frac{\cos \Omega t - \cos \omega t}{\omega^2 - \Omega^2}$$

$$\frac{1}{(s^2 + \Omega^2)(s^2 + \omega^2)} \rightarrow \frac{(1/\Omega) \sin \Omega t - (1/\omega) \sin \omega t}{\omega^2 - \Omega^2}$$

$$\bar{u} \rightarrow \frac{P_0}{M} \frac{1}{\omega^2 - \Omega^2} \left[\cos \Omega t - \cos \omega t + i\Omega \left(\frac{1}{\Omega} \sin \Omega t - \frac{1}{\omega} \sin \omega t\right)\right]$$

$$\therefore u = \frac{P_0}{k} \frac{1}{1 - \Omega^2/\omega^2} \left[e^{i\Omega t} - \left(\cos \omega t + i \frac{\Omega}{\omega} \sin \omega t\right)\right]$$

Only the first of the functions in brackets represents the steady-state response. The terms containing $\sin \omega t$ and $\cos \omega t$ are the transient part of the response and are ignored accordingly. Using the symbol $u_{st} = P_0/k$, we write

$$u = \frac{u_{st}}{1 - \Omega^2/\omega^2} e^{i\Omega t} = \frac{u_{st}}{1 - \Omega^2/\omega^2} (\cos \Omega t + i \sin \Omega t)$$

The result is a set of two independent response functions induced by the corresponding forcing terms.

EXERCISES

2-19 In a basic oscillator without damper the mass M is pulled away by a distance u_0 from its initial position and then released. What should u_0 be if the maximum velocity is to attain the prescribed value v_{max}?

2-20 Using Eq. 2.5 instead of Eq. 2.4 is desirable, because it gives us a briefer, more meaningful expression. The disadvantage, however, is that some thought must be given to the choice of α if we define this angle by $\tan \alpha = v_0/(\omega u_0)$. For example, if $\tan \alpha = 0.12$, we can put

$$\alpha = 6.8° \quad \text{or} \quad \alpha = 6.8° - 180° = -173.2°$$

Also, when $v_0 = 0$, we can put

$$\alpha = 0° \quad \text{or} \quad \alpha = 180°$$

One way to solve the problem is to see whether a selected value of α is going to give the correct initial conditions and if not, use the other value differing by 180°. A better way is to improve the definition of the angle α. Develop an alternative definition and check on it for several combinations of u_0 and v_0.

2-21 The basic oscillator is subjected to an exciting force $P = P_0 \sin \Omega t$. The steady-state component of response is $u = A \sin \Omega t$, where A is the amplitude. The work-energy equation of the system is

$$\Delta \Pi + \Delta T = U_p$$

where $\Delta \Pi$ and ΔT are the changes of the strain and kinetic energy, respectively, and U_p is the work of force P during the time of interest. Express the terms of the equation by the amplitude of displacement A.

2-22 In Prob. 2-4 the complete solution was found for the vibration due to the applied harmonic force. Suppose now that the forcing frequency is only slightly different from the natural frequency, so that $\Omega - \omega = 2\Delta$, where $\Delta \ll \Omega$. Analyze the shape of the displacement response.

2-23 The following observations can be made about the steady-state component of the basic oscillator when subjected to harmonic forcing $P = P_0 \sin \Omega t$. (1) When $\Omega \ll \omega$, the displacement is spring controlled. (2) At $\Omega = \omega$, the displacement is damper controlled. (3) When $\Omega \gg \omega$, the displacement is mass controlled. Using Eqs. 2.15, to 2.17, write the expressions justifying those observations.

2-24 The basic oscillator is acted on by a composite harmonic force

$$P = P_1 \sin \Omega_1 t + P_2 \sin \Omega_2 t + P_3 \sin \Omega_3 t$$

where $P_1 = 0.8 P_0$ $\Omega_1 = 0.8 \omega$

$\qquad\quad P_2 = 1.2 P_0$ $\Omega_2 = 1.2 \omega$

$\qquad\quad P_3 = 1.8 P_0$ $\Omega_3 = 1.8 \omega$

The natural frequency $f_n = 10\,\text{Hz}$ and the damping ratio $\zeta = 0.1$. Find the largest possible displacement response. Consider only the steady-state components.

2-25 The exciting force acting on an oscillator with damping $\zeta = 0.2$ is $P = P_0 \sin \Omega t$, where $\Omega = 0.75 \omega$. Suppose that damping is reduced to $\zeta = 0.02$. What the new amplitude P_1 of the exciting force will give the same maximum displacement of the steady-state component?

2-26 A mechanical system with a relatively large damping $\zeta = 0.2$ has a natural frequency of 100 Hz. Find the forcing frequencies for which the displacement, velocity, and acceleration responses reach their maxima.

2-27 A basic oscillator, $W = 80\,\text{Lb}$, is subjected to a sinusoidally varying force. What will be the total maximum spring load (including gravity) at resonance if the amplitude of forcing is $P_0 = 6\,\text{Lb}$ and the damping ratio $\zeta = 0.04$?

2-28 Calculate the harmonic components of the forcing function with a triangular pattern.

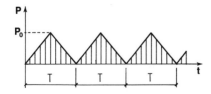

2-29 Calculate the harmonic components of the forcing function shown in the figure.

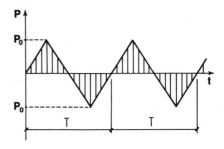

2-30 Calculate the steady-state response due to forcing shown in Fig. 2-28. Assume that a forcing function $P_0 \cos \Omega t$ produces a response proportional to $\cos(\Omega t - \theta)$. Ignore damping.

2-31 The basic oscillator is subjected to a force growing in proportion to time as shown in the figure. Calculate displacement as a function of time, assuming zero initial conditions and ignoring damping. *Hint.* Assume the particular solution of the equation of motion as $at + b$, where a and b are the coefficients to be determined.

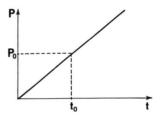

2-32 The basic oscillator is subjected to a load pattern in (a), which may be treated as a superposition of two linear functions for the appropriate time intervals. Derive the formula for deflections for $t > t_0$. What is the value of t_0 for which maximum possible displacement can be attained? What happens when $t_0 \to 0$ while P_0 remains unchanged? For what value of t_0 are there no oscillations when $t > t_0$?

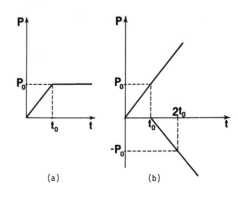

(a)   (b)

2-33 The basic oscillator without a damper is subjected to a sine-pulse force with amplitude P_0 and period T as shown in (a). In (b) the superposition of two sine waves giving such a pulse is shown. The second one, indicated by the dashed line, begins $T/2$ later than the first. Using the results of Prob. 2-4 with $T = 2\pi/\Omega$, develop the equations for the displacement response for $t > t_0$.

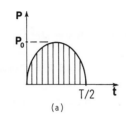

(a)

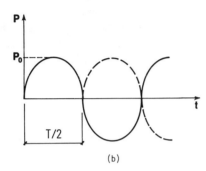

(b)

2-34 The disc of a torsional oscillator in Fig. 2.1b is subjected to rectangular pulses of torque, as in Fig. 2-14. The shaft stiffness $K = 42,000$ Lb-in./rad and damping $\zeta = 0.05$. The period of the forcing function is about $8\tau_d$, where τ_d is the damped natural period. Within each forcing cycle the disc is loaded for $2.5\tau_d$ and unloaded for the remaining $5.5\tau_d$. The magnitude of torque is such that it causes the shaft to twist by $0.25°$ when applied statically. Calculate the maximum amplitude of the dynamic twisting moment in the shaft.

2-35 Calculate the response of an undamped oscillator following the sequence of two impulses shown in the figure. What value of t_0 will produce the maximum response?

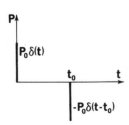

2-36 Calculate the displacement of an undamped torsional oscillator subjected to a pulse shown in the figure.

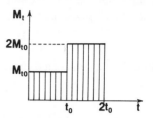

2-37 Weight W is suspended on a cable so that the spring below the weight is unstrained. If the cable suddenly breaks, what will be the maximum downward deflection of the weight? How high will it go after rebound? What will be its maximum velocity? Use the work-energy method.

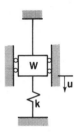

2-38 An impulse that is applied to the basic oscillator at time t_0 has the magnitude P_0 Lb-s and is of infinitely short duration. Determine the subsequent motion, assuming zero initial conditions.

2-39 The basic oscillator is subjected to a force growing linearly with time as in Fig. 2-31. Calculate the displacement response, assuming zero initial conditions and no damping.

2-40 A torque acting on the undamped torsional oscillator has the form of a pulse with an initial magnitude M_{t0} and the time of decay t_1. Find the angular displacement α for $t \geq t_1$. What is the magnification factor μ for $t_1/\tau = 0.1$? τ is the natural period of vibration.

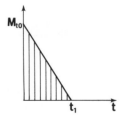

2-41 Using Laplace transforms, calculate the response of an undamped oscillator to a force that instantaneously reaches the value P_0 and then exponentially decays. *Hint.* Represent $u(s)$ as a sum of partial fractions.

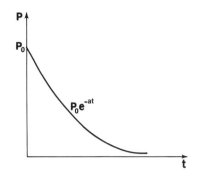

CHAPTER

3

Nonlinear Vibrations

When the relationship between the external load and the resistance offered by a deformable system cannot be represented by a straight line, the system is said to be nonlinear. The natural frequency of such a structure is most often amplitude dependent, unlike the case of the perfectly linear system, where frequency remains constant regardless of amplitude. The analysis is mostly limited to the nonlinearly elastic structures, for which there is only one value of resistance for a particular displacement. The problems of determining the natural frequency as well as response to a harmonic forcing of damped and undamped systems are the main topics of this chapter.

3A. General Remarks About Nonlinearities. The equation of motion of the basic oscillator

$$M\ddot{u} + c\dot{u} + ku = P(t)$$

is an example of a linear differential equation, because the unknown function $u(t)$ as well as its derivatives appear in the first power. When an equation of motion does not satisfy that condition, we call it a nonlinear equation and the system it describes is referred to as a nonlinear system. Most engineering objects show some deviation from linearity, but this is usually ignored for the practical reason of keeping computations simple. When deflection amplitudes are small in comparison with dimensions of a structure, it is often quite realistic to assume proportionality be-

tween deflections and the elastic resistance. Yet, when those displacements significantly change the manner in which the external loads act on a structure, the nonlinearity of the system may not be ignored.

A common example of a nonlinear system is a body, say a wing of an airplane, subjected to aerodynamic forces that are proportional to the square of velocity. Another large class of problems pertains to systems with a general resisting force $R(u)$ so that we have

$$M\ddot{u} + c\dot{u} + R(u) = P(t) \qquad (3.1)$$

instead of the previous equation. The function $R(u)$ is also referred to as a *characteristic* of a system. The first example of an element with a nonlinear characteristic is shown in Fig. 3.1. It is a cantilever that is slightly curved upward in its unstressed state. When subjected to a vertical downward force, the free length of the cantilever diminishes because the beam is gradually leaning against the rigid plane. In consequence, for every successive increment of displacement, a larger and larger increment of force P is

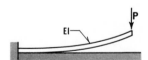

Figure 3.1 Hardening system.

36

Figure 3.2 Softening system.

needed. Such elements in which the resistance grows faster than displacements are called *hardening* elements. Figure 3.2 gives an example of a *softening* element. This is the same cantilever as before, but loaded with a compressive force and having no horizontal restraining plane. As the deflection grows, the process of deforming becomes easier, at least up to a point. This is because the effective arm of force P with respect to the deflected axis is increasing. The characteristics of both types of element are plotted in Fig. 3.3.

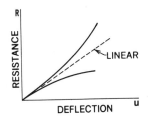

Figure 3.3 Characteristics of nonlinear systems.

This example has illustrated an important principle, namely, that not only the element itself, but also its manner of loading and the magnitude of deflections may cause nonlinear effects to appear.

Most structural materials begin to yield at a certain stress level. Once yielding starts, it makes the characteristic of an element deviate from the initial direction. This is called a *material nonlinearity*, as opposed to *geometrical nonlinearity* caused by the change of shape during the loading process. A problem of two bodies impacting each other often involves both types of nonlinearity, because the change in contact area may be accompanied by yielding.

There are many other types of nonlinear behavior, but they are outside the scope of this book. Unlike the case of the linear problems, very few exact, closed-form solutions are available, and various approximations must be employed. That makes the field of nonlinear vibrations a diversified and difficult branch of engineering science.

An important computational aspect of a nonlinear system is that the principle of superposition does not in general apply. When a system is subject to several dynamic loads acting at the same time, the entire loading must be simultaneously considered.

3B. Elastic, Nonlinear Systems. A system or a body is called elastic if there is a 1:1 relationship between displacement and load. If in addition to that, the resistance-deflection curve is a straight line, the body is called *linearly elastic*; otherwise it is *nonlinearly elastic*. The latter type of elements are described in this section.

Among the plots in Fig. 3.4, only the first one does not show symmetry. All are made of lines with two distinct slopes, but the term *bilinear* is usually applied only to the characteristic in Fig. 3.4b. Except for the first, all plots are antisymmetric about the origin, which means that if the direction of displacement changes, so does the direction of the resisting force, but the magnitude of the latter remains unchanged. Such systems are said to possess elastic symmetry, and it is sufficient to define their resistance-deflection curve for positive values of displacement.

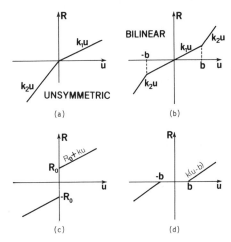

Figure 3.4 Simple nonlinear systems.

Figure 3.5 shows a general shape of the amplitude-frequency relationship in free vibration for three different types of spring. It is assumed that all three of them have the same slope of resistance-deflection curve at the origin. The numerical difference between the three natural frequencies may be made arbitrarily small if the amplitude is made small enough.

When we load an elastic body to some point (R_1, u_1) on its R–u curve in Fig. 3.4 and then begin unloading, a point determining the condition of that body begins to move back toward the center of plot along the same path. This is the essential feature of elasticity, which is characterized by the fact that there is only one value of the resistance force for a given value of displacement. Not so for inelastic elements, where the loading and unloading paths are not in general the same.

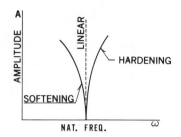

Figure 3.5 Amplitude-frequency relations.

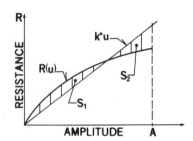

Figure 3.6 Equivalent linear oscillator.

3C. Determination of Natural Frequency by an Equivalent Linear Oscillator.

Consider mass M attached to a nonlinear elastic spring and vibrating with amplitude A. Imagine an associated linear system with a spring of stiffness k^*, so chosen that at the maximum displacement $u_m = A$ the strain energy is the same in both springs (Fig. 3.6). The natural period of that nonlinear system, which is characterized by $R(u)$, can now be approximated by

$$\tau = 2\pi \left(\frac{M}{k^*} \right)^{1/2} \tag{3.2}$$

The equivalent linear stiffness k^* is valid only for a given extreme displacement $u_m = A$. The deviation of curve $R(u)$ from a straight line must be small to make Eq. 3.2 reasonably accurate.

3D. Harmonic Approximation for Systems with Symmetric Resistance.

When a system is symmetric about the neutral point, the plot of the resisting force is an odd function of displacement u. In detail, we have $+R$ for positive u and $-R$ for negative values of u. (See Fig. 3.7). A good approximation may often be obtained by assuming that displacement is a harmonic function of time:

$$u = A \cos \omega t \tag{3.3}$$

The resisting function is written as

$$R(u) = k_0 f(u) \tag{3.4}$$

where k_0 is a positive constant. The equation of motion is

$$M\ddot{u} + k_0 f(u) = 0 \tag{3.5}$$

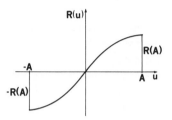

Figure 3.7 Symmetry of resistance versus deflection.

According to the Ritz method, the assumed displacement (Eq. 3.3) is substituted in Eq. 3.5 and the resulting expression is multiplied by $\cos \omega t$. Integration with respect to ωt gives the relation between the natural frequency ω and amplitude A

$$\omega^2 = \frac{k_0}{M} F(A) \tag{3.6}$$

in which

$$F(A) = \frac{1}{\pi A} \int_0^{2\pi} f \cdot \cos \omega t \, d(\omega t) \tag{3.7}$$

The difficulty now is to evaluate this integral. If we select the value of the independent variable ωt, say $\omega t = (\omega t)_1$, then $\cos(\omega t)_1$ and $f_1 = f[A \cos(\omega t)_1]$ can be calculated, the latter also depending on A. A plot of the function under the integral sign is shown in Fig. 3.8. This plot is generated by dividing one quarter of a period into four segments. It can be shown that when Simpson's integration formula is employed, we obtain

$$F(A) = \frac{1}{6A} \left(f_0 + 3.6955 f_1 + 1.4142 f_2 + 1.5308 f_3 \right) \tag{3.8}$$

(The indices of f in this equation correspond to the indices of y in Fig. 3.8). If the plot of $f \cos \omega t$ does not have a large slope change, even a simpler formula, based on two rather than four segments, can be used:

$$F(A) = \frac{1}{3A} \left(f_0 + 2.8284 f_c \right) \tag{3.9}$$

in which $f_0 = f(A)$ and $f_c = f(0.7071A)$ are the values of f at $\omega t = 0$ and $\omega t = \pi/4$, respectively. [Function f indicates an operation on the displacement u. Writing $f(A)$ means that the value of f is to be calculated for u equal to the vibration amplitude A.]

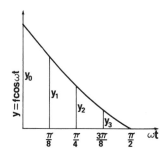

Figure 3.8 Plot for determination of frequency.

Notice that Eq. 3.3 implies a displacement prescribed as the initial condition. If the initial velocity is given instead, one can find the equivalent initial displacement from the energy balance.

3E. Response of Undamped System to Harmonic Forcing. When a forcing function $P = P_0 \cos \Omega t$ is applied to an undamped system, the equation of motion is

$$M\ddot{u} + k_0 f(u) = P_0 \cos \Omega t. \qquad (3.10)$$

The solution is assumed to be:

$$u = A \cos \Omega t \qquad (3.11)$$

That is, the system will vibrate with the forcing frequency. The Ritz method applied to the system gives a relation between Ω and A:

$$\frac{\Omega^2}{\omega_0^2} = F(A) - \frac{P_0}{Ak_0} \qquad (3.12)$$

in which $\omega_0^2 = k_0/M$. As we see from Eq. 3.6, the function $F(A)$ is always positive. The value of A itself may be either positive or negative, depending on whether the response defined by Eq. 3.11 is in phase or out-of-phase with the forcing.

To construct a graph of amplitude versus forcing frequency, we assume a value for A and calculate the corresponding Ω. Figures 3.9 and 3.10 show typical plots for a hardening and a softening system, respectively. Although the sign is lost in plotting the absolute value of the amplitude, one can easily find with the help of Eq. 3.12 which of the two branches corresponds to positive A. The dashed line between

the branches is the natural frequency ω as a function of Ω.

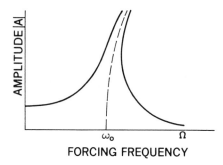

Figure 3.9 Response of hardening system.

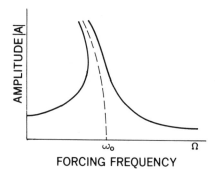

Figure 3.10 Response of softening system.

3F. Response of Damped System to Harmonic Forcing. Figure 3.11 illustrates such a response, using a hardening system as an example. The plot is obtained as a result of solving the equation of motion

$$M\ddot{u} + c\dot{u} + k_0 f(u) = P_0 \cos \Omega t \qquad (3.13)$$

in which the only difference from Eq. 3.10 is the additional viscosity term. The exact solution is rather laborious, so an approximate method is outlined.

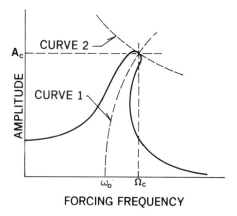

Figure 3.11 Damped response of hardening system.

The method is based on the observation that the influence of damping is important only in the vicinity of resonance, that is, near the maximum possible amplitude of response. To plot the response in Fig. 3.11 we must first draw the undamped amplitude as in Fig. 3.9 using Eq. 3.12. The dashed line, which is the characteristic of free vibrations and is defined by Eq. 3.6 should also be included. In the second step the envelope of maximum response, defined by

$$A = \frac{P_0}{c\Omega} \qquad (3.14)$$

and marked as "curve 2" in Fig. 3.11 is drawn. The intersection of the latter with the free-vibration characteristic (curve 1) approximately determines the resonance point. The anticipated damped response curve may then be inscribed as illustrated in detail in Fig. 3.12. If only the resonance amplitude A_c and frequency Ω_c are desired, it is sufficient to locate the point at which the two curves intersect.

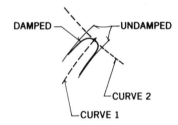

Figure 3.12   Approximate determination of damped response.

At the resonance point, displacement response lags 90° behind forcing. This is another similarity in comparison with a corresponding linear case. The decisive difference is that the resonant frequency will vary with the magnitude of forcing in a nonlinear system, while there is only a single such frequency in a linear system.

SOLVED PROBLEMS

3-1   When the mass in (a) is in the neutral position the springs are unstrained. (If the block is deflected from the neutral position, only one spring is in contact with it.) The right spring is stiffer than the left, $k_2 > k_1$. Determine (1) the largest spring force, (2) the largest deflection, and (3) the natural period of vibration after the block starts with the initial velocity v_0 from the neutral position. Given: M, k_1, and k_2.

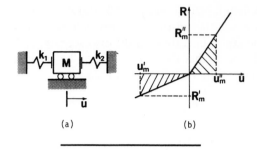

(a)   (b)

Suppose that the initial velocity is directed to the right. The initial kinetic energy is $T = \frac{1}{2} M v_0^2$. At the extreme right deflection designated by u_m'' in (b), the kinetic energy is zero and the strain energy is $\Pi = \frac{1}{2} k_2 (u_m'')^2$. Comparing both values gives

$$M v_0^2 = k_2 (u_m'')^2 \qquad \text{and} \qquad u_m'' = \sqrt{\frac{M}{k_2}}\, v_0$$

After the block returns to the initial position with its velocity reversed, it begins to compress the left spring and reaches the extreme left deflection u_m', which is

$$u_m' = \sqrt{\frac{M}{k_1}}\, v_0$$

The area under the resistance-deflection plot $R-u$ in (b) represents the strain energy accumulated in a spring. The hatched areas must be equal to conform to the principle of energy conservation. From this we may recognize that the largest deflection will occur in the softer spring k_1 and the largest force in the stiffer spring k_2. In conclusion:

1. The largest force occurs in spring k_2,

$$R_m'' = k_2 u_m'' = \sqrt{k_2 M}\, v_0$$

2. The largest deflection is that of the spring k_1,

$$u_m' = \sqrt{\frac{M}{k_1}}\, v_0$$

The period of vibrations of a linear oscillator is

$$\tau = 2\pi \sqrt{\frac{M}{k}}$$

The movement of the block from the initial to the extreme right position and back takes one-half the period of a linear system with spring k_2 or $\pi \sqrt{M/k_2}$. Repeating the same movement with the other spring and adding both time intervals gives the final part of the answer:

3. The period is $\tau = \pi \left(\sqrt{M/k_1} + \sqrt{M/k_2} \right)$.

3-2   Find the natural period of vibrations for a mass restrained by a rigid elastic system, whose characteristic is shown in Fig. 3.4c. Assume $u_0 = 0$ and $v_0 \neq 0$.

For positive deflections the resisting force is

$$R = R_0 + ku.$$

The general equation of motion (Eq. 3.1) becomes here

$$M\ddot{u} = -R_0 - ku \qquad \text{or} \qquad M\ddot{u} + ku = -R_0$$

Noting that it is analogous to Eq. 2.7, we can use Eq. 2.8 as a solution. One can assume $u_p = C$, where C is a constant, and put it in the equation of motion in place of u. The result is $u_p = -R_0/k$ and the complete solution is

$$u = B_1 \cos \omega_1 t + B_2 \sin \omega_1 t - \frac{R_0}{k}$$

where $\qquad \omega_1^2 = k/M$.

The first initial condition is $u(0) = 0$;

$$0 = B_1 - \frac{R_0}{k} \qquad \therefore B_1 = \frac{R_0}{k}$$

The second condition, $\dot{u}(0) = v_0$ is used with Eq. 2.8a:

$$\dot{u} = -B_1 \omega_1 \sin \omega_1 t + B_2 \omega_1 \cos \omega_1 t + \dot{u}_p$$

$$v_0 = 0 + B_2 \omega_1 + 0 \qquad \therefore B_2 = \frac{v_0}{\omega_1}$$

The expression for the velocity is

$$\dot{u} = -\frac{R_0 \omega_1}{k} \sin \omega_1 t + v_0 \cos \omega_1 t$$

The extreme point $u = u_m$ is reached at some $t = t_1$ for which the velocity becomes zero or

$$\tan \omega_1 t_1 = \frac{k v_0}{\omega_1 R_0}$$

The entire period of motion is equal to $4t_1$:

$$\tau = \frac{4}{\omega_1} \arctan \frac{k v_0}{\omega_1 R_0} \qquad \text{with} \qquad \omega_1^2 = \frac{k}{M}$$

Note that the expression for $u(t)$ derived above is valid only for positive u. A separate calculation is needed for negative u.

3-3 The beam stiffness measured with respect to the deflection of mass M is k Lb/in. If the magnitude of deflection exceeds b, the mass comes in contact with one of the springs. Construct the resistance-deflection plot of the system. What will be the natural period of vibrations initiated by giving mass M velocity v_0 sufficient to cross the gap b? Put $k = 1000$ Lb/in., $M = 10$ Lb-s^2/in., and $b = 0.5$ in. The initial velocity is $v_0 = 10$ in./s at the neutral point.

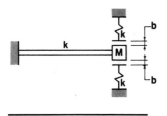

The resistance-deflection plot is shown in Fig. 3.4b. Here we have $k_1 = k$ and $k_2 = 2k$, but it is just as easy to find the answer using general notation. From the conditions of the problem we know that the mass will be oscillating between u_m and $-u_m$, where $u_m > b$. As long as $u \leq u_1 = b$ we have the harmonic motion described by Eq. 2.4,

$$u = \frac{v_0}{\omega_1} \sin \omega_1 t$$

where $\omega_1^2 = k_1/M$. Also

$$\dot{u} = v_0 \cos \omega_1 t$$

The first equation gives us t_1 at which b is reached,

$$t_1 = \frac{1}{\omega_1} \arcsin \left(\frac{b \omega_1}{v_0} \right)$$

The velocity at this instant is

$$\dot{u}_1 = v_0 \left(1 - \sin^2 \omega_1 t_1 \right)^{1/2} = v_0 \left[1 - \left(\frac{b \omega_1}{v_0} \right)^2 \right]^{1/2}$$

Let us think now about point u_1 as the origin of a coordinate system and velocity u_1 as the initial velocity. We want to find the time interval t_2 that is needed to get to the extreme point u_m (i.e., to travel the distance $u_m - b$). The solution of Prob. 3-2 provides the answer. After adjusting the notation, we have

$$\tan \omega_2 t_2 = \frac{k_2 \dot{u}_1}{\omega_2 R_1} \qquad \text{with} \qquad R_1 = k_1 b$$

or

$$t_2 = \frac{1}{\omega_2} \arctan \left(\frac{k_2 \dot{u}_1}{\omega_2 k_1 b} \right)$$

where $\qquad \omega_2^2 = k_2/M$

The total time the mass travels from the center to the extreme point is therefore $(t_1 + t_2)$, so we have

$$\tau = 4t_1 + 4t_2$$

Substitute the numerical values:

$$\omega_1^2 = \frac{k_1}{M} = \frac{1000}{10} \therefore \omega_1 = 10 \text{ rad/s}$$

$$\omega_2^2 = \frac{k_2}{M} = \frac{2000}{10} \therefore \omega_2 = 14.142 \text{ rad/s}$$

$$\frac{b \omega_1}{v_0} = \frac{0.5 \times 10}{10} = 0.5$$

$$\dot{u}_1 = 10 (1 - 0.5^2)^{1/2} = 8.6603 \text{ in./s}$$

$$\tau = 0.2094 + 0.3347 = 0.5441 \text{ s}$$

is the natural period of interest.

3-4 A body of mass $M = 5$ Lb-s^2/in. is restrained by a system with a resisting function shown in Fig. 3.4c.

Put $R_0 = 500$ Lb and $k = 1000$ lb/in. and calculate the maximum velocity when the body is released at a distance $u_m = 3$ in. from the neutral position. Using the result of Prob. 3-2, calculate the natural frequency of vibration.

Prior to release the strain energy (area under the $R–u$ curve) is

$$\Pi = \tfrac{1}{2}(R_0 + R_m)u_m = \tfrac{1}{2}(R_0 + R_0 + ku_m)u_m$$

$$= R_0 u_m + \tfrac{1}{2}ku_m^2 = 6000 \text{ Lb-in.}$$

At the neutral point the kinetic energy is

$$T = \tfrac{1}{2}Mv_m^2 = 2.5v_m^2$$

Setting $\quad\quad T = \Pi$ gives $v_m = 48.99$ in./s

We can reverse the problem and assume that $v_0 = v_m$, that is, the system was started with a velocity v_0 from the neutral position, which makes it possible to use the result of Prob. 3-2,

$$\omega_1^2 = k/M = 1000/5; \quad \therefore \omega_1 = 14.142 \text{ rad/s} \quad\quad v_0 = 48.99 \text{ in./s}$$

$$\tau = \frac{4}{14.142} \arctan \frac{1000 \times 48.99}{14.142 \times 500} = 0.4037 \text{ s}$$

$$\therefore \omega = \frac{2\pi}{\tau} = 15.562 \text{ rad/s}$$

3-5 Consider again Prob. 3-3. What is the maximum displacement u_m when the initial velocity is $v_0 = 10$ in./s? If the initial condition is specified as the displacement $u_0 = u_m = 1.5$ in. with zero initial velocity, what will the velocity be at the neutral point?

The area under the resistance plot, which represents the strain energy, may be broken down into three parts.

$$S_1 = \tfrac{1}{2}R_1 u_1 = \tfrac{1}{2}k_1 u_1^2$$

$$S_2 = R_1(u_m - u_1) = k_1 u_1 u_m - k_1 u_1^2$$

$$S_3 = \tfrac{1}{2}(R_m - R_1)(u_m - u_1)$$

$$= \tfrac{1}{2}k_2\left(u_m^2 - 2u_m u_1 + u_1^2\right)$$

Total strain energy:

$$\Pi = S_1 + S_2 + S_3 = \tfrac{1}{2}k_2 u_m^2 - (k_2 - k_1)u_1 u_m + \tfrac{1}{2}(k_2 - k_1)u_1^2$$

In this case $k_1 = k$ and $k_2 = 2k$, therefore,

$$\Pi = ku_m^2 - ku_1 u_m + \tfrac{1}{2}ku_1^2$$

$$k = 1000 \text{ Lb/in.} \quad\quad u_1 = b = 0.5 \text{ in.}$$

$$\Pi = 1000\left(u_m^2 - 0.5u_m + 0.125\right) \quad\quad\quad (*)$$

This is to be equated with the maximum kinetic energy

$$T = \tfrac{1}{2}Mv_0^2 = \tfrac{1}{2} \times 10 \times 10^2 = 500 \text{ Lb-in.}$$

From $\Pi = T$ we get

$$u_m^2 - 0.5u_m - 0.375 = 0 \quad\quad \therefore u_m = 0.9114 \text{ in.}$$

Equation ($*$) is also used in finding the second part of the answer, which is the maximum velocity resulting from the initial displacement $u_0 = 1.5$ in.

$$\Pi = 1000(1.5^2 - 0.5 \times 1.5 + 0.125) = 1625$$

$$\tfrac{1}{2}Mv_m^2 = \tfrac{1}{2} \times 10v_m^2 = 1625 \quad\quad \therefore v_m = 18.028 \text{ in./s}$$

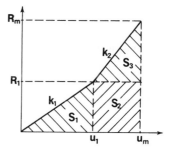

3-6 The attachment brackets of a rigid body with mass $M = 12$ Lb-s^2/in. are fixed to the base by bolts. Because of unavoidable misalignment between the bolt and the bracket axes, the assembly stiffness is $k_1 = 48{,}000$ Lb/in. when pushed down and only $k_2 = 6000$ Lb/in. when pulled up. What is the natural frequency of vertical oscillation? What would be the computed frequency if the analyst treated the stiffness as symmetric and used only the higher figure based on simple axial stiffness of brackets?

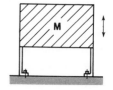

The directional character of stiffness is illustrated by Fig. 3-1b. Using the formula developed therein we have

$$\tau = \pi\left[\left(\frac{M}{k_1}\right)^{1/2} + \left(\frac{M}{k_2}\right)^{1/2}\right] = \pi\left[\left(\frac{12}{48{,}000}\right)^{1/2} + \left(\frac{12}{6000}\right)^{1/2}\right]$$

$$= 0.1902 \text{ s}$$

$$\therefore \omega = 33.04 \text{ rad/s}$$

For a linear system with k_1 only,

$$\omega' = \left(\frac{k_1}{M}\right)^{1/2} = 63.25 \text{ rad/s}$$

which would be a gross error. One may notice that including the influence of gravity would have a twofold effect. First, it would make the frequency amplitude dependent. Second, for small amplitudes ω would increase and might even reach ω'.

3-7 A rigid block, $W = 3000$ Lb, is restrained by a single spring from above and a plate with two springs from below. The lower two springs are preloaded, each to $P_0 = 10,000$ Lb. The block is pressed down so that it deflects by 0.5 in., then released. As soon as the plate hits the stop on the way back, the block separates from the plate and later it engages two impact-absorbing springs, each of stiffness k. Assuming that there are no losses in the system, calculate the amplitude of vibrations. Put $k = 8000$ Lb/in. and $b = 0.75$ in. Ignore the effect of gravity.

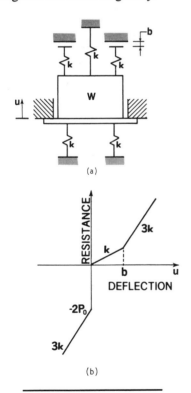

(a)

(b)

The characteristic in (b) shows that deflecting the system down by $\delta = 0.5$ in. induces the resisting force

$$R'_m = 2P_0 + 3k\delta = 2 \times 10,000 + 3 \times 8000 \times 0.5 = 32,000 \text{ Lb}$$

The strain energy prior to release is

$$\Pi = \tfrac{1}{2}(2P_0 + R'_m)\delta = \tfrac{1}{2}(2 \times 10,000 + 32,000)0.5 = 13,000 \text{ Lb-in.}$$

Once the weight returns to the initial position, this energy is converted into kinetic energy. When the upper stationary point is reached, the same level of Π must be attained. The amount of strain energy needed for the initial closing of gap b is

$$\Pi_1 = \tfrac{1}{2}kb^2 = \tfrac{1}{2} \times 8000 \times 0.75^2 = 2250 \text{ Lb-in.}$$

The remaining kinetic energy is

$$\Pi_2 = \Pi - \Pi_1 = 13,000 - 2250 = 10,750 \text{ Lb-in.}$$

Denoting by u_m the maximum upward deflection, we note that the resistance force changes from

$$R_b = kb = 8000 \times 0.75 = 6000 \text{ Lb} \qquad \text{at} \quad u = b$$

to

$$R_m = R_b + 3k(u_m - b) = 24,000u_m - 12,000 \qquad \text{at} \quad u = u_m$$

The latter is found from

$$\tfrac{1}{2}(6000 + 24,000u_m - 12,000)(u_m - 0.75) = \Pi_2 = 10,750$$

$$\therefore u_m = 1.4789 \text{ in.}$$

The stationary points are therefore 0.5 in. below and 1.4789 in. above $u = 0$ level, respectively. Defining the amplitude as half the entire range, we get

$$A = \tfrac{1}{2}(0.5 + 1.4789) = 0.9895 \text{ in.}$$

3-8 The resistance-deflection curve of the system shown in (a) is $R = k_0(u + 0.75u^3)$. Calculate and plot the natural frequency as a function of amplitude for the values of the independent variable up to $u_m = 2$ in. Put $k_0 = 16$ Lb/in. and $M = 1.0$ Lb-s^2/in.

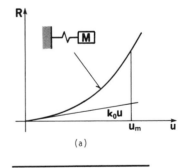

(a)

If only the u-term had been present in the equation for R, the system would have been linear and for $u = 2$ in. the resistance would have been $R_m = 2k_0$. The actual resisting force at this maximum amplitude is $R_m = k_0(2 + 0.75 \times 2^3) = 8k_0$ or four times the linear value. This indicates we are dealing with a strongly nonlinear problem. Yet, in spite of it, we will try a fairly crude approximation offered by Eq. 3.9. For maximum amplitude $A = 2$ in., $f_0 = 8$, and

$$f_c = f(1.4142) = 1.4142 + 0.75 \times 1.4142^3 = 3.5355$$

and

$$F(A) = \frac{1}{3A}(f_0 + 2.8284f_c) = 3.0$$

From Eq. 3.6:

$$\omega^2 = \frac{16}{1.0} \times 3.0 \quad \text{and} \quad \omega = 6.928 \text{ rad/s}$$

When this procedure is repeated for smaller amplitudes, we obtain the following table, which is plotted in (b).

A (in.)	ω (rad/s)
2.00	6.928
1.75	6.364
1.50	5.831
1.25	5.338
1.00	4.899
0.50	4.243

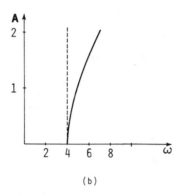

(b)

When A tends to zero, the spring characteristic may be approximated by $u \approx 16x$, which corresponds to $\omega = 4$ rad/s. A more accurate approximation, which can be found on page 174 of Ref. 2, gives $\omega = 7.498$ rad/s at the amplitude of 2 in. The error of our solution is therefore about 8%.

Note that this problem could be solved more easily by developing a formula for $F(A)$. The purpose of the solution, however, was to illustrate the general approach.

3-9   Two identical springs, each with a characteristic

$$R = k(u + 2u^3)$$

are joined in a series. What is the resistance-deflection curve $\bar{R}-\bar{u}$ for the assembly?

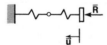

If one of the springs is compressed by u, the assembly deflection is $\bar{u} = 2u$. The resisting force in each single spring is the same as that of the assembly. We can rewrite the given equation as

$$R = k\left[\frac{\bar{u}}{2} + 2\left(\frac{\bar{u}}{2}\right)^3 \right]$$

and since $\bar{R} = R$, we obtain

$$\bar{R} = \frac{k}{2}\left(\bar{u} + \frac{1}{2}\bar{u}^3 \right)$$

It is interesting that even in this simple case the assembly equation does not have the same form as the characteristic of a component.

3-10   Two masses placed on a frictionless surface are connected with an elastic element whose characteristic $R(u)$ is nonlinear. The relative displacement of the ends of the element is $u = u_1 + u_2$. Assume that the system is in a vibratory motion, which can be approximated by

$$u_1 = A_1 \cos \omega t \quad \text{and} \quad u_2 = A_2 \cos \omega t$$

where both amplitudes are frequency dependent. Using Eqs. 3.6 and 3.9, determine the frequency ω in terms of the system properties. As an aid, refer to Prob. 1-5, noting, however, a different sign convention for the displacements.

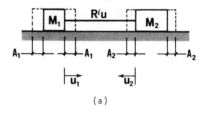

(a)

Equation 3.6 may be written for either of the two masses,

$$\omega^2 = \frac{k_{01}}{M_1} F(A_1) = \frac{k_{01}}{M_1} \frac{1}{3A_1}(f_{01} + 2.8284 f_{c1})$$

where the added index shows that M_1 is chosen for the following development. Since the total momentum of an unconstrained system in a vibratory motion is zero, the velocities are inversely proportional to the masses. The same is true with respect to the displacement amplitudes, because of the harmonic nature of motion,

$$\frac{A_1}{A_2} = \frac{M_2}{M_1}$$

Denoting by A the total amplitude, we get

$$A_1 + A_2 = A \quad \text{and} \quad A_1 = \frac{AM_2}{M_1 + M_2}$$

The equation for the frequency becomes now

$$\omega^2 = \frac{1}{M^*} \frac{1}{3A}(k_{01}f_{01} + 2.8284 k_{01}f_{c1})$$

in which M^* is the effective mass calculated from

$$\frac{1}{M^*} = \frac{1}{M_1} + \frac{1}{M_2}$$

The resistance-deflection equation for the element is given in the form

$$R = k_0 f(u)$$

The deflection u may be thought of as a sum of two components, each coming from one of two springs connected in series. A resolution of the total displacement amplitude is shown in (b). As previously stated, these amplitudes are inversely proportional to the respective masses. The reactions for all three items being equal, we can write

$$R_1 = k_{01} f(u_1) \qquad R_2 = k_{02} f(u_2) \qquad R_1 = R_2 = R$$

At full amplitude we have

$$k_{01} f(A_1) = k_0 f(A)$$

that is,

$$k_{01} f_{01} = k_0 f_0$$

We cannot make the same claim about the second term in the frequency equation, but it would be reasonable to assume

$$k_{01} f(0.7071 A_1) \approx k_0 f(0.7071 A)$$

that is,

$$k_{01} f_{c1} \approx k_0 f_c$$

which gives us the final approximation for the natural frequency:

$$\omega^2 = \frac{k_0}{M^*} F(A)$$

where

$$F(A) = \frac{1}{3A}(f_0 + 2.8284 f_c)$$

If instead of the combined characteristic $R(u)$ we are given a separate expression for each component spring, it may take much work to obtain an analytical relation for $R(u)$. This is seldom necessary, however, because all we need are the reactions for some selected deflections.

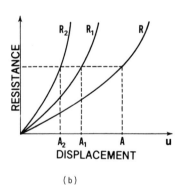

(b)

3-11 In Prob. 3-10 it was determined that the natural frequency of two masses joined with an elastic link is found from

$$\omega^2 = \frac{k_0}{M^*} F(A)$$

where k_0 is a stiffness parameter appearing in $R = k_0 f(u)$, M^* is the effective mass found from $1/M^* = 1/M_1 + 1/M_2$, and $F(A)$ is a function of the amplitude,

$$F(A) = \frac{1}{3A}(f_0 + 2.8284 f_c)$$

where $f_0 = f(A)$ and $f_c = f(0.7071A)$. Check the accuracy of this formulation by using it to find the frequency of the system shown in the figure when amplitude $A = 4.0$ in. The resistance is

$$R_1 = 16(u_1 + 0.75 u_1^3)$$

for each of the two springs and $M = 1.0$ Lb-s^2/in. for each mass. Compare the results with Prob. 3-8, noting the identical behavior of both systems due to symmetry of this system.

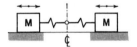

If the relative displacement of masses during vibratory motion is u, then $u_1 = u/2$ and

$$R = R_1 = 16\left(\frac{u}{2} + 0.75 \times \frac{u^3}{8}\right) = 8\left(u + \frac{3}{16} u^3\right).$$

$$\frac{1}{M^*} = \frac{1}{M} + \frac{1}{M} = \frac{2}{M} \qquad \therefore M^* = 0.5$$

$$f_0 = f(A) = 4 + \frac{3}{16} \times 4^3 = 16$$

$$f_c = f(0.7071A) = 2.8284 + \frac{3}{16} \times 2.8284^3 = 7.071$$

$$F(A) = \frac{1}{3 \times 4}(16 + 2.8284 \times 7.071) = 3.0$$

$$\omega^2 = \frac{8}{0.5} \times 3 = 48 \qquad \therefore \omega = 6.928 \text{ rad/s}$$

The result is the same as in Prob. 3-8 at the amplitude of a single mass $A_1 = 2.0$ in. (Notice that $k_0 = 8$ because this is a constant multiplier in the expression for R.)

3-12 Calculate the natural frequency of a mathematical pendulum of length L oscillating between two horizontal positions.

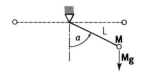

The equation of motion in the direction tangent to the path is

$$ML\ddot{\alpha} = -Mg\sin\alpha \quad \text{or} \quad M\ddot{\alpha} + M\frac{g}{L}\sin\alpha = 0$$

According to notation of Sect. 3D,

$$k_0 = \frac{Mg}{L} \quad \text{and} \quad f(\alpha) = \sin\alpha$$

Using Eq. 3.9, we have

$$A = \pi/2 \quad f_0 = f(A) = \sin\frac{\pi}{2} = 1.0$$

$$f_c = f\left(\frac{0.7071\pi}{2}\right) = \sin 1.1107 = 0.896$$

$$F(A) = \frac{1}{3(\pi/2)}(1.0 + 2.8284 \times 0.896) = 0.75$$

Equation 3.6 yields the approximate frequency,

$$\omega^2 = \frac{k_0}{M}F(A) = 0.75\frac{g}{L} \quad \therefore \omega = 0.866(g/L)^{1/2}$$

If we used a small-deflection approximation and replaced $\sin\alpha$ by α, we would obtain $\omega_0 = (g/L)^{1/2}$, regardless of the amplitude. A different and more precise approach using tables of elliptical integrals gives $\omega = 0.8472(g/L)^{1/2}$, which indicates an error of just over 2% in our calculation.

3-13   A linear mass-spring system is subjected to a harmonic force $P = P_0\cos\Omega t$. Discuss the relationship between the amplitude and the forcing frequency. Compare the resulting expression with Eq. 3.12, which is valid for a more general, nonlinear system.

The results of Sect. 2G may be used if $\sin\Omega t$ is replaced by $\cos\Omega t$. The steady-state component of response is (Eqs. 2.24 and 2.25)

$$u = \frac{P_0/k}{(1 - \Omega^2/\omega^2)}\cos\Omega t = A\cos\Omega t$$

in which A denotes the amplitude. One can write

$$\frac{\Omega^2}{\omega^2} = 1 - \frac{P_0}{Ak}$$

Thus the only essential difference between this and Eq. 3.12 is that 1 is replaced by $F(A)$ in the latter.

Treating the amplitude of the forcing function P_0, stiffness k, and the natural frequency ω as fixed values, we can make a plot of amplitude A as a function of forcing frequency Ω.

Note that at $\Omega/\omega = 1.0$ the amplitude A has negative values, which means that the displacement is out of phase with the forcing function. The ordinate of the plot is the absolute value of the nondimensional amplitude. Except for the sign there is no difference between this plot and Fig. 2-4.

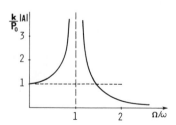

3-14   The system with a nonlinear spring described in Prob. 3-8 is subjected to a force $P = P_0\cos\Omega t$ in which $P_0 = 2$ Lb. Calculate the amplitude of forced vibrations as a function of the forcing frequency Ω.

We have $k_0 = 16$ Lb/in. and $\omega_0^2 = 16$ (rad/s)^2 and Eq. 3.12 gives us

$$\Omega^2 = 16F - \frac{2}{A}$$

where F is a function of amplitude A. Employing Eq. 3.9:

$$F = \frac{1}{3A}(f_0 + 2.8284f_c)$$

with $f_0 = f(A)$ and $f_c = f(0.7071A)$, while $f = u + 0.75u^3$. After some algebraic manipulations we get

$$F = \frac{1}{3A}(3A + 1.5A^3) = 1 + 0.5A^2$$

We choose $A = 3$ in. to be the maximum amplitude of interest. According to the method described in Sect. 3E, the following table of values is calculated.

A	$F(A)$	Ω^2
-3.0	5.5	88.67
-2.0	3.0	49.0
-1.0	1.5	26.0
-0.1	1.005	36.08
0.	1.0	∞
0.1	—	—
0.2	1.02	6.32
1.0	1.5	22.0
2.0	3.0	47.0
3.0	5.5	87.33

Note that there is no value of Ω corresponding to $A = 0.1$ in., because the smallest positive amplitude is equal to the static deflection $P_0/k_0 = 2/16 = 0.125$ in. The plot of the absolute value of the amplitude is shown in the figure.

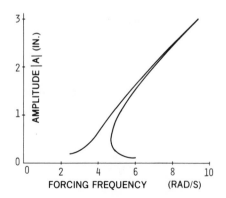

EXERCISES

3-15 When the mass M is in the neutral position, it is given the initial velocity v_0. The restraining system has a resisting force defined in Fig. 3.4c and the quantity $\omega = k/M$ is also given. Determine the extreme displacement u_m and the maximum resisting force R_m.

3-16 Weight W is suspended on a spring of stiffness k_1. The initial velocity v_0 at the neutral position is large enough that the weight may cross the gap b and engage the lower spring. The total stiffness of both springs is k_2. Calculate the period of vibration.

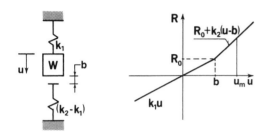

3-17 An isolation spring exhibits a stiffer characteristic during loading (k_1) than during unloading (k_2). Calculate the natural frequency of vibration of mass M attached to such a spring. (Notice that according to our definitions this is an inelastic spring, because the loading and unloading paths do not coincide).

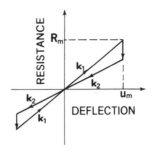

3-18 The resistance-deflection curve of a mass-spring system is shown in Fig. 3.4c. Take the amplitude as $u_m = A = 3$ in. and calculate the corresponding frequency using the harmonic approximation with Eq. 3.8. Solve for $R_0 = 500$ Lb, $M = 5$ Lb-s^2/in., and $k = 1000$ Lb/in. (The same data as in Prob. 3-4).

3-19 Find the answer to Prob. 3-18 by using a simpler approximation in the form of Eq. 3.9 rather than Eq. 3.8. Compare that answer with the exact value of $\omega = 15.562$ rad/s.

3-20 Consider again the system in Prob. 3-8, now with the added viscous damping c. The magnitude of c is such that when the system is treated as linear, it has damping ratio $\zeta = 0.12$. Forcing is defined by $P = P_0 \cos \Omega t$ with $P_0 = 6$ Lb. What is the resonant amplitude of the system, and how does it compare with the amplitude of the system treated as linear?

CHAPTER

4

Damping

Besides the fundamental concept of viscous damping, the other types of energy dissipation like frictional damping, structural damping, and velocity-squared damping are considered. All these other types are reducible to viscous damping as long as the equivalent ratio ζ is small. The dynamic response of a damped SDOF system to an applied force can be performed in the same manner as in Chapter 2 as long as ζ is known. A direct approach to certain problems with nonviscous damping is also a practical proposition.

4A. Viscous Damping was introduced in Chapter 2 in the form of a damper parallel to a spring in an SDOF system (Fig. 2.1). Here is the summary of its properties and influence.

1. A damper with a constant c generates a resistance force of magnitude $c\dot{u}$, where $\dot{u}$ is the velocity of one end of the damper with respect to the other.
2. The damping ratio is defined as $\zeta = c/c_c$, where $c_c = 2\sqrt{kM}$.
3. The amplitude of free vibrations gradually diminishes according to Eq. 2.11.
4. When the system is under harmonic forcing, the magnification factor at resonance is

$$\mu_{max} = \frac{1}{2\zeta}$$

When the basic oscillator is subjected to the harmonic force $P = P_0 \sin \Omega t$, it is useful to know the energy dissipated (lost) by the damper in one cycle of motion:

$$D = \pi A^2 c \Omega \qquad (4.1)$$

where A is the amplitude of deflection. The work performed by the driving force with amplitude P_0 per cycle of motion is

$$L_p = \pi A P_0 \sin \theta \qquad (4.2)$$

The phase angle θ is defined in terms of damping ratio ζ by Eq. 2.17.

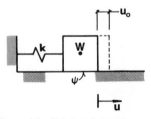

Figure 4.1 Frictional damping.

4B. Frictional Damping, also called *Coulomb damping*, is a result of dry friction between two parts whose surfaces slide against each other. Figure 4.1 shows a simple example of this type of damping. The weight W subjected to the force of gravity rests on a plane with respect to which it can slide. The kinetic coefficient of friction between both surfaces is ψ. The spring with stiffness k provides the restraining force. The equation of motion of the weight is

$$M\ddot{u}+ku=-F\,\mathrm{sgn}\,\dot{u} \tag{4.3}$$

where F is the friction force, $F=\psi W$. The function $\mathrm{sgn}\,\dot{u}$ is read "sign of $\dot{u}$." It has the value of $+1$ when velocity is positive (the weight moving to the right) and the value of -1 for negative velocity. In effect, the direction of the friction force is opposite to that of the velocity. The equation of motion is non-linear because of the presence of $\mathrm{sgn}\,\dot{u}$ function. If we give the weight an initial displacement u_0 and release it without an initial velocity, Eq. 4.3 becomes

$$M\ddot{u}+ku=F \tag{4.4}$$

and is valid until the weight reaches the opposite extreme point. This is a linear equation of a weight subjected to a constant force and may be solved by the methods of Chapter 2. Note that for the motion to take place, the force in the spring must be larger than the friction force, that is,

$$ku_0>F \quad \text{or} \quad u_0>a=\frac{F}{k}$$

It can be easily demonstrated that after each half-cycle of motion (i.e., after a movement from one to the next stationary point) the amplitude diminishes by $2a$. The oscillations come to a stop after the amplitude becomes less than a. This is an interesting case in comparison with a viscous damping, where the motion lasts indefinitely, at least in theory.

Also note the second important difference. The frequency of friction-damped motion does not differ from the undamped frequency of the system. The influence of viscous damping, on the other hand, is to reduce the natural frequency.

4C. Forced Vibrations with Frictional Damping. Consider steady-state vibrations of a system in Fig. 4.1 subjected to a horizontal force $P=P_0\sin\Omega t$. One of the simplest ways of looking at the problem is to ignore any phase difference between the applied forcing and the displacement response. The plot of the driving force P and frictional resistance F in Fig. 4.2 is based on the response $u=A\sin\Omega t$.

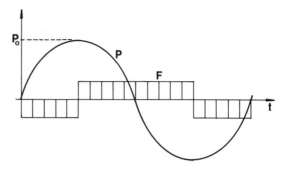

Figure 4.2 Simplified relation between driving force P and frictional resistance F.

In most technical applications F is much smaller than P_0 and a reasonable approximation for the magnitude of response is obtained when one assumes that the cumulative effect of P and F is such that one should apply the force

$$P'=P_{ef}\sin\Omega t \tag{4.5}$$

to the undamped system. The amplitude of this force is

$$P_{ef}=\left[P_0^2-\left(\frac{4}{\pi}F\right)^2\right]^{1/2} \tag{4.6}$$

and the response, by Eq. 2.24, is

$$u=\frac{P_{ef}}{k}\frac{1}{1-\Omega^2/\omega^2}\sin\Omega t \tag{4.7}$$

For Ω approaching ω, the amplitude of vibration tends to infinity. This is another important difference in comparison with viscous damping.

4D. Viscous Equivalent of Frictional Damping may be introduced by considering the amount of dissipated energy during one cycle of motion. If the friction force is F and the displacement amplitude A, the work performed by F in one cycle is $4FA$ and it is equal to the dissipated energy. Equating this with the viscous dissipation of energy expressed by Eq. 4.1, we get

$$c'=\frac{4F}{\pi A\Omega} \tag{4.8}$$

as the constant of the equivalent viscous damper. The equivalent damping ratio becomes

$$\zeta'=\frac{c'}{c_c}=\frac{2F}{\pi Ak}\frac{\omega}{\Omega} \tag{4.9}$$

When this value is substituted for ζ, one can use Eqs. 2.15 and 2.16 to derive the amplitude of response in Eq. 4.7.

4E. Viscous Equivalent of Structural Damping. If a sample of material is subjected to a tensile or a compressive stress, which grows very slowly from zero to its maximum value, the characteristic is a straight line, as shown in Fig. 4.3. After a few load cycles applied with some finite velocity, we notice that the characteristic forms a narrow loop (solid line in Fig. 4.3). The area under the loading line is the work performed by the external forcing on the system (positive work), while the area under the unloading curve counts as negative. The area within the loop is equal to the work U_s that must be supplied from the outside to carry out the loading cycle. Alternatively, we may say that U_s is the energy dissipated by the sample in one cycle. The damping associated with this material behavior is called *structural* or *material damping*. It is thought to be caused by the internal friction within the volume of material. The energy dissipated in one cycle is assumed to be proportional to the square of displacement amplitude A:

$$D = U_s = sA^2 \qquad (4.10)$$

where s is a proportionality constant determined from an experiment. Equating this expression with Eq. 4.1 we obtain the equivalent viscous damping constant

$$c'' = \frac{s}{\pi\Omega} = \frac{\eta k}{\Omega} \qquad (4.11)$$

The new constant η introduced by the second equality is called the *loss factor*. The equivalent damping ratio is

$$\zeta'' = \frac{\eta}{2}\frac{\omega}{\Omega} \qquad (4.12)$$

The magnification factor for harmonic forcing is obtained by substituting ζ'' in place of ζ in Eq. 2.16:

$$\mu = \frac{1}{\left[\left(1 - \Omega^2/\omega^2\right)^2 + \eta^2\right]^{1/2}} \qquad (4.13)$$

At resonance when $\Omega = \omega$ we have $\mu = 1/\eta$.

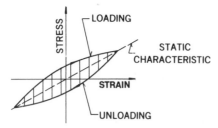

Figure 4.3 Loading and unloading curves when structural damping is present.

4F. Viscous Equivalent of Velocity-Squared Damping. It can be shown that when the resisting force is proportional to the square of velocity, the energy dissipated by a velocity-squared damper in one cycle of motion is

$$D_a = \tfrac{8}{3}c_a A^3\Omega^2$$

where c_a is the proportionality constant, A is the amplitude of the assumed harmonic motion, and Ω is the forcing frequency. (See Prob. 4-10). If this is equated with D from Eq. 4.1, the equivalent viscous damping constant becomes

$$c''' = \frac{8c_a A\Omega}{3\pi} \qquad (4.14)$$

When the amplitude of a harmonic force is P_0, we can find the deflection amplitude A by inserting $\zeta''' = c/c_c$ in place of ζ in Eqs. 2.15 and 2.16. The equation thus obtained will have A on both sides. To find A we rewrite the expression to

$$\left(\frac{8c_a\Omega^2}{3\pi}\right)^2 A^4 + k^2\left(1 - \frac{\Omega^2}{\omega^2}\right)^2 A^2 - P_0^2 = 0 \qquad (4.15)$$

where A^2 is the unknown.

SOLVED PROBLEMS

4-1 A basic oscillator consisting of mass M, spring k, and damper c (Fig. 2.1a) is put in motion by an initial displacement u_0 with no initial velocity. Calculate the energy dissipated by the damper during one cycle of motion. Express the result in terms of the damping ratio ζ.

The equation of motion 2.10 is

$$u = Ae^{-\omega\zeta t}\cos(\omega_d t - \alpha_d)$$

Let us choose some initial time t_1 so that $\omega_d t_1 - \alpha_d = 0$ and

$$u_1 = Ae^{-\omega\zeta t_1}$$

One period later, for $t_2 = t_1 + \tau_d$:

$$u_2 = Ae^{-\omega\zeta(t_1 + \tau_d)} \quad \text{and} \quad \frac{u_2}{u_1} = e^{-\omega\zeta\tau_d}$$

Time t_1 was chosen so that both deflections are at the stationary points of the mass, where the only energy the system has is the strain energy of spring. The difference

$$D = \Pi_1 - \Pi_2 = \tfrac{1}{2}ku_1^2 - \tfrac{1}{2}ku_2^2$$

is the energy dissipated by the damper in one cycle of motion.

$$D = \tfrac{1}{2} k u_1^2 \left(1 - \frac{u_2^2}{u_1^2} \right) = \Pi_1 \left(1 - e^{-2\omega\zeta\tau_d} \right)$$

$$2\omega\zeta\tau_d = 2\omega\zeta \frac{2\pi}{\omega_d} = \frac{4\pi\zeta}{\left(1 - \zeta^2\right)^{1/2}}$$

Even for small damping ratios the loss of energy in one cycle is noticeable. For $\zeta = 0.01$, $D = 0.1181\Pi_1$, while for $\zeta = 0.1$, $D = 0.7172\Pi_1$, where Π_1 is the strain energy in the beginning of a cycle.

4-2 Weight W is suspended on a light rigid rod of length $2L$. At the neutral position both springs are unstretched. The natural frequency of the system was experimentally determined to be f_d. What is the damping ratio ζ caused by the liquid into which the weight is submerged? Solve with $W = 20$ Lb, $L = 8$ in., $k = 200$ Lb/in., and $f_d = 6.32$ Hz.

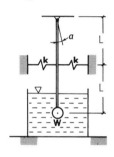

In Chapter 1 a similar problem (1-13) was solved using the energy method. A more direct approach is employed here. Imagine the rod deflected by α. The elastic resistance is then $2kL\alpha$ and the gravity force W is offset by $2L\alpha$. If the effective stiffness is denoted by K, we have the total moment of resistance equal to

$$K\alpha = (2kL\alpha)L + 2L\alpha W \qquad \therefore K = 2kL^2 + 2WL$$

The moment of inertia about the pivot point is $J = 4(WL^2/g)$ and the undamped frequency ω:

$$\omega^2 = \frac{K}{J} = \frac{kg}{2W} + \frac{g}{2L}$$

The damped frequency is defined by

$$\omega_d^2 = \omega^2 \left(1 - \zeta^2 \right) \qquad \therefore \zeta^2 = 1 - \frac{\omega_d^2}{\omega^2}$$

Substituting the numerical values:

$$\omega^2 = \left(\frac{200}{2 \times 20} + \frac{1}{2 \times 8} \right) 386 = 1954.1$$

$$\omega_d = 2\pi f_d = 2\pi \times 6.32 = 39.71$$

$$\zeta^2 = 1 - \frac{39.71^2}{1954.1} \qquad \therefore \zeta = 0.4394$$

4-3 A rigid block, which can pivot about a point, has a moment of inerita $J = 24$ Lb-s^2-in. with respect to that point. The block is restrained by a spring, $k = 10,500$ Lb/in., and three identical viscous dampers, $c = 2.1$ Lb-s/in. Calculate the damping ratio ζ using $a = 12$ in.

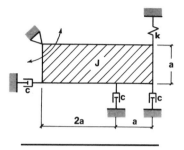

The equation of motion (Eq. 2.1) is written here as

$$J\ddot{\alpha} + c'\dot{\alpha} + K\alpha = 0$$

where α is the angle of small rotation about the pivot point. It is easy to determine that $K = (3a)^2 k = 13.608 \times 10^6$ Lb-in./rad, but to calculate the effective damping coefficient c' we must first know the forces applied by the dampers to the block. The velocities along the axes of dampers, going from left to right, are

$$a\dot{\alpha} \qquad 2a\dot{\alpha} \qquad 3a\dot{\alpha}$$

The moments of damper forces about the pivot point:

$$a^2 c\dot{\alpha} \qquad (2a)^2 c\dot{\alpha} \qquad (3a)^2 c\dot{\alpha}$$

The total moment of viscous resistance,

$$c'\dot{\alpha} = 14 a^2 c\dot{\alpha} = 4233.6\dot{\alpha}$$

$$\therefore c' = 4233.6 \text{ Lb-in.-s/rad}$$

Critical damping:

$$c'_c = 2\sqrt{KJ} = 36,144$$

$$\therefore \zeta = \frac{c'}{c'_c} = 0.1171$$

4-4 The figure shows a fragment of a plot of the magnification factor μ for an SDOF system. Two values of the forcing frequency, Ω_1 and Ω_2, for which μ is the same are given, and so is the natural frequency ω. Determine the viscous damping ratio ζ.

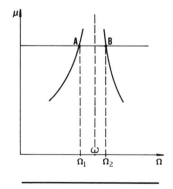

We set $x = \Omega/\omega$ in Eq. 2.16 for the sake of brevity and equate the two magnification factors, obtaining:

$$\left(1 - x_1^2\right)^2 + \left(2\zeta x_1\right)^2 = \left(1 - x_2^2\right)^2 + \left(2\zeta x_2\right)^2$$

or equivalently,

$$2\left(1 - 2\zeta^2\right)\left(x_2^2 - x_1^2\right) - \left(x_2^4 - x_1^4\right) = 0$$

This may easily be reduced to

$$\zeta^2 = \tfrac{1}{4}\left(2 - x_2^2 - x_1^2\right)$$

and on returning to the original notation,

$$\zeta^2 = \tfrac{1}{4}\left(2 - \frac{\Omega_2^2 + \Omega_1^2}{\omega^2}\right)$$

4-5 A basic oscillator is put into motion by an initial velocity v_0 at zero initial deflection. Derive an approximate expression for the energy dissipated by the damper on the path between the neutral position and the extreme deflected position. Assume damping to be small.

The displacement due to initial velocity is given by Eq. 2.9,

$$u = \frac{v_0}{\omega_d} \sin \omega_d t \cdot \exp(-\omega\zeta t)$$

and the velocity may be obtained by differentiation:

$$\dot{u} = \left(v_0 \cos \omega_d t - \frac{v_0}{\omega_d}\omega\zeta \sin \omega_d t\right) \exp(-\omega\zeta t)$$

The maximum displacement is reached when $\dot{u} = 0$ or

$$\tan \omega_d t_m = \frac{\omega_d}{\omega}\frac{1}{\zeta} \approx \frac{1}{\zeta}$$

the latter near-equality being a good approximation for small ζ. Since the right-hand side of the equation is a fairly large number, we expect the angle $\omega_d t_m$ to be only slightly less than $\pi/2$. In fact we may notice that up to $\zeta = 0.2$ the last equality is quite well satisfied by

$$\omega_d t_m \approx \frac{\pi}{2} - \zeta$$

The maximum displacement is then given by

$$u_m \approx \frac{v_0}{\omega_d}\exp\left[\left(\zeta - \frac{\pi}{2}\right)\zeta\right]$$

after assuming again $\omega/\omega_d \approx 1$ and $\sin(\pi/2 - \zeta) \approx 1.0$. After the maximum displacement is reached, there is only the strain energy

$$\Pi_m = \tfrac{1}{2}ku_m^2$$

in the system. If damping were not involved, the strain energy would be

$$\Pi_m' = \frac{1}{2}ku_m'^2 = \frac{1}{2}k\left(\frac{v_0}{\omega}\right)^2$$

which would be the same as the initial kinetic energy T_0. The energy lost between the neutral point and the stationary point is therefore

$$D_1 = \Pi_m' - \Pi_m = T_0\left(1 - \frac{u_m^2}{u_m'^2}\right) \quad \text{or} \quad D_1 \approx T_0\left[1 - \exp\left(2\zeta^2 - \pi\zeta\right)\right]$$

4-6 Weight W restrained by spring k can slide along a surface with a friction coefficient ψ (see Fig. 4.1). The initial displacement is u_0 and the weight is released without an initial velocity. Describe how the amplitude of oscillations changes during motion. How many cycles will the weight go through before it stops if $W = 100$ Lb, $k = 500$ Lb/in., $\psi = 0.3$, and $u_0 = 1.25$ in.?

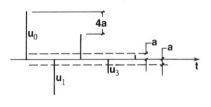

We have the following work-energy equation to describe the movement of weight from one extreme position to the other.

$$\tfrac{1}{2}ku_0^2 - F(u_0 + u_1) = \tfrac{1}{2}ku_1^2$$

The second term on the left is the work of the friction force F over the segment of interest; u_0 and u_1 denote the absolute values of the extreme deflections. Rewriting, we get

$$\tfrac{1}{2}k(u_0 - u_1)(u_0 + u_1) = F(u_0 + u_1)$$

or

$$u_1 = u_0 - \frac{2F}{k} = u_0 - 2a$$

During the travel from the first to the second stationary point, the amplitude is reduced by $2a$. This can be proved to take place for every half-cycle that follows. The diagram of motion is shown in the figure.

The motion stops when the amplitude falls below a. This is because the force in the spring is not sufficiently large to "restart" the mass from a stationary point. In our example, $F = W\psi = 100 \times 0.3 = 30$ Lb, and $a = F/k = 30/500 = 0.06$ in. Since the amplitude diminishes by $4a$ after each full cycle, there is a possibility of $1.25/(4 \times 0.06) = 5.208$ cycles. After 5 cycles the amplitude is

$$1.25 - 5 \times 4 \times 0.06 = 0.05 \text{ in.}$$

As this is less than $a = 0.06$ in., the weight will stop after 5 cycles.

4-7 Mass $M = 1.0$ Lb-s²/in. is restrained by a spring $k = 100$ Lb/in. and a frictional damper developing the force $F = 20$ Lb. (The role of that damper is identical with that of the ground in Fig. 4.1.) As a result of an impulse, the mass acquires the initial velocity v_0. Calculate the maximum displacement u_m for $v_0 = 10$ and $v_0 = 50$ in./s.

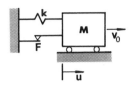

The initial kinetic energy must be equal to the strain energy at $u = u_m$ plus the work performed by the friction force:

$$\tfrac{1}{2}Mv_0^2 = \tfrac{1}{2}ku_m^2 + Fu_m$$

or

$$u_m^2 + \frac{2F}{k}u_m - \frac{M}{k}v_0^2 = 0$$

$$\therefore u_m = \left[\left(\frac{F}{k}\right)^2 + \frac{M}{k}v_0^2\right]^{1/2} - \frac{F}{k}$$

Inserting the numbers:

$$u_m = \left(0.04 + \frac{v_0^2}{100}\right)^{1/2} - 0.2$$

For $v_0 = 50$ in./s, $u_m = 4.804$ in.

Notice that for fairly large initial velocities, such that u_m is, say, five times the value $a = F/k$, one can use the approximation

$$u_m \approx \left(\frac{M}{k}\right)^{1/2}v_0 - a$$

which is the same as saying that friction can be at first ignored and then the value $a = F/k$ subtracted from the calculated displacement. This statement may be useful in more involved problems.

4-8 Plot a resistance-deflection characteristic for a frictional damper as in Prob. 4-7. The deflection starts at the neutral position (unstretched spring), grows to the maximum value A, diminishes to $-A$, and returns to the neutral position. The friction force is F and the spring constant is k. Calculate the energy that must be expended to execute this cycle of movement.

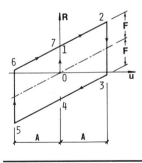

The initial status of the system is defined by point 0 in the plot. During one full cycle of motion the resistance changes according to the polygonal curve 0-1-2-3-4-5-6-7. This curve is sometimes called a hysteresis loop. The area it encloses represents the energy that must be supplied from outside to perform this cycle of motion, and it is equal to $4FA$. The broken line represents the characteristic of the spring itself.

4-9 A body subjected to the force of gravity can pivot about a support point. The moment of friction forces opposing the motion at the support is $\mathfrak{M}_f$. The body is initially deflected by a small angle α_0 and then released without an initial velocity. Develop a formula for a decrease of the deflection angle after each cycle of motion.

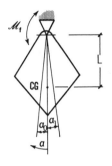

When the body travels from the initial position to the lowest point, the work of gravity forces is

$$U_g' = Mg(L - L\cos\alpha_0) \approx \tfrac{1}{2}MgL\alpha_0^2$$

From the lowest point to the next extreme at $\alpha = -\alpha_1$ the work of gravity is negative:

$$U_g'' = -\tfrac{1}{2}MgL\alpha_1^2$$

The net work performed on the system by the gravity is equal to the work of the friction moment:

$$\tfrac{1}{2}MgL(\alpha_0^2 - \alpha_1^2) = \mathfrak{M}_f(\alpha_0 + \alpha_1)$$

$$\therefore \alpha_1 = \alpha_0 - \frac{2\mathfrak{M}_f}{MgL}$$

It can be shown that the next half-cycle also causes a decrease in the angle by the same amount. After each cycle the amplitude will decrease by

$$\frac{4\mathfrak{M}_f}{MgL}$$

and the motion will cease once the amplitude falls below $\mathfrak{M}_f/(MgL)$. To complete the analogy with Prob. 4-6, note that the presence of gravity may be replaced by an angular spring with $K = MgL$.

4-10 A body moving in air is subjected to a damping force of magnitude $c_a\dot{u}^2$ directed opposite to velocity. Assume that the motion is a harmonic function of time, $u = A \sin \Omega t$ and calculate the energy D_a that is dissipated by damping in one cycle of motion.

The formula for the damping force may be written as $R = -c_a \dot{u} |\dot{u}|$ to correctly show its direction. The increment of work performed by R is

$$dU = R\dot{u}\, dt = -c_a \dot{u}^2 |\dot{u}|\, dt$$

The dissipated energy is equal to the work performed by the force opposite to R, therefore $dD_a = -dU$ and

$$D_a = c_a \int_0^T \dot{u}^2 |\dot{u}|\, dt = c_a A^3 \Omega^3 \int_0^T \cos^2 \Omega t |\cos \Omega t|\, dt$$

The absolute value sign may be dropped if we limit ourselves to the first quarter of the period,

$$D_a = 4 c_a A^3 \Omega^3 \int_0^{T/4} \cos^3 \Omega t\, dt$$

The value of the integral is $2/(3\Omega)$

$$D_a = \tfrac{8}{3} c_a A^3 \Omega^2$$

4-11 When loaded and unloaded, an instrument shows a hysteresis loop drawn with a solid line in the figure. The dashed line is the same loop translated to the u-axis in a manner that preserves the area enclosed. Assuming the projected loop to be a doubly symmetric line consisting of two second-order parabolas, calculate the damping ratio ζ'' at resonance. Use the average, linear stiffness as a reference.

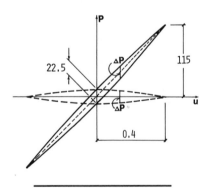

One may easily check that for an area under a second-order parabola the filling factor (as defined in Appendix I) is $2/3$. Applying this to calculate the area within the hysteresis loop in the figure, we have

$$D = 4\left(0.4 \times \frac{22.5}{2} \times \frac{2}{3}\right) = 12 \text{ Lb-in.}$$

Our amplitude $A = 0.4$ in. From the identity $D = sA^2$ we obtain

$$s = \frac{12}{0.4^2} = 75 \text{ Lb-in.}$$

The linear stiffness is

$$k = \frac{115}{0.4} = 287.5 \text{ Lb/in.}$$

Combining Eqs. 4.11. and 4.12 at resonance ($\Omega = \omega$) gives us

$$\zeta'' = \frac{\eta}{2} = \frac{s}{2\pi k} = \frac{75}{2\pi \times 287.5} = 0.0415$$

which is the equivalent damping ratio.

EXERCISES

4-12 A machine subjected to a harmonic force is mounted on steel springs. The only form of damping in the system is the material damping in springs, which may be treated as a viscous damper, parallel to the springs and having $\zeta = 0.005$. If the force applied to the foundation by the springs is of concern, will it be effective to increase damping to $\zeta = 0.05$ by means of installing some additional devices? Estimate the change in the transmitted force if (1) $\Omega/\omega = 1.2$ and (2) $\Omega/\omega = 1.6$.

4-13 As a result of a harmonic response test, a plot of displacement amplitude versus forcing frequency was obtained. Show that when the damping ratio ζ is small, it can be expressed by only Ω_1 and Ω_2, which are the frequencies at so called *half-power points*; *Hint.* Note that $\omega \approx \frac{1}{2}(\Omega_1 + \Omega_2)$ and that when u_0 is the static displacement, we have

$$u_{max}/u_0 \approx 1/2\zeta$$

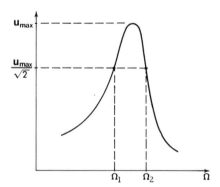

4-14 The rectangular frame, together with the load resting on it (not shown) is supported by four identical springs each with $k = 1200$ Lb/in. and four dampers, each with $c = 11.2$ Lb-s/in. Three separate components of motion are considered: (1) rocking about the x-axis, (2) rocking about the y-axis, and (3) vertical translation. These components are treated as separate SDOF motions, independent from one another. The respective natural frequencies for undamped oscillations are

$$\omega_1 = 9.5 \qquad \omega_2 = 16.0 \qquad \omega_3 = 28.4 \text{ rad/s}$$

Find the corresponding damping ratios. *Hint.* Note that the ratio of the effective stiffness and the effective damping constant is the same for each component of motion.

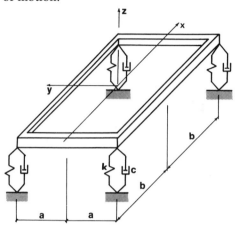

4-15 The natural frequency of the system in Fig. 4.1 is 36 rad/s. When the weight was displaced by 1.5 in. from the initial position and released, the motion stopped after 11 cycles because of friction. The displacement was then found to be 0.09 in., in the same direction as at the beginning. Determine the friction coefficient ψ.

4-16 A cylinder restrained by spring $k = 300$ Lb/in. is subjected to a harmonic force with amplitude $P_0 = 10$ Lb and frequency $\Omega = 155$ rad/s. The drag (damping) force of the surrounding air has the amplitude $R_a = \frac{1}{2}\rho A_0 C_d \dot{u}^2$, in which $\rho g = 43.6 \times 10^{-6}$ Lb/in.^{3} is the specific weight of air, $A_0 = 360$ in.^{2} is the cross-sectional area of cylinder normal to the direction of motion, and $C_d = 0.85$ is the drag coefficient for this cylinder. The weight of the cylinder is $W = 4.5$ Lb. What is the maximum spring force?

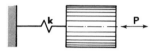

CHAPTER

5

Vibration Isolation

Consider again the basic oscillator from Chapter 2, with some time-dependent force applied to the mass. The determination of the load applied to the ground by the spring-damper assembly is called an *isolation* problem, because that assembly is thought to isolate the ground from the dynamic action of the mass. A reverse problem would be, for example, calculating the acceleration of the mass when the foundation (ground) is moving. This is called a *passive* isolation as opposed to *active* isolation, as previously described. Only the passive type is discussed in this chapter; there is enough information in Chapters 2 and 4 to permit the solution of parallel problems pertaining to active isolation. Only the basic types of isolator are considered here: a spring, a spring with a viscous damper, and a spring with a frictional damper. Both translational and angular motion are considered.

5A. Types of Isolators. Figure 5.1 shows a *simple viscous isolator*. The illustration does not in itself differ from the basic oscillator in Fig. 2.1*a*, except that the base can perform a prescribed motion $u_b(t)$. The parameters defining this isolator (i.e., viscous constant c and the spring constant k) are so chosen that the response of mass M is kept within the prescribed limits.

A different device called a *simple frictional isolator* is symbolically shown in Fig. 5.2. Instead of a dashpot it has a frictional damper working parallel with

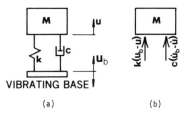

Figure 5.1 Simple viscous isolator and its internal forces.

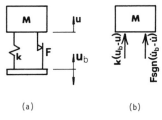

Figure 5.2 Simple frictional isolator and its internal forces.

the spring. The friction force F, which characterizes the damper, is assumed to be relatively small in comparison with the largest inertia force the mass can exert, so that the resistance of the damper against sliding can be easily overcome.

The rigid block referred to as mass M may be an assembly or a machine that is to be protected from the motion of the base by means of the isolator. Usually the concern is that the acceleration a to which this block is subjected be not excessive. Another way of putting it is to request that the overload factor

n, defined as

$$n = \frac{a}{g}$$

where g is the acceleration of gravity and a is the maximum acceleration to which the assembly is subjected, does not exceed the permissible value.

Another concern is the strength of the isolator itself. In the case of a spring element, the applied load depends on a relative displacement of the base and the mass. For a viscous damper the relative velocity determines the applied load. In either case there is a limited magnitude of the load that can be safely applied.

5B. Effect of Foundation Motion in Mass-Spring System. Consider an isolator consisting of a single spring, as in Fig. 5.3. The difference between the deflections of both ends of the spring results in the force $k(u_b - u)$ applied to the mass. The equation of motion is

$$M\ddot{u} = k(u_b - u), \quad \text{or} \quad M\ddot{u} + ku = ku_b \quad (5.1)$$

This equation is similar to Eq. 2.7, except that the forcing function is expressed in terms of foundation displacement $u_b(t)$. The complete solution is given by Eq. 2.8. Suppose that the forcing is prescribed as $u_b = u_0 \sin \Omega t$, then the particular solution is

$$u = \frac{u_0}{1 - \Omega^2/\omega^2} \sin \Omega t \quad (5.2)$$

(Eqs. 2.24 and 2.25). Thus we may say that prescribing a displacement u_b on the foundation gives the same result that would be obtained if the force ku_b were applied to the mass, while the foundation remained stationary.

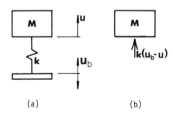

(a)   (b)

Figure 5.3 Spring Isolator.

Another manner of looking at the problem is this: if we are interested only in the *relative displacement* $u - u_b$ of the mass with respect to the foundation, we can introduce a new coordinate, $\bar{u} = u - u_b$. Then

$$u = \bar{u} + u_b \quad \text{and} \quad \ddot{u} = \ddot{\bar{u}} + \ddot{u}_b$$

while the equation of motion becomes

$$M\ddot{\bar{u}} + k\bar{u} = -M\ddot{u}_b \quad (5.3)$$

If the spring were rigid, the function on the right side of the equals sign would be the inertia force applied to the mass M. This is a known function of time, and this equation can be solved in the same way as Eq. 5.1. When $\bar{u}$ is calculated, the spring force $k\bar{u}$ also becomes known. The merit of using Eq. 5.3 is that we can think of a mass connected to a fixed rather than moving foundation.

When the harmonic motion $u_b = u_0 \sin \Omega t$ is imposed at the base, the acceleration at this point is $\ddot{u}_b = -\Omega^2 u_0 \sin \Omega t$. If we are interested in the acceleration response of the mass, we can differentiate Eq. 5.2. Equivalently, we may say that the magnitude of the acceleration response may be obtained by multiplying the displacement response by Ω^2.

5C. Simple Viscous Isolator. Transmissibility. The action of a viscous damper c is similar to that of spring k, except that velocities replace displacements. The equation of motion of mass M supported by the isolator in Fig. 5.1b is

$$M\ddot{u} = k(u_b - u) + c(\dot{u}_b - \dot{u}),$$

or

$$M\ddot{u} + c\dot{u} + ku = ku_b + c\dot{u}_b \quad (5.4)$$

This equation is analogous to Eq. 2.1. The forcing function is expressed by the foundation displacement u_b and the velocity $\dot{u}_b$. If this displacement is given as a harmonic function, we have:

$$u_b = u_0 \sin \Omega t; \quad \dot{u}_b = u_0 \Omega \cos \Omega t$$

$$P = ku_0 \sin \Omega t + cu_0 \Omega \cos \Omega t$$

With the help of trigonometric identities this becomes

$$P = P_b \sin(\Omega t + \phi) \quad (5.5)$$

in which

$$P_b = (k^2 + c^2 \Omega^2)^{1/2} u_0 \quad \text{and} \quad \tan \phi = \frac{c\Omega}{k}$$

If the amplitude of the displacement response

(steady-state part only) is denoted by A, we have, after Eqs. 2.15 and 2.16,

$$\frac{A^2}{u_0^2} = \frac{1 + (2\zeta\Omega/\omega)^2}{(1 - \Omega^2/\omega^2)^2 + (2\zeta\Omega/\omega)^2} \quad (5.6)$$

where $\omega = \sqrt{k/M}$ is the natural frequency. The ratio A/u_0 is called *transmissibility* of the isolator, and it is a measure of how well the oscillator transmits the deflection from the foundation to the mass. For some values of Ω/ω we get a reduction of motion; for others a magnification is to be expected. We usually design an isolator so that the transmissibility has the least possible value.

In Eq. 5.6 A denotes the mass displacement with respect to some fixed reference frame. A displacement relative to the foundation, $\bar{u} = u - u_b$ may also be of interest. Acting as in Sect. 5B, we can develop the equation of motion in terms of $\bar{u}$:

$$M\ddot{\bar{u}} + c\dot{\bar{u}} + k\bar{u} = -M\ddot{u}_b \quad (5.7)$$

This is treated as an equation of motion of a mass attached to a fixed foundation and acted on by an external force $P = -M\ddot{u}_b$. If the isolator were rigid, P would be the inertia force applied to the mass as a result of the whole system moving with an acceleration $\ddot{u}_b$.

In the particular case of a harmonic forcing $u_b = u_0 \sin \Omega t$,

$$P = M\Omega^2 u_0 \sin \Omega t$$

and on the basis of Eqs. 2.15 and 2.16:

$$\frac{\bar{A}^2}{u_0^2} = \frac{(\Omega/\omega)^4}{(1 - \Omega^2/\omega^2)^2 + (2\zeta\Omega/\omega)^2} \quad (5.8)$$

where $\bar{A}$ is the amplitude of the steady-state motion of the mass relative to the base. For this reason the ratio $\bar{A}/u_0$ is called the *relative transmissibility* of the isolator.

When the base of a simple oscillator is subjected to a forcing acceleration $a = a_0 \sin \Omega t$, the amplitude of the absolute acceleration of the mass $\ddot{u}_{max}$ is found from

$$\frac{\ddot{u}_{max}^2}{a_0^2} = \frac{1 + (2\zeta\Omega/\omega)^2}{(1 - \Omega^2/\omega^2)^2 + (2\zeta\Omega/\omega)^2} \quad (5.9)$$

This *acceleration transmissibility* (squared) is expressed

by the same formula as the transmissibility of the absolute displacement, Eq. 5.6 (see Prob. 5-9).

5D. Simple Frictional Isolator. The forces acting on mass M supported by a frictional isolator are shown in Fig. 5.2b. The equation of motion is

$$M\ddot{u} = k(u_b - u) + F\,\text{sgn}(\dot{u}_b - \dot{u})$$

and when the relative displacement $\bar{u} = u - u_b$ is introduced, we get

$$M\ddot{\bar{u}} + k\bar{u} = -F\,\text{sgn}\,\dot{\bar{u}} - M\ddot{u}_b \quad (5.10)$$

When the base motion is a harmonic function $u_b = u_0 \sin \Omega t$, the last term of the equation becomes

$$M\ddot{u}_b = -M\Omega^2 u_0 \sin \Omega t$$

Equation 5.10 is analogous to Eq. 4.3 if a forcing term is added to the latter. The solution of forced vibrations was developed in Sect. 4C and it is applicable here if we put $P_0 = M\Omega^2 u_0$. From Eq. 4.6 the effective net force to be applied is

$$P_{ef} = \left[\left(M\Omega^2 u_0 \right)^2 - \left(\frac{4}{\pi} F \right)^2 \right]^{1/2}$$

$$= u_0 k \left[\left(\frac{\Omega}{\omega} \right)^4 - \left(\frac{4a}{\pi u_0} \right)^2 \right]^{1/2}$$

where $a = F/k$ is the parameter characterizing the friction element. Equation 4.7 gives the amplitude of displacement response, which for us is the amplitude of the relative motion $\bar{A}$:

$$\frac{\bar{A}^2}{u_0^2} = \frac{\left(\dfrac{\Omega}{\omega} \right)^4 - \left(\dfrac{4a}{\pi u_0} \right)^2}{(1 - \Omega^2/\omega^2)^2} \quad (5.11)$$

The ratio $\bar{A}/u_0$ is called the relative transmissibility. In motion with respect to a fixed reference frame, the amplitude of displacement u is again denoted by A and is expressed as

$$\frac{A^2}{u_0^2} = \frac{1 + \left(\dfrac{4a}{\pi u_0} \right)^2 \left(1 - \dfrac{2\omega^2}{\Omega^2} \right)}{(1 - \Omega^2/\omega^2)^2} \quad (5.12)$$

The ratio A/u_0 is referred to as the absolute transmissibility.

SOLVED PROBLEMS

5-1 The base of the cantilever beam is in a vertical motion described by $u = 2\sin 10t$, where u is in inches if t is in seconds. Find the amplitude of displacement of mass $M = 0.2$ Lb-s^2/in. with respect to the moving and fixed reference frames. The vertical stiffness of the cantilever is 800 Lb/in.

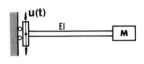

The natural frequency of mass supported by the cantilever is

$$\omega = \left(\frac{k}{M}\right)^{1/2} = (800/0.2)^{1/2} = 63.25 \text{ rad/s}$$

while the forcing frequency $\Omega = 10$ rad/s. Equation 5.2 gives the absolute amplitude

$$A = \frac{u_0}{1 - \Omega^2/\omega^2} = \frac{2}{1 - 10^2/63.25^2} = 2.051 \text{ in.}$$

The acceleration of the base is

$$\ddot{u}_b = -2 \times 10^2 \sin 10t = -200 \sin 10t$$

and Eq. 5.3 may be written as

$$M\ddot{\bar{u}} + k\bar{u} = -0.2(-200 \sin 10t) = 40 \sin 10t$$

Notice that the right-hand side of Eq. 5.1 for this case is $ku_0 \sin \Omega t$. If we put $ku_0 = 40$ or $800u_0 = 40$, which gives $u_0 = 0.05$, the differential equation for $\bar{u}$ will not differ from Eq. 5.1 except for the magnitude of the forcing function. Equation 5.2 may be used again, this time to calculate the relative amplitude:

$$\bar{A} = \frac{0.05}{1 - 10^2/63.25^2} = 0.051 \text{ in.}$$

We see that because of the absence of damping, the amplitudes of the components are directly additive:

$$A = u_0 + \bar{A}$$

5-2 The rigid, prismatic cantilever beam of mass M and length $L = L_1 + L_2$ is driven by a spring k, which is attached to a block oscillating according to $u_b = u_0 \sin \Omega t$. Develop the equation for the angular deflection α of the cantilever.

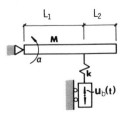

The mass moment of inertia about the pivot point is $J = \frac{1}{3}ML^2$. The force resulting from the compression of spring is $k(u_b - L_1\alpha)$ acting upward on the beam. The equation of the angular motion:

$$J\ddot{\alpha} = \mathfrak{M} = L_1 k (u_b - L_1\alpha)$$

or

$$J\ddot{\alpha} + L_1^2 k\alpha = L_1 k u_b = L_1 k u_0 \sin \Omega t$$

This equation shows the angular stiffness to be $K = L_1^2 k$, therefore the natural frequency is

$$\omega^2 = \frac{K}{J} = \frac{L_1^2 k}{\frac{1}{3}ML^2} = 3\frac{kL_1^2}{ML^2}$$

To solve the differential equation for α, we can use the analogy with Eq. 5.1, or we may get the particular solution by substituting $\alpha = A\sin \Omega t$ into the equation of motion,

$$J(-A\Omega^2 \sin \Omega t) + L_1^2 kA \sin \Omega t = L_1 k u_0 \sin \Omega t$$

$$\therefore \alpha = \frac{u_0/L_1}{1 - \Omega^2/\omega^2} \sin \Omega t$$

5-3 The base of the mass-spring system is given a constant acceleration a_0. Calculate the displacement u of mass M with respect to a fixed coordinate frame. The initial displacement is zero, as is the velocity.

If the upward movement of the base starts at $t = 0$, we have

$$u_b = \frac{1}{2}a_0 t^2$$

The equation of motion is in this case

$$M\ddot{u} + ku = \frac{1}{2}ka_0 t^2$$

The particular solution in Eq. 2.8 is assumed as

$$u_p = C_1 t^2 + C_2 t + C_3$$

and then $\ddot{u}_p = 2C_1$. Substituting, we obtain

$$M(2C_1) + k(C_1 t^2 + C_2 t + C_3) = \frac{1}{2}ka_0 t^2$$

Equating the coefficients,

$$C_1 = \frac{1}{2}a_0; \qquad C_2 = 0; \qquad C_3 = -2C_1\frac{M}{k} = -a_0\frac{M}{k}$$

and finally

$$u = B_1 \cos \omega t + B_2 \sin \omega t + \frac{1}{2}a_0 t^2 - \frac{a_0 M}{k}$$

Using the initial conditions

$$u(0)=0=B_1-\frac{a_0 M}{k} \qquad \therefore B_1=\frac{a_0 M}{k}$$

$$\dot{u}(0)=B_2\omega=0 \qquad \therefore B_2=0$$

The complete solution is thus

$$u=\tfrac{1}{2}a_0 t^2-(1-\cos\omega t)\frac{a_0 M}{k}$$

We clearly see that the first part of the solution is the same as the movement of the base, while the second is the displacement relative to the base.

5-4 The base of the oscillator vibrates according to $u_b=u_0\cos\Omega t$, which gives rise to a reaction at the fixed point B. Let us designate the amplitude of this reaction by R_d and introduce the *force transmissibility* $\mu=R_d/R_{st}$, where R_{st} is the static reaction due to slow application of u_0. Develop the equation for μ as a function of Ω, M, k_1, and k_2.

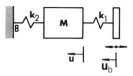

Equation of motion of mass M is

$$M\ddot{u}=k_1(u_b-u)-k_2 u$$

or

$$M\ddot{u}+(k_1+k_2)u=k_1 u_b$$

This corresponds to a simple oscillator, with a spring stiffness k_1+k_2, acted on by a force $k_1 u_b=k_1 u_0\cos\Omega t$. From Eqs. 2.24 and 2.25 (replacing sin by cos), we have

$$u=\frac{1}{1-\Omega^2/\omega^2}\frac{k_1 u_0}{k_1+k_2}\cos\Omega t=A\cos\Omega t$$

The amplitude of force transmitted to B is $k_2 A$, or

$$R_d=\frac{1}{1-\Omega^2/\omega^2}\frac{k_1 k_2 u_0}{k_1+k_2}=\frac{1}{1-\Omega^2/\omega^2}k^* u_0$$

When u_0 is statically applied,

$$R_{st}=k^* u_0$$

and the transmissibility of force is

$$\mu=\frac{R_d}{R_{st}}=\frac{1}{1-\Omega^2/\omega^2}$$

where $\omega^2=(k_1+k_2)/M$. According to this definition μ is negative for $\Omega>\omega$.

5-5 A heavy valve with 42 in. radius has a constant, 8 in. thickness. It can be rotated about its center so that the hollow, 17 in. circle may be positioned where it is needed. The valve is driven by a small gear, which is prevented from rotating while at a fixed position by an elastic angular restraint K. The axle of the valve is attached to the base, which may undergo acceleration in any direction in the plane of paper. The acceleration is a step function in time and has magnitude of $2.5g$. What is the maximum torque that can be applied to the elastic restraint K? The valve material is steel, $\gamma=0.284$ Lb/in.[3] The small gear may be considered weightless.

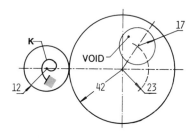

Notice that a second hole in the valve, located on the same radius, but 180° away from the first, would balance the system. A translational acceleration in the plane of paper would not be then capable of rotating the valve. This indicates that the source of unbalance as the mass M,

$$Mg=\pi\times 17^2\times 8\times 0.284=2062.8 \text{ Lb}$$

located at $r_0=23$ in. When $2.5g$ acceleration is applied, the torque about the valve axis is

$$M_t=2.5Mgr_0=2.5\times 2062.8\times 23=118{,}611 \text{ Lb-in.}$$

This torque is resisted by the driving gear with 12 in. radius. Including the dynamic factor of 2.0, the maximum torque on the restraint is

$$M_t'=2.0\times\frac{12}{42}M_t=67{,}778 \text{ Lb-in.}$$

To induce so large a torque, the acceleration must act normal to the radius joining the center of the valve and the center of hole.

5-6 A simple viscous isolator as in Fig. 5.1 has the natural frequency $\omega=12$ Hz. The base motion is described by $u_b=2.7\sin 12.5t$, where u_b is in inches if t in seconds. It was determined by a measurement that the amplitude of the motion of M with respect to the base was 5.2 in. What is the damping ratio $\zeta=c/c_c$ of the viscous damper?

When the problem data is inserted into Eq. 5.8, we get

$$\frac{5.2^2}{2.7^2} = \frac{(12.5/12)^4}{\left(1-12.5^2/12^2\right)^2 + \left(2\zeta \times 12.5/12\right)^2} = \frac{1.1773}{0.0072 + 4.3403\zeta^2}$$

$$\therefore \ \zeta = 0.2673$$

5-7 The system is similar to a simple viscous isolator, except that the spring and the damper are in series rather than in parallel. The motion of the end of spring is given by $u_b = u_0 \sin \Omega t$. Treating Ω as a variable calculate the maximum damper force and estimate the upper bound of the spring force. Solve for $u_0 = 1.5$ in., $M = 12$ Lb-s^2/in., $k = 100$ Lb/in., and $c = 11$ Lb-s/in.

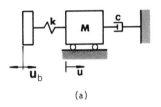

(a)

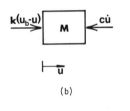

(b)

Isolating the mass, we find the forces applied as in (b). The equation of motion is

$$M\ddot{u} = k(u_b - u) - c\dot{u}$$

or

$$M\ddot{u} + c\dot{u} + ku = ku_b$$

This expression is the same as Eq. 2.1 when $P = ku_0 \sin \Omega t$. The displacement solution is given by Eqs. 2.15 and 2.16. The damping coefficient is

$$\zeta = \frac{c}{2\sqrt{kM}} = \frac{11}{2\sqrt{100 \times 12}} = 0.1588$$

while the resonance takes place for

$$\omega = \Omega = \left(\frac{k}{M}\right)^{1/2} = \left(\frac{100}{12}\right)^{1/2} = 2.8868 \text{ rad/s}$$

The maximum deflection is

$$u_{max} = \frac{1}{2\zeta}\frac{P_0}{k} = \frac{u_0}{2\zeta} = \frac{1.5}{2 \times 0.1588} = 4.7229 \text{ in.}$$

at resonance, $\Omega = \omega$. From Eq. 2.20, the resonance velocity is

$$\dot{u}_{max} = \frac{1}{2\zeta}\frac{\Omega P_0}{k} = \frac{\Omega u_0}{2\zeta} = 2.8868 \times 4.7229 = 13.634 \text{ in./s}$$

The damper force is

$$c\dot{u}_{max} = 11 \times 13.634 = 149.97 \text{ Lb}$$

The largest possible spring force may be estimated by noting that the extension will not exceed

$$u_0 + u_{max} = 1.5 + 4.7229 = 6.2229 \text{ in.}$$

This is a conservative estimate, because it implies not only that extreme displacement of both ends is attained at the same time, but also that the signs are such that the spring undergoes maximum elongation. The spring force will not exceed

$$N_{max} = k(u_0 + u_{max}) = 100 \times 6.2229 = 622.3 \text{ Lb}$$

5-8 The figure shows a single-axis trailer moving along an uneven road whose surface is described by $u_b = u_0 \sin \pi y/L$. Find the constant velocity v of the vehicle at which the maximum absolute displacement amplitude of the weight W is reached. Put $W = 1500$ Lb, $k = 300$ Lb/in. and the damping factor $\zeta = 0.3$ ($c = 20.49$ Lb-s/in.). The road shape parameters are $u_0 = 2$ in. and $L = 20$ in.

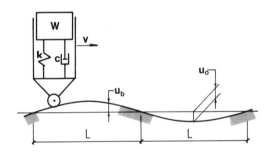

The only essential difference between this scheme and Fig. 5.1 is that instead of being directly imposed, the base motion must be deduced from some other movement. Leaving this simple task for the end, first find that value of the forcing frequency Ω at which the maximum of the right side of Eq. 5.6 is reached. We have

$$\omega^2 = \frac{kg}{W} = \frac{300 \times 386}{1500} = 77.2 \quad \therefore \ \omega = 8.7864 \text{ rad/s}$$

When we set $x = \Omega/\omega$, Eq. 5.6 becomes

$$\frac{A^2}{u_0^2} = \frac{1 + (0.6x)^2}{(1-x^2)^2 + (0.6x)^2}$$

Since our damping is not very small, we can get the maximum response for a value of Ω differing somewhat from ω (see Sect. 2G for comparison). Instead of searching for a maximum by means of differentiation, let us simply calculate the numerical value of the transmissibility for a few values of x close to 1.

x	$\dfrac{A^2}{u_0^2}$
0.90	3.941
0.93	3.978
0.96	3.941
0.99	3.830
1.02	3.654

The maximum is between $x=0.9$ and $x=0.96$. For $x=0.92$ and $x=0.94$ we get $A^2/u_0^2=3.974$ and 3.974, respectively. We stop the search here and decide that the maximum is reached for $x=\Omega/\omega=0.93$, or

$$\Omega_c=0.93\times8.7864=8.1741 \text{ rad/s}$$

and the value of transmissibility at that point is

$$\frac{A}{u_0}=\sqrt{3.978}=1.994$$

Therefore

$$A_{max}=1.994\times2=3.988 \text{ in.}$$

When the vehicle moves along the road with constant velocity v, its position is determined by $y=vt$ and the vertical location of the wheel may be found from

$$u_b=u_0\sin\left(\frac{\pi y}{L}\right)=u_0\sin\left(\frac{\pi vt}{L}\right)$$

We can treat this expression as $u_0\sin\Omega t$, where Ω is the circular frequency given by

$$\Omega=\frac{\pi v}{L}$$

This allows us to find the critical value of velocity v_c:

$$8.1741=\frac{\pi v_c}{20} \qquad \therefore\ v_c=52.04 \text{ in./s}=2.957 \text{ mph}$$

5-9 The base of the system in Fig. 5.1 is subjected to a forcing acceleration $a=a_0\sin\Omega t$. Find the acceleration amplitude $\ddot{u}_{max}$ of the mass M. Present the answer in the nondimensional form $\ddot{u}_{max}^2/a_0^2$, which is called the *acceleration transmissibility* (squared).

As long as the forcing motion is harmonic, it does not really matter whether a displacement or a forcing amplitude is prescribed. We can put $u_0=a_0/\Omega^2$ and use the formula for the amplitude of the applied force from Sect. 5C,

$$P_b=(k^2+c^2\Omega^2)^{1/2}u_0=k\frac{a_0}{\Omega^2}\left(1+\frac{4\zeta^2\Omega^2}{\omega^2}\right)^{1/2}$$

Equation 2.21 tells us that the amplitude of the response accelera-

tion, herein denoted as $\ddot{u}_{max}$, is

$$\ddot{u}_{max}=\mu\left(\frac{\Omega^2}{\omega^2}\right)\frac{P_b}{M}=\mu\left(\frac{\Omega^2}{\omega^2}\right)\frac{k}{M}\frac{a_0}{\Omega^2}\left(1+\frac{4\zeta^2\Omega^2}{\omega^2}\right)^{1/2}$$

$$\therefore\ \frac{\ddot{u}_{max}^2}{a_0^2}=\frac{1+(2\zeta\Omega/\omega)^2}{(1-\Omega^2/\omega^2)^2+(2\zeta\Omega/\omega)^2}$$

Thus we see that the transmissibility of the absolute acceleration is the same as the transmissibility of the absolute displacement, as expressed by Eq. 5.6.

5-10 A basic oscillator performs a harmonic motion $u=u_0\sin\Omega t$. Find the expression for the resultant force applied to the foundation by the spring and the damper and present the result graphically as a rotating vector. Discuss the analogy between a stationary-base and a moving-base oscillator.

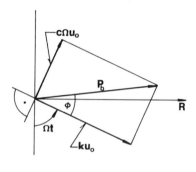

The resultant force R applied to the foundation of the basic oscillator is

$$R=ku+c\dot{u}=u_0(k\sin\Omega t+c\Omega\cos\Omega t)$$

The figure gives the geometric interpretation of this equation, and R can be written as

$$R=P_b\sin(\Omega t+\phi) \qquad \text{with} \quad \tan\phi=\frac{c\Omega}{k}$$

where $\qquad P_b=(k^2+c^2\Omega^2)^{1/2}u_0$

The analogy between these equations and those developed in Sect. 5C illustrates how the problem of the absolute response of the mass attached to the moving base is reduced to that of the motion of the basic oscillator:

1. Hold the mass stationary and calculate the amplitude of the force transmitted to the mass when the base is moving.
2. Apply the above-calculated force to the mass of the stationary-base oscillator at the same frequency Ω and compute the response using Eqs. 2.15 and 2.16.

5-11 A single-axis trailer moving with a constant horizontal velocity v rolls over a discontinuity in the road depicted in (*a*). Assuming the wheel axis to be moving strictly parallel to the road surface, find the upper bound of the acceleration experienced by the weight W. Use $d=6$ in., $h=1.5$ in., $v=650$ in./s, and the other constants as in Prob. 5-8.

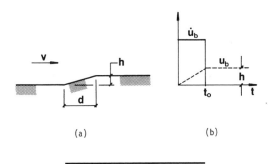

(a)   (b)

The displacement u_b and velocity $\dot{u}_b$ of the wheel axis along the vertical are plotted as a function of time in (*b*). Assuming that the horizontal component of velocity is constant, $\dot{u}_b$ may be found from the velocity vector triangle as

$$\dot{u}_b = \frac{vh}{d}$$

The forcing terms in Eq. 5.4, ku_b and $c\dot{u}_b$ vary in time according to (*b*). That equation visualizes the fact that the road discontinuity is in this case equivalent to ku_b and $c\dot{u}_b$ applied to the mass of a system with fixed base. The time needed by the wheel to traverse distance d is

$$t_0 = \frac{d}{v} = \frac{6}{650} = 0.00923 \text{ s}$$

We can decide whether this is an event of short or long duration only by comparing t_0 with the period of free vibrations. The undamped frequency, from the reference problem, is $\omega = 8.7864$ rad/s. The actual, damped period of vibration is

$$\tau_d = \frac{2\pi}{\omega(1-\zeta^2)^{1/2}} = \frac{2\pi}{8.7864(1-0.3^2)^{1/2}} = 0.7496 \text{ s}$$

We can use the fact that $t_0 \ll \tau_d$ to treat the traversing of the discontinuity as a set of initial conditions of the system. The damper force is

$$c\dot{u}_b = \frac{cvh}{d} = 20.49 \times 650 \times \frac{1.5}{6} = 3330 \text{ Lb}$$

while the spring force may reach

$$ku_b = 300 \times 1.5 = 450 \text{ Lb}$$

The total acceleration (assuming no displacement of mass until the slope is passed) may amount to

$$\left(c\dot{u}_b + ku_b\right)\frac{g}{W} = 2.52g$$

On the other hand, we may look at the phenomenon from the viewpoint of energy imparted to the system in the vertical direction. As a result of the damper force being applied during t_0, the weight acquires the initial velocity

$$v_0 = \frac{c\dot{u}_b t_0 g}{W} = 7.909 \text{ in./s}$$

while $h=1.5$ in. is the initial base deflection. The total initial energy is

$$\Pi_0 + T_0 = \frac{1}{2}kh^2 + \frac{1}{2}\frac{W}{g}v_0^2$$

$$= \frac{1}{2} \times 300 \times 1.5^2 + \frac{1}{2} \times \frac{1500}{386} \times 7.909^2 = 459.0 \text{ Lb-in.}$$

If, from that moment on, the damper is ignored, the maximum upward deflection measured from the new neutral position (after passing the slope) will be u_m

$$\tfrac{1}{2}ku_m^2 = 459 \qquad \therefore \ u_m = 1.7494 \text{ in.}$$

The acceleration associated with this figure would be

$$\ddot{u} = \frac{ku_m}{W}g = 0.35g$$

The actual figure would be less than that because of the damper action. The acceleration previously computed (2.52g) must therefore be treated as an upper bound.

5-12   A disc with moment of inertia $J=20$ Lb-s^2-in. is placed in the cavity of a rigid body. The axis about which the body rotates coincides with the disc axis. The steady-state angular motion of that rigid body is described by $\alpha = \alpha_0 \cos\Omega t$, where Ω is a slowly varying excitation frequency. The cavity is filled with a viscous liquid, such that $\zeta = 0.08$. What is the maximum twisting moment the shaft can experience when $\alpha_0 = 0.02$ rad? The shaft data are: $L=28$ in., $d=4$ in., and $G=11\times10^6$ psi.

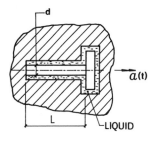

Denoting by $\bar{\alpha}$ the amplitude of disc rotation relative to the rigid body, we can take advantage of Eq. 5.8, placing $(\bar{\alpha}/\alpha_0)^2$ on its left side. When damping is small, the maximum of that expression occurs near $\Omega = \omega$ and then

$$\frac{\bar{\alpha}}{\alpha_0} = \frac{1}{2\zeta} = 6.25$$

The spring constant of the shaft is

$$K = \frac{GC}{L} = \frac{G}{L}\frac{\pi d^4}{32} = 9.8736 \times 10^6 \text{ Lb-in./rad}$$

The twisting moment in the shaft,

$$M_t = K\bar{\alpha} = 9.8736 \times 10^6 \times 6.25 \times 0.02 = 1.2342 \times 10^6 \text{ Lb-in.}$$

5-13 A simple frictional isolator is subjected to a sinusoidal forcing of the base, $u_b = u_0 \sin \Omega t$. For a given amplitude u_0 determine the value of frequency Ω at which the frictional damper begins to slide. Compare the result from an equilibrium equation to that from Eq. 5.11. The mass M and the frictional resistance F are known.

———————

The mass begins to move with respect to the base when the inertia force exceeds the friction force F. The acceleration of the mass repeating the base motion is

$$\ddot{u}_b = -\Omega^2 u_0 \sin \Omega t$$

and the maximum of the inertia force is $M\Omega^2 u_0$. The equilibrium equation

$$F = M\Omega_b^2 u_0$$

gives us the desired "break-loose" frequency:

$$\Omega_b^2 = \frac{F}{Mu_0}$$

Another way is to use Eq. 5.11, which is valid only for values of Ω that make the numerator of that equation positive. The limiting frequency is found from

$$\left(\frac{\Omega_b}{\omega}\right)^2 = \frac{4a}{\pi u_0} \quad \text{or} \quad \Omega_b^2 = \frac{k}{M} \frac{4}{\pi} \frac{F}{k} \frac{1}{u_0}$$

$$\therefore \; \Omega_b^2 = \frac{4}{\pi} \frac{F}{Mu_0}$$

This second method gives us the forcing frequency, which is $\sqrt{4/\pi} = 1.128$ larger than the true value. Employing Eq. 5.12 instead and setting $A/u_0 = 1$ would give the identical result. This is because both equations are based on the energy approach, which involves some inaccuracies.

EXERCISES

5-14 What is the maximum acceleration to which the mass in Prob. 5-1 is subjected? What is the maximum shear force V and the maximum bending moment $\mathfrak{M}$ if the length of cantilever is 20 in.?

5-15 A light block is moving according to $u_b = u_0 \sin \Omega t$ and is driving the rigid body having mass M through a viscous damper c. The body is attached to the ground with two springs having total stiffness k. Find the maximum force in the springs and estimate the upper bound of the damper force. Solve for the same parameters as in Prob. 5-7, that is, $u_0 = 1.5$ in., $M = 12$ Lb-s²/in., $k = 100$ Lb/in., and $c = 11$ Lb-s/in.

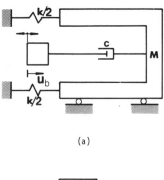

(a)

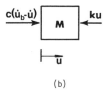

(b)

5-16 What is the overload factor due to the vertical motion of the wheel in Prob. 5-8 at $\Omega = \omega$ and $\Omega = 2\omega$? To what travel velocity v do these frequencies correspond?

5-17 A simple frictional isolator as in Fig. 5.2 supports the weight $W = 150$ Lb. The spring constant is $k = 5000$ Lb/in. The frequency of the base vibrations can be anywhere from 28 to 40 Hz. If the base oscillates with the amplitude 0.6 in., what is the minimum friction force F necessary in the isolator to keep the amplitude $\bar{A}$ within 1 in.? ($\bar{A}$ is the amplitude of the mass moving with respect to the base.)

5-18 A frictional isolator is so designed that its break-loose frequency Ω_b, resulting from Eq. 5.11 equals 1.1ω. Determine the maximum relative displacement transmissibility $\bar{A}/u_0$ as a function of Ω/ω.

CHAPTER

6

Shock and Impact

A *shock load* is characterized by a short duration of applied force—usually less than one natural period of vibration. *Impact* is the sudden contact of a moving body with a motionless barrier, or with a body of much larger size. This chapter analyzes shock and impact of objects with linear, piecewise-linear, or nonlinear characteristics. A distinction is made between local and overall deformation and their relative influence on response is compared. The *plastic hinge*, which is a useful concept in estimating the resistance capacity of structures, is introduced. A continuation of shock and impact problems into MDOF systems is presented in Chapter 13.

6A. Central Impact Against Rigid Wall. This is probably the simplest form of impact and as such it provides a good illustration of the basic concepts involved. Let us consider a free fall and rebound of a ball against the ground and separate this event into four distinct stages as shown in Fig. 6.1.

Stage 1 lasts from the moment the ball is released at height h_0 until it first touches the ground. At the end of this stage all particles of the falling body have velocity v, called the impact velocity. During Stage 2 the deformation process takes place as a result of inertia forces acting on the ball. At the end of this stage all (or nearly all) particles have zero velocity, which means that the initial kinetic energy has been converted into strain and other forms of energy. Stage 3 is predominantly the recovery process during

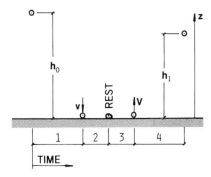

Figure 6.1 Four phases of impact.

which the ball (nearly) returns to its original shape and gains the upward velocity. At the end of Stage 3 the ball barely touches the ground and all its particles have velocity V, called the rebound velocity. Stage 4 lasts from the time the ball leaves the ground, the initial velocity being V, until the peak point of rebound h_1 above the ground is reached. The time of impact proper (Stages 2 and 3) is very short in comparison with the free fall (1) and rebound (4).

The impulse of contact forces is a vector quantity, which shows itself through the change of velocity of the impacted body. In general the impulse $\mathbf{S}_{12}$ is

$$\mathbf{S}_{12} = M\mathbf{v}_2 - M\mathbf{v}_1 \qquad (6.1)$$

where $\mathbf{v}_1$ and $\mathbf{v}_2$ are, respectively, the velocities before and after the impulse is imparted to the body. In this particular case we choose the positive direction upward, along the z-axis and write the impulse of Stage 2 as

$$S' = M \cdot 0 - (-Mv) = Mv$$

while at Stage 3

$$S'' = MV - M \cdot 0 = MV$$

The total impulse applied to the ball is therefore

$$S = S' + S'' = Mv + MV \qquad (6.2)$$

There are two components, S' which stops the ball, and S'' which makes it rebound. The ratio of the rebound velocity to the impact velocity is called the coefficient of restitution,

$$\kappa = \frac{V}{v} = \sqrt{\frac{h_1}{h_0}} \qquad (6.3)$$

The second equality is written on the basis of the free fall formulas

$$V = \sqrt{2gh_1} \quad \text{and} \quad v = \sqrt{2gh_0}$$

When the kinetic energy is changed into strain energy, and vice versa, during the process of deformation of the ball, some energy losses take place, which may be measured by the loss of velocity:

$$\Delta T = \tfrac{1}{2}Mv^2 - \tfrac{1}{2}MV^2 = \tfrac{1}{2}Mv^2(1 - \kappa^2) \qquad (6.4)$$

When the velocities of rebound and impact are the same, the coefficient of restitution $\kappa = 1.0$ and there are no energy losses. This ideal event is called a perfectly elastic impact and, needless to say, it cannot be attained in practice. The other extreme is the perfectly plastic impact, for which $\kappa = 0$ and the entire amount of kinetic energy is lost. This can be quite well approximated in the real world if the impacting body is made of easily and permanently deformable material (e.g., a bag of sand). Most likely is an intermediate type of impact, for which $0 < \kappa < 1$. When the loss of energy is associated with viscous damping so that we can use the model in Fig. 6.2, the following relation exists between the damping ratio and the coefficient of restitution:

$$\kappa = e^{-\pi\zeta} \approx 1 - \pi\zeta \qquad (6.5)$$

in which the last near-equality may be used when ζ does not exceed a few percent.

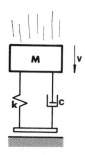

Figure 6.2 Impact with viscoelastic bumper.

As far as the impact against a rigid wall is considered, the coefficient of restitution is understood to characterize the impacting body. For a deformable surface the formulas would not change; however, the coefficient would also depend on the properties of the wall material.

The term *central impact* is used when the velocity of the center of gravity (CG) of the impacting body lies on the same line as the reaction at the impacted surface.

6B. Local *Versus* General Deformation. Consider a beam on two supports with the load concentrated over a small portion of the length. Apart from bending deflection, there is always some deformation of the beam cross section. For a very slender beam the latter may usually be ignored when compared with the former; that is, the *local deformation* is negligible in comparison with the *general deformation*. When the length, however, is not much larger than the depth, the preceding statement will not be true. It is a rule that for compact bodies (all three dimensions of comparable magnitude), the local deformations are important and may even predominate. The latter is the case when a compact body strikes a barrier and the contact area is small.

6C. Piecewise-Linear Impact. Figure 6.3 shows a projectile impacting a barrier, both bodies being deformable. The barrier is so large that its overall motion is not noticeable. Because of compactness of the projectile and a small contact area only the local deformations are of interest. A better visual approach to the problem is to show the combined local deformations of both bodies, as a deflection δ of a fictitious spring, as in Fig. 6.4. The length of spring shown as L_0 usually does not enter the calculations. Introducing that imaginary spring allows us to treat both the projectile and the barrier as rigid.

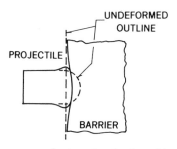

Figure 6.3 Deformation of projectile and barrier.

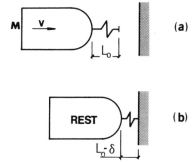

Figure 6.4 Equivalent "bumper" spring to simulate contact deformation.

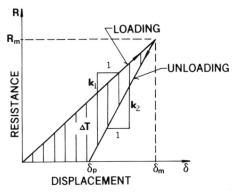

Figure 6.5 Loading and unloading in piecewise-linear impact.

Typically, the characteristic of such a "bumper spring" is nonlinear. This section discusses the linearized theory of impact, which means that the actual resistance-deflection plot is replaced by a straight line. To keep calculations simple, yet not unrealistic, different loading and unloading properties are assumed (Fig. 6.5). This type of characteristic is called piecewise linear. By equating the kinetic energy of the impacting body with the strain energy at maximum deflection δ_m we get

$$\delta_m = v\sqrt{\frac{M}{k_1}} \qquad (6.6)$$

The maximum impact force $R_m = k_1\delta_m$ is

$$R_m = v\sqrt{Mk_1} \qquad (6.7)$$

The impact ends when the contact force reaches zero value during the unloading phase. This corresponds to a point with coordinates $R=0$ and $\delta=\delta_P$ in Fig. 6.5. One of the results of the spring being stiffer during unloading, $k_2 > k_1$, is the permanent deformation δ_P. The shaded area represents the kinetic energy ΔT lost during impact:

$$\Delta T = \frac{1}{2}k_1\delta_m^2\left(1 - \frac{k_1}{k_2}\right) \qquad (6.8)$$

This energy loss may also be expressed in terms of the coefficient of restitution κ, (Eq. 6.4). Since $Mv^2 = k_1\delta_m^2$, we have

$$\kappa^2 = \frac{k_1}{k_2} \qquad (6.9)$$

Let t_1 be the time of the loading phase of impact. It lasts from the first instant of contact until the impacting body reaches the stationary point. The motion during this phase is described by Eq. 2.4 with zero initial displacement, $u_0 = 0$, and initial velocity v_0. We identify t_1 as one-quarter of the period of a harmonic motion:

$$t_1 = \frac{1}{4}\frac{2\pi}{\omega} = \frac{\pi}{2}\left(\frac{M}{k_1}\right)^{1/2} \qquad (6.10)$$

A similar relation holds for the unloading phase, except that the spring constant is k_2.

In the presence of viscous damping, the impact force is less than predicted by Eq. 6.7. Using the model in Fig. 6.2 for the loading phase, one may find that

$$R_m = v\left(\frac{Mk}{1-\zeta^2}\right)^{1/2}\exp\left[\left(2\zeta - \frac{\pi}{2}\right)\zeta\right] \quad (6.11)$$

is a very close approximation for $\zeta \leqslant 0.2$. The peak impact force and the maximum deflection do not coincide exactly when damping is involved (see Prob. 6-3).

6D. Local Deformation Formulas for Elastic Contact. Consider two bodies with spherical surfaces that are in contact. As we slowly increase the load R, the displacement of the moving plane a–a (Fig. 6.6) increases according to the Hertz formula:

$$\delta = \left\{ \left[\frac{3}{4} \left(\frac{1-\nu_1^2}{E_1} + \frac{1-\nu_2^2}{E_2} \right) \right]^2 \left(\frac{1}{r_1} + \frac{1}{r_2} \right) \right\}^{1/3} R^{2/3}$$

(6.12)

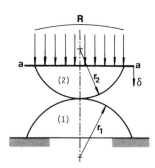

Figure 6.6 Contact of two curved surfaces.

A few important special cases may be deduced from this equation. When r_1 or r_2 is infinitely large—that is, when $1/r_1$ or $1/r_2$ is zero—we are dealing with a sphere pressed into an elastic half-space. When the larger radius (r_1 in this case) is negative, it means that the upper sphere is pressing on the bottom of a spherical cavity having the radius r_1. If either of the two bodies is assumed to be rigid, we treat the appropriate E as infinitely large.

When a rod of diameter d_1 is pressed against another rod with a diameter d_2, the axial deformation that results from these two diameters being unequal, $d_1 < d_2$, is

$$\delta_t = \frac{1-\nu^2}{E d_1} \left(1 - \frac{d_1}{d_2} \right) R$$

(6.13)

This approximate expression treats the thinner rod as rigid. When we put $d_2 = \infty$, the equation for a rod pressed into a half-space is obtained.

6E. Parameters of Impact According to Hertz Theory. Although Eq. 6.12 is based on the elastic material properties, the relation between the resistance force R and local deflection δ is nonlinear. A briefer form is,

$$R = k_0 \delta^{3/2} \quad \text{or} \quad \delta = \left(\frac{R}{k_0} \right)^{2/3}$$

(6.14)

as illustrated in Fig. 6.7. The increase of contact area during the loading process is responsible for this geometric nonlinearity.

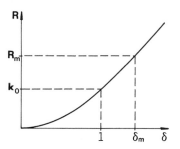

Figure 6.7 Resistance-deflection plot for local deformation.

Keeping in mind that the area under the curve R–δ represents the strain energy, we can write the following energy balance equations,

$$\Pi_m = \int_0^{\delta_m} R \, d\delta = \int_0^{\delta_m} k_0 \delta^{3/2} \, d\delta = \frac{2}{5} k_0 \delta_m^{5/2}$$

But $\Pi_m = T_0 = \frac{1}{2} M v^2$ (initial kinetic energy)

$$\therefore \delta_m = \left(\frac{5}{4} \frac{M v^2}{k_0} \right)^{0.4}$$

(6.15)

Equation 6.14 allows us to calculate the maximum force of impact:

$$R_m = \left(\frac{5}{4} M v^2 \right)^{0.6} k_0^{0.4}$$

(6.16)

The time of the loading phase of impact (i.e., between the instant at which the object first touches the ground and the stationary point) is calculated by integrating the equation of motion of mass M. As a result,

$$t_1 = 1.609 \left(\frac{M}{k_0} \right)^{0.4} \left(\frac{1}{v} \right)^{0.2}$$

(6.17)

The Hertz theory of impact is based on the use of local deformation only. The stress induced by impact must remain within the elastic range.

6F. Impact of a Compact Body with a Nonlinear Material Characteristic. More often than not, an engineer performing an impact analysis will find the impact stress to be above the yield point of material. Since the local characteristic does not conform to the Hertz formula outside the elastic range, a more general relation must be used. A theoretical approach presents considerable difficulties, and testing is an expense that an engineering office cannot always afford. Computer simulation is probably the best way to obtain a resistance-deflection curve because programs offering the capability of nonlinear analysis are widely available.

Once the local characteristic is known, we can find the maximum displacement δ_m and the accompanying impact force R_m by means of the energy method. Although the details of the procedure may vary, the approach is essentially the same as presented in Sect. 6E. The time of the first phase of impact t_1 may be estimated by using a harmonic approximation

$$\delta = \delta_m \sin \omega t$$

where ω is a fictitious frequency of vibration and the equation is valid only for the interval $t_1 = \pi/(2\omega)$. To use the results of Sect. 3D, where the approximation was first developed, we imagine observing the unloading phase of impact, which is identical to the loading phase, except that the sign of the velocity is reversed. If we choose to set $t=0$ at the stationary point, the equation of motion will be $\delta = \delta_m \cos \omega t$, the same as in Sect. 3D. Writing

$$R(\delta) = k_0 f(\delta)$$

and changing the notation from δ_m to A we can use Eqs. 3.6 and 3.8 (or 3.9) to find the fictitious frequency ω.

6G. Response of Basic Oscillator to Shock Load.
A load that acts for a short time, say less than the natural period of vibration, will be referred to as a shock load. If we plot the applied force versus time, the hatched area in Fig. 6.8 represents the impulse acting on the system. Although the force itself may be very large, if the impulse duration t_0 is quite small, the net effect in terms of deflection may be insignificant. When a shock is of a very short duration, say $t_0 \leqslant 0.1\tau$, where τ is the natural period, we can assume that mass M acquires its initial velocity v_0 instantly, and then we have

$$S = Mv_0$$

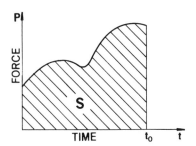

Figure 6.8 Applied force versus time.

This is, in effect, specifying the initial velocity S/M and we can apply Eq. 2.4, which gives us the maximum displacement

$$u_{\max} = \frac{v_0}{\omega} \qquad \text{or} \qquad u_{\max} = \frac{S\omega}{k} \qquad (6.18)$$

When a shock is of somewhat longer duration, $0.1\tau < t_0 \leqslant 0.25\tau$, a more accurate formula must be used:

$$u_{\max} = \frac{S\omega}{k\left[1 - (\omega t_0/\pi)^2\right]} \cos \frac{\omega t_0}{2} \qquad (6.19)$$

This equation results from approximating the actual P–t plot with a half-sinusoid, which contains the same impulse as the given shock. For $t_0 > 0.25\tau$ this relation ceases to be reasonably accurate.

6H. Response of Systems with Piecewise-Linear Characteristics.
Figure 6.9 shows a spring, or more generally a structural element, connecting mass M to the ground. The resistance-deflection curve of that element can take one of three forms: *bilinear, elastic–perfectly plastic* and *rigid-plastic*. (Note that according to the terminology introduced in Sect. 3B the elements are not elastic, because the loading and unloading paths do not coincide.)

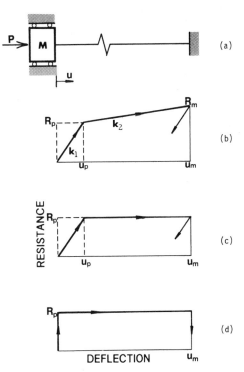

Figure 6.9 Resistance-deflection plots for inelastic elements.

The maximum force in the structural element and maximum displacement u_m induced by a shock load can often be found by applying the energy balance equation. Suppose that a system characterized by a bilinear curve in Fig. 6.9b is subjected to a step load $P = P_0 H(t)$ (see Appendix III). We know from Chapter 2 that for a linear system under this type of loading the magnification factor $\mu = 2.0$, which means

$$u_m = 2u_{st} = \frac{2P_0}{k_1}$$

if k_1 is the spring constant. This will also be true in our case as long as $P_0 \leqslant R_p/2$. When u_m is calculated according to this equation and is larger than u_p, this tells us that the deflection is in the plastic range. Equating the strain energy with the work of the external force P_0, we get the maximum deflection u_m.

A deflection cannot be determined in case of a rigid–perfectly plastic system in Fig. 6.9d, when the step load is applied. No deflection occurs for $P_0 < R_p$ and when $P_0 > R_p$, u_m becomes unbounded. The latter merely means that the system is incapable of resisting the applied load.

Another type of external loading, which is fairly easy to deal with, is the velocity shock (i.e., imposing an initial velocity on the system). Again equating the initial kinetic energy with the maximum strain energy allows us to calculate maximum deflection.

The plots of the type shown in Fig. 6.9 were originally developed to approximate the stress-strain curves of various types of material. It is a trivial matter to obtain from any of those the resistance-deflection curve for an axial bar. The task becomes much more difficult for a member such as a beam under lateral load, because the bending moment changes along the length. A very helpful concept in that respect is the *plastic hinge*. When the bending moment attains a certain limiting value $\mathfrak{M} = \mathfrak{M}_p$ in a particular section, we assume the yielding starts abruptly at this location and a hinge forms at that point. Regardless of how large the deflection is, the bending moment may not exceed $\mathfrak{M}_p$. This limiting value of bending moment capacity is selected on the basis of the actual moment-curvature relation for the cross section.

SOLVED PROBLEMS

6-1 Mass M with an elastic bumper of stiffness k and length L_0 is dropped from the height h_0. Calculate the maximum spring force and the height of rebound if the coefficient of restitution is κ. Solve for $M = 1$ Lb-s^2/in., $k = 10,000$ Lb/in., $L_0 = 5$ in., $h_0 = 100$ in., and $\kappa = 0.9$.

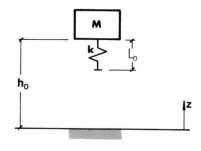

Net drop height:

$$h_0' = h_0 - L_0 = 100 - 5 = 95 \text{ in.}$$

Impact velocity:

$$v = (2gh_0')^{1/2} = (2 \times 386 \times 95)^{1/2} = 270.8 \text{ in./s}$$

To determine the maximum deflection, use the energy balance equation (Eq. 1.7), which includes the work of gravity forces U_g:

$$\Delta \Pi + \Delta T = U_g$$

From the undeflected to deflected position we have

$$\Delta \Pi = \tfrac{1}{2} k \delta^2 - 0$$

where δ is unknown deflection.

$$\Delta T = 0 - \tfrac{1}{2} M v^2 \qquad \text{and} \qquad U_g = Mg\delta$$

Substituting:

$$\tfrac{1}{2} k \delta^2 - \tfrac{1}{2} M v^2 = Mg\delta$$

$$\delta^2 - \frac{2}{k} Mg\delta - \frac{M}{k} v^2 = 0$$

$$\therefore \delta = \frac{Mg}{k} + \left[\left(\frac{Mg}{k} \right)^2 + \frac{Mv^2}{k} \right]^{1/2} = 2.747 \text{ in.}$$

The maximum force in spring:

$$P_{max} = k\delta = 10,000 \times 2.747 = 27,470 \text{ Lb}$$

The rebound velocity

$$V = \kappa v = 0.9 \times 270.8 = 243.72 \text{ in./s}$$

Vertical travel from level $z = L_0$:

$$h_1' = \frac{V^2}{2g} = \frac{243.72^2}{2 \times 386} = 76.94 \text{ in.}$$

Total height:

$$h_1 = 5.0 + 76.94 = 81.94 \text{ in.}$$

6-2 Mass M with a spring and a damper (Fig. 6.2) is impacting a surface with velocity v_0. What is the coefficient of restitution κ if the damping ratio is ζ?

———————

Once the contact is established, the mass can be treated as performing vibratory motion with the initial velocity v_0 and no initial displacement. According to Eq. 2.9 we have

$$u = \frac{v_0}{\omega_d} \sin \omega_d t \cdot \exp(-\omega \zeta t)$$

The velocity is obtained by differentiation,

$$\dot{u} = \left(v_0 \cos \omega_d t - \frac{v_0}{\omega_d} \omega \zeta \sin \omega_d t \right) \cdot \exp(-\omega \zeta t)$$

At $t = \tau_d/2 = \pi/\omega_d$, the displacement returns to zero and a rebound begins. Designating the rebound velocity by V_0, we have

$$V_0 = v_0 \cos \pi \cdot \exp\left(-\omega \zeta \frac{\pi}{\omega_d} \right) \approx -v_0 \exp(-\pi \zeta)$$

According to our convention,

$$\kappa = \frac{-V_0}{v_0} = e^{-\pi \zeta}$$

When ζ is only a few percent or less, we can replace the exponential by the first two terms in its series expansion and then write

$$\kappa \approx 1 - \pi \zeta$$

6-3 Develop the relationship between the impact velocity and maximum contact force for the model in Fig. 6.2. Evaluate the effect of damping for $\zeta = 0.2$.

———————

When the impacting mass is being slowed by a spring and a damper in parallel, the resultant force R in the two is the impact force. By virtue of Eq. 2.2 we can write

$$R = -M\ddot{u} = M(\omega^2 u + 2\omega \zeta \dot{u})$$

Using the expression for u and $\dot{u}$, which were developed in Prob. 6-2,

$$\frac{R}{M} = \left[\frac{\omega}{\omega_d} (1 - 2\zeta^2) \sin \omega_d t + 2\zeta \cos \omega_d t \right] \omega v_0 \exp(-\omega \zeta t)$$

With a trigonometric identity this can be presented as

$$\frac{R}{M} = \varepsilon^{1/2} \omega v_0 \sin(\omega_d t + \beta) \cdot \exp(-\omega \zeta t)$$

where

$$\varepsilon = \left(\frac{\omega}{\omega_d} \right)^2 (1 - 2\zeta^2)^2 + 4\zeta^2$$

Noting that $\omega^2/\omega_d^2 = 1/(1 - \zeta^2)$, we obtain $\varepsilon = 1/(1 - \zeta^2)$ and

$$R = \frac{M\omega v_0}{(1 - \zeta^2)^{1/2}} \sin(\omega_d t + \beta) \cdot \exp(-\omega \zeta t)$$

The angle β makes R reach its maximum a little before the end of one-fourth of the damped period, namely, at $t_m = (\pi/2 - \beta)1/\omega_d$. At this instant

$$R_m = v_0 (Mk)^{1/2} \frac{\exp(-\omega \zeta t_m)}{(1 - \zeta^2)^{1/2}}$$

The first part of this expression agrees with Eq. 6.7, while the second is the correction factor showing the influence of damping. The angle β can be recovered from the expression for R containing both sine and cosine terms:

$$\sin \beta = \frac{2\zeta}{\varepsilon^{1/2}} = \frac{2\zeta}{(1 - \zeta^2)^{1/2}}$$

Substituting $\zeta = 0.2$, one obtains:

$$\sin \beta = 0.4082 \qquad \beta = 0.4205 \qquad \frac{\pi}{2} - \beta = 1.1503$$

$$\omega \zeta t_m = \left(\frac{\pi}{2} - \beta \right) \frac{\zeta}{(1 - \zeta^2)^{1/2}} = 0.2348$$

and finally $R_m = 0.8071 v_0 (Mk)^{1/2}$. A close approximation of R_m may be obtained by putting

$$\beta \approx 2\zeta \qquad \text{and} \qquad \omega \zeta t_m \approx \left(\frac{\pi}{2} - 2\zeta \right) \zeta$$

$$\therefore R_m = v_0 \left(\frac{Mk}{1 - \zeta^2} \right)^{1/2} \exp\left[-\left(\frac{\pi}{2} - 2\zeta \right) \zeta \right]$$

6-4 At the end of a light cantilever beam with stiffness $k_b = 120$ Lb/in. there is a lumped mass $M = 0.002$ Lb-s^2/in. This mass suddenly acquires velocity $v_0 = 75$ in./s and after crossing the gap $b = 0.15$ in. strikes a spring support, $k_s = 80$ Lb/in. What is the maximum deflection and the impact force?

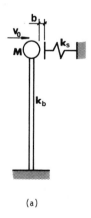

(a)

———————

When the beam is undeflected, the kinetic energy is

$$T_0 = \tfrac{1}{2} M v_0^2 = \tfrac{1}{2} \times 0.002 \times 75^2 = 5.625 \text{ Lb-in.}$$

The strain energy accumulated in the beam after crossing the gap is only

$$\Pi_0 = \tfrac{1}{2} k_b b^2 = \tfrac{1}{2} \times 120 \times 0.15^2 = 1.35 \text{ Lb-in.}$$

therefore the motion will continue upon crossing the gap. Once the mass reaches the maximum deflection u_m, the strain energy can be visualized as an area under the bilinear characteristic in (b). The formula for this energy may be found from Prob. 3-5 after replacing k_1 by k_b, k_2 by $(k_b + k_s)$ and u_1 by b:

$$\Pi_m = \tfrac{1}{2}(k_b + k_s) u_m^2 - k_s b u_m + \tfrac{1}{2} k_s b^2 = 100 u_m^2 - 12 u_m + 0.9$$

From the equality of T_0 and Π_m we can calculate $u_m = 0.2855$ in. The impact force

$$R_m = k_b b + (k_b + k_s)(u_m - b) = 45.1 \text{ Lb}$$

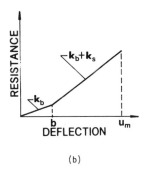

(b)

6-5 Solve Prob. 6-4 when there is a viscous damper with $c = 0.22$ Lb-s/in. working in parallel with the spring.

The phase of motion from the time the mass touches the support until the maximum deflection is reached will be treated as a harmonic motion with some initial velocity v_0. (This approximate approach ignores the fact that there already is some resisting force in the beam when the support is touched.) When the mass is in contact with the spring support, the fictitious critical damping of the system is

$$c_c = 2[(k_b + k_s)M]^{1/2} = 2[(120 + 80)0.002]^{1/2} = 1.2649$$

in accordance with Sect. 2A. The damping ratio

$$\zeta = \frac{c}{c_c} = 0.1739$$

As may be found in the reference problem, the maximum undamped deflection in the phase of motion in which we are interested is

$$u_m - b = 0.2855 - 0.15 = 0.1355 \text{ in.}$$

Reviewing Prob. 4-5 we note that when an oscillator starts from a neutral position with a certain initial velocity, the effect of damping is to reduce the maximum deflection by the factor

$$\exp[(\zeta - \pi/2)\zeta] = 0.7843,$$

which gives

$$u_{m1} - b = 0.1355 \times 0.7843 = 0.1063$$

The maximum deflection is therefore

$$u_{m1} = 0.15 + 0.1063 = 0.2563 \text{ in.}$$

and the maximum force of impact:

$$R_{m1} = 120 \times 0.15 + 200 \times 0.1063 = 39.26 \text{ Lb}$$

6-6 In a machine for crushing ore, two drums are pressed against each other with a system of springs having stiffness $k = 3000$ Lb/in. and a preload of $N_0 = 4000$ Lb. When a piece to be crushed, with diameter $d = 1.5$ in., drops down and is drawn between the drums, it makes them separate by the distance d. At this moment crushing takes place and the moving drum, assumed to be instantaneously released, returns to the original position impacting the stationary drum. Treating the process of separation of the drums as static, calculate the magnitude of impulse caused by the return of the moving drum ($W = 2500$ Lb) to its original position. Reduce the impulse to the CG of the system, with $h = 46$ in.

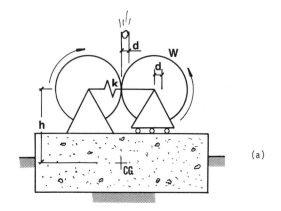

(a)

The drums are pressing each other with the force $P_0 = 4000$ Lb, which is the same as the spring preload. When the drums are separated by $d = 1.5$ in., the spring force grows to

$$N_{\max} = 4000 + 1.5 \times 3000 = 8500 \text{ Lb}$$

As long as the piece is not crushed, the stationary drum is under the action of two equal and opposite forces. When the idealized crushing occurs, one of the forces suddenly disappears and the spring force applies the impulse to the support. The magnitude of this impulse may be calculated from the velocity v attained by the moving drum prior to striking the stationary drum. Equating the change of strain energy and kinetic energy we have

$$\frac{1}{2}(N_0 + N_{\max})d = \frac{1}{2}\frac{Wv^2}{g}$$

that is,

$$\frac{1}{2}(4000+8500)1.5=\frac{1}{2}\times\frac{2500}{386}v^2 \qquad \therefore v=53.805 \text{ in./s}$$

The impulse of the spring force is therefore

$$S=\frac{W}{g}v=348.48 \text{ Lb-s}$$

This impulse pulls the stationary support to the right. When the drums collide, an impulse of the same magnitude, but opposite sense, is applied to the stationary support. The time needed for the drum to cross the gap may be obtained by solving the equation of motion or by the following simplified reasoning. The average force acting on a drum is

$$N_{av}=\frac{1}{2}(N_0+N_{max})=6250 \text{ Lb}$$

while the impulse is

$$S=N_{av}t_1=348.48 \text{ Lb-s} \qquad \therefore t_1=0.05576 \text{ s}$$

The approximate time-force plot for the stationary support is shown in (b). The force during the collision of drums is undetermined, as long as the drums are treated as rigid. The CG experiences not only the impulse S, but also the moment of impulse

$$Sh=348.48\times46=16,030 \text{ Lb-in.-s}$$

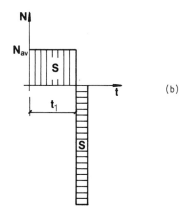

(b)

6-7 The cantilever beam of length $L=16$ in. shown in (a) has a circular cross section with diameter $D=2.5$ in. The material is steel, having $E=29\times10^6$ psi and the yield point $\sigma_p=45,000$ psi. Mass $M=0.5$ Lb-s^2/in. impacts the tip of beam with the velocity $v_0=85$ in./s and does not rebound. Calculate the maximum displacement u_m assuming the following behavior: elastic action until the maximum bending moment $\mathfrak{M}$ reaches the value of $\mathfrak{M}_p=\frac{1}{6}D^3\sigma_p$ (rectangular stress distribution). At this instant the plastic hinge forms and further increase of displacement comes from that hinge only. Do not include the mass of the beam in the calculation.

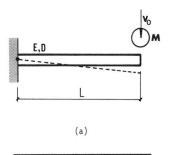

(a)

The resistance of the plastic hinge is

$$\mathfrak{M}_p=\frac{1}{6}\times2.5^3\times45,000=117,188 \text{ Lb-in.}$$

This corresponds to the tip load of

$$P_p=\frac{\mathfrak{M}_p}{L}=117,188/16=7324 \text{ Lb}$$

The force-deflection characteristic of the tip of beam is shown in (b). The energy, which may be absorbed by the elastic action, is

$$\Pi_e=\frac{1}{2}P_pu_p=\frac{1}{2}P_p\left(\frac{P_pL^3}{3EI}\right)$$

For a round section $I=\pi D^4/64=1.9175$ in.^{4}, $u_p=0.1798$ in. and then $\Pi_e=658.5$ Lb-in. The total energy to be absorbed is

$$T=\frac{1}{2}Mv_0^2=\frac{1}{2}\times0.5\times85^2=1806.3 \text{ Lb-in.}$$

and the portion of it left for plastic range is

$$\Pi_p=T-\Pi_e=1806.3-658.5=1147.8 \text{ Lb-in.}$$

But $\Pi_p=\mathfrak{M}_p\vartheta=\mathfrak{M}_p(u_m-u_p)/L$, therefore

$$1147.8=\frac{117,188(u_m-0.1798)}{16}$$

$$\therefore u_m=0.3365 \text{ in.}$$

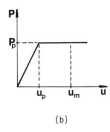

(b)

6-8 A long cantilever beam with a lumped mass $M=2.1$ Lb-s^2/in. and a distributed mass $m=1/14.96$ Lb-s^2/in.^{2} is subjected to a suddenly applied tip load, $P=P_0H(t)$ with $P_0=344,000$ Lb. The plastic moment capacity of the cross section is $\mathfrak{M}_p=11.04\times10^6$ Lb-in. Calculate the distance L_1 at which a plastic hinge will form assuming the material to be rigid-plastic.

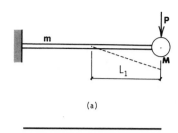

(a)

The fact that the material is rigid-plastic means that the deformation, if any, will take place at the plastic hinge only. Normally, the hinge will form at the fixed section, but that is not always the case. If the applied force is large enough, the hinge may form earlier and closer to the tip. The dynamic equilibrium of such an event is depicted in (b). The reaction R_1 is the resultant of the distributed inertia forces

$$R_1 = \tfrac{1}{2} L_1 ma_t$$

where a_t is the tip acceleration and ma_t is the inertia force per unit length near the tip. The vertical equilibrium equation is

$$P_0 - Ma_t - \tfrac{1}{2} L_1 ma_t = 0$$

and the moment equilibrium about the tip:

$$\left(\tfrac{1}{2} L_1 ma_t \right) \frac{L_1}{3} - \mathfrak{M}_p = 0$$

Using the second equation to eliminate a_t, we get

$$P_0 m L_1^2 - 3\mathfrak{M}_p m L_1 - 6M\mathfrak{M}_p = 0$$

and upon substitution,

$$22{,}995 L_1^2 - 2.214 \times 10^6 L_1 - 139.1 \times 10^6 = 0$$

$$\therefore L_1 = 139.61 \text{ in.}$$

Our solution will be meaningful only when L_1 is less than the total beam length. The larger the applied force, the more likely it is to take place.

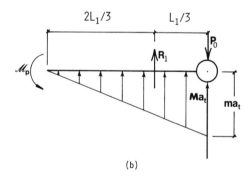

(b)

EXERCISES

6-9 A solid ball impacts a heavy plate and experiences contact force of magnitude R_m. How will R_m change if the mass of the ball is doubled but the material remains unchanged? Use the Hertz theory.

6-10 A solid ball impacts a thick steel plate. How will the impact force R_m change (1) if the initial velocity increases four times and (2) if instead of a steel plate a concrete slab is used? For steel: $E = 29 \times 10^6$ psi, $\nu = 0.3$; for concrete: $E = 3.5 \times 10^6$ psi, and $\nu = 0.15$. How much will the time of the loading phase of impact change in both cases? Use the Hertz theory.

6-11 The rod in (a) has a spherical ending. For the purpose of calculating the parameters of its impact against the elastic half-space, determine the characteristic of the equivalent bumper spring in (b): $E = 29 \times 10^6$ psi, $\nu = 0.3$, $L = 12$ in., $d = 1$ in., and $\gamma = 0.284$ Lb/in.^{3} Assume the impact to be elastic and include only local deformation. Calculate the impact duration if the rod is dropped from the height of 6 in. and the axis remains vertical.

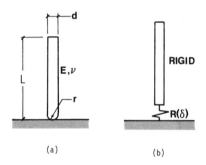

(a)   (b)

6-12 A bar of length L_1 moving with velocity v impacts another bar, which is fixed at the opposite end. Calculate the maximum impact force, ignoring the mass of the fixed bar. The mass of the moving bar is to be lumped at the left end, and the local deformations due to the diameters being unequal must be included. The material is aluminum, $E = 10.4 \times 10^6$ psi, $\nu = 0.32$, and $\gamma = 0.1$ Lb/in.^{3} The cross sections are circular, $d_1 = 5$ in. and $d_2 = 15$ in., $L_1 = 100$ in., $L_2 = 20$ in., and $v = 150$ in./s.

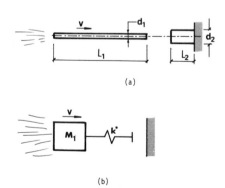

(a)

(b)

6-13 Solve Prob. 6-4 using the step load $P = 20H(t)$ instead of the initial velocity v_0.

6-14 In Prob. 6-11 the loading phase of impact calculated according to the Hertz theory is $t_1 = 0.1275 \times 10^{-3}$ s. Using the same local characteristic

$$R = 15.023 \times 10^6 \delta^{3/2}$$

(R in pounds if δ in inches), calculate the time from Eqs. 3.6 and 3.8. Impact velocity $v_0 = 68.06$ in./s and mass $M = 0.006934$ Lb-s^2/in.

6-15 In Prob. 6-14 a time of loading phase of impact was calculated. Repeat the calculation using Eq. 3.9 instead of 3.8 and see what error it gives.

6-16 The figure shows a rectangular impulse (a) and a half-sine impulse (b) acting on the basic undamped oscillator. The duration is $t_0 = 0.1\tau$, where τ is the natural period of vibration. Calculate the true magnification factor μ' for each case and compare it with the factor μ'' obtained from Eq. 6.18. Use the displacement equations developed in Probs. 2-13 and 2-33.

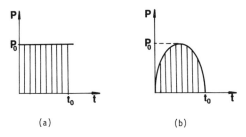

(a) (b)

6-17 Figure 6-16a shows a rectangular pulse, which is applied to the basic oscillator. The duration of the pulse is $t_0 = 0.25\tau$, where τ is the natural period, and the magnification factor may be calculated as

$$\mu' = 2\left| \sin\left(\frac{\omega t_0}{2} \right) \right|$$

The triangular pulse shown in this figure also lasts $t_0 = 0.25\tau$. Calculate the errors if Eq. 6.19 is used in place of the exact relations for those pulses.

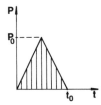

6-18 The figure shows an outline of an electric motor on a block foundation. The length of the block, normal to the paper, is 312 in. The soil on which the block rests is not directly visualized, but is represented by the angular spring $K = 6.5 \times 10^9$ Lb-in./rad and a fixed pivot point. The moment of inertia of the system about the pivotal axis is $J = 27.46 \times 10^6$ Lb-s^2-in. Because of malfunctioning of the motor, a torsional step load with magnitude $M_{t0} = 8.4 \times 10^6$ Lb-in. lasting for 0.1 s is applied. Calculate the maximum dynamic stress applied by the block to the soil, assuming linear distribution of this stress.

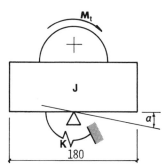

6-19 A frame consists of two elastic columns and a rigid horizontal beam. Each column is made up of a set of parallel, wide-flange beams with total $I = 10,420$ in.^{4} and section depth $H = 16.32$ in. The frame is impacted by the weight $W_1 = 2000$ Lb moving with the velocity 120 mph. The impact is perfectly plastic and from the first contact W_1 and W_2 move as one object. Calculate the maximum bending stress in columns for $W_2 = 50,000$ Lb, $h = 300$ in., and $E = 29 \times 10^6$ psi.

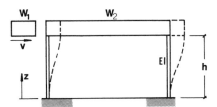

6-20 An object of mass $M = 0.01$ Lb-s^2/in. is elastically attached to its enclosure with two springs of stiffness $k = 10$ Lb/in. each. The entire assembly is dropped from the height $H = 50$ in. vertically and the enclosure does not rebound. (1) What is the maximum deflection h? (2) What is the overload factor if the springs are replaced with a different set, allowing $h = 1.0$ in.?

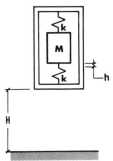

6-21 The mass $M=2$ Lb-s^2/in. is isolated from the foundation by a spring $k=10,000$ Lb/in. The foundation experiences a constant acceleration $a_0=5g$ that lasts 0.2τ, where τ is the natural period of vibration. Calculate the overload factor of the mass and the largest spring deflection. At $t=0$ both the displacement and the velocity are zero.

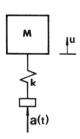

6-22 The system is the same as in Prob. 6-21 except that a damper has been added. The viscous constant c is such that $\zeta=0.12$. If the same acceleration shock is applied to the base, what is the overload factor and the maximum spring deflection? *Hint.* Ignore damping at first and then use a reduction factor for displacement developed in Prob. 4-5.

6-23 A heavy object with $M=2.1$ Lb-s^2/in. is placed inside a container having a flexible element $k=1',000$ Lb/in. on each side. When the container is moved horizontally, the object is allowed to slide freely with a total clearance of 0.25 in. Before the motion under consideration begins, the entire clearance is on the right-hand side. A constant acceleration $a_0=12g$ is

applied in the form of a step load at some instant. Calculate the maximum force of the first impact against a flexible element.

6-24 Suppose that the acceleration in Prob. 6-23 is applied in the opposite sense so that there is no gap involved. Calculate the new value of the impact force and comment on the desirability of the gap.

6-25 Both ends of the beam are clamped. When the step load $P=P_0H(t)$ is applied, it causes three plastic hinges to form and the beam deflects as shown by the dashed line. The center of beam is supported by a spring of stiffness $k=4000$ Lb/in. The resistance of the plastic hinge is $\mathfrak{M}_p=12,000$ Lb-in. Find the maximum spring force when $P_0=8000$ Lb and $L=40$ in.

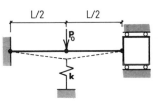

6-26 The frame is identical to that in Prob. 6-19, except that now we consider the pattern of plastic deformation. Assume the impacting weight W_1 to be large enough to cause the plastic hinges to appear at the ends of columns. The plastic hinge moment is $\mathfrak{M}_p=62\times10^6$ Lb-in. What is the largest weight W_1 that can impact the beam with $v=120$ mph if the allowable horizontal displacement is 8 in.? Ignore the elastic component of displacement.

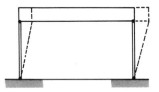

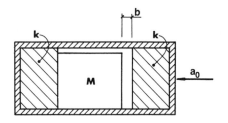

CHAPTER

7

Miscellaneous Problems

A new topic in this chapter is fatigue strength, which is introduced to enable the reader to relate the applied, time-dependent stress pattern to possible fatigue failure of the element under consideration. The remaining problems are associated with the ideas developed in previous chapters.

7A. Fatigue Strength. A structural element subjected to repeated loading and unloading will fail under a lower stress level than the same element under static loading. This phenomenon, known as a *fatigue failure*, may be of concern wherever vibrations or repeated shocks take place. The curve in Fig. 7.1 illustrates a stress level at some point of a structure subjected to a combination of a constant and a sinusoidally varying load. Two numbers are needed to characterize this stress distribution, the *mean stress* $\bar{\sigma}$ and the *alternating stress* σ_a. When introducing these definitions, one has in mind a uniaxial type of stress, because most of the fatigue data available are related to this condition.

Figure 7.2 shows a typical result of a test in which the applied stress varies between $-\sigma_a$ and $+\sigma_a$. (According to Fig. 7.1, this implies zero mean stress, $\bar{\sigma}=0$.) The objective of such a test is to establish the number of cycles to failure $\tilde{n}$ as a function of the alternating stress σ_a. For some materials (e.g., steel) the curve begins to flatten out after a number of cycles (solid line in Fig. 7.2). The stress level, which can be repeated infinitely many times without break-

ing the part in fatigue, is called the *endurance limit* F_f. The dashed branch of the curve applies to materials that do not exhibit an endurance limit, as is the case for nonferrous metals and alloys.

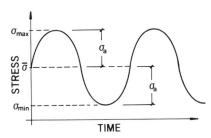

Figure 7.1 Parameters of sinusoidal stress level.

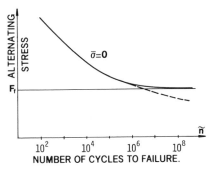

Figure 7.2 Alternating stress level as a function of number of cycles before failure.

77

In most practical applications, the mean applied stress is different from zero and the knowledge of F_f (or the entire curve in Fig. 7.2) is not sufficient to perform a fatigue strength check. If this is the case, we calculate the *damage ratio*

$$D = \frac{\bar{\sigma}}{F_u} + \frac{\sigma_a}{F_f} \qquad (7.1)$$

in which F_u is the ultimate static strength. If $D<1$, the element under investigation is safe, while $D \geq 1$ indicates a fatigue failure. When the relationship

$$\frac{\bar{\sigma}}{F_u} + \frac{\sigma_a}{F_f} = 1 \qquad (7.2)$$

is plotted, as in Fig. 7.3, this plot is called a Goodman diagram. For a combination of $\bar{\sigma}$ and σ_a marked by point B_1 we have damage ratio D less than unity, while for B_2 this ratio is larger than unity. If we are interested in the element withstanding only $\tilde{n}$ cycles, not an infinite number of cycles, we replace F_f in Fig. 7.3 (as well as in Eqs. 7.1 and 7.2) by σ_a' corresponding to that $\tilde{n}$. The desired value of σ_a' may be found from a plot of the type presented in Fig. 7.2 or from an empirical relationship

$$\tilde{n}(\sigma_a')^b = r \qquad (7.3)$$

which is a close approximation for many materials. (If there is a definite limit F_f for a particular material, the equation holds only for $\sigma_a' > F_f$). This relationship represents a straight line, defined by constants b and r, when plotted on a log-log scale. Although the established literature calls it the S–N curve, we will refer to it as $\tilde{n}$–σ curve to be consistent with our notation.

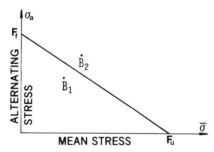

Figure 7.3 Goodman diagram.

When an applied stress is calculated using a simple formula like

$$\sigma = \frac{N}{A} \quad \text{or} \quad \sigma = \frac{\mathfrak{M}c}{I}$$

we speak of *nominal stress*. Manual calculations and computer printouts usually give nominal stress levels in members, unless special steps are taken to include the effect of *stress concentrations* (often called the stress risers) in the form of holes, fillets, and other deviations from an ideal shape. A factor by which the nominal stress must be multiplied to obtain the local peak stress is called a stress concentration factor k_t or a *fatigue notch factor* k_f. (The first is found on a theoretical basis, while the second is obtained from experimental results. Typically, $k_f < k_t$). It is only the fatigue, not the static strength, that is affected by stress concentrations (at least for ductile materials), and for this reason k_f is applied only to the alternating stress, not to the mean stress level. If, in Eq. 7.2, F_f is given with reference to a nearly perfect specimen ($k_f = 1.0$), while the actual part under consideration has a notch ($k_f > 1$), we use

$$\frac{\bar{\sigma}}{F_u} + \frac{k_f \sigma_a}{F_f} = 1 \qquad (7.4)$$

as a criterion of failure instead of Eq. 7.2 (σ_a is the nominal alternating stress). When k_f is the same for the test sample and in the part under consideration, Eq. 7.2 still holds.

When a structural element is subjected to dynamic loads with amplitudes varying in time, the damage ratio for the history of loading is determined as follows. Let n_1 be the number of cycles experienced at a load level corresponding to a failure after $\tilde{n}_1$ cycles, similarly for n_2 versus $\tilde{n}_2$, and so on. The damage ratio is then

$$D = \frac{n_1}{\tilde{n}_1} + \frac{n_2}{\tilde{n}_2} + \cdots + \frac{n_n}{\tilde{n}_n} \qquad (7.5)$$

According to Miner's hypothesis, the failure takes place when $D = 1.0$. It is known from experiments that D may be anywhere from 0.3 to 3.0 when failure occurs.

All the foregoing is applicable only to uniaxial stress and specifically to tension, because experimental evidence shows that there is no failure under direct compression. If the state of stress has several components (e.g., tension and shear), we must first calculate an equivalent tension. This is done using the *combined stress hypotheses*, the most prominent being the energy of distortion theory and the maximum shear theory.

SOLVED PROBLEMS

7-1 In Sect. 1E the equations for a series and a parallel elastic assembly were given in which the effective stiffness is expressed as a combination of spring constants of component elements. Develop the equations for the natural frequencies when a mass M is supported by one or the other type of assembly.

———————

When both sides of Eq. 1.9 have been multiplied by mass M, the series connection is defined by

$$\frac{1}{\omega^2} = \frac{1}{\omega_1^2} + \frac{1}{\omega_2^2} + \cdots + \frac{1}{\omega_n^2}$$

This means that the natural frequency ω is less than any component frequency $\omega_r = (k_r/M)^{1/2}$. The component frequency ω_r will be exhibited when all springs except k_r have become rigid. When both sides of Eq. 1.10 are divided by M, we have another definition of a parallel connection,

$$\omega^2 = \omega_1^2 + \omega_2^2 + \cdots + \omega_n^2$$

A fictitious component frequency ω_r can be realized if M is restrained only by k_r. The natural frequency is larger than any fictitious component frequency.

7-2 The cantilever beam forms one continuous element with its base. Determine the tip deflection, taking into account not only flexure and shear, but also the angular deformability of the base material. (This deformability shows itself in the slope of the deflected axis at point B.) Also find the natural frequency by placing one-fourth of the beam weight at the tip, while treating the beam itself as weightless. Use the following data:

$$L = 8 \text{ in., } h = 2 \text{ in., } r = 0.5 \text{ in.,}$$

$$E = 29 \times 10^6 \text{ psi, } G = 11 \times 10^6 \text{ psi,}$$

$$\gamma = 0.284 \text{ Lb/in.}^3, \nu = 0.3.$$

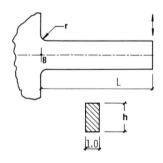

———————

From page 157 of Ref. 7 we can find the angular flexibility of the base as

$$\frac{\alpha}{PL} = \frac{16.67}{\pi E h_1^2} + \frac{1-\nu}{ELh_1}$$

where α is the slope at point B and

$$h_1 = h + 1.5r = 2.75 \text{ in.}$$

The translational tip flexibility corresponding to this,

$$\frac{\alpha L}{P} = \frac{1}{E}\left[\frac{16.67}{\pi}\frac{L^2}{h_1^2} + (1-\nu)\frac{L}{h_1}\right] = 1.619 \times 10^{-6} \text{ in./Lb}$$

The remaining components:

$$\frac{u_b}{P} + \frac{u_s}{P} = \frac{L^3}{3EI} + \frac{L}{GA_s} = (8.827 + 0.436)10^{-6} \text{ in./Lb}$$

where $A_s = A/1.2 = 1.667 \text{ in.}^2$ and $I = 0.6667 \text{ in.}^4$

Summation of the three components gives us the total flexibility. The stiffness is $k = 91,895 \text{ Lb/in.}$

The calculations show that in a beam of these proportions the influence of shear is quite small in comparison with bending. The effect of base deflection, on the other hand, is quite significant and may not be ignored. The lumped mass and frequency are, respectively,

$$M_0 = \frac{1}{4} \times 8 \times 2 \times 1 \times \frac{0.284}{386} = 2.943 \times 10^{-3}$$

$$\omega^2 = \frac{91,895}{2.943} \times 10^3 \qquad \therefore \omega = 5588 \text{ rad/s}$$

7-3 A cantilever with a distributed mass m may be treated as a weightless beam if a lumped mass equal to about $mL/4$ is placed at the tip. That gives, however, a good approximation for frequency only if the base of the cantilever is truly fixed. With this in mind, obtain a better answer to Prob. 7-2 using the reasoning from Prob. 7-1.

———————

Let us think about two parallel flexibilities, one of the beam itself and the other of the rotational base restraint. The first component, from the reference problem,

$$\frac{u_b}{P} + \frac{u_s}{P} = 9.263 \times 10^{-6} \text{ in./Lb}$$

The corresponding fictitious frequency, with $mL/4$ at the tip,

$$\omega_1^2 = \frac{1}{2.943 \times 10^{-3}} \times \frac{10^6}{9.263} = 36.682 \times 10^6$$

The angular flexibility

$$\frac{\alpha}{PL} = \frac{1.619 \times 10^{-6}}{8^2} = 25.3 \times 10^{-9} \text{ rad/(Lb-in.)}$$

The mass moment of inertia about the base end is

$$J = \frac{1}{3}(4 \times 2.943 \times 10^{-3})8^2 = 0.2511 \text{ Lb-s}^2\text{-in.}$$

The fictitious frequency is

$$\omega_2^2 = \frac{1}{0.2511} \times \frac{1}{25.3 \times 10^{-9}} = 157.41 \times 10^6$$

Combining the frequencies we get

$$\frac{10^6}{\omega^2} = \frac{1}{36.682} + \frac{1}{157.41} \qquad \therefore \omega = 5454 \text{ rad/s}$$

The answer differs only slightly from 5588 rad/s obtained in the reference problem.

7-4 Construct a diagram representing the displacement u, the velocity $\dot{u}$, and the acceleration $\ddot{u}$ of an SDOF system under a steady-state harmonic force $P = P_0 \sin \Omega t$. All quantities of interest are to be shown as components of a vector rotating with forcing frequency Ω.

From Sect. 2G we can determine that if $u = A \sin(\Omega t - \theta)$, then

$$\dot{u} = A\Omega \sin(\Omega t - \theta + 90^0)$$

$$\ddot{u} = A\Omega^2 \sin(\Omega t - \theta + 180^0)$$

which makes it easy to construct the diagram shown. The system of three amplitude vectors rotates with the angular velocity Ω. The projections of these vectors on the vertical axis give the desired quantities.

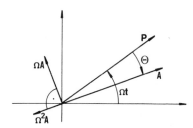

7-5 The system is as described in Prob. 1-3. The forcing frequency is $\Omega = 62$ rad/s while the amplitude $P_0 = 1800$ Lb. Calculate the maximum vibratory stress.

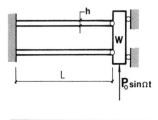

When P_0 is applied statically, the maximum stress at the root of beam is

$$\sigma_b = \frac{6\mathfrak{M}}{bh^2} = \frac{6(P_0 L/2)}{bh^2} = \frac{3 \times 1800 \times 50}{3 \times 2^2} = 22,500 \text{ psi}$$

The dynamic magnification factor is

$$\mu = \frac{1}{1 - \Omega^2/\omega^2} = \frac{1}{1 - 62^2/105.4^2} = 1.5291$$

Maximum vibratory stress:

$$\mu \sigma_b = 1.5291 \times 22,500 \approx 34,400 \text{ psi}$$

7-6 Horizontal force $P_0 H(t)$ is applied to the top of vertical arm of the mechanism in Prob. 1-23. What is the maximum spring moment and maximum acceleration of the horizontal member? (Note that everything except the spring is treated as rigid.)

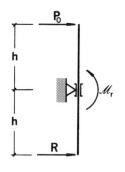

When P_0 is statically applied, the only resistance to rotation of the vertical arm comes from spring K. The moment of resistance is equal to $K\alpha$, where α is the angle of rotation. When the applied moment is $P_0 h$, the static angle is

$$\alpha_{st} = \frac{P_0 h}{K}$$

When P_0 is imposed as a step load, we can apply the magnification factor $\mu = 2$ and obtain the maximum spring moment as

$$\mathfrak{M}_{max} = 2K\alpha_{st} = 2P_0 h$$

Since both the vertical arm and the horizontal member are rigid and there is no gap in the joint connecting them, motion of any point in the system determines the motion of the remaining points. When the acceleration of the horizontal member is a, the dynamic equilibrium (neglecting the spring) is shown in the figure. The reaction R is due to inertia of the horizontal member,

$$R = a(M_1 + M_2)$$

while $\mathfrak{M}_r$ is caused by the angular inertia of the vertical arm,

$$\mathfrak{M}_r = J\ddot{\alpha} = \frac{Ja}{h}$$

The equilibrium equation

$$P_0 h - \mathfrak{M}_r - Rh = 0$$

allows us to compute the acceleration:

$$P_0 h = \frac{Ja}{h} + a(M_1 + M_2)h$$

$$\therefore a = \frac{P_0}{M_1 + M_2 + J/h^2}$$

This relation is valid as long as the spring is unstrained, that is, only at the initial position of the arm. Tilting of the (initially) vertical arm will activate the spring. The resistance moment of the spring will have the same sign as $\mathfrak{M}_r$ in the figure and thus it will work to diminish a. Conclusion: the acceleration formula developed above gives maximum a for step loading.

7-7 The natural frequency of a square, simply supported plate with side length L and thickness h, is given by

$$\omega = \frac{5.7}{L^2}\left[\frac{Eh^2}{\rho(1-\nu^2)}\right]^{1/2}$$

while the maximum bending stress due to a static force P_0 applied at the center is

$$\sigma = \frac{3P_0}{2\pi h^2}\left[(1+\nu)\log\frac{L}{2r_0}+0.75\right]$$

where r_0 is a radius of a small circle over which P_0 is distributed. The dynamic load is a step function cut off at $t=t_0$ (Fig. 2-13a):

$$P(t)=P_0 H(t-t_0)$$

Calculate the maximum bending stress in the plate for $L=12$ in., $h=0.04$ in., $r_0=0.2$ in., $E=29\times10^6$ psi, $\nu=0.3$, $g\rho=0.282$ Lb/in.^{3}, $P_0=50$ Lb, and $t_0=0.007$ s.

The natural frequency

$$\omega=\frac{5.7}{12^2}\left[\frac{29\times10^6\times0.04^2}{(0.282/386)(1-0.3^2)}\right]^{1/2}=330.7 \text{ rad/s}$$

If the force P_0 were applied statically, the bending stress would be

$$\sigma=\frac{3\times50}{2\pi\times0.04^2}\left(1.3\times\log\frac{12}{2\times0.2}+0.75\right)=77,164 \text{ psi}$$

Treating the plate as an SDOF system, the dynamic magnification factor may be deduced from Prob. 2-13,

$$\mu=\frac{u_{\max}}{u_{\text{st}}}=2\sin\frac{\pi t_0}{\tau}$$

τ is the natural period, $\tau=2\pi/\omega=0.019$ s, which is much longer than t_0,

$$\mu=2\sin\frac{\pi\times0.007}{0.019}=1.8315$$

The maximum bending stress,

$$\sigma_{\max}=1.8315\times77,164=141,326 \text{ psi}$$

7-8 A pipe is suspended on a row of hangers such as the one shown. When constructing a computer model of the system, the hanger is considered to be a vertical restraint only. Yet, our intuition tells us the hanger will also offer some resistance to a horizontal force acting on the pipe. What element should be used in the model to simulate this effect? (W and L are given.)

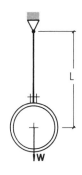

The lateral resistance comes from the geometric stiffening effect due to gravity as discussed in Chapter 1. From Prob. 1-12, for example, we find that the natural frequency of a pendulum of length L is

$$\omega=\left(\frac{g}{L}\right)^{1/2}$$

when no lateral spring is present. Writing

$$\omega^2=\frac{W}{ML}$$

we realize that the effect of gravity is equivalent to a spring with stiffness W/L, where W is the weight carried by the hanger. To simulate this effect in a computer model, we must attach the center of pipe to a fixed base using a horizontal spring with stiffness W/L. Although this additional restraint will be relatively flexible, it may have an appreciable effect on the lateral load response.

7-9 A simple oscillator described in Prob. 2-5 is vibrating at $\Omega/\omega=0.9$. The amplitude of forcing is $P_0=4500$ Lb and this is superposed on the static load of $P_1=3200$ Lb. The cross-sectional area of the elastic element of the oscillator is $A=0.82$ in.^{2}, and it is subjected to a direct stress. Can the element withstand an infinite number of vibratory cycles if the endurance limit of the material is $F_f=22,000$ psi, while the static strength $F_u=85,000$ psi?

The direct stress due to P_0 applied in a static manner is

$$\frac{P_0}{A}=5488 \text{ psi}$$

Because of dynamic application of the load, the alternating stress is

$$\sigma_a=\mu\frac{P_0}{A}=20,960 \text{ psi}$$

in which $\mu=3.82$ is the magnification factor taken from the reference problem for the prescribed ratio Ω/ω. The sustained or mean stress is

$$\bar\sigma=\frac{P_1}{A}=3902 \text{ psi}$$

Now we are ready to use Eq. 7.1:

$$D = \frac{\bar{\sigma}}{F_u} + \frac{\sigma_a}{F_f} = \frac{3902}{85,000} + \frac{20,960}{22,000} = 0.9986$$

Since D is (barely) less than unity, the elastic element will be able to withstand an infinite number of applied stress cycles.

7-10   The tube is subjected to pressure oscillating between 0 and 5400 psi as well as a torque varying from 0 to 63,000 Lb-in.; the frequency of both components is the same, $f = 1.5$ Hz. The tube is to be treated as open ended (i.e., no axial stress is induced). The material is an alloy steel having $F_u = 180$ ksi and the fatigue strength σ_a' varying according to

$$\bar{n}(\sigma_a')^{5.6276} = 3.66 \times 10^{31}$$

How long can this loading be applied to the tube before fatigue failure takes place? No dynamic magnification of the applied load is to be considered here.

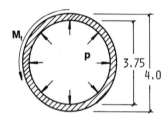

The approximate circumferential stress is

$$\sigma_h = \frac{pr}{t} = 5400 \left(\frac{3.75 + 4.0}{4} \right) \left(\frac{2}{4.0 - 3.75} \right) = 83,700 \text{ psi}$$

Maximum shear stress when twisting a tube with inside radius $r_0 = 1.875$ in. and outside radius $r_1 = 2.0$ in. is

$$\tau = \frac{2 M_t r_1}{\pi (r_1^4 - r_0^4)} = 22,035 \text{ psi}$$

The effective tensile stress, according to the energy of distortion theory, is

$$\sigma_{ef} = (\sigma_h^2 + 3\tau^2)^{1/2} = 91,990 \text{ psi}$$

Since both the load and the stress oscillate between zero and the maximum value, we have

$$\bar{\sigma} = \sigma_a = \frac{91,990}{2} \approx 46,000 \text{ psi}$$

Equation 7.2, when written for $\bar{n}$ cycles, is

$$\frac{46,000}{180,000} + \frac{46,000}{\sigma_a'} = 1.0 \qquad \therefore \sigma_a' = 61,790 \text{ psi}$$

Inserting this required value of σ_a' in the equation given in the problem statement we get the number of cycles to failure, $\bar{n} = 40,005$. The maximum time during which the tube can be exposed to the prescribed load is

$$t = \frac{\bar{n}}{f} = \frac{40,005}{1.5} = 26,670 \text{ s} \approx 7.4 \text{ hr}$$

7-11   The system and the loading are the same as in Prob. 7-5, except that the amplitude P_0 varies in a stepwise manner,

$P_{01} = 500$ Lb	applied for	$t_1 = 200$ min
$P_{02} = 900$ Lb	applied for	$t_2 = 80$ min
$P_{03} = 2000$ Lb	applied for	$t_3 = 12$ min

Evaluate the damage ratio D resulting from this loading program using the material characterized by $b = 5.6276$ and $r = 36.6 \times 10^{30}$. The fatigue constants are based on a test with a notch factor $k_{f1} = 2.0$, while in our case $k_{f2} = 3.3$.

The frequency of load application is

$$\Omega = 62 \text{ rad/s} = 9.8676 \text{ Hz}$$

Converting time in minutes to cycles, we have

118,410	cycles with	500 Lb
47,365	cycles with	900 Lb
7,105	cycles with	2000 Lb

$P_0 = 1800$ Lb applied with this frequency induces the vibratory stress of 34,400 psi, as was calculated in the reference problem. Consequently, the load amplitudes above correspond to vibratory stress of 9556, 17,200, and 38,220 psi, respectively. These, however, are nominal stress levels, which must be multiplied by the approximate ratio of the notch factors involved:

$$\frac{k_{f2}}{k_{f1}} = \frac{3.3}{2.0} = 1.65$$

All this results in the following levels of alternating stress:

118,410	cycles with	15,770 psi
47,365	cycles with	28,380 psi
7,105	cycles with	63,060 psi

Equation 7.3 gives us the number of cycles to failure at 15,770 psi,

$$\bar{n}_1 (15,770)^{5.6276} = 36.6 \times 10^{30} \qquad \therefore \bar{n}_1 = 87.05 \times 10^6$$

After finding $\bar{n}_2$ and $\bar{n}_3$, the damage ratio is obtained from Eq. 7.5:

$$D = \frac{118,410}{87.05 \times 10^6} + \frac{47,365}{3.189 \times 10^6} + \frac{7105}{35,677} = 0.2154 < 1.0 \text{ O.K.}$$

EXERCISES

7-12 Calculate the natural frequencies of weightless beams with lumped masses.

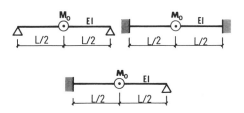

7-13 The moment of inertia of a rigid body is J, when measured about pivot point B. Find the expression for the natural frequency.

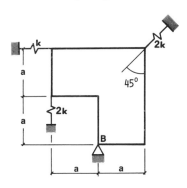

7-14 When the center load P is applied to a round plate with thickness h, the center deflection is

$$u = \frac{(3+\nu)PR^2}{16\pi(1+\nu)D} \qquad \text{(simply supported)}$$

$$u = \frac{PR^2}{16\pi D} \qquad \text{(clamped)}$$

where $D = \dfrac{Eh^3}{12(1-\nu^2)}$

What are the corresponding natural frequencies when the plate is weightless and there is a lumped mass M at the center? What is the ratio of frequencies for these boundary conditions when $\nu = 0.3$?

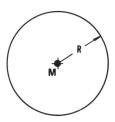

7-15 In the equation of motion

$$\ddot{u} + 2\omega\zeta\dot{u} + \omega^2 u = \frac{P}{M}$$

u with its derivatives as well as P may be treated as complex functions of t. (This means there are two numbers defining u at any time point, not just one. Those numbers are called the real part and the imaginary part of u, respectively.) Develop a formula for steady-state response when forcing is a complex harmonic function:

$$P(t) = P_0 e^{i\Omega t} = P_0(\cos\Omega t + i\sin\Omega t)$$

7-16 Using all data and assumptions from Prob. 1-4, calculate maximum bending stress in the beam and maximum spring force. The harmonic load $P(t)$ is at resonance with the system and has the amplitude $P_0 = 180$ Lb. The damping ratio is $\zeta = 0.08$.

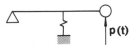

7-17 In some engineering applications the force P grows to its limiting value P_0 in asymptotic fashion. How can such a function be obtained from simpler shapes in Chapter 2? What is the response of the basic oscillator to such forcing?

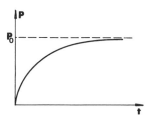

7-18 Using the graphical representation in Prob. 7-4 as an aid, discuss how to obtain a steady-state response of the basic oscillator to the forcing function:

$$P(t) = P_{01}\sin\Omega t + P_{02}\sin(\Omega t - \alpha) + P_{03}\sin(\Omega t - \beta)$$

7-19 The system in Fig. 6.2 drops and impacts a rigid surface with a velocity v_0. For a given k and M determine the damping ratio so that the acceleration of the mass is minimum. Express the minimum acceleration as a function of v_0, k, and M. *Hint.* Use the approximate expression 6.11 for the maximum impact force.

7-20 A weight is restrained by a spring, a damper, and the effect of a friction force due to gravity. The undamped natural frequency is $f_n = 3.1$ Hz, the viscous damping ratio $\zeta_v = 0.08$, and the friction coefficient $\psi = 0.06$. The initial amplitude is $A = 2.0$ in. After how many cycles will the amplitude be reduced to less than 0.2 in.? *Hint.* Use an average amplitude when computing the equivalent damping ratio.

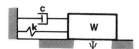

7-21 An aircraft-arresting system consists of a cable, four pulleys, and a friction brake B. The aircraft, weighing W, impacts the cable with a velocity v_0 and begins to pull the cable, which causes the brake to move in the opposite direction. The brake friction force F is set before impact and remains constant thereafter. Assuming the cable to be rigid and weightless, calculate the maximum displacement of the aircraft and its acceleration at the instant it reaches that maximum displacement. Put $W = 25,000$ Lb, $v_0 = 1440$ in./s, $b = 540$ in., and $F = 75,000$ Lb. (See illustration, upper right.)

7-22 An aircraft-arresting system is similar to the one described in Prob. 7-21, except that the friction brake is fixed in place and the kinetic energy of impact is absorbed by the cable. The length of one-half of the cable is $L = 2460$ in. and the axial stiffness

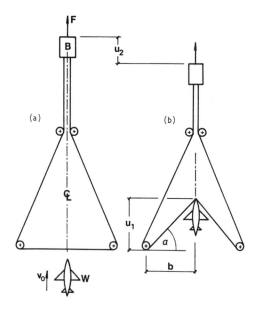

parameter $EA = 370,000$ Lb/in. Using the same data as in the reference problem, determine the overload factor for an arrested aircraft.

7-23 A turbine blade is subjected to a steady centrifugal stress, 17,460 psi in tension, at the root area. At the same location periodic steam pressure results in the application of a pulsating stress of 8920 psi. Both the figures above are nominal levels, while the notch factor is 2.6. The endurance limit of the material is 55 ksi (unnotched specimen) and ultimate strength 87 ksi. Calculate the damage ratio in fatigue.

PART
II

MULTIPLE DEGREE-OF-FREEDOM
SYSTEMS

Stiffness Properties of Elastic Systems

Stiffness and flexibility matrices are defined and the process of their formation is described in detail. The direct approach based on definitions is employed, as well as the Castigliano theorem. The relation between individual element matrices and the assembly matrix is introduced by means of coordinate transformation. The unit load method, one of the most efficient tools for manual derivation of flexibility coefficients, is described. Symmetry properties are formulated and their use discussed with regard to modeling. This chapter is devoted entirely to the static properties of elastic structures.

8A. Definition of Stiffness and Flexibility Matrices.

Consider an arbitrary elastic system, supported in a statically determinate or indeterminate manner. Suppose there are three points at which forces may be applied or deflections have to be calculated (Fig. 8.1).

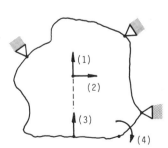

Figure 8.1 Directions of interest or displacement coordinates.

The directions of interest are numbered from 1 to 4. Although numbers 1 and 3 are associated with the arrows on the same vertical line, we still treat them as two separate directions, because the points of interest do not coincide. The stiffness matrix is generated by applying a unit displacement to each direction in turn and not allowing any other deflections at the same time.

Let us apply a displacement of a unit magnitude in the direction 1, $u_1 = 1$, while the remaining three displacements are prevented. The force, which must be applied in the direction 1 for this purpose, is designated by k_{11} while the reactions (i.e., the forces that prevent deflections) in the remaining directions are k_{21}, k_{31}, and k_{41}, respectively. For u_1 having some arbitrary value, the set of forces associated with the imposed deformation pattern is

$$\begin{Bmatrix} P_1 \\ P_2 \\ P_3 \\ P_4 \end{Bmatrix} = \begin{Bmatrix} k_{11} \\ k_{21} \\ k_{31} \\ k_{41} \end{Bmatrix} u_1$$

(Note that there is no need to introduce separate symbols to distinguish moments from forces and rotations from translations.)

The system is now restored to the original unstrained position and the displacement u_2 is applied. The active force is k_{22}, while the remaining three are the reactions,

$$\begin{Bmatrix} P_1 \\ P_2 \\ P_3 \\ P_4 \end{Bmatrix} = \begin{Bmatrix} k_{12} \\ k_{22} \\ k_{32} \\ k_{42} \end{Bmatrix} u_2$$

When the whole set of displacements, u_1 through u_4, has been applied simultaneously, the forces associated with the directions of interest are superposed and the result is presented as:

$$\begin{Bmatrix} P_1 \\ P_2 \\ P_3 \\ P_4 \end{Bmatrix} = \begin{bmatrix} k_{11} & k_{12} & k_{13} & k_{14} \\ k_{21} & k_{22} & k_{23} & k_{24} \\ k_{31} & k_{32} & k_{33} & k_{34} \\ k_{41} & k_{42} & k_{43} & k_{44} \end{bmatrix} \begin{Bmatrix} u_1 \\ u_2 \\ u_3 \\ u_4 \end{Bmatrix}$$

The symbolic form of this equation is

$$\vec{P} = k\vec{u} \qquad (8.1)$$

where $\vec{P}$ is called the force vector, k the stiffness matrix, and $\vec{u}$ the displacement vector. An entry k_{ij} of k has the physical meaning of a force applied in ith direction as a result of a unit displacement in jth direction only. Column number j is the set of external forces associated with the deformed pattern in which the unit displacement in jth direction is the only nonzero deflection.

An important property of stiffness matrix is its symmetry about the main diagonal. This means that the rows can take the place of columns, and vice versa, without changing the value of the matrix. In terms of indices we have

$$k_{ij} = k_{ji}$$

which is another way of saying that the order of indices is immaterial. Therefore it is enough to calculate the entries above (or below) the main diagonal only.

When the words "deflection" and "load" are interchanged, the development of the stiffness matrix can be used as a pattern for introducing the flexibility matrix. Referring to the picture described before, we apply a unit load only in direction 1, with all four directions unconstrained. The deflection in the ith direction due to that load is designated by a_{i1}. If a load $P_1 \neq 1$ is applied, the deformed pattern is de-

scribed by the following set of displacements.

$$\begin{Bmatrix} u_1 \\ u_2 \\ u_3 \\ u_4 \end{Bmatrix} = \begin{Bmatrix} a_{11} \\ a_{21} \\ a_{31} \\ a_{41} \end{Bmatrix} P_1$$

The system is then unloaded and the force P_2 is applied. The deflections are

$$\begin{Bmatrix} u_1 \\ u_2 \\ u_3 \\ u_4 \end{Bmatrix} = \begin{Bmatrix} a_{12} \\ a_{22} \\ a_{32} \\ a_{42} \end{Bmatrix} P_2$$

When all the forces act at once, the displacements are superposed and the result is

$$\begin{Bmatrix} u_1 \\ u_2 \\ u_3 \\ u_4 \end{Bmatrix} = \begin{bmatrix} a_{11} & a_{12} & a_{13} & a_{14} \\ a_{21} & a_{22} & a_{23} & a_{24} \\ a_{31} & a_{32} & a_{33} & a_{34} \\ a_{41} & a_{42} & a_{43} & a_{44} \end{bmatrix} \begin{Bmatrix} P_1 \\ P_2 \\ P_3 \\ P_4 \end{Bmatrix}$$

In symbolic form:

$$\vec{u} = a\vec{P} \qquad (8.2)$$

where a is called the flexibility matrix. Entry a_{ij} of a is the displacement in the ith direction due to a unit force applied in the jth direction. Column number j is the set of displacements induced by a unit load applied in jth direction. Matrix a is also symmetric about the main diagonal, $a_{ij} = a_{ji}$. This property is known as the Maxwell reciprocity principle.

In Chapter 1 flexibility is defined as a reciprocal of stiffness. The analogous relationship exists between the corresponding matrices:

$$a = k^{-1} \qquad \text{or} \qquad ak = I \qquad (8.3)$$

where I designates the unit matrix.

A displacement in a direction of interest is often referred to as a *displacement coordinate* or a *degree of freedom*.

8B. Work of External Forces and Strain Energy.

The strain energy accumulated in a linear spring of stiffness k and flexibility $a = 1/k$ can be expressed either by the force P or by the deflection u:

$$\Pi = \tfrac{1}{2}Pu = \tfrac{1}{2}aP^2 = \tfrac{1}{2}ku^2 \qquad (8.4)$$

where P and u are the current values (i.e., the values at the time we want to calculate Π). If we consider a linear elastic system with displacement coordinates $u_1, u_2, \ldots, u_n$, the strain energy accumulated during the deformation process is:

$$\Pi = \tfrac{1}{2}(P_1 u_1 + P_2 u_2 + \ldots + P_n u_n)$$

Using matrix notation, this is the same as

$$\Pi = \tfrac{1}{2}\vec{P}^T \vec{u} \qquad (8.5)$$

From Eq. 8.2, we get

$$\Pi = \tfrac{1}{2}\vec{P}^T \mathbf{a} \vec{P} \qquad (8.6)$$

To express Π in terms of stiffness, rather than flexibility, Eq. 8.1 may be transposed:

$$\vec{P}^T = \vec{u}^T \mathbf{k}$$

which, when substituted into Eq. 8.5, gives

$$\Pi = \tfrac{1}{2}\vec{u}^T \mathbf{k} \vec{u} \qquad (8.7)$$

This energy is equal to work of forces overcoming the elastic resistance. Note that Eqs. 8.6 and 8.7 are similar to Eq. 8.4, when the latter is written as $\Pi = \tfrac{1}{2}PaP = \tfrac{1}{2}uku$.

8C. Castigliano's Theorem.

The strain energy can be represented as a second-order polynomial in terms of applied forces or deflections, Eq. 8.6 or 8.7, respectively. There are two parts of Castigliano's theorem that correspond to those forms.

$$\frac{\partial \Pi}{\partial u_r} = P_r \qquad \text{(part I)} \qquad (8.8)$$

$$\frac{\partial \Pi}{\partial P_r} = u_r \qquad \text{(part II)} \qquad (8.9)$$

Part I states that strain energy differentiated with respect to deflection u_r gives the force P_r applied in that direction. Part II is similar, but the roles of force and displacement are interchanged.

The practical use of this theorem is to facilitate the calculation of stiffness and flexibility matrices. When Eq. 8.1 is written in a developed form, the force associated with rth direction is

$$P_r = k_{r1} u_1 + k_{r2} u_2 + \cdots + k_{rn} u_n$$

On the other hand, Part I tells us the expression we get as a result of differentiation of Π with respect to u_r is also equal to P_r. A comparison of coefficients of u's in both expressions determines the entries of the rth row of the stiffness matrix. Performing the differentiation with respect to the other displacements gives us the entire stiffness matrix.

When Π is expressed by the applied forces, it may be differentiated with respect to some selected P_r and the resulting expression is equal to u_r by virtue of Part II of the theorem. On the other hand, Eq. 8.2 tells us that u_r may be written as

$$u_r = a_{r1} P_1 + a_{r2} P_2 + \cdots + a_{rn} P_n$$

Comparing the coefficients of P's in both expressions yields the entries of the rth row in the flexibility matrix.

The use of Castigliano's theorem in this book is limited to *unconstrained displacements*, that is, to displacements that are independent of each other.

8D. Stiffness Matrices of Structural Elements.

As stated in Chapter 1, a pin-ended rod (Fig. 8.2) has axial stiffness EA/L. If it is preloaded with a stretching force N_0, it also has lateral stiffness N_0/L. For the direction shown in Fig. 8.2, the stiffness matrix is

$$\mathbf{k} = \begin{bmatrix} \dfrac{EA}{L} & 0 \\ 0 & \dfrac{N_0}{L} \end{bmatrix} \qquad (8.10)$$

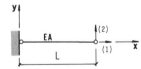

Figure 8.2 Pin-ended bar.

For a more general case in Fig. 8.3 it is convenient to show the stiffness matrix as a sum of geometric and elastic components

$$\mathbf{k} = \begin{bmatrix} c_x^2 & c_x c_y \\ c_y c_x & c_y^2 \end{bmatrix} \frac{EA}{L} + \begin{bmatrix} 1 - c_x^2 & -c_x c_y \\ -c_y c_x & 1 - c_y^2 \end{bmatrix} \frac{N_0}{L}$$

$$(8.11)$$

where $c_x = \cos\alpha$ and $c_y = \cos\beta$. This matrix is valid not only for rods, but also for other axial elements like cables and springs, provided they remain straight during the loading cycle. (One of the conditions for a cable to be straight is that it is subject to tension. An apparent compression, which is a decrease in tension, is acceptable if no resultant compression takes place.)

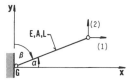

Figure 8.3   Pin-ended bar.

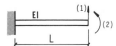

Figure 8.4   Constrained beam.

For a beam with 2 degrees of freedom (Fig. 8.4),

$$\mathbf{k} = \begin{bmatrix} \dfrac{12}{L^2} & \dfrac{-6}{L} \\ \dfrac{-6}{L} & 4 \end{bmatrix} \dfrac{EI}{L} \qquad (8.12)$$

Using the equilibrium equations and Fig. 8.5, we can develop a 4×4 beam matrix,

$$\mathbf{k} = \begin{bmatrix} \dfrac{12}{L^2} & \dfrac{-6}{L} & \dfrac{-12}{L^2} & \dfrac{-6}{L} \\ \dfrac{-6}{L} & 4 & \dfrac{6}{L} & 2 \\ \dfrac{-12}{L^2} & \dfrac{6}{L} & \dfrac{12}{L^2} & \dfrac{6}{L} \\ \dfrac{-6}{L} & 2 & \dfrac{6}{L} & 4 \end{bmatrix} \dfrac{EI}{L} \quad (8.13)$$

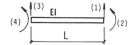

Figure 8.5   Unconstrained beam.

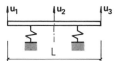

Figure 8.6   Constrained beam.

Another generalization of Fig. 8.4 is to include a shear deflection component. To have all meaningful stiffness coefficients in the matrix, the axial and the twisting directions are added (Fig. 8.6). The stiffness matrix is

$$\mathbf{k} = \begin{bmatrix} \dfrac{12}{L^2(1+\psi)} & \dfrac{-6}{L(1+\psi)} & 0 & 0 \\ \dfrac{-6}{L(1+\psi)} & \dfrac{4+\psi}{1+\psi} & 0 & 0 \\ 0 & 0 & \dfrac{A}{I} & 0 \\ 0 & 0 & 0 & \dfrac{GC}{EI} \end{bmatrix} \dfrac{EI}{L}$$

$$(8.14)$$

The effect of shear deflections is accounted for by the factor ψ:

$$\psi = \frac{12EI}{GA_s L^2} \qquad (8.15)$$

where G is the shear modulus and A_s is the effective shear area of the cross section. This factor will be significantly larger than zero only when the beam is short (say the length is less than twice the depth of beam) or when the cross section is of thin-wall type. Symbol C denotes the torsional constant, which in case of a circular section is equal to the polar moment of inertia of the area.

The stiffness matrices presented above were written for a beam whose principal plane of bending coincides with the plane of the paper. In a three-dimensional case the two displacements associated with bending in the second principal plane have to be added. This increases the size of the matrix to 6×6. When both ends are allowed to move, the matrix becomes 12×12.

8E.   Change of Displacement Coordinates.   A set of displacements initially chosen to describe deformation of an elastic body is shown in Fig. 8.7. Suppose

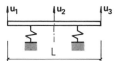

Figure 8.7   Selected displacement coordinates.

we want to use an alternative set, q_1, q_2, and q_3, defined by a transformation

$$u_1 = q_1 + q_3$$

$$u_2 = q_2$$

$$u_3 = q_1 - q_3$$

or, in matrix form

$$\begin{Bmatrix} u_1 \\ u_2 \\ u_3 \end{Bmatrix} = \begin{bmatrix} 1 & 0 & 1 \\ 0 & 1 & 0 \\ 1 & 0 & -1 \end{bmatrix} \begin{Bmatrix} q_1 \\ q_2 \\ q_3 \end{Bmatrix}$$

This change of (displacement) coordinates is written briefly as

$$\vec{u} = C\vec{q} \tag{8.16}$$

where C is called the *transformation matrix*. If there are forces P_1, P_2, and P_3 associated with the u's, there is also a set of corresponding forces Q acting along the displacements q. If one system is to be equivalent to the other, the work performed in both must be the same:

$$P_1 u_1 + P_2 u_2 + P_3 u_3 = Q_1 q_1 + Q_2 q_2 + Q_3 q_3$$

which is, in matrix notation:

$$\vec{P}^T \vec{u} = \vec{Q}^T \vec{q} \tag{8.17}$$

From which one may obtain

$$\vec{Q} = C^T \vec{P} \tag{8.18}$$

where C^T is the transpose of the transformation matrix.

Denote by k_u the stiffness matrix related to the first set of displacements:

$$\vec{P} = k_u \vec{u}$$

In the second set we have

$$\vec{Q} = k_q \vec{q}$$

The second stiffness matrix may be obtained from the first by performing two multiplications:

$$k_q = C^T k_u C \tag{8.19}$$

When the vector $\vec{q}$ in Eq. 8.16 has the minimum number of entries needed to describe a deformed

pattern, the set of q's is called the *generalized coordinates*. The initial coordinates represented by vector $\vec{u}$ are usually chosen for convenience and need not be independent of one another. Typically, there are more u's than q's.

8F. Generation of Stiffness Matrix for Assembly of Elements. Beginning with the known stiffness matrices of elements, we can construct the matrix of the assembly or the global matrix by merely adding the contributions of elements along the appropriate directions. This is especially easy when the elements have the same orientation with respect to some reference coordinate system. When the arrangement of elements is somewhat complicated and the problem is more suitable for computer-aided analysis, the manual superposition is replaced by matrix operations. The displacement coordinates characterizing individual elements are numbered consecutively $u_1, u_2, \ldots, u_n$, even though some of them coincide. A set of displacements $q_1, q_2, \ldots, q_n$, characterizing the deformation of the structure as a whole, is also designated. The relation between both sets is then written with the form of Eq. 8.16, and matrix C is used to calculate the global stiffness matrix by means of Eq. 8.19.

If in a stiffness matrix an off-diagonal term k_{ij} equals zero, we say that directions i and j are *elastically uncoupled*. If the stiffness matrix is diagonal, the whole system is considered to be elastically uncoupled and all directions of displacement may be treated independently.

8G. Singularity of Stiffness and Flexibility Matrices. A body that is not suitably constrained, such as the beam in Fig. 8.5, has a singular stiffness matrix. This means that the determinant of that matrix is zero. This is a normal situation for an airborne vehicle. It also occurs intentionally in some other systems (e.g., rotors). The other reasons for this condition are design errors and omissions that result in incomplete constraining of a structure. Also, there may be an insufficient number of structural elements in a portion of a system. This singularity of a stiffness matrix, which is not always evident at the outset of calculation, manifests itself in infinitely large deflections, which occur when external forces are applied. In computer analyses these deflections usually have very large but finite values.

A flexibility matrix may also be singular. This takes place when a relationship

$$d_1 u_1 + d_2 u_2 + \cdots + d_n u_n = 0$$

between displacements $u_1, u_2, \ldots, u_n$ exists, while the matrix was constructed treating these variables as independent (d_i's are constants). The singularity of flexibility matrix causes the absence of a stiffness matrix for a given system.

The beam in Fig. 8.8a is an example of what will be called a *properly constrained body*, which has a 2×2 stiffness matrix. When the right support is removed, as in Fig. 8.8b, a displacement with no elastic resistance is possible and the stiffness matrix becomes singular. This will be called an *underconstrained* body, or, when all supports are removed, an *unconstrained* body. Finally, a requirement $u_1 = u_2$ for the beam in Fig. 8.8a would cause the 2×2 flexibility matrix to be singular and the beam itself to be *overconstrained*.

Figure 8.8 (a) Properly constrained beam; (b) underconstrained beam.

8H. Unit Load Method is an efficient tool for calculating deflections when forming a flexibility matrix. Although the great majority of structural computer programs are based on stiffness approach, the flexibility method is quite often easier to use in manual calculations.

When the strain energy in a three-dimensional beam element is expressed by the internal forces and differentiation is carried out according to Part II of the Castigliano theorem, one may show that a deflection u along a direction of interest is

$$u = \int \frac{Nn}{EA} dx + \int \frac{\mathfrak{M}m}{EI} dx + \int \frac{Vv}{GA_s} dx + \int \frac{M_t m_t}{GC} dx$$

(8.20)

in which N, $\mathfrak{M}$, V, and M_t are the internal forces caused by the external load imposed on the structure and n, m, v, and m_t are the internal forces induced by a unit load applied in the direction of interest. The integration is extended over the whole structure. If bending and shear take place in two planes, two additional integrals will appear in the expression. For many special cases some integrals vanish; for example, an axial bar and a shaft would be represented by only the first and the last integral, respectively, in Eq. 8.20. Most practical structures have segments with

constant cross-sections (or can be approximated by such); hence the stiffness constants may be taken outside the integration symbol. The calculation of integrals is greatly facilitated by the tables like those on page 454 of Ref. 14. One may also notice that the flexural displacements of many simple beams and frames are tabulated in several standard references and that often voids the need for an actual calculation of the second integral in Eq. 8.20.

Sometimes it is more convenient to generate the stiffness matrix by inverting the flexibility matrix than in a direct process. Consider, for example, the beam in Fig. 8.8a. The direct computation of stiffness using the general beam matrix given by Eq. 8.13 would involve introducing two angles of rotation and generating a 4×4 matrix first. The two rotations would then be eliminated by stating that the moments applied from outside are zero, and in effect we would get a 2×2 stiffness matrix. It is much simpler, however, to form a 2×2 flexibility matrix to invert it.

The flexibility method is convenient for statically determinate structures, and its efficiency becomes worse with a growing number of redundancies. When it is not possible to quickly estimate the redundant forces, the stiffness method may be a better choice.

8J. Symmetry Properties. A structure is said to be symmetric with respect to a plane if one half of it is a mirror image of the other half. When the material properties are also symmetrically distributed, we speak of the *elastic symmetry*. An arbitrary loading system applied to a structure can be resolved into a symmetric and an antisymmetric part. To achieve this, each individual force is treated as schematically shown in Fig. 8.9. The formulas used for this operation are

$$P'' = \tfrac{1}{2}(P_1 + P_2) \quad \text{(symmetric)}$$
$$P' = \tfrac{1}{2}(P_1 - P_2) \quad \text{(antisymmetric)}$$

(8.21)

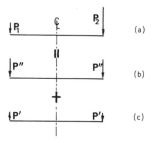

Figure 8.9 (a) General loading; (b) symmetric forces; (c) antisymmetric forces.

With the loading so resolved we may apply the following theorem:

When a structure, symmetric about a plane, is subjected to a symmetric set of forces, its response is also symmetric. Similarly, the antisymmetric load would produce only the antisymmetric response.

By the term *response* we mean here displacements, internal forces, reactions, and so on. Figure 8.10 shows the symmetric and the antisymmetric parts of the internal forces in cross sections of a three-dimensional beam and a two-dimensional beam. (Note that when applying the theorem above to a nonplane problem, one should draw the moments in the form of circular arrows.)

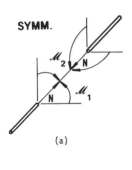

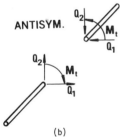

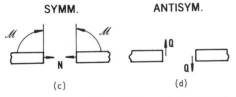

Figure 8.10 Resolution of internal forces: (*a*) and (*b*) in space; (*c*) and (*d*) in plane.

Structural symmetry may be used to reduce the size of a problem, as shown in Figure 8.11. The arbitrary loading acting on a frame is resolved into a symmetric and an antisymmetric component. Each is analyzed separately with only one-half of the frame modeled,

the other half being simulated by the boundary condition at the plane of cut. The static response is obtained by a superposition of both subcases. In spite of the additional operations involved, reducing the number of Eqs. 8.1 by one-half may often be advantageous.

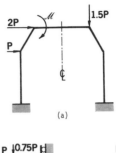

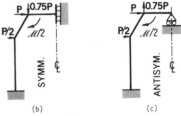

Figure 8.11 Resolution of original loading into symmetric and antisymmetric parts.

In real structures there seldom is a perfect symmetry, whether with respect to elastic properties or the applied load. In such cases a near-symmetry may be used to conclude about near-symmetry of response.

SOLVED PROBLEMS

8-1 Calculate the stiffness matrix for the system of blocks connected with springs.

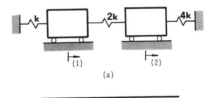

Impose a unit displacement in the direction 1 only. To keep the deflected system in equilibrium, the external forces P_1 and P_2 must be applied,

$$P_1 = 3k \quad \text{and} \quad P_2 = -2k$$

The system is then restored to the unstrained position and the unit translation is applied only in the direction 2. The reactions of springs are shown in (*c*). The blocks will remain displaced if the following forces are applied to them:

$$P_1 = -2k \quad \text{and} \quad P_2 = 6k$$

If the displacements are not unit values but u_1 and u_2, respectively, and if they are applied simultaneously, we have to exert the following forces to hold the system in the deflected position:

$$P_1 = 3ku_1 - 2ku_2 \quad \text{and} \quad P_2 = -2ku_1 + 6ku_2$$

In matrix form:

$$\begin{Bmatrix} P_1 \\ P_2 \end{Bmatrix} = \begin{bmatrix} 3k & -2k \\ -2k & 6k \end{bmatrix} \begin{Bmatrix} u_1 \\ u_2 \end{Bmatrix}$$

The 2×2 square array is the stiffness matrix. It is symmetric with respect to the main diagonal, as mentioned before. The first column represents the forces that are applied from outside to enforce deflection $u_1 = 1$, as shown in (b). Similarly, the second column is associated with deflection u_2 equal to unity.

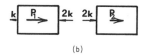

(b)

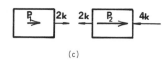

(c)

8-2   Calculate the flexibility matrix for the system of blocks connected with springs shown in Fig. 8.1a.

Apply a unit force in the direction 1. The spring forces and the block deflections may be obtained using the techniques presented in Chapter 1. Springs $2k$ and $4k$ are treated as one with effective stiffness k_1^*:

$$\frac{1}{k_1^*} = \frac{1}{2k} + \frac{1}{4k} \qquad \therefore k_1^* = \frac{4}{3}k$$

The unit force in the direction 1 is distributed between springs k and k_1^*. The force in spring k is

$$\frac{k}{k + (4/3)k} = \frac{3}{7}$$

while in the assembly of the other two springs it has a value of $4/7$. This means that the deflections of blocks are as shown in (a).

The system is then unloaded and the unit force is applied only in the direction 2. A similar reasoning gives the displacements as shown in (b). If, instead of unit forces, there are P_1 and P_2 applied simultaneously, the resultant displacements are obtained by superposition,

$$\begin{Bmatrix} u_1 \\ u_2 \end{Bmatrix} = \begin{bmatrix} \dfrac{3}{7k} & \dfrac{1}{7k} \\ \dfrac{1}{7k} & \dfrac{3}{14k} \end{bmatrix} \begin{Bmatrix} P_1 \\ P_2 \end{Bmatrix}$$

This 2×2 square array is the flexibility matrix. It is symmetric with respect to the main diagonal. The first column represents the displacements that result from applying only $P_1 = 1$ to the system.

Similarly, the second column is associated with the displacements caused by $P_2 = 1$.

The system considered is statically indeterminate, and calculating the flexibility matrix is relatively complicated.

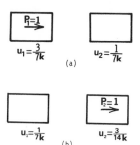

(a)

(b)

8-3   The system consists of two shafts, each with a gear at both ends. The independent angles of rotation in the assembly are α_1, α_2, and α_3. Angle α_4 is related to α_2 by a constraint equation $\alpha_2 r_2 = \alpha_4 r_4$. Calculate the 3×3 stiffness matrix.

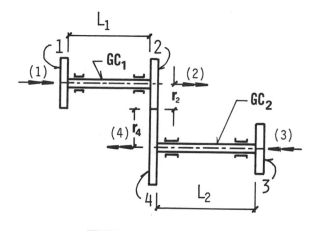

Shaft stiffness is $K = GC/L$ and notation $K_1 = GC_1/L_1$, $K_2 = GC_2/L_2$ will be used. To rotate wheel 1 by $\alpha_1 = 1$, the moment $k_{11} = K_1$ is required. The reaction at the other end is $k_{21} = -K_1$. We have $k_{31} = 0$, since no torque is transmitted to wheel 3. When the rotation $\alpha_2 = 1$ is imposed, wheel 4 turns by

$$\alpha_4 = \alpha_2 \frac{r_2}{r_4}$$

The torque in the shaft 3-4 is then $k_{32} = -K_2(r_2/r_4)$, and the contact force between wheels 2 and 4 is $K_2(r_2/r_4)(1/r_4) = P$.

The total resistance to turning of wheel 2 is

$$k_{22} = Pr_2 + K_1 = K_2\left(\frac{r_2}{r_4}\right)^2 + K_1$$

The moment needed to hold wheel 1 is $k_{12} = -K_1$. Finally, when wheel 3 is turned by the unit angle, we have $k_{33} = K_2$. Using the symmetry property, we can now write the entire matrix,

$$\mathbf{K} = \begin{bmatrix} K_1 & -K_1 & 0 \\ -K_1 & K_1 + K_2 \left(\dfrac{r_2}{r_4}\right)^2 & -K_2 \left(\dfrac{r_2}{r_4}\right) \\ 0 & -K_2 \left(\dfrac{r_2}{r_4}\right) & K_2 \end{bmatrix}$$

8-4 Calculate the stiffness matrix with respect to the directions shown, using the data in the figure.

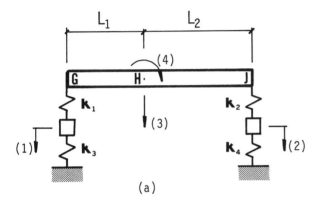

(a)

The equilibrium of the beam after each unit displacement is individually applied is shown in (b).

$$k_{11} = k_1 + k_3 \qquad k_{21} = 0 \qquad k_{31} = -k_1 \qquad k_{41} = k_1 L_1$$

$$k_{22} = k_2 + k_4 \qquad k_{32} = -k_2 \qquad k_{42} = -k_2 L_2$$

$$k_{33} = k_1 + k_2 \qquad k_{43} = k_2 L_2 - k_1 L_1$$

$$k_{44} = k_2 L_2^2 + k_1 L_1^2$$

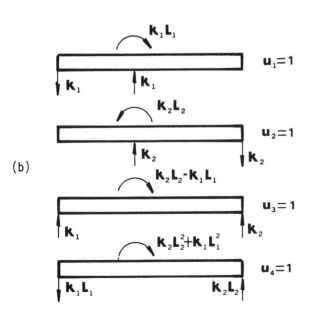

(b)

and

$$\mathbf{k} = \begin{bmatrix} k_1 + k_3 & 0 & -k_1 & k_1 L_1 \\ 0 & k_2 + k_4 & -k_2 & -k_2 L_2 \\ -k_1 & -k_2 & k_1 + k_2 & k_2 L_2 - k_1 L_1 \\ k_1 L_1 & -k_2 L_2 & k_2 L_2 - k_1 L_1 & k_2 L_2^2 + k_1 L_1^2 \end{bmatrix}$$

8-5 Using the formulas for translation (i.e., lateral deflection) and slope at the end of a cantilever beam, demonstrate the symmetry of stiffness and flexibility matrices and discuss the Maxwell reciprocity principle.

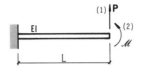

From Strength of Materials:

$$u = \frac{PL^3}{3EI} + \frac{\mathcal{M}L^2}{2EI}$$

$$\vartheta = \frac{PL^2}{2EI} + \frac{\mathcal{M}L}{EI}$$

where the translation u and the rotation ϑ are measured in the directions of P and $\mathcal{M}$, respectively. Numbering the directions as shown in parentheses and modifying the notation, we put the equations in matrix form:

$$\begin{Bmatrix} u_1 \\ u_2 \end{Bmatrix} = \begin{bmatrix} \dfrac{L^2}{3} & \dfrac{L}{2} \\ \dfrac{L}{2} & 1 \end{bmatrix} \begin{Bmatrix} P_1 \\ P_2 \end{Bmatrix} \frac{L}{EI}$$

Inversion of the flexibility matrix gives

$$\begin{Bmatrix} P_1 \\ P_2 \end{Bmatrix} = \begin{bmatrix} \dfrac{12}{L^2} & -\dfrac{6}{L} \\ -\dfrac{6}{L} & 4 \end{bmatrix} \begin{Bmatrix} u_1 \\ u_2 \end{Bmatrix} \frac{EI}{L}$$

and the symmetry of both matrices is thus demonstrated. Noting the relationship between the translation caused by the bending moment and the slope induced by the lateral force, we can state the Maxwell principle:

If a unit load applied in the direction i causes a displacement in the direction j, then an independently applied unit load in the direction j gives the numerically identical displacement in the direction i.

8-6 The formula for the lateral deflections of a cantilever beam, when there is a force P applied to the tip, is:

$$u = \frac{PL^3}{6EI} \left(\frac{x^3}{L^3} - 3\frac{x}{L} + 2 \right)$$

Determine the flexibility matrix with respect to the directions shown.

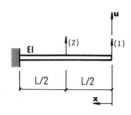

Put $P=1$ along (1). For $x=0$, $u=a_{11}=L^3/3EI$. For $x=L/2$, $u=a_{21}=5L^3/(48EI)$. If P is applied in direction 2, the same general formula for u may be used if L is replaced by $l=L/2$. (The formula will be valid for the left half of cantilever.) Thus, $a_{22}=l^3/(3EI)=L^3/(24EI)$. Finally,

$$\mathbf{a}=\frac{L^3}{3EI}\begin{bmatrix}1 & \dfrac{5}{16}\\[2mm] \dfrac{5}{16} & \dfrac{1}{8}\end{bmatrix}$$

8-7 A thin, inextensible ring, with radius R and bending stiffness EI, may be subjected to two equal, opposite forces acting along a diameter. If those forces are applied as shown, the horizontal diameter expands by $0.149PR^3/EI$, while the vertical diameter shrinks by $0.137PR^3/EI$. Derive the stiffness matrix of the system with respect to the generalized coordinates 1 and 2.

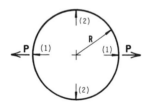

A generalized deflection in this problem consists of two opposite, radial deflections along the same diameter. Using the reciprocity principle, we have

$$u_1=(0.149P_1-0.137P_2)\frac{R^3}{EI}$$

in which P_1 is the system of forces in the figure and P_2 is the corresponding system along the vertical diameter. If u_2 is the expansion of the vertical diameter, we can write the flexibility matrix as

$$\begin{Bmatrix}u_1\\u_2\end{Bmatrix}=\frac{R^3}{EI}\begin{bmatrix}0.149 & -0.137\\-0.137 & 0.149\end{bmatrix}\begin{Bmatrix}P_1\\P_2\end{Bmatrix}$$

Inverting the flexibility matrix gives us the stiffness equation:

$$\begin{Bmatrix}P_1\\P_2\end{Bmatrix}=\frac{EI}{R^3}\begin{bmatrix}43.415 & 39.918\\39.918 & 43.415\end{bmatrix}\begin{Bmatrix}u_1\\u_2\end{Bmatrix}$$

8-8 Two cantilever beams with box sections are tied at the front ends to a plate, which may be considered infinitely rigid in its own plane, but perfectly flexible as far as warping is concerned. The back ends are built in. Develop the 2×2 stiffness matrix for a point of reference located midway between the elastic axes of the cantilevers. The wall thickness of beams is constant, $t=0.1$ in., while $L=60$ in., $d=40$ in., $E=10.6\times10^6$ psi, and $G=4\times10^6$ psi. Ignore the shear deflection component.

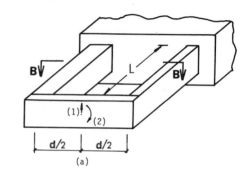

(a)

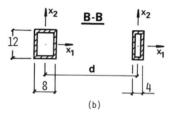

(b)

The section properties of the cantilevers are:

$$I_1'=84.03,\qquad C'=89.27\text{ in.}^4\qquad\text{(left)}$$
$$I_1''=55.71,\qquad C''=27.26\text{ in.}^4\qquad\text{(right)}$$

When the u_1 displacement is enforced, the ends of cantilevers can rotate freely because of the lack of warping stiffness of the plate. Therefore

$$k_{11}=\frac{3E}{L^3}(I_1'+I_1'')=12,371+8202=20,573$$

From the free-body sketch of the plate in (c):

$$k_{21}=83,380$$

When $u_2=1$ rad is applied, each beam end is subjected to 1 rad rotation and a 20 in. translation. The associated shears and torques are:

$$Q'=20\times\frac{3EI_1'}{L^3}=247,420\text{ Lb}$$

$$M_t'=\frac{GC'}{L}=5.951\times10^6\text{ Lb-in.}$$

$$Q''=20\times\frac{3EI_1''}{L^3}=164,040\text{ Lb}$$

$$M_t''=\frac{GC''}{L}=1.817\times10^6\text{ Lb-in.}$$

The torque needed to enforce the rotation (see (d)) is

$$k_{22} = 16 \times 10^6$$

Assemblying the matrix, we have

$$\mathbf{k} = \begin{bmatrix} 20{,}573 & 83{,}380 \\ 83{,}380 & 16 \times 10^6 \end{bmatrix}$$

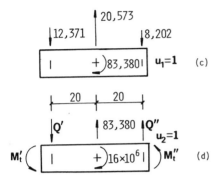

(c)

(d)

8-9 A rigid platform is placed on four pipe columns, two of them different from the other two. Each column has one end fixed in the ground and the other in the platform, the net column height being $L = 165$ in. Develop a 3×3 stiffness matrix. For smaller columns $D_1 = 18$ in., $t = 0.5$ in., $A = 27.49$ in.², and $I = 1053$ in.⁴ For larger columns $D_2 = 30$ in., $t = 0.5$ in., $A = 46.3$ in.², and $I = 5042$ in.⁴ Also, $B = 150$ in., $E = 29 \times 10^6$ psi, and $G = 11 \times 10^6$ psi.

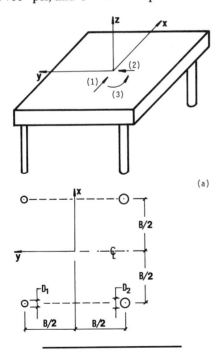

(a)

In solving this problem we use the 4×4 cantilever-beam matrix, Eq. 8.14, and Fig. 8.6. For a thin-wall tube $A_s = 0.5A$, which gives

us the ψ factors as follows:

$$\psi = \frac{12EI}{GA_s L^2} = \frac{12 \times 29 \times 10^6 \times 1053}{11 \times 10^6 (0.5 \times 27.49) \times 165^2} = \frac{1}{11.23} \quad \text{(smaller tube)}$$

$$\psi = 0.2531 \quad \text{(larger tube)}$$

A unit translation produces a resistance of a single tube equal to

$$k_{11}' = \frac{12EI}{L^3(1+\psi)} = 74{,}905 \quad \text{(smaller tube)}$$

$$k_{11}' = 311{,}706 \text{ Lb/in.} \quad \text{(larger tube)}$$

Summing the effects of components, we get

$$k_{11} = 2(74{,}905 + 311{,}706) = 773{,}222 = k_{22}$$

Since we are bending the columns along their principal directions, (as is always the case for tubes), $k_{21} = 0$. Also $k_{32} = 0$, because yz is the plane of symmetry. However, plane xz is not, and we get

$$k_{31} = 2(311{,}706 - 74{,}905)\frac{150}{2} = 35.52 \times 10^6$$

The resisting forces when the platform rotates by an angle α about its center are shown in (b). The tube section also rotates by α (with the whole platform) and translates by 106.07α. For a circular section, the torsional constant is $C = 2I$. The stiffness matrix, Eq. 8.14, gives us

$$k_{33}' = \frac{GC}{L} = \frac{11 \times 10^6 (2 \times 1053)}{165} = 140.4 \times 10^6 \quad \text{(smaller tube)}$$

$$k_{33}' = 672.27 \times 10^6 \quad \text{(larger tube)}$$

When the angle α is unity, the twisting moment applied to the platform and overcoming this resistance is

$$M_t + 106.07P = 140.4 \times 10^6 + 106.07^2 \times 74{,}905$$

$$= 983.14 \times 10^6 \quad \text{(smaller tube)}$$

$$M_t + 106.07P = 4179.2 \times 10^6 \quad \text{(larger tube)}$$

where $\qquad P = 106.07 k_{11}'$

Finally, $k_{33} = 2(983.14 + 4179.2)10^6 = 10{,}324.7 \times 10^6$, and we can assemble the matrix as

$$\mathbf{k} = \begin{bmatrix} 0.77322 & 0 & 35.52 \\ 0 & 0.77322 & 0 \\ 35.52 & 0 & 10{,}324.7 \end{bmatrix} 10^6$$

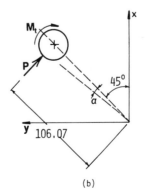

(b)

8-10 The concept of a propulsion system of a surface-effect ship is illustrated in (a). The turbine (J_3) is driving the water-jet pump consisting of the first- and second-stage rotors, J_1 and J_2, respectively. The shafts of pump rotors are concentric and can twist independently of each another. The motion from turbine to pump is transmitted through a set of four reducing gears. Develop a 3×3 stiffness matrix. Assume the following: $K_1 = 20.1 \times 10^6$, $K_2 = 8.72 \times 10^6$, $K_3 = 0.34 \times 10^6$ Lb-in./rad, with gear ratios $g_1 = r_1'/r_1'' = 0.2242$ and $g_2 = r_2'/r_2'' = 0.4926$.

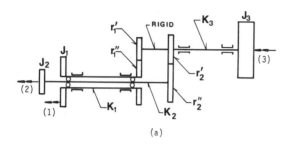

(a)

First, let us solve the auxiliary problem in (b). Suppose that the external torque is applied to the subassembly consisting of two gears with a rigid shaft, while no rotation is permitted at the three degrees of freedom, as in (a). This external torque M_{t4} causes the subassembly to rotate by angle α_4. The equilibrium equation for (b) is

$$M_{t4} = g_1 M_{t1} + g_2 M_{t2} + M_{t3}$$

(Note that if there is a resisting torque M_{t1} in shaft K_1, the circumferential force at the mating point of gears r_1' and r_1'' is M_{t1}/r_1'' and the torque applied to the subassembly is $M_{t1}r_1'/r_1'' = g_1 M_{t1}$, and similar for the other pair of gears.) Expressing the torques by the angles of rotation, we have

$$M_{t4} = K_4 \alpha_4 \qquad M_{t1} = K_1 g_1 \alpha_4$$
$$M_{t2} = K_2 g_2 \alpha_4 \qquad M_{t3} = K_3 \alpha_4$$

The combined stiffness of the system K_4 with all three shafts resisting the rotation of the subassembly is thus

$$K_4 = K_1 g_1^2 + K_2 g_2^2 + K_3 \qquad (*)$$

Let us now apply a torque M_{t1}' to J_1 while holding J_2 and J_3 fixed. The torque acting on the subassembly is then $g_1 M_{t1}'$ and the subassembly rotates by

$$\frac{g_1 M_{t1}'}{K_2 g_2^2 + K_3}$$

according to (*) when only K_2 and K_3 are resisting. The total rotation of J_1 is then

$$\alpha_1 = \frac{M_{t1}'}{K_1} + \frac{g_1^2 M_{t1}'}{K_2 g_2^2 + K_3}$$

When M_{t1} is selected so that $\alpha_1 = 1$ rad, it becomes equal to the stiffness coefficient k_{11},

$$\frac{1}{k_{11}} = \frac{1}{K_1} + \frac{g_1^2}{K_2 g_2^2 + K_3}$$

After substitution, $k_{11} = 14.241 \times 10^6$. The torque applied to the subassembly is $g_1 k_{11}$. From Eq. (*), in which we omit the term $K_1 g_1$, we find that the fraction

$$\frac{K_3}{K_2 g_2^2 + K_3} = 0.1384$$

of this torque is reacted by K_3, which gives

$$k_{31} = 0.1384 \times 0.2242 \times 14.241 \times 10^6 = 0.4420 \times 10^6$$

The remaining $1 - 0.1384 = 0.8616$ of that torque is reacted by K_2, therefore

$$k_{21} = -0.8616 \frac{1}{g_2} g_1 k_{11} = -5.585 \times 10^6$$

Next, we fix J_1 and J_3 while applying a torque to J_2. The distribution of internal loading is then found by removing the second term on the right-hand side of Eq. (*). The stiffness coefficients are found from similar formulas:

$$\frac{1}{k_{22}} = \frac{1}{K_2} + \frac{g_2^2}{K_1 g_1^2 + K_3} \qquad \therefore k_{22} = 3.397 \times 10^6$$

$$k_{32} = \frac{K_3 k_{22} g_2}{K_1 g_1^2 + K_3} = 0.4213 \times 10^6$$

Finally, the torque is applied to J_3, while J_1 and J_2 are fixed. This time we have

$$\frac{1}{k_{33}} = \frac{1}{K_3} + \frac{1}{K_1 g_1^2 + K_2 g_2^2} \qquad \therefore k_{33} = 0.3067 \times 10^6$$

The assembled matrix is presented below. Note that introducing the fourth DOF (α_4) would make the computation of the coefficients much simpler, even though the size of the matrix would increase.

$$k = \begin{bmatrix} 14.241 & -5.585 & 0.442 \\ -5.585 & 3.397 & 0.4213 \\ 0.442 & 0.4213 & 0.3067 \end{bmatrix} 10^6$$

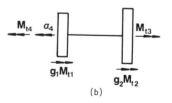

(b)

8-11 The frame consists of slender, rigidly joined beams. Specifying two translation components and a rotation at each node, at which a force may be applied, defines a deformed shape. As the figure shows, there are six independent displacements involved, but owing to a certain property of the system not all the displacements have to be used. The component beams may be treated as inextensible, because it is known that the influence of direct stress is usually negligible in comparison with bending. Using this property, reduce the number of displacements to a necessary minimum and write the appropriate transformation equation.

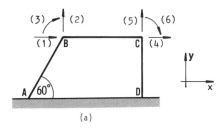

(a)

If u_A and u_B stand for translation of ends, the condition that the length remain unchanged is, in general:

$$u_B \cos \beta - u_A \cos \alpha = 0$$

where α and β denote the angles between the respective displacements and the member axis, (b). At the end A of member AB the displacement is zero, while the axial component at B is

$$u_1 \cos 60° + u_2 \sin 60°$$

The condition of axial rigidity is

$$u_1 + u_2 \tan 60° = 0$$

Similarly, for member BC, $u_4 - u_1 = 0$, and for CD, $u_5 = 0$.

As three equations of constraints were written, the number of independent displacement components needed to describe the deformation pattern is now

$$6 - 3 = 3$$

From the first two equations we see that u_1 and u_4 can be defined in terms of u_2. The following u's are chosen as independent and redesignated:

$$\begin{Bmatrix} u_2 \\ u_3 \\ u_6 \end{Bmatrix} = \begin{Bmatrix} q_2 \\ q_3 \\ q_6 \end{Bmatrix}$$

and the transformation equation, $\vec{u} = C\vec{q}$ becomes

$$\begin{Bmatrix} u_1 \\ u_2 \\ u_3 \\ u_4 \\ u_5 \\ u_6 \end{Bmatrix} = \begin{bmatrix} -\tan 60° & 0 & 0 \\ 1 & 0 & 0 \\ 0 & 1 & 0 \\ -\tan 60° & 0 & 0 \\ 0 & 0 & 0 \\ 0 & 0 & 1 \end{bmatrix} \begin{Bmatrix} q_2 \\ q_3 \\ q_6 \end{Bmatrix}$$

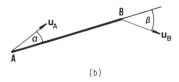

(b)

8-12 Using Castigliano's theorem, calculate the stiffness matrix for the system in Prob. 8-1.

Part I of the theorem must be used after the strain energy is expressed by displacements. For a single spring in the sketch we have

$$\Pi = \tfrac{1}{2} k (u_B - u_A)^2$$

Summing the contributions of each of three springs in the problem, we have

$$\Pi = \tfrac{1}{2} k u_1^2 + \tfrac{1}{2}(2k)(u_2 - u_1)^2 + \tfrac{1}{2}(4k)(-u_2)^2$$

Differentiate this successively with respect to u_1 and u_2:

$$\frac{\partial \Pi}{\partial u_1} = k u_1 + 2k(u_2 - u_1)(-1) = 3k u_1 - 2k u_2 = P_1$$

$$\frac{\partial \Pi}{\partial u_2} = 2k(u_2 - u_1) + 4k u_2 = -2k u_1 + 6k u_2 = P_2$$

We can write the result in the matrix form as

$$\begin{bmatrix} 3k & -2k \\ -2k & 6k \end{bmatrix} \begin{Bmatrix} u_1 \\ u_2 \end{Bmatrix} = \begin{Bmatrix} P_1 \\ P_2 \end{Bmatrix}$$

The stiffness matrix obtained is the same as in Prob. 8-1.

8-13 Two rigid pulleys with fixed centers are connected with an elastic cable. Calculate the stiffness matrix assuming that the pulleys will not slip, nor will the cable slacken. The projected effective length is to be used as in Prob. 1-14. Apply Castigliano's theorem.

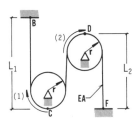

There are three segments of cable, which are considered separately: BC, CD, and DF. Strain energy in each is expressed by

$$\Pi = \tfrac{1}{2} \frac{EA}{L}(u_B - u_A)^2$$

Using projected length and putting $u = \theta r$ we have

$$BC : \Pi = \frac{EA}{2L_1}(\theta_1 r)^2$$

$$CD : \Pi = \frac{EA}{2L_2}(\theta_1 r + \theta_2 r)^2$$

because $\theta_1 r + \theta_2 r$ is the extension of this cable.

$$DF : \Pi = \frac{EA}{2L_2}(\theta_2 r)^2$$

Summing these components and differentiating yields:

$$\frac{\partial\Pi}{\partial\theta_1} = \frac{EA}{L_1}r^2\theta_1 + \frac{EA}{L_2}(\theta_1 r + \theta_2 r)r = EAr^2\left(\frac{\theta_1}{L_1} + \frac{\theta_1}{L_2} + \frac{\theta_2}{L_2}\right)$$

$$\frac{\partial\Pi}{\partial\theta_2} = \frac{EA}{L_2}(\theta_1 r + \theta_2 r)r + \frac{EA}{L_2}r^2\theta_2 = EAr^2\left(\frac{\theta_1}{L_2} + \frac{\theta_2}{L_2} + \frac{\theta_2}{L_2}\right)$$

This stiffness matrix is shown in the equilibrium equation:

$$EAr^2\begin{bmatrix}\left(\dfrac{1}{L_1}+\dfrac{1}{L_2}\right) & \dfrac{1}{L_2} \\[2mm] \dfrac{1}{L_2} & \dfrac{2}{L_2}\end{bmatrix}\begin{Bmatrix}\theta_1 \\ \theta_2\end{Bmatrix} = \begin{Bmatrix}\mathfrak{M}_1 \\ \mathfrak{M}_2\end{Bmatrix}$$

The vector on the right-hand side represents the external moments applied to the pulleys.

8-14 The system consists of a "T" member, connected to the ground by spring k_a, a short beam, spring k_s, and mass M. Develop the stiffness matrix for the system, where $k_a = 57,370$, $k_s = 39,042$ Lb/in., $h = 52.3$ in., $H = 61.07$ in., $L = 25$ in., and $a = 11.29$ in.

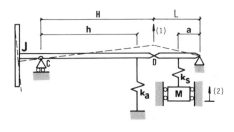

The elongations of springs are

$$\frac{h}{H}u_1 \quad \text{and} \quad \frac{a}{L}u_1 - u_2$$

The strain energy is

$$\Pi = \tfrac{1}{2}k_a\left(\frac{h}{H}\right)^2 u_1^2 + \tfrac{1}{2}k_s\left(\frac{a}{L}u_1 - u_2\right)^2$$

After differentiating and rearranging, we get

$$\begin{Bmatrix}P_1 \\ P_2\end{Bmatrix} = \begin{bmatrix}\dfrac{h^2}{H^2}k_a + \dfrac{a^2}{L^2}k_s & -\dfrac{a}{L}k_s \\[3mm] -\dfrac{a}{L}k_s & k_s\end{bmatrix}\begin{Bmatrix}u_1 \\ u_2\end{Bmatrix}$$

which defines the stiffness matrix.

8-15 Calculate the flexibility matrix for this beam and obtain from it the stiffness matrix. Use $L =$

60 in., $I_1 = 100$ in.^{4}, $I_2 = 125$ in.^{4}, and $E = 29 \times 10^6$ psi.

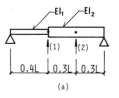

(a)

The bending moments induced by the unit forces applied in the directions of interest appear in (b) and (c). Using the unit load method, we express the flexibility coefficients as

$$a_{ij} = \sum_k \frac{1}{EI_k}\int m_i m_j\,dx$$

where the integration is extended on segments over which I_k is constant. The integrals are calculated with the help of the table on page 454 of Ref. 14.

$$Ea_{11} = \frac{1}{I_1}\frac{0.4L}{3}(0.24L)^2 + \frac{1}{I_2}\frac{0.6L}{3}(0.24L)^2 = \frac{L^3}{5918.6}$$

$$Ea_{22} = \frac{1}{I_1}\frac{0.4L}{3}(0.12L)^2 + \frac{1}{I_2}\frac{0.3L}{3}(0.12^2 + 0.12\times0.21 + 0.21^2)L^2$$

$$+ \frac{1}{I_2}\frac{0.3L}{3}(0.21L)^2 = \frac{L^3}{8234.5}$$

$$Ea_{12} = \frac{1}{I_1}\frac{0.4L}{3}(0.12L)(0.24L) + \frac{1}{I_2}\frac{0.3L}{6}[0.12(2\times0.24+0.12)$$

$$+ 0.21(2\times0.12+0.24)]L^2 + \frac{1}{I_2}\frac{0.3L}{3}(0.12L)(0.21L)$$

$$= \frac{L^3}{7832.1}$$

Assemblying these into the flexibility matrix, we have:

$$\mathbf{a} = \begin{bmatrix}1.2585 & 0.9510 \\ 0.9510 & 0.9045\end{bmatrix}10^{-6}$$

and finally

$$\mathbf{k} = \mathbf{a}^{-1} = \begin{bmatrix}3.8668 & -4.0656 \\ -4.0656 & 5.3802\end{bmatrix}10^6$$

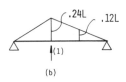

(b)

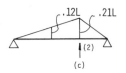

(c)

8-16 Resolve the arbitrary load applied to a symmetric beam into symmetric and antisymmetric components.

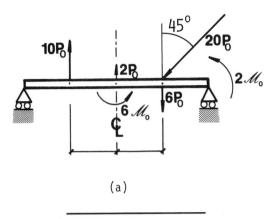

(a)

The resolution shown in (b) and (c) was made in accordance with Eq. 8.21. One can easily check that superposition of (b) and (c) gives us the original load pattern in (a).

The force $2P_0$ and the moment $6\mathfrak{M}_0$ applied in (a) just *at* the plane of symmetry are each resolved into two equal components applied at infinitesimally short distances away from that plane.

(b)

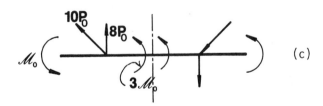

(c)

8-17 An applied load is antisymmetric with respect to either plane of symmetry. Construct a simplified model of this frame. What can be said about the internal forces? The frame is in equilibrium.

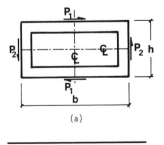

(a)

A cross section located at a plane of symmetry cannot undergo a symmetric displacement when the external loading is antisymmetric. This means that the translation along said plane is prohibited, which is illustrated in (b). According to the theorem in Sect. 8J, only a shear force can be applied to the cut at the plane of symmetry. We can therefore replace the supports in (b) by the shear forces:

$$V_1 = \frac{P_1}{2} \qquad \text{in the vertical beam}$$

$$V_2 = \frac{P_2}{2} = \frac{P_1 h}{2b} \qquad \text{in the horizontal beam}$$

Under this loading, the frame is statically determinate.

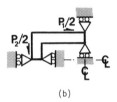

(b)

8-18 A frame has two planes of symmetry, xz and yz. The supports, which are also doubly symmetric, can resist only forces, not moments. The principal axes of the cross section are parallel to either the x and z or the y and z coordinates. An arbitrary out-of-plane loading can be resolved into the following subcases:

1. Symmetric about xz, symmetric about yz.
2. Symmetric about xz, antisymmetric about yz.
3. Antisymmetric about xz, symmetric about yz.
4. Antisymmetric about xz, antisymmetric about yz.

The response to each subcase can then be evaluated using only one-fourth of the entire frame. Discuss the boundary conditions for each of the subcases.

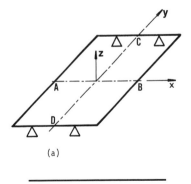

(a)

For a frame of this type the out-of-plane loading causes only out-of-plane deformations, and vice versa. This means that the following displacement components are not possible: u_x, u_y, and α_z. This is true for any point along the elastic frame axis. When the points B and C located in the planes of symmetry are chosen to isolate one-fourth of the frame, some of the remaining three components (u_z, α_x, and α_y) may also vanish, as illustrated in the sketch. Only subcases 1 and 4 are shown in (*b*); the other two may be easily deduced.

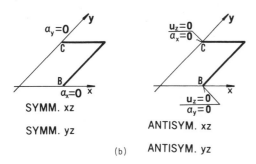

(b)

EXERCISES

8-19　Generate stiffness matrix for the assembly of rods with a rigid block. The modulus of elasticity is E.

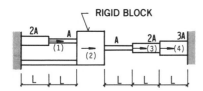

8-20　The left end of the beam can only translate, while the right end can both translate and rotate. Calculate the stiffness matrix.

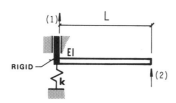

8-21　The left end of the beam can only translate; the right end is simply supported. Calculate the stiffness matrix with respect to the directions shown.

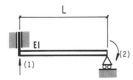

8-22　The deflected line of a cantilever subjected to tip load is given in Prob. 8-6. With that as an aid, derive a 3×3 flexibility matrix for the beam shown in figure.

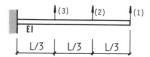

8-23　In this scheme of a helicopter rotor drive system there are nine independent coordinates. The pitch radii of mating gears (r_{ij}) and the torsional spring constants of shafts (K_i) marked in the figure are given. Derive the stiffness matrix.

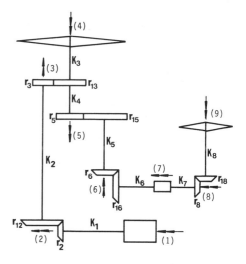

8-24　The stiffness matrix for the unconstrained axial member in (*a*) is

$$\mathbf{k}_u = \begin{bmatrix} \overset{(1)}{1} & \overset{(2)}{-1} \\ -1 & 1 \end{bmatrix} \frac{EA}{L}$$

Using a transformation matrix, develop a 1×1 stiffness matrix for the assembly of two bars in (*b*).

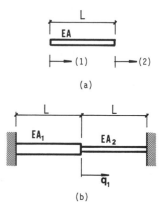

8-25　If the system of coordinates aligned with a rod, as in Fig. 8.2, is designated by $\vec{\mathbf{u}}$ and the system in Fig. 8.3 by $\vec{\mathbf{q}}$, the transformation of displacement components is expressed by

$$\left\{ \begin{array}{c} u_1 \\ u_2 \end{array} \right\} = \left[\begin{array}{cc} c_x & c_y \\ -c_y & c_x \end{array} \right] \left\{ \begin{array}{c} q_1 \\ q_2 \end{array} \right\}$$

where c_x and c_y are $\cos\alpha$ and $\cos\beta$ in accordance with Fig. 8.3. Using this transformation, derive the matrix in Eq. 8.11 from the one in Eq. 8.10.

8-26 Using the individual stiffness matrix, Eq. 8.13, derive the global matrix for the cantilever made up of two segments. Use a matrix transformation from the individual to the global displacement coordinates.

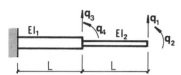

8-27 Using Castigliano's theorem, calculate the flexibility matrix for the beam in Prob. 8-6. The strain energy of a beam element of length l, expressed by bending moments is

$$\Pi = \int_0^l \frac{\mathfrak{M}^2 \, dx}{2EI}$$

8-28 Solve Prob. 8-13 again, choosing a different arrangement of pulleys.

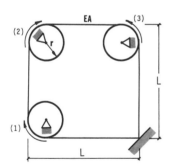

8-29 Using Castigliano's theorem, derive the stiffness matrix for the system in Prob. 8-3.

8-30 A heavy cylinder is attached to a dolly, which can travel in the direction normal to the paper. The system is sketched in (a), while (b) shows its model for analysis of lateral deflections, in the plane of the paper. The top portion, segment 1–2 of the cylinder, has the second area moment $I_1 = 10,500$ in.[4] while the rest has $I_2 = 22,600$ in.[4] The spring constants are $k = 4.808 \times 10^6$ Lb-in. and $K = 20.2 \times 10^9$ Lb-in./rad, respectively. Calculate the flexibility matrix if $E = 29 \times 10^6$ psi.

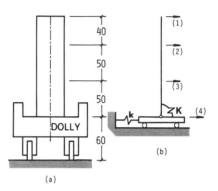

8-31 The cross section of an L-shaped frame has the following bending and twisting properties: $I_{xx} = 800$ in.[4], $I_{yy} = 37.2$ in.[4], and $C = 2.45$ in.[4] Calculate the flexibility matrix and the stiffness matrix with respect to the displacements shown. Ignore axial and shearing deformation components. Use $L_1 = 50$ in., $L_2 = 150$ in., $E = 29 \times 10^6$ psi, and $G = 11 \times 10^6$ psi. *Hint.* Make plots of in-plane and out-of-plane bending as well as torsion before using the unit load method.

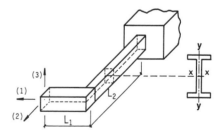

CHAPTER

9

Free Vibrations

When there is more than one displacement component associated with inertia properties, we speak of multiple degree-of-freedom (MDOF) systems. Equations of motion for MDOF systems are developed in matrix form using inertia and stiffness properties. The use of Lagrange equations for the purpose of mass matrix generation is presented. The mathematical concepts of *eigenvalues* and *eigenvectors* described in Appendix II are used here to deduce the *natural frequencies* and *mode shapes* of a vibrating structure. The *normal mode* method, which allows one to describe the motion of an nth degree system by a set of n equations of motion independent of one another is developed and used to calculate the response to the initial conditions. The vibratory effects of structural symmetry are discussed. The basic concepts in converting a system with a continuous mass distribution to one with lumped masses are outlined.

9A. Equations of Motion. For a simple SDOF system consisting of a mass and a spring, the equation of motion in the absence of damping and external forces is

$$M\ddot{u} = -ku \qquad (9.1)$$

according to Chapter 2. It is merely a statement of the second Newton's law in which only the force of elastic resistance acts on mass M. If the mass is displaced by a distance u from the neutral position,

the spring applies to the mass a reaction of magnitude ku in the direction opposite to u.

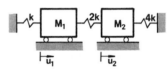

Figure 9.1 Example two-DOF system.

In MDOF systems the equations of motion may be constructed using the same principle. To be more specific, consider a system in Prob. 8-1 with the masses assigned to both blocks (Fig. 9.1). Let us return for a moment to Chapter 8, where the equation

$$\vec{\mathbf{P}} = \mathbf{k}\vec{\mathbf{u}} \qquad (9.2)$$

was derived. It shows the relation between the deformed pattern defined by $\vec{\mathbf{u}}$ and the vector of forces $\vec{\mathbf{P}}$, applied from outside to induce this pattern. The masses at the nodal points deflected by $\vec{\mathbf{u}}$ from their respective neutral positions are subjected to the elastic reactions $\vec{\mathbf{R}} = -\mathbf{k}\vec{\mathbf{u}}$. The equations of motion of the nodal points, in our case the blocks in Fig. 9.1, may be written as

$$\begin{aligned} M_1\ddot{u}_1 &= -k_{11}u_1 - k_{12}u_2 \\ M_2\ddot{u}_2 &= -k_{21}u_1 - k_{22}u_2 \end{aligned} \qquad (9.3)$$

104

or, in matrix form:

$$
\begin{bmatrix} M_1 & 0 \\ 0 & M_2 \end{bmatrix} \begin{Bmatrix} \ddot{u}_1 \\ \ddot{u}_2 \end{Bmatrix} = - \begin{bmatrix} k_{11} & k_{12} \\ k_{21} & k_{22} \end{bmatrix} \begin{Bmatrix} u_1 \\ u_2 \end{Bmatrix}
$$

and symbolically:

$$
\mathbf{M}\ddot{\vec{u}} = -\mathbf{k}\vec{u}
$$

This expression is analogous to Eq. 9.1. Here **M** is called the *mass matrix* and $\ddot{\vec{u}}$ is the vector of acceleration. This equation is usually written as

$$
\mathbf{M}\ddot{\vec{u}} + \mathbf{k}\vec{u} = \vec{0} \tag{9.4}
$$

in which $\vec{0}$ is a column matrix of zeros. It is stressed here that Eqs. 9.3 may be derived without any reference to matrices, as is often done in elementary texts. Matrix algebra is merely a bookkeeping tool, which makes the task easier.

Using the concept of an inertia force, there is an alternative way of developing the equations of motion. In a SDOF system the inertia force is $-M\ddot{u}$, while in a MDOF structure the equivalent expression is $-\mathbf{M}\ddot{\vec{u}}$. Treating the inertia forces as the external load vector we put $-\mathbf{M}\ddot{\vec{u}}$ in place of $\vec{P}$ in Eq. 9.2, and we again get Eq. 9.4. For a system where the angular displacements are predominant, Eq. 9.4 is written as

$$
\mathbf{J}\ddot{\vec{\alpha}} + \mathbf{K}\vec{\alpha} = \vec{0} \tag{9.5}
$$

Here **J** is the matrix of moments of inertia, **K** the angular stiffness matrix, and $\vec{\alpha}$ the vector of rotation. When translations as well as rotations take place, the symbols in Eq. 9.4 are used for both types of degrees of freedom.

9B. Mass Matrix. Lagrange's Equations. When a structural system is composed of masses lumped into points and connected by elastic elements, each equation of motion has only one inertia term. When those equations are presented in a matrix form, the mass matrix is diagonal, as was seen in Sect. 9A. This situation, however, is an exception rather than a rule, because in many engineering problems there are off-diagonal terms in a mass matrix.

A convenient method of calculating a mass matrix is by the use of Lagrange's equations. For the rth degree of freedom the simplest form of this relationship is

$$
\frac{d}{dt}\left(\frac{\partial T}{\partial \dot{u}_r}\right) = Q_r \tag{9.6}
$$

where T is the kinetic energy, which normally is a function of velocities $\dot{u}_1, \dot{u}_2, \ldots, \dot{u}_n$ of a system with n degrees of freedom. (The unusual situations when T is also a function of displacements are not considered here.) The expression in the parentheses is the partial derivative of T with respect to the rth velocity, and Q_r is the resultant of all forces, elastic reactions included, acting along the rth displacement. When the operations indicated by Eq. 9.6 are completed, we obtain an equation of motion in rth direction as follows:

$$
M_{r1}\ddot{u}_1 + M_{r2}\ddot{u}_2 + M_{r3}\ddot{u}_3 = Q_r \tag{9.7}
$$

if a three-DOF system is used as an example. The coefficients of the acceleration terms in the rth equation constitute the rth row of a mass matrix of the system. Writing Eq. 9.6 for the remaining two degrees of freedom, a 3×3 mass matrix is obtained.

The elastic reactions are just the opposite of the statically applied forces, $Q_r = -P_r$, the latter defined in Sect. 8C. Using Part I of the Castigliano theorem, the Lagrange's equations become

$$
\frac{d}{dt}\left(\frac{\partial T}{\partial \dot{u}_r}\right) + \frac{\partial \Pi}{\partial u_r} = 0 \tag{9.8}
$$

(The second term was used in Sect. 8C to calculate the entries of the rth row of the stiffness matrix.) Equation 9.7 may now be written fully as:

$$
M_{r1}\ddot{u}_1 + M_{r2}\ddot{u}_2 + M_{r3}\ddot{u}_3 + k_{r1}u_1 + k_{r2}u_2 + k_{r3}u_3 = 0 \tag{9.9}
$$

If there is a zero mass associated with any degree of freedom, the mass matrix becomes singular and the number of natural frequencies is diminished by the number of missing masses.

9C. Natural Frequencies and Mode Shapes. A careful review of Sect. 2B shows that a free undamped vibration of an SDOF elastic system may always be presented as $u = u_{max}\sin\omega t$, provided time t is counted not necessarily from the beginning of a free motion, but from some suitably chosen instant. The amplitude of motion u_{max} depends on the initial conditions. In structures having many degrees of freedom, we assume that each DOF performs a simple harmonic motion and that the only difference between the individual degrees is their amplitudes. For a three-DOF system we thus have

$$
u_1 = x_1\sin\omega t \qquad u_2 = x_2\sin\omega t \qquad u_3 = x_3\sin\omega t
$$

where x_1, x_2, and x_3 are the amplitudes in the respective directions. In matrix form:

$$\begin{Bmatrix} u_1 \\ u_2 \\ u_3 \end{Bmatrix} = \begin{Bmatrix} x_1 \\ x_2 \\ x_3 \end{Bmatrix} \sin \omega t \quad \text{or} \quad \vec{\mathbf{u}} = \vec{\mathbf{x}} \sin \omega t \tag{9.10}$$

Double differentiation with respect to time gives

$$\ddot{\vec{\mathbf{u}}} = -\omega^2 \vec{\mathbf{x}} \sin \omega t = -\omega^2 \vec{\mathbf{u}} \tag{9.11}$$

When this is substituted into Eq. 9.4, we get

$$(\mathbf{k} - \omega^2 \mathbf{M})\vec{\mathbf{u}} = \vec{\mathbf{0}} \tag{9.12}$$

This is similar to Eq. A.2.9b in Sect. A2.6 of Appendix II. Natural frequency squared is an eigenvalue and $\vec{\mathbf{u}}$ is the eigenvector of the system. When eigenvectors are discussed in connection with a vibrating system, they are usually called *mode shapes*. This is because each natural frequency of a system is associated with a certain deflected shape, also called the mode of vibration. That mode is described by specifying the entries of a vector of amplitudes in Eq. 9.10. Everything said about eigenvalues and eigenvectors in Appendix II is applicable to natural frequencies (squared) and the mode shapes. In particular, the characteristic equation

$$|\mathbf{k} - \omega^2 \mathbf{M}| = 0 \tag{9.13}$$

is called here the *frequency equation*. If a vibrating body is properly constrained (see Sect. 8G), the stiffness matrix is nonsingular. If the mass matrix is also free from singularities (see Sect. 9B), we can expect that a system with n DOF will have n natural frequencies and a like number of mode shapes. Although some frequencies may repeat, the mode shapes are distinct. The smallest frequency is called the *first* or the *fundamental frequency*, and the associated mode shape is the first mode. This approach to free-vibration problems is sometimes called the *stiffness method* because it involves the direct use of the stiffness matrix.

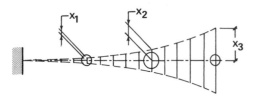

Figure 9.2 Beam vibrating in natural mode.

Figure 9.2 shows a beam vibrating in a natural mode with the frequency satisfying Eq. 9.13 and the amplitudes x_1, x_2, and x_3 determined by Eq. 9.12. The thick line depicts the neutral, undeflected position, while the dashed lines are the upper and the lower extreme positions, respectively. All point masses reach the uppermost deflection at the same instant of time, at which point the entire beam is motionless as the velocity changes its sign. The deflections of all points in the structure are described by the same time function, when vibrating in a natural mode. The only difference is the amplitude, as previously stated.

9D. Normal Coordinates. Response to Initial Conditions. Let us perform a change of displacement coordinates, which is defined by:

$$\vec{\mathbf{u}} = \mathbf{\Phi}\vec{\mathbf{s}} \tag{9.14a}$$

where $\vec{\mathbf{u}}$ is a set of displacements with respect to which the stiffness and mass matrices are determined. This set will be referred to as the *physical coordinates*. The term $\mathbf{\Phi}$ is a square modal matrix (see Appendix II), while the new displacements $\vec{\mathbf{s}}$ are called *normal coordinates*. Differentiations of Eq. 9.14a with respect to time gives velocities and accelerations:

$$\dot{\vec{\mathbf{u}}} = \mathbf{\Phi}\dot{\vec{\mathbf{s}}} \tag{9.14b}$$

$$\ddot{\vec{\mathbf{u}}} = \mathbf{\Phi}\ddot{\vec{\mathbf{s}}} \tag{9.14c}$$

Substitution of Eqs. 9.14 into the equation of motion (Eq. 9.4) plus premultiplication by $\mathbf{\Phi}^T$ results in

$$\overline{\mathbf{M}}\ddot{\vec{\mathbf{s}}} + \overline{\mathbf{k}}\vec{\mathbf{s}} = \vec{\mathbf{0}} \tag{9.15}$$

where $\overline{\mathbf{M}}$ and $\overline{\mathbf{k}}$ are diagonal matrices according to Appendix II:

$$\overline{\mathbf{M}} = \mathbf{\Phi}^T \mathbf{M} \mathbf{\Phi}$$
$$\overline{\mathbf{k}} = \mathbf{\Phi}^T \mathbf{k} \mathbf{\Phi} \tag{9.16}$$

The set of equations of motion from Sect. 9A, describing the motion of blocks in Fig. 9.1 is therefore equivalent to

$$\overline{M}_1 \ddot{s}_1 + \overline{k}_1 s_1 = 0$$

$$\overline{M}_2 \ddot{s}_2 + \overline{k}_2 s_2 = 0$$

when written in the normal coordinates, in accordance with Eq. 9.15. The rth mode frequency is

$$\omega_r = \left(\frac{\overline{k}_r}{\overline{M}_r} \right)^{1/2} \tag{9.17}$$

We have thus succeeded in converting an MDOF system of nth order into n independent SDOF systems. The structure shown as an example in Fig. 9.1 is now equivalent to what is depicted in Fig. 9.3, namely, two separate SDOF systems.

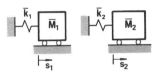

Figure 9.3 Two oscillators, equivalent to system in Fig. 9.1.

Once the modal matrix of a system is known, we can find the response to the initial conditions that in the most general case, are given as the deflection and velocity vectors at $t=0$, $\vec{u}_0$ and $\dot{\vec{u}}_0$. The corresponding values in normal coordinates are found by solving Eqs. 9.14a and 9.14b for $\vec{s}_0$ and $\dot{\vec{s}}_0$:

$$\vec{u}_0 = \mathbf{\Phi}\vec{s}_0 \qquad (9.18a)$$

$$\dot{\vec{u}}_0 = \mathbf{\Phi}\dot{\vec{s}}_0 \qquad (9.18b)$$

The rth mode of vibration is described by Eq. 9.15 as

$$\overline{M}_r \ddot{s}_r + \overline{k}_r s_r = 0 \qquad (9.19)$$

Equation 2.4 tells us that if the initial displacement and velocity for this elementary system are, respectively, s_{r0} and $\dot{s}_{r0}$, the displacement response is

$$s_r = s_{r0}\cos\omega_r t + \frac{\dot{s}_{r0}}{\omega_r}\sin\omega_r t \qquad (9.20)$$

In actual computations of *modal masses* $\overline{M}_r$ and *modal stiffnesses* $\overline{k}_r$ we employ an alternative form of Eq. 9.16:

$$\begin{aligned}\overline{M}_r &= \vec{\Phi}_r^T \mathbf{M} \vec{\Phi}_r \\ \overline{k}_r &= \vec{\Phi}_r^T \mathbf{k} \vec{\Phi}_r\end{aligned} \qquad (9.21)$$

In a particular case of a diagonal mass matrix

$$\mathbf{M} = \lfloor M_1 \ M_2 \cdots M_n \rfloor$$

we have

$$\overline{M}_r = M_1 x_1^2 + M_2 x_2^2 + \cdots + M_n x_n^2 \quad (9.21a)$$

where $x_1, x_2, \ldots, x_n$ are the displacement components of the rth mode, $\vec{\Phi}_r$.

Once the response has been found in terms of normal coordinates, we use Eqs. 9.14 to return to the physical system. The above-described procedure is called the *normal-mode method*.

9E. Free Vibrations of Underconstrained Systems. As noted in Chapter 8, when some displacements are possible without an elastic resistance, the stiffness matrix is singular. This lack of a sufficient number of constraints may be internal (components of a system not adequately tied together) or external (the system as a whole is not completely attached to a fixed base). For example, an elastic body in space has in general six unconstrained degrees of freedom. This means that the stiffness matrix of that body will be singular and a free-vibration analysis will show six zero frequencies. The modes of vibration associated with zero frequencies degenerate to kinematic conditions of free motion and are referred to as *rigid-body modes*. When using the stiffness method to calculate frequencies and mode shapes, no deviation from regular procedure is needed for the underconstrained systems. The stiffness and mass matrices are set up, the frequency equation is solved, and the modes of vibration are found in the same manner as for a fully constrained body. The only difference occurs in the response to the initial conditions. In a rigid-body mode we have stiffness $\overline{k}_r = 0$, which reduces Eq. 9.19 to

$$\overline{M}_r \ddot{s}_r = 0$$

Since $\overline{M}_r \neq 0$, we have, in effect, a constant acceleration motion, $\ddot{s}_r = 0$. Integrating this twice with respect to time gives

$$s_r = s_{r0} + \dot{s}_{r0}t \qquad (9.22)$$

which should be used for the rigid-body modes as a counterpart of Eq. 9.20.

9F. Free Vibrations of Symmetric Systems. Section 8J discussed the static aspects of structural symmetry, and its consequences. If in addition to elastic symmetry there also is a symmetric mass distribution, the findings of Sect. 8J also becomes meaningful for dynamic behavior. In a symmetric structure the modes of vibration are either symmetric or antisymmetric. An arbitrary initial condition may be resolved into a symmetric component and an antisymmetric component, the response to each of the two may be found separately, and the total vibratory response determined by superposition. Only one-half of the entire structure needs to be analyzed when there is one plane of symmetry. For two such planes, only a quarter of the structure has to be evaluated.

9G. Mass Lumping in Dynamic Models.   In all real, deformable bodies the mass is continuously distributed along the physical dimensions, and there are no "massless" elements. In contrast to that rather obvious truth, most problems in this book employ discrete or *lumped* masses connected by weightless, elastic elements. A process of reducing a structural system to such a lumped-mass model will be illustrated below with some simple examples. A beam in Fig. 9.4a has seven degrees of freedom, designated by arrows at the nodes marked by dots. A mass is associated with each coordinate by assigning to it a portion of beam inertia. For example, M_1 is the part of the beam comprising half of length L_1 and half of L_2, the total being called the *tributary length* of M_1. The tributary length of M_3 is taken as $L_4 + L_3/2$. The hatch lines indicate how the remaining portions of the beam are assigned to their respective points of reference. The vertical segment of length L_5 is used in computing mass M_5 or M_6 and it also gives rise to the angular inertia at coordinate 7. The short segment $L_1/2$ adjacent to the fixed end is not contributing to the moving inertia elements. When the lumping process is complete, we have the model as in Fig. 9.4b.

with each node. The hatched pattern shows the mass assigned to the center node is twice as big as that at the tip (presuming a uniform mass distribution along the axis). The inertia terms associated with the respective DOFs are

$$M_1 = \tfrac{1}{2} A \rho L \qquad\qquad M_3 = \tfrac{1}{4} A \rho L$$

$$M_2 \equiv J_2 = \tfrac{1}{2} I \rho L \qquad M_4 \equiv J_4 = \tfrac{1}{4} I \rho L$$

where A is the cross-sectional area (in.²), I is the second area moment (in.⁴), and ρ is the mass density of the material. Notice that rotatory inertia J used here is not the actual mass moment of inertia of a respective segment, but only the part of it that is associated with the dimension normal to the axis of the beam. Not including the rotatory inertia in a model is justified only for slender beams.

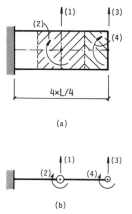

(a)

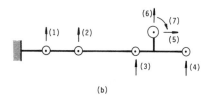

(b)

Figure 9.5 Mass lumping including rotatory inertia: (*a*) original system; (*b*) lumped model.

This section provides only a preliminary description of the mass-lumping process and neglects many important questions. A more thorough treatment of the problem is presented in Chapter 19.

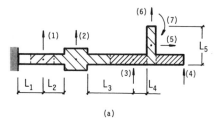

(a)

(b)

Figure 9.4 Mass lumping on geometric basis: (*a*) original system; (*b*) lumped model.

The second example, Fig. 9.5, is a deep beam for which the rotatory inertia of the cross section is important. The nodal points are located at the tip and midlength and there are two DOFs associated

SOLVED PROBLEMS

9-1 The gears in Prob. 8-3 have mass moments of inertia J_1, J_2, J_3, and J_4. There are only three independent rotations, as previously stated. Determine the mass matrix by means of Lagrange's equations.

The kinetic energy is

$$T = \tfrac{1}{2} J_1 \dot{\alpha}_1^2 + \tfrac{1}{2} J_2 \dot{\alpha}_2^2 + \tfrac{1}{2} J_3 \dot{\alpha}_3^2 + \tfrac{1}{2} J_4 \dot{\alpha}_4^2$$

Before Lagrange's equations may be used, the constraint equation $\alpha_4 = \alpha_2 r_2 / r_4$ must be employed to remove the dependent coordinate α_4. In terms of velocities:

$$\dot{\alpha}_4 = \frac{\dot{\alpha}_2 r_2}{r_4}$$

and

$$T = \frac{1}{2} J_1 \dot{\alpha}_1^2 + \frac{1}{2}\left(J_2 + J_4 \frac{r_2^2}{r_4^2} \right) \dot{\alpha}_2^2 + \frac{1}{2} J_3 \dot{\alpha}_3^2$$

Differentiate with respect to each velocity in turn:

$$\frac{\partial T}{\partial \dot{\alpha}_1} = J_1 \dot{\alpha}_1$$

$$\frac{\partial T}{\partial \dot{\alpha}_2} = \left(J_2 + J_4 \frac{r_2^2}{r_4^2} \right) \dot{\alpha}_2$$

$$\frac{\partial T}{\partial \dot{\alpha}_3} = J_3 \dot{\alpha}_3$$

When differentiated with respect to time, the expressions above represent the inertia terms in the equations of motion. The mass matrix is

$$\mathbf{J} = \begin{bmatrix} J_1 & 0 & 0 \\ 0 & \left(J_2 + J_4 \dfrac{r_2^2}{r_4^2} \right) & 0 \\ 0 & 0 & J_3 \end{bmatrix}$$

9-2 The system consists of two rigid beams that are connected to each other, as well as to the fixed base, by hinges and angular springs. The displacement coordinates are angles α_1 and α_2. Moments of inertia J_1 and J_2 are calculated with respect to the centers of masses M_1 and M_2, which coincide with the geometrical centers of the respective beams. Using Lagrange's equations, calculate the mass matrix.

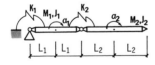

In the case of a body with finite dimensions, the kinetic energy has both translational and rotational components. The velocities of gravity centers are:

$$v_1 = \dot{\alpha}_1 L_1$$

$$v_2 = 2\dot{\alpha}_1 L_1 + \dot{\alpha}_2 L_2$$

The kinetic energy is

$$T = \tfrac{1}{2} M_1 \left(\dot{\alpha}_1 L_1 \right)^2 + \tfrac{1}{2} J_1 \dot{\alpha}_1^2 + \tfrac{1}{2} M_2 \left(2\dot{\alpha}_1 L_1 + \dot{\alpha}_2 L_2 \right)^2 + \tfrac{1}{2} J_2 \dot{\alpha}_2^2$$

Differentiating this with respect to velocities, we have

$$\frac{\partial T}{\partial \dot{\alpha}_1} = M_1 L_1^2 \dot{\alpha}_1 + J_1 \dot{\alpha}_1 + M_2 \left(2\dot{\alpha}_1 L_1 + \dot{\alpha}_2 L_2 \right)(2 L_1)$$

$$\frac{\partial T}{\partial \dot{\alpha}_2} = M_2 \left(2\dot{\alpha}_1 L_1 + \dot{\alpha}_2 L_2 \right) L_2 + J_2 \dot{\alpha}_2$$

Differentiating with respect to time gives

$$\frac{d}{dt} \frac{\partial T}{\partial \dot{\alpha}_1} = \left(J_1 + M_1 L_1^2 + 4 M_2 L_1^2 \right) \ddot{\alpha}_1 + 2 M_2 L_1 L_2 \ddot{\alpha}_2$$

$$\frac{d}{dt} \frac{\partial T}{\partial \dot{\alpha}_2} = 2 M_2 L_1 L_2 \ddot{\alpha}_1 + \left(M_2 L_2^2 + J_2 \right) \ddot{\alpha}_2$$

The mass matrix is:

$$\mathbf{M} = \begin{bmatrix} \left(J_1 + M_1 L_1^2 + 4 M_2 L_1^2 \right) & 2 M_2 L_1 L_2 \\ 2 M_2 L_1 L_2 & \left(J_2 + M_2 L_2^2 \right) \end{bmatrix}$$

9-3 Find the natural frequencies and mode shapes of a system in Fig. 9.1, putting $k = 1000$ Lb/in., $M_1 = 0.5$ Lb-s^2/in. and $M_2 = 1.5$ Lb-s^2/in. Refer to Prob. 8-1 for the stiffness matrix.

With the use of the stiffness matrix developed in the reference problem, the frequency equation (9.13) becomes:

$$\begin{vmatrix} 3k - \omega^2 M_1 & -2k \\ -2k & 6k - 3\omega^2 M_1 \end{vmatrix} = 0$$

Dividing by k and putting $\tilde{\omega}^2 = \omega^2 M_1 / k$ gives

$$(3 - \tilde{\omega}^2)(6 - 3\tilde{\omega}^2) - 4 = 0$$

(The purpose of this notation is to avoid introducing the numerical values for as long as possible, to prevent frequent division and multiplication by the same number.) After rearranging terms we have

$$3\tilde{\omega}^4 - 15\tilde{\omega}^2 + 14 = 0$$

$$\therefore \omega_1^2 = 1.2417 \frac{k}{M_1} \qquad \omega_2^2 = 3.7584 \frac{k}{M_1}$$

Now ω_1^2 is inserted into Eq. 9.12:

$$\begin{bmatrix} 1.7583 & -2 \\ -2 & 2.2749 \end{bmatrix} \begin{Bmatrix} u_1 \\ u_2 \end{Bmatrix} = \begin{Bmatrix} 0 \\ 0 \end{Bmatrix}$$

The coefficient matrix is singular, so according to Appendix II we set $u_1 = 1.0$ and calculate u_2 from one of the equations:

$$1.7583 \times 1.0 - 2u_2 = 0 \qquad \therefore u_2 = 0.8792$$

The second mode shape is obtained using $\omega^2 = \omega_2^2$ in Eq. 9.12:

$$\begin{bmatrix} -0.7584 & -2 \\ -2 & -5.2752 \end{bmatrix} \begin{Bmatrix} u_1 \\ u_2 \end{Bmatrix} = \begin{Bmatrix} 0 \\ 0 \end{Bmatrix}$$

Setting $u_1 = 1.0$ we get $u_2 = -0.3792$. Both mode shapes are now arranged in a modal matrix:

$$\Phi = \begin{bmatrix} 1.0 & 1.0 \\ 0.8792 & -0.3792 \end{bmatrix}$$
$$\quad (1) \qquad (2)$$

The frequencies are: $f_1 = 7.9313$ Hz and $f_2 = 13.799$ Hz.

As we see from the modal matrix, the first mode is associated with the simultaneous movement of blocks in the same direction, while in the second mode blocks are moving opposite each other.

9-4   A heavy weight W is attached to the end of the cantilever beam. The stiffness matrix was developed in Prob. 8-5. Determine the natural frequencies, mode shapes, and modal masses. Put $E = 10 \times 10^6$ psi, $I = 4$ in.^{4}, $L = 30$ in., $h = 15$ in., and $W = 100$ Lb. Disregard the weight of the beam itself.

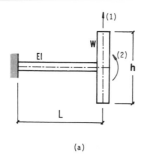

(a)

The diametral mass moment of inertia about the axis through the CG of the weight is

$$J = \frac{1}{12} \frac{W}{g} h^2 = \frac{1}{12} \times \frac{100}{386} \times 15^2 = 4.8575 \text{ Lb-s}^2\text{-in.}$$

The mass matrix is thus

$$\mathbf{M} = \begin{bmatrix} 0.2591 & 0 \\ 0 & 4.8575 \end{bmatrix}$$

Substituting the data into the stiffness matrix expression in Prob. 8-5, we obtain

$$\mathbf{k} = \begin{bmatrix} 17.78 \times 10^3 & -266.7 \times 10^3 \\ -266.7 \times 10^3 & 5333 \times 10^3 \end{bmatrix}$$

The frequency equation is

$$\begin{vmatrix} (17.78 \times 10^3 - 0.2591\omega^2) & -266.7 \times 10^3 \\ -266.7 \times 10^3 & (5333 \times 10^3 - 4.8575\omega^2) \end{vmatrix} = 0$$

This gives

$$\omega_1^2 = 1.6366 \times 10^4 \quad \text{or} \quad f_1 = 20.36 \text{ Hz}$$

$$\omega_2^2 = 1.1501 \times 10^6 \quad \text{or} \quad f_2 = 170.68 \text{ Hz}$$

Using Eq. 9.12 twice we obtain

$$\Phi = \begin{bmatrix} 1.0 & 1.0 \\ 0.05077 & -1.051 \end{bmatrix}$$

Both mode shapes are shown in (b). It is seen that in the second mode the end rotation is considerably more pronounced than in the first one.

The modal masses can be determined with the aid of Eq. 9.21. The first mode shape gives the first modal mass:

$$\overline{M}_1 = \vec{\Phi}_1^T \mathbf{M} \vec{\Phi}_1 =$$

$$= [1.0 \quad 0.05077] \begin{bmatrix} 0.2591 & 0 \\ 0 & 4.8575 \end{bmatrix} \begin{Bmatrix} 1.0 \\ 0.05077 \end{Bmatrix} = 0.2716$$

Similarly $\overline{M}_2 = 5.6247$.

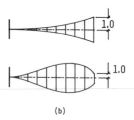

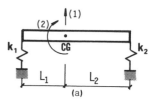

(b)

9-5   A rigid beam supported by two springs has two degrees of freedom, as shown. Determine the natural frequencies and the mode shapes. The mass is $M = 0.25$ Lb-s^2/in. and the moment of inertia about the axis through the CG and perpendicular to the plane of the paper is $J = 6.2$ Lb-s^2-in. Also, $L_1 = 6$ in., $L_2 = 10$ in., $k_1 = 2000$ Lb/in., and $k_2 = 1800$ Lb/in.

(a)

Considering the equilibrium after each of the unit displacements is separately applied allows us to immediately write the stiffness matrix. Noting that the mass matrix is diagonal, our frequency equation is

$$\begin{vmatrix} k_1 + k_2 - \omega^2 M & k_1 L_1 - k_2 L_2 \\ k_1 L_1 - k_2 L_2 & k_1 L_1^2 + k_2 L_2^2 - \omega^2 J \end{vmatrix} = 0$$

Inserting the numbers, we obtain

$$1.55\omega^4 - 86{,}560\omega^2 + 0.9216 \times 10^9 = 0$$

$$\therefore \omega_1^2 = 14{,}318 \qquad \omega_2^2 = 41{,}527$$

Substitution of ω_1^2 in Eq. 9.12 gives

$$\begin{bmatrix} 220.5 & -6000 \\ -6000 & 163{,}228 \end{bmatrix} \begin{Bmatrix} u_1 \\ u_2 \end{Bmatrix} = \begin{Bmatrix} 0 \\ 0 \end{Bmatrix}$$

If $u_1 = 1$, $u_2 = 0.03675$.

Similarly, the second mode is defined by $u_1 = 1$, $u_2 = -1.097$. The end deflections corresponding to these modes are shown in (b). Note that the first mode is predominantly translation, while the second is essentially rotation.

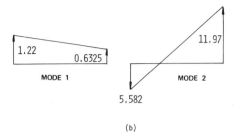

(b)

9-6 A disc with mass $M = 0.2627$ Lb-s^2/in. and the mass moment of inertia $J' = 4.86$ Lb-s^2-in. (about the axis perpendicular to the paper, through the CG) is mounted on the shaft made of steel, $E = 29 \times 10^6$ psi, $\gamma = 0.284$ Lb/in.^{3}. Disregarding the weight of the shaft calculate the natural frequencies and the mode shapes. Assume simple supports at the bearings.

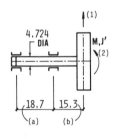

When we apply load P_1 in direction 1 and separately load P_2 (moment) in direction 2, we obtain the flexibility coefficients for the system, which allows us to write the formulas for translation u_1 and rotation u_2:

$$\left\{ \begin{array}{c} u_1 \\ u_2 \end{array} \right\} = \left[\begin{array}{cc} \dfrac{b^2}{3}(a+b) & b\left(\dfrac{a}{3}+\dfrac{b}{2}\right) \\ b\left(\dfrac{a}{3}+\dfrac{b}{2}\right) & \left(\dfrac{a}{3}+b\right) \end{array} \right] \left\{ \begin{array}{c} P_1 \\ P_2 \end{array} \right\} \dfrac{1}{EI}$$

The second area moment of the shaft section is

$$I = \frac{\pi D^4}{64} = 24.446 \text{ in.}^4$$

and $EI = 708.94 \times 10^6$ Lb-in.^{2} After inserting the figures, we obtain the flexibility matrix as

$$\mathbf{a} = \left[\begin{array}{cc} 3.742 & 0.2996 \\ 0.2996 & 0.03037 \end{array} \right] \times 10^{-6}$$

The inversion procedure gives us the stiffness matrix

$$\mathbf{k} = \left[\begin{array}{cc} 1.2715 & -12.544 \\ -12.544 & 156.67 \end{array} \right] \times 10^6$$

The mass matrix is written by inspection:

$$\mathbf{M} = \left[\begin{array}{cc} 0.2627 & 0 \\ 0 & 4.86 \end{array} \right]$$

The frequency equation is

$$\left| \begin{array}{cc} (1.2715 \times 10^6 - 0.2627\omega^2) & (-12.544 \times 10^6) \\ (-12.544 \times 10^6) & (156.67 \times 10^6 - 4.86\omega^2) \end{array} \right| = 0$$

from which

$$\omega_1 = 952.01 \text{ rad/s} \qquad \omega_2 = 6014.3 \text{ rad/s}$$

Using Eq. 9.12 with ω_1 and ω_2, we obtain

$$\Phi = \left[\begin{array}{cc} 1.0 & 1.0 \\ 0.08238 & -0.6560 \end{array} \right]$$
$$\quad (1) \qquad (2)$$

9-7 A system is called *elastically coupled* when the stiffness matrix is not diagonal. Similarly, the term *inertially coupled* is used for a system whose mass matrix is not diagonal. Show how to change the displacement coordinates in Prob. 9-5 to uncouple the system from stiffness viewpoint and develop a new mass matrix.

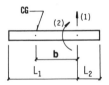

The reference point for translation and rotation in Prob. 9-5 was chosen at the CG, which gave rise to the off-diagonal term

$$k_1 L_1 - k_2 L_2$$

in the stiffness matrix. To reduce this term to zero, we must choose the reference point according to

$$\frac{k_1}{k_2} = \frac{L_2}{L_1}$$

This is shown in the figure for $k_2 > k_1$. The kinetic energy expressed by the new displacements is

$$T = \tfrac{1}{2} M (\dot{u}_1 + b\dot{u}_2)^2 + \tfrac{1}{2} J \dot{u}_2^2$$

in which J is with respect to the CG. After differentiation we find

$$\mathbf{M} = \left[\begin{array}{cc} M & Mb \\ Mb & J + Mb^2 \end{array} \right]$$

The stiffness matrix remains unchanged except that the off-diagonal term vanishes. (The numerical values will change, however, because L_1 and L_2 are different now.) The new reference point for displacements is called the *elastic center*.

9-8 A rigid beam of length L, with uniformly distributed mass M is elastically supported at both ends. Develop the formulas for the natural frequencies and determine whether it is possible to make all frequencies the same by adjusting the supporting spring constants.

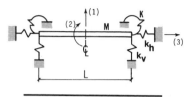

Because symmetry is complete, the three directions in the plane are uncoupled and may be treated separately.

1. Vertical motion: $k_{11} = 2k_v$,

$$\omega_1^2 = \frac{2k_v}{M}$$

2. Angular motion: $k_{22} = 2K + \frac{1}{2}k_v L^2$ (cf. k_{22} in Prob. 9-5). The moment of inertia is $J = \frac{1}{12}ML^2$,

$$\omega_2^2 = \frac{24K + 6k_v L^2}{ML^2}$$

3. Horizontal motion: $k_{33} = 2k_h$,

$$\omega_3^2 = \frac{2k_h}{M}$$

Horizontal frequency is the same as vertical frequency when $k_h = k_v$. To make the rocking frequency identical, we would have to put

$$\frac{24K}{L^2} + 6k_v = 2k_v$$

which is not possible, because the spring constants must be positive.

9-9 A simplified model of a three-story building consists of elastic, weightless columns and rigid, horizontal beams. No rotation takes place at either end of a column. The multiples of k in (a) indicate the horizontal stiffness of all columns in a story. Calculate the natural frequencies and mode shapes with $k = 4 \times 10^6$ Lb/in. and $W = 2 \times 10^6$ Lb.

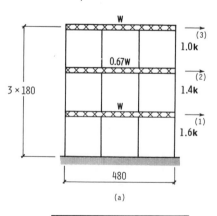

(a)

The stiffness and mass matrices are

$$\mathbf{k} = \begin{bmatrix} 3.0 & -1.4 & 0 \\ -1.4 & 2.4 & -1.0 \\ 0 & -1.0 & 1.0 \end{bmatrix} k \qquad \mathbf{M} = \begin{bmatrix} W & 0 & 0 \\ 0 & 0.67W & 0 \\ 0 & 0 & W \end{bmatrix} \frac{1}{g}$$

Substituting the figures, we obtain the frequency equation,

$$\begin{vmatrix} \left(12 - \dfrac{\omega^2}{193}\right) & -5.6 & 0 \\ -5.6 & \left(9.6 - \dfrac{\omega^2}{288.1}\right) & -4 \\ 0 & -4 & \left(4 - \dfrac{\omega^2}{193}\right) \end{vmatrix} = 0$$

The natural frequencies are

$$\omega_1 = 15.587 \text{ rad/s} \qquad \omega_2 = 39.561 \text{ rad/s} \qquad \omega_3 = 63.603 \text{ rad/s}$$

and the modal matrix

$$\mathbf{\Phi} = \begin{bmatrix} 1.0 & 1.0 & 1.0 \\ 1.9181 & 0.6948 & -1.6000 \\ 2.7989 & -0.6763 & 0.3774 \end{bmatrix}$$

The mode shapes are illustrated in (b). A problem with similar stiffness and inertia parameters may be found on page 309 of Ref. 13.

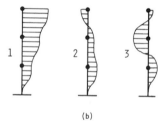

(b)

9-10 A heavy cylinder is attached to a dolly, which travels in the direction normal to the paper. The system is sketched in (a), while (b) shows its model for the dynamic analysis of lateral motion, in the plane of the paper. Calculate the natural frequencies and mode shapes, treating the cylinder as rigid and using $W_1 = 40,000$ Lb, $W_2 = 13,000$ Lb, $W_3 = 21,200$ Lb, $W_4 = 76,800$ Lb, $k = 4.808 \times 10^6$ Lb/in., and $K = 20.2 \times 10^9$ Lb-in./rad.

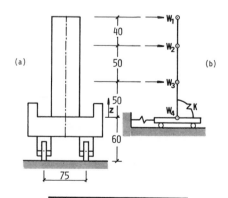

(a)   (b)

The total weight is:

$$W = (40 + 13 + 21.2 + 76.8)1000 = 151,000 \text{ Lb}$$

Location of the CG:

$$z_c = \frac{40 \times 140 + 13 \times 100 + 21.2 \times 50}{151} = 52.72 \text{ in.}$$

The stick in (b) representing the cylinder in (a) is a rigid body and its motion can be completely described by translation of the CG and rotation about it, in accordance with (c).

To construct the mass matrix we need the moment of inertia about the CG:

$$\frac{Jg}{1000} = 40(140 - 52.72)^2 + 13(100 - 52.72)^2 + 21.2(50 - 52.72)^2$$

$$+ 76.8(0 - 52.72)^2$$

$$\therefore J = 1.418 \times 10^6 \text{ Lb-s}^2\text{-in.}$$

The coefficients of the stiffness matrix are read from the sketches in (d):

$$\mathbf{k} = \begin{bmatrix} k & -kz_c \\ -kz_c & (K + kz_c^2) \end{bmatrix} = \begin{bmatrix} 4.808 & -253.48 \\ -253.48 & 33,563 \end{bmatrix} 10^6$$

The frequency equation is:

$$\begin{vmatrix} \left(4.808 - \dfrac{\omega^2}{2556.3}\right) & -253.48 \\ -253.48 & (33,563 - 1.418\omega^2) \end{vmatrix} = 0$$

From this we obtain

$$\omega_1 = 76.2 \text{ rad/s} \qquad \omega_2 = 173.65 \text{ rad/s}$$

$$\Phi = \begin{bmatrix} 1.0 & 1.0 \\ 0.01 & -0.02757 \end{bmatrix}$$

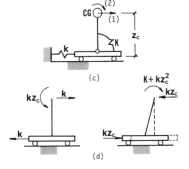

(c)

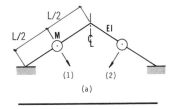

(d)

If the beams are treated as inextensible, the frame corner located on the centerline cannot translate and the only motion the mass can undergo is the translation normal to the frame axis. (Rotation of masses is not precluded, but it is not of interest here either because there are no angular masses present.) One mode of vibration occurs when the sign and the magnitude of displacement are the same for both masses, as shown in (b). Owing to symmetry, there is no angular displacement of the corner and each beam acts as fixed against rotation at both ends. The center deflection of such beam under the load P in the middle of length is

$$u = \frac{PL^3}{192EI} \qquad \therefore k = \frac{192EI}{L^3}$$

The natural frequency $\omega^2 = 192EI/ML^3$.

Another possibility is presented in (c), where the signs of displacement are opposite and the magnitudes are equal. The antisymmetry of the deflected shape causes an absence of symmetric interaction (bending moment) at a fictitious cut at the plane of symmetry. Consequently, each beam acts as if it were simply supported at one end. Proceeding as before, we find

$$\omega^2 = \frac{109.71EI}{ML^3}$$

The second frequency is smaller and therefore it is the fundamental frequency of the system. In our modal matrix, which we can construct by inspection,

$$\Phi = \begin{bmatrix} 1 & 1 \\ -1 & 1 \end{bmatrix}$$

we show the deformed pattern from (c) in the first column.

(b)

(c)

9-11 Using symmetry properties, identify the mode shapes and write the formulas for the natural frequencies of bending vibrations of a two-beam frame. Assume the beams to be inextensible.

9-12 A heavy block with mass $M = 50$ Lb-s^2/in. is in the middle of a thin-wall beam supported at both ends and having wall thickness $t = 0.25$ in. The beam material properties are $E = 29 \times 10^6$ psi and $G = 11 \times 10^6$ psi. The moments of inertia of the block are $J_x = 5100$ Lb-s^2-in., $J_y = 2700$ Lb-s^2-in., and $J_z = 5100$ Lb-s^2-in. The CG of the block, the geometric center of the beam section, and the origin of the coordinates are at the same point. Ignoring the weight of the beam material, calculate the natural frequencies.

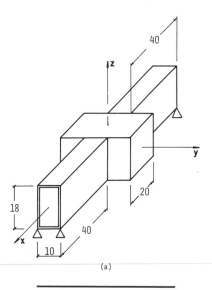

(a)

First of all, it is understood that the design of the supports and the beam ends precludes local deformation, which might otherwise be significant for a beam of these proportions. The section properties are

$$I_y = 617.2 \qquad I_z = 249.7 \qquad C = 544.6 \text{ in.}^4$$
$$A_{sz} = 9 \qquad A_{sy} = 5 \qquad A = 13.75 \text{ in.}^2$$

(Shear area A_{sz} pertaining to the action of a vertical force was assumed to consist only of vertical walls. A similar assumption was used for the other direction.) When a force is applied along the x-axis, there are two axial elements, working in parallel, which resist it. The stiffness and frequency are, respectively,

$$k_x = \frac{2EA}{L} = 2 \times 29 \times 10^6 \times \frac{13.75}{40} = 19.938 \times 10^6 \text{ Lb/in.}$$

$$\omega_x^2 = \frac{k_x}{M} \qquad \therefore \omega_x = 631.5 \text{ rad/s}$$

Along the y-axis there are bending and shear deflections,

$$\frac{1}{k_y} = \frac{L^3}{2 \times 3EI_z} + \frac{L}{2 \times GA_{sy}} = (1.473 + 0.3636)10^{-6}$$

$$k_y = 544,474 \text{ Lb/in.}$$

$$\omega_y^2 = \frac{k_y}{M} \qquad \therefore \omega_y = 104.4 \text{ rad/s}$$

In a similar computation for the z-axis:

$$\frac{1}{k_z} = (0.5959 + 0.202)10^{-6} \qquad \therefore k_z = 1.253 \times 10^6 \text{ Lb/in.}$$

$$\omega_z = 158.32 \text{ rad/s}$$

When the block is twisted about the x-axis, the two beams work in parallel to restrain it:

$$k_{yz} = \frac{2GC}{L} = 2 \times 11 \times 10^6 \times \frac{544.6}{40} = 299.53 \times 10^6 \text{ Lb-in./rad}$$

$$\omega_{yz}^2 = \frac{k_{yz}}{J_x} \qquad \therefore \omega_{yz} = 242.3 \text{ rad/s}$$

The beam subjected to a unit torque acting about the y-axis is presented in (b). The support reaction is $1/100$ Lb at either end. Applying the unit load method, we have

$$\alpha_y = 2 \int_0^L \frac{m^2 \, dx}{EI_y} + 2 \int_0^L \frac{v^2 \, dx}{GA_{sz}}$$

where m is a linear function growing from zero to 0.4 Lb-in. and $v = 1/100 = $ constant. Evaluating the integrals gives

$$\alpha_y = \frac{2}{EI_y}\left(\frac{40}{3} \times 0.4^2\right) + \frac{2}{GA_{sz}}\left(40 \times \frac{1}{100^2}\right) = 0.3192 \times 10^{-9}$$

$$k_{xz} = \frac{1}{\alpha_y} = 3.133 \times 10^9 \text{ Lb-in./rad}$$

$$\omega_{xz}^2 = \frac{k_{xz}}{J_y} \qquad \therefore \omega_{xz} = 1077.2 \text{ rad/s}$$

Similarly for rotation about the z-axis:

$$\alpha_z = 0.7347 \times 10^{-9} \qquad k_{xy} = 1.3612 \times 10^9 \qquad \omega_{xy} = 516.6 \text{ rad/s}$$

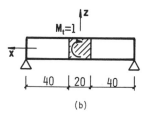

(b)

9-13 Calculate the symmetric mode frequency (left) and the antisymmetric mode frequency (right) of a frame with lumped rotary inertias J. Every frame segment has the same bending stiffness EI and is considered to be inextensible.

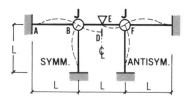

Owing to the assumption of inextensibility, the node points where the inertias are placed cannot translate. The only difference between the symmetric and the antisymmetric shape is the deflected form of the center segment. To evaluate the stiffness associated with the symmetric pattern (left), apply a unit rotation at joint B. The three beams converging at this node resist the rotation like three springs in parallel. The contribution from either BA or BC is found from the matrix 8.12 as $k_{22} = 4EI/L$. The resistance of the center beam BD is determined by setting

$$u_1 = 0 \qquad u_2 = \theta \qquad u_3 = 0 \qquad u_4 = -\theta$$

in Eq. 8.13. (It is more convenient, at this point, to think of the entire center beam, rather than about its half.) The matrix multiplication gives us the end moments,

$$\mathfrak{M}_2 = -\mathfrak{M}_4 = 2\frac{EI}{L}\theta$$

which means the angular stiffness $2EI/L$. The total for the joint is

$$K_s = \frac{(2\times4+2)EI}{L} = \frac{10EI}{L}$$

For the antisymmetric pattern the center half-beam EF may be treated as simply supported at both ends, while a moment is being applied at node F. It is known from strength of materials that the angle of rotation and the corresponding stiffness are, respectively,

$$\theta = \frac{\mathfrak{M}(L/2)}{3EI} \quad \text{and} \quad K_1 = \frac{6EI}{L}$$

The total joint stiffness is

$$K_a = \frac{(2\times4+6)EI}{L} = \frac{14EI}{L}.$$

We can now write formulas for the natural frequency ω_1 and ω_2:

$$\omega_1^2 = \frac{K_s}{J} = \frac{10EI}{JL} \quad \text{(symmetric)}$$

$$\omega_2^2 = \frac{K_a}{J} = \frac{14EI}{JL} \quad \text{(antisymmetric)}$$

9-14 A continuous beam consists of n identical spans. Using symmetry principles, show that the structure has a natural frequency equal to that of a single, isolated span

$$\omega_1 = \left(\frac{48EI}{ML^3}\right)^{1/2}$$

Also prove that this is the fundamental frequency of the system.

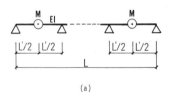

(a)

A particular feature of simple supports is that only the bending moment may be transmitted between the adjacent spans. A shear force may exist at either side of a support, but it is zero in the cross section coinciding with the support point. To begin with, let us consider a two-span beam in (b), vibrating in an antisymmetric mode. The conditions of symmetry do not allow the bending moment to exist in the center cross section. This means that the two spans vibrate without affecting each other, hence the frequency is equal to ω_1. If we attach the third span on the right, convex upward at the instant depicted in (b), it will not influence the second span and the frequency of the system will remain equal to ω_1. This process of adding spans with opposite curvatures that do

not interfere with one another may be continued as long as necessary, which demonstrates the existence of a natural frequency ω_1 in the continuous beam.

When several structural members that may freely vibrate become interconnected, it usually becomes more difficult for them to deform and the fundamental frequency of the assembly is larger than the smallest frequency of a single component. If our multi-span beam is viewed as an assembly of single spans, we find one natural frequency (ω_1) in the assembly equal to the frequency of each component when separated. The act of joining the beams together obviously creates some frequencies larger than ω_1, but it cannot cause a new frequency smaller than ω_1 to appear. For this reason we conclude that ω_1 is the fundamental frequency of the assembly.

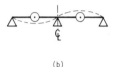

(b)

9-15 Consider a system of two blocks identical to that analyzed in Prob. 9-3. Initially, the first block is fixed, while the second is displaced by $u_2 = 0.1$ in. from the neutral position. Both blocks are released at the same time. Calculate the displacements resulting from this initial condition as well as the maximum force in spring k joining the left block with the ground. Assume $k=1000$ Lb/in. and $M=0.5$ Lb-s^2/in.

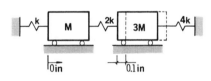

The initial condition may be stated as

$$\begin{Bmatrix} u_1 \\ u_2 \end{Bmatrix}_0 = \begin{Bmatrix} 0 \\ 0.1 \end{Bmatrix}$$

at time $t=0$. The natural frequencies and the mode shapes are known from the reference problem, and the initial displacements in the normal coordinates are found by solving Eq. 9.18a for s's,

$$\begin{bmatrix} 1.0 & 1.0 \\ 0.8792 & -0.3792 \end{bmatrix} \begin{Bmatrix} s_1 \\ s_2 \end{Bmatrix}_0 = \begin{Bmatrix} 0 \\ 0.1 \end{Bmatrix}$$

$$\therefore \begin{Bmatrix} s_1 \\ s_2 \end{Bmatrix}_0 = \begin{Bmatrix} 0.07947 \\ -0.07947 \end{Bmatrix}$$

Equation 9.20 gives us displacements in normal coordinates:

$$s_1 = s_{10}\cos\omega_1 t = 0.07947\cos 49.83t$$

$$s_2 = s_{20}\cos\omega_2 t = -0.07947\cos 86.70t$$

The physical displacement components are now found from Eq. 9.14a,

$$u_1 = 0.07947\cos\omega_1 t - 0.07947\cos\omega_2 t$$

$$u_2 = 0.06987\cos\omega_1 t + 0.03014\cos\omega_2 t$$

As a check, we put $t=0$ in the equations above, obtaining $u_1=0$ and $u_2=0.10001$ in., which agree quite well with the actual initial condition. The force in the spring joining the left block with the ground is maximum when u_1 reaches its extreme value. Instead of resorting to a mathematical procedure to determine that extremum, we will simply note that the maximum absolute value of u_1 may not be larger than $0.07947+0.07947=0.1589$ in. This gives the largest possible spring force as

$$N_{max}=k(u_1)_{max}=1000\times0.1589=158.9 \text{ Lb}$$

9-16 The masses in Prob. 9-11 are forced to deflect, respectively, by

$$u_1=\delta \quad \text{and} \quad u_2=0.5\delta$$

and then are released at the same time. Describe the subsequent motion of the system and evaluate the bending moment at the center point of the frame.

The total initial deflection may be treated as a sum of two components:

$$u_1=0.75\delta \quad u_2=0.75\delta \quad \text{(symmetric)}$$

$$u_1=0.25\delta \quad u_2=-0.25\delta \quad \text{(antisymmetric)}$$

Owing to symmetry, we can limit our investigation to the left half, whose right end is the center point of the frame. According to the reference problem, the antisymmetric deflection pattern does not induce any bending at the center point. The frequency of this mode is

$$\omega_1^2=\frac{109.71 EI}{ML^3}$$

Equation 2.4 gives us the motion following the release

$$u=0.25\delta\cos\omega_1 t$$

In the symmetric mode we have a fixity at the plane of symmetry. The motion after the release is

$$u=0.75\delta\cos\omega_2 t \quad \omega_2^2=\frac{192 EI}{ML^3}$$

Displacing the mass by 0.75δ is accomplished by the force

$$P=0.75\delta k=0.75\delta\times\frac{192 EI}{L^3}=\frac{144\delta EI}{L^3}$$

It is known from statics that this force is associated with the bending moment $PL/8$ at the frame center point. Our moment amplitude at this location is therefore

$$\mathfrak{M}_0=\frac{144\delta EI}{L^3}\frac{L}{8}=\frac{18\delta EI}{L^2}$$

The resultant motion of the left mass is

$$u=0.25\delta\cos\omega_1 t+0.75\delta\cos\omega_2 t$$

The mass on the right moves according to

$$u=-0.25\delta\cos\omega_1 t+0.75\delta\cos\omega_2 t$$

9-17 Find the natural frequencies and the mode shapes of an unconstrained system consisting of three masses connected with elastic links. Put $M_1=0.5M$; $M_2=M$; $M_3=0.5M$, and $k_1=k_2=k$.

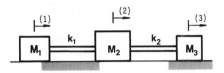

The stiffness and mass matrices are

$$\mathbf{k}=\begin{bmatrix} k & -k & 0 \\ -k & 2k & -k \\ 0 & -k & k \end{bmatrix} \quad \mathbf{M}=\begin{bmatrix} \dfrac{M}{2} & 0 & 0 \\ 0 & M & 0 \\ 0 & 0 & \dfrac{M}{2} \end{bmatrix}$$

The frequency equation may be written as

$$\begin{vmatrix} \left(1-\dfrac{\tilde{\omega}^2}{2}\right) & -1 & 0 \\ -1 & (2-\tilde{\omega}^2) & -1 \\ 0 & -1 & \left(1-\dfrac{\tilde{\omega}^2}{2}\right) \end{vmatrix}=0$$

where $\tilde{\omega}^2=\omega^2 M/k$. This can be developed into

$$\left(-\frac{\tilde{\omega}^4}{4}+\frac{3\tilde{\omega}^2}{2}-2\right)\tilde{\omega}^2=0$$

$$\therefore \tilde{\omega}_1^2=0 \quad \tilde{\omega}_2^2=2.0 \quad \tilde{\omega}_3^2=4.0$$

Equation 9.12 is now written with $\tilde{\omega}_1^2=0$,

$$\begin{bmatrix} k & -k & 0 \\ -k & 2k & -k \\ 0 & -k & k \end{bmatrix}\begin{Bmatrix} u_1 \\ u_2 \\ u_3 \end{Bmatrix}=\begin{Bmatrix} 0 \\ 0 \\ 0 \end{Bmatrix}$$

which yields $u_1=u_2=u_3=1$. This is the rigid-body mode. The other two mode shapes are shown in the modal matrix,

$$\mathbf{\Phi}=\begin{bmatrix} 1 & 1 & 1 \\ 1 & 0 & -1 \\ 1 & -1 & 1 \end{bmatrix}$$
$$\quad (1) \quad (2) \quad (3)$$

9-18 Rework Prob. 9-17 using symmetry principles.

The system exhibits both elastic and inertial symmetry with respect to the center of block M_2. There is one rigid-body mode (of no concern here) and two elastic modes. The choice of displacements for the symmetric mode is easy:

$$u_1=1 \quad u_2=0 \quad u_3=-1$$

The center block is motionless and it serves as a base of the other two; therefore

$$\omega^2 = \frac{k_1}{M_1} = \frac{2k}{M}$$

The other elastic mode is antisymmetric, which tells us only that the displacements of the end masses will have the same sense:

$$u_1 = A_1 \qquad u_2 = A_2 \qquad u_3 = A_1$$

The ratio of amplitudes A_1 and A_2 may be determined from the equilibrium condition. The sum of inertia forces of a system vibrating in a natural mode must be zero, that is,

$$M_1 A_1 \omega^2 + M_2 A_2 \omega^2 + M_3 A_1 \omega^2 = 0$$

or

$$0.5MA_1 + MA_2 + 0.5MA_1 = 0$$

$$\therefore A_1 = -A_2$$

and the mode shape is given by

$$u_1 = 1 \qquad u_2 = -1 \qquad u_3 = 1$$

Since this mode corresponds to antisymmetric deformation pattern, there may not be any symmetric interaction at the plane of symmetry. This means that the two halves of the system obtained by cutting the center block in two along the plane of symmetry do not exert any force on one another while vibrating in this mode and may thus be treated separately. Referring to Prob. 1-5, in which we put $M_1 = M_2 = 0.5M$ (i.e., we model the left half of our system using only a half the center mass) we find the natural frequency:

$$\omega^2 = k\left(\frac{1}{0.5M} + \frac{1}{0.5M}\right) = \frac{4k}{M}$$

9-19 A beam with length L and has two identical discs at the ends. The mass of one disc is M, its diametral moment of inertia is J' and the beam itself is weightless. Identify the mode shapes of the system.

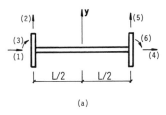

(a)

There is a total of six degrees of freedom. The first three modes are the rigid-body motions characterized by the following nonzero displacements.

1. $u_1 = u_4 = 1$

2. $u_2 = u_5 = 1$

3. $u_3 = u_6 = 1$ $\qquad u_2 = \dfrac{L}{2}; \qquad u_5 = -\dfrac{L}{2}$

The remaining three modes are elastic,

4. $u_1 = 1 \qquad u_4 = -1 \qquad$ (axial)

5. $u_3 = 1 \qquad u_6 = -1 \qquad$ (symmetric bending)

6. $u_2 = 1 \qquad u_3 = -\dfrac{ML}{2J'} \qquad u_5 = -1 \qquad u_6 = -\dfrac{ML}{2J'} \equiv -\dfrac{1}{c}$

To understand how the sixth mode is obtained, see (b). First of all, the mode is antisymmetric, which sets $u_5 = -u_2$ and $u_6 = u_3$. Second, the statement that the resultant inertia force applied to the system is zero dictates the mutual orientation of translation u_2 and rotation u_3. The equilibrium gives us $PL = 2\mathfrak{M}$. Noting that these inertia forces may be expressed as $P = Mu_2\omega^2$ and $\mathfrak{M} = -J'u_3\omega^2$, we have

$$-\frac{u_3}{u_2} = \frac{\mathfrak{M}M}{J'P} = \frac{L}{2}\frac{M}{J'}$$

The displacement patterns are now assembled into a modal matrix.

$$\Phi = \begin{bmatrix} 1 & 0 & 0 & 1 & 0 & 0 \\ 0 & 1 & \dfrac{L}{2} & 0 & 0 & 1 \\ 0 & 0 & 1 & 0 & 1 & -\dfrac{1}{c} \\ 1 & 0 & 0 & -1 & 0 & 0 \\ 0 & 1 & -\dfrac{L}{2} & 0 & 0 & -1 \\ 0 & 0 & 1 & 0 & -1 & -\dfrac{1}{c} \end{bmatrix}$$

$\quad$ (1) (2) $\quad$ (3) $\quad$ (4) $\quad$ (5) $\quad$ (6)

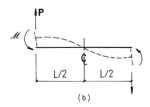

(b)

EXERCISES

9-20 Find natural frequencies and mode shapes for the three-DOF system using $k = 1000$ Lb/in. and $M = 0.5$ Lb-s^2/in. Note that what we have here may be viewed as a modification of Prob. 9-3 with the right end freed and a mass attached to it.

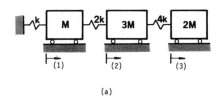

(a)

9-21 Find the natural frequency of bending vibrations and the associated mode shapes for a shaft supporting two discs with the effective lumped masses

$$M_1 = 0.8 \text{ Lb-s}^2/\text{in.} \quad \text{and} \quad M_2 = 2.0 \text{ Lb-s}^2/\text{in.}$$

The angular inertia of discs is not to be included. The stiffness matrix was calculated in Prob. 8-15 as

$$k = \begin{bmatrix} 3.8668 & -4.0656 \\ -4.0656 & 5.3802 \end{bmatrix} \times 10^6$$

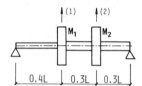

9-22 A simplified model of a two-story building consists of elastic, weightless columns and rigid, horizontal beams. No rotation takes place at either end of a column. Calculate the natural frequencies and mode shapes with $k = 75{,}000$ Lb/in. and $W = 500{,}000$ Lb.

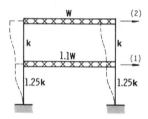

9-23 Construct the mass matrix for Prob. 8-14, and determine the natural frequencies and the mode shapes, where $M = 0.07583$ Lb-s^2/in. and $J = 358$ Lb-in.-s^2. (The latter is the moment of inertia of the T-member measured with respect to the pivot point C.)

9-24 Improve the accuracy of the bending frequency calculation in Prob. 9-6 by lumping a portion of the shaft mass at the CG of the disc. The mass of shaft and its mass moment of inertia are

$$\frac{\rho \pi D^2}{4} \quad \text{and} \quad \frac{\rho \pi D^4}{64}$$

per unit length, respectively. The shaft diameter is D, and ρ is the mass density. Lump one-fourth of the overhanging part of the shaft with the disc.

9-25 Calculate the natural frequency of the symmetric and the antisymmetric mode shape of the two-mass beam.

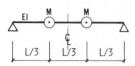

9-26 Determine the natural frequencies and the mode shapes of a cantilever with three lumped masses. Use $L = 600$ in., $E = 29 \times 10^6$ psi, $I = 919.54$ in.^{4}, and $W = 20{,}000$ Lb. The flexibility matrix may be found in the answer to Prob. 8-22.

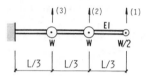

9-27 In the plan view the platform in Prob. 8-9 is a square with a side 180 in. long. The weight is 38,000 Lb. Ignoring the weight of the columns, calculate the frequencies and the mode shapes relative to the three directions.

9-28 The three-mass system analyzed in Prob. 9-20 is subjected to the initial displacement vector

$$\vec{u}_0^T = [0.1 \quad 0.0185 \quad -0.05874]$$

and released without the initial velocity. Find the displacement response of the system.

9-29 The entire system illustrated in Prob. 9-20 is moving toward the left with respect to a more general reference frame. The base is also moving, but the motion does not induce any stress. The velocity is 10 in./s. At time $t = 0$ the base is suddenly stopped. What is the subsequent motion of the system?

9-30 Find the natural frequencies and the mode shapes of a system consisting of two masses connected with an elastic link. Use the matrix approach.

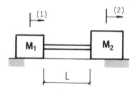

9-31 Calculate the natural frequencies and mode shapes of the propulsion system in Prob. 8-10, using the following inertia data:

$$J_1 = 1780 \text{ Lb-s}^2\text{-in.} \quad J_2 = 423.5 \text{ Lb-s}^2\text{-in.}$$

$$J_3 = 722.5 \text{ Lb-s}^2\text{-in.}$$

9-32 The shaft segment on the left is initialiy twisted with a torque of magnitude M_{t0} applied to the center wheel in the positive direction and to the left wheel in the negative direction. At time $t=0$ both torques are suddenly removed and the system is allowed to rotate freely. Determine the twisting moments in both shafts upon release.

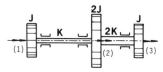

9-33 A beam with length L and a circular cross section has two identical discs at the ends (M, J about the axis of the system and J' about diametral axis). Identify the mode shapes of the system.

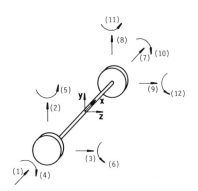

Dynamic Response

The *normal-mode method*, which in effect allows the analyst to treat a MDOF system as a set of independent, simple oscillators, is the main working tool of this chapter. The effect of viscous damping and its handling by matrix operations are illustrated by a number of problems. The response to harmonic forcing is one of the major topics. In the absence of damping, the harmonic solution is formulated in terms of physical coordinates. The use of symmetry principles is illustrated for the purpose of simplifying the solution process. The *upper-bound estimate* of results is often used to eliminate the time variable when combining the modal components. The models analyzed in this chapter are kept simple to allow the reader to concentrate on methods without being distracted by physical complexities.

10A. Forced, Undamped Motion. When there is a set of forces $P_i(t)$ applied to an MDOF system, the equations of motion have the form

$$\mathbf{M}\ddot{\vec{u}} + \mathbf{k}\vec{u} = \vec{\mathbf{P}}(t) \qquad (10.1)$$

This is similar to Eq. 9.4 except that the time-dependent vector on the right-hand side replaces the vector of zeros. For a three-DOF system, Eq. 10.1

becomes:

$$\begin{bmatrix} M_{11} & M_{12} & M_{13} \\ M_{21} & M_{22} & M_{23} \\ M_{31} & M_{32} & M_{33} \end{bmatrix} \begin{Bmatrix} \ddot{u}_1 \\ \ddot{u}_2 \\ \ddot{u}_3 \end{Bmatrix} + \begin{bmatrix} k_{11} & k_{12} & k_{13} \\ k_{21} & k_{22} & k_{23} \\ k_{31} & k_{32} & k_{33} \end{bmatrix} \begin{Bmatrix} u_1 \\ u_2 \\ u_3 \end{Bmatrix}$$

$$= \begin{Bmatrix} P_1(t) \\ P_2(t) \\ P_3(t) \end{Bmatrix}$$

Figure 10.1 shows an example of such a system.

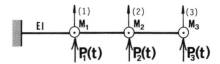

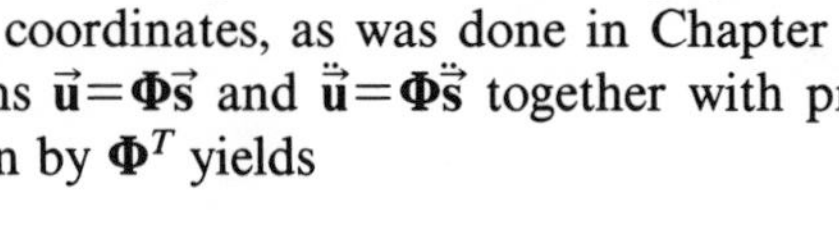

Figure 10.1   General, time-dependent loading.

To solve Eq. 10.1 for a displacement response, we transform the equations from the physical to the normal coordinates, as was done in Chapter 9. Substitutions $\vec{u} = \mathbf{\Phi}\vec{s}$ and $\ddot{\vec{u}} = \mathbf{\Phi}\ddot{\vec{s}}$ together with premultiplication by $\mathbf{\Phi}^T$ yields

$$\overline{\mathbf{M}}\ddot{\vec{s}} + \overline{\mathbf{k}}\vec{s} = \mathbf{\Phi}^T \vec{\mathbf{P}}(t) \qquad (10.2)$$

The square matrices on the left are diagonal. The equation of motion in the rth normal coordinate is

$$\bar{M}_r \ddot{s}_r + \bar{k}_r s_r = \vec{\Phi}_r^T \vec{P}(t) \qquad (10.3a)$$

where $\vec{\Phi}_r^T$ is the transposed rth mode shape. If, for example, the second mode in a three-DOF system is determined by the numbers x_1, x_2, and x_3, Eq. 10.3a will give

$$\bar{M}_2 \ddot{s}_2 + \bar{k}_2 s_2 = x_1 P_1(t) + x_2 P_2(t) + x_3 P_3(t)$$

The forcing function on the right is a combination of nodal forces, and as such it may be treated as a simple generalized force $Q_r(t)$ associated with the rth normal coordinate. A briefer alternative of Eq. 10.3a is

$$\bar{M}_r \ddot{s}_r + \bar{k}_r s_r = Q_r(t) \qquad (10.3b)$$

This method of calculating the response is called the *normal-mode method* and the force $Q_r(t)$ will be referred to as the *modal force*. A transformation of the original, coupled system of Eqs. 10.1 into a set of independent Eqs. 10.3b allows us, for example, to treat the system in Fig. 10.1 as three separate SDOF systems depicted in Fig. 10.2.

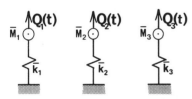

Figure 10.2 Reduction of three-DOF system to three oscillators.

Once the equations in the normal coordinates are obtained, we proceed to calculate the response exactly in the same manner used for SDOF systems. According to Eq. 2.8 we have, for the rth coordinate

$$s_r = B_{1r} \cos \omega_r t + B_{2r} \sin \omega_r t + s_{pr}(t) \qquad (10.4)$$

where B_{1r} and B_{2r} are the constants dependent on the boundary conditions and $s_{pr}(t)$ is the particular solution, which depends on the forcing function $Q_r(t)$. The means of finding the particular solution were discussed in Sect. 2C. One may also solve Eq. 10.3b by other means, for instance, by using Laplace transforms. To get the displacement in terms of the

physical coordinates, we again use the transformation equation $\vec{u} = \Phi \vec{s}$.

10B. Viscous Damping in MDOF Systems. This most common form of damping is characterized by the presence of forces obstructing the motion that are proportional to the velocities. For the basic translational SDOF system in Fig. 2.1a, this force of viscous resistance is $-c\dot{u}$, where c is a constant characterizing the damper and the negative sign indicates that the direction of the force is opposite to that of the velocity $\dot{u}$. The expression for the elastic resistance has a similar form, namely, $-ku$. This similarity between the types of resistance is also manifested in Fig. 10.3. The conclusion is that if a damping constant is replaced by a spring constant and velocity by displacement, the substitute spring acts in the same manner as the original damper. It is therefore logical to introduce the term *viscous stiffness* with reference to a damper and to use it in the same way as the elastic stiffness of a spring.

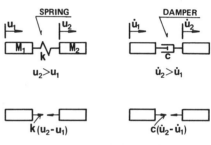

Figure 10.3 Spring-damper analogy.

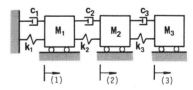

Figure 10.4 Example of damped, three-DOF system.

Consider the example in Fig. 10.4, where the elastic and viscous elements are parallel. The stiffness matrix, by inspection, is

$$\mathbf{k} = \begin{bmatrix} (k_1 + k_2) & -k_2 & 0 \\ -k_2 & (k_2 + k_3) & -k_3 \\ 0 & -k_3 & k_3 \end{bmatrix}$$

By Sect. 8A, an entry of this matrix in the ith row

and jth column is the external force that must be applied in direction i as a result of a unit displacement in direction j. Similarly, when a vector of unit velocities $\dot{\vec{u}}$ is imposed on the system and the associated force vector $\vec{P}_v$ is calculated, we have an equation $\vec{P}_v = \mathbf{c}\dot{\vec{u}}$, where $\mathbf{c}$ is the viscous stiffness matrix:

$$\mathbf{c} = \begin{bmatrix} (c_1 + c_2) & -c_2 & 0 \\ -c_2 & (c_2 + c_3) & -c_3 \\ 0 & -c_3 & c_3 \end{bmatrix}$$

Analogously, an entry in the ith row and jth column is the external force, which must be applied in the ith direction as a result of a unit velocity in the jth direction. The symmetry property exhibited by an elastic stiffness matrix also holds for a viscous matrix,

$$c_{ij} = c_{ji}$$

In the example above both matrices have exactly the same form. This is an exception rather than a rule, and it is possible only when each elastic element has a parallel, viscous counterpart.

10C. Equations of Damped Motion in Normal Coordinates. With the viscous stiffness included in the equations of motion, we have, in place of Eq. 10.1,

$$\mathbf{M}\ddot{\vec{u}} + \mathbf{c}\dot{\vec{u}} + \mathbf{k}\vec{u} = \vec{P}(t) \tag{10.5}$$

When this is transformed to normal coordinates, as was done in Sect. 10A, we encounter the expression $\mathbf{\Phi}^T\mathbf{c}\mathbf{\Phi}$. Under certain conditions, outlined in Sect. 10E, this expression becomes a diagonal matrix,

$$\bar{\mathbf{c}} = \mathbf{\Phi}^T\mathbf{c}\mathbf{\Phi} \tag{10.6}$$

Consequently, Eq. 10.5 takes the form of

$$\overline{\mathbf{M}}\ddot{\vec{s}} + \overline{\mathbf{c}}\dot{\vec{s}} + \overline{\mathbf{k}}\vec{s} = \mathbf{\Phi}^T\vec{P}(t) \tag{10.7}$$

This makes the equations of the system independent of each other. For the rth normal direction one can write

$$\overline{M}_r\ddot{s}_r + \bar{c}_r\dot{s}_r + \bar{k}_r s_r = Q_r(t) \tag{10.8}$$

which is, in effect, Eq. 10.3b with a damping term added. It is also possible to write it in a form analogous to Eq. 2.13:

$$\ddot{s}_r + 2\omega_r\zeta_r\dot{s}_r + \omega_r^2 s_r = \frac{Q_r(t)}{\overline{M}_r} \tag{10.9}$$

where

$$\omega_r^2 = \frac{\bar{k}_r}{\overline{M}_r}$$

$$\zeta_r = \frac{\bar{c}_r}{\bar{c}_{rc}} \tag{10.10}$$

and ζ_r is called the modal damping ratio. Note that the modal matrix $\mathbf{\Phi}$ is found from the properties of an undamped system and then used to decouple (or make independent of one another) the equations in the system represented by Eq. 10.5.

The normal mode method is limited to lightly damped systems. This means that ζ_r must be small, say $\zeta_r \leq 0.2$, for all modes that significantly contribute to the response.

10D. Free, Damped Motion is described by Eqs. 10.5 if we set $\vec{P}(t) = 0$. When the equations are decoupled by the procedure described in the previous section, each normal coordinate is treated as if it belonged to an SDOF system. The motion is described by the homogeneous part of Eq. 10.9,

$$\ddot{s}_r + 2\omega_r\zeta_r\dot{s}_r + \omega_r^2 s_r = 0 \tag{10.11}$$

Except for the change in notation, there is no difference between this and Eq. 2.2. The solution may therefore be found in Sect. 2D and the damped natural frequency, which will appear in the response equation, is expressed as

$$\omega_{rd} = \omega_r\left(1 - \zeta_r^2\right)^{1/2} \tag{10.12}$$

This solution is valid up to $\zeta_r = 1$ (i.e., for periodic motion only). As was mentioned before, the actual limit of applicability of the normal mode method ends at much smaller values of ζ_r.

The ratio of the initial maximum deflection to the local maximum after n cycles is given for each mode by Eq. 2.11b. It is seen that this ratio is likely to be larger, let us say, for the third mode than for the first, because there are more cycles of motion in the third mode than there are in the first for any selected time interval.

10E. General Case of Forced, Damped Motion. The developments in Sect. 10C allow any system, together with the forces applied to it, to be transformed into a set of separate basic SDOF systems. All statements and equations given in Sect. 2E apply here, with some adjustment in notation. A response of an rth basic system to the generalized force

$Q_r(t)$ is

$$s_r = (B_{1r}\cos\omega_{dr}t + B_{2r}\sin\omega_{dr}t)\exp(-\omega_r\zeta_r t) + s_{pr}(t)$$

$$(10.13)$$

The term containing the exponential function is the transient component, which is caused by putting the system in motion and decays with time. The steady-state component of displacement $s_{pr}(t)$ lasts as long as the exciting force is applied.

Once the response equation in each normal coordinate has been obtained from Eq. 10.13, the transformation to the physical coordinates is made by means of Eq. 9.14a.

This chapter is concerned with two types of problems associated with this general type of motion.

1. Steady-state component of response that involves only the functions $s_{pr}(t)$. This is practically all the response remaining after a sufficiently large number of cycles.
2. Total response that takes place just after the excitation is applied. To avoid the excessive computational difficulties, one should include damping only in the steady-state part of the solution. Instead of Eq. 10.13, we again use Eq. 10.4, which is a result of setting $\zeta_r = 0$ in the transient part of the response.

The approximate procedure described is reasonable only for lightly damped systems and then only for a short interval of time after the motion starts. Outside these limitations, this method tends to overestimate the response. Under most circumstances, however, using Eq. 10.13 as a solution in its full form is not practical in manual computation, even for SDOF systems.

The key assumption in applying the normal-mode method to analysis of damped systems is that the transformation to normal coordinates diagonalizes the damping matrix. This takes place, for example, when the damping and the stiffness matrices are proportional, $\mathbf{c} = b\mathbf{k}$, where b is a constant. This leads directly to the proportionality of the corresponding modal constants,

$$\bar{c}_r = b\bar{k}_r \quad \text{or} \quad \zeta_r = \tfrac{1}{2}b\omega_r \quad (10.14)$$

If we now assume the value of ζ_r in the first mode, this establishes b and allows us to calculate the damping ratios for all remaining modes. Higher frequency modes are damped more strongly than the lower ones.

One may also easily find out that Eq. 10.6 gives a diagonal matrix if the damping matrix is proportional to the mass matrix. Under this condition the damping ratio is smaller for the higher modes. It is also possible to decouple the modes when the damping matrix is a linear combination of $\mathbf{M}$ and $\mathbf{k}$.

Since damping is the least known of all structural properties, we are unlikely to know exactly the form of the damping matrix, and the situations described above are somewhat idealized. For this reason another approach is frequently employed, in which we use the decoupled Eqs. 10.8 without presuming any direct relationship between the modal damping ratios ζ_r. Those ratios are selected for a particular problem from the experience of an analyst or on the basis of tests. Quite often the same ζ is used for all the modes of vibration.

All forms of damping discussed thus far are known as *proportional damping*. Sometimes, however, we know enough about a physical system under consideration to be certain that Eq. 10.6 will not give us a diagonal matrix and the equations of motion will therefore be coupled. In this event the normal-mode method presented here will not be applicable and may be used only as an approximation.

10F. Response to Harmonic Forcing is calculated based on equations in Sect. 10C. For a three-DOF system the applied load is defined as

$$\vec{\mathbf{P}}(t) = \begin{Bmatrix} P_{01} \\ P_{02} \\ P_{03} \end{Bmatrix} \sin\Omega t = \vec{\mathbf{P}}_0 \sin\Omega t \quad (10.15)$$

This means that all the load components have the same frequencies and differ only with respect to their amplitudes. In the principal coordinates, we again get Eq. 10.9, with the right side more precisely defined as

$$\frac{Q_r(t)}{\bar{M}_r} = \frac{Q_{0r}}{\bar{M}_r}\sin\Omega t \quad (10.16)$$

where Q_{0r} is the amplitude of the rth mode

$$Q_{0r} = \vec{\mathbf{\Phi}}_r^T \vec{\mathbf{P}}_0 = x_1 P_{01} + x_2 P_{02} + x_3 P_{03} \quad (10.17)$$

in which x_1, x_2, and x_3 are the numbers defining the rth mode shape. The steady-state solution of Eq. 10.9 is given by Eqs. 2.15, 2.16, and 2.17 with suitably modified notation,

$$s_r = \mu_r \frac{Q_{0r}}{\bar{k}_r}\sin(\Omega t - \theta_r) \quad (10.18)$$

and μ_r and θ_r are, respectively, the magnification factor and the phase angle for the rth mode,

$$\frac{1}{\mu_r} = \left[\left(1 - \frac{\Omega^2}{\omega_r^2} \right)^2 + \left(\frac{2\zeta_r\Omega}{\omega_r} \right)^2 \right]^{1/2} \quad (10.19)$$

$$\tan\theta_r = \frac{2\zeta_r\Omega/\omega_r}{1 - \Omega^2/\omega_r^2} \quad (10.20)$$

Once the modal responses are known, a transformation to the physical coordinates can be made. If the forcing function is given as a multiple of $\cos\Omega t$ rather than $\sin\Omega t$, the solution remains unchanged, except that cos replaces sin.

When damping is absent, one may easily solve the problem using only physical coordinates. In the general equation of motion 10.1 we put

$$\vec{u} = \vec{A}\sin\Omega t \quad \text{and} \quad \ddot{\vec{u}} = -\vec{A}\Omega^2\sin\Omega t$$

which gives us

$$(\mathbf{k} - \mathbf{M}\Omega^2)\vec{A} = \vec{P}_0 \quad (10.21)$$

This is a set of equations that can be solved for the unknown amplitudes A_n. There is no need to know the natural frequencies of the system, however; if the given forcing frequency Ω is equal to a natural frequency ω_r, the infinitely large response will be obtained.

Calculating the physical amplitudes of response in the presence of damping is, by comparison, a much more involved process. Even when the response is known in terms of modal components, those components may not be algebraically added because of phase differences.

10G. Upper Bound of Dynamic Response.
When a structure performs steady-state, harmonic vibrations with a frequency Ω, the displacement at some point of interest may be written as

$$u = u_1\sin(\Omega t - \theta_1) + \cdots + u_n\sin(\Omega t - \theta_n)$$

where $u_1, \ldots, u_n$ are the amplitudes and $\theta_1, \ldots, \theta_n$ are the phase angles of the n modal components. The expression

$$\text{ub}(u) = |u_1| + |u_2| + \cdots + |u_n|$$

will be called the *upper bound* of the displacement response at that particular point, because it is the limiting value the displacement cannot exceed. If another quantity (e.g., a shear force) is to be estimated, we again find the amplitude of that quantity in each individual mode and then sum the absolute values of the modal contributions.

The reader may note that a harmonic forcing with a single frequency is not necessarily a very difficult case, because the amplitude of response may always be exactly determined. When two frequencies Ω_1 and Ω_2 are present, the response is no longer a harmonic function and a systematic search for maxima must be conducted. In this latter case, as in many others, the upper bound estimate is a very useful first approximation.

10H. Response of Symmetric Structures.
The discussion of the response of symmetric structures to the applied dynamic forces or displacements is based on what was said in Sects. 8J (stiffness properties) and 9F (free vibrations). Any arbitrary dynamic loading can be resolved into a symmetric part and an antisymmetric part. Equations 8.21 may again be used to achieve this resolution, except that now the forces involved are the functions of time. Symmetric loading excites only the symmetric modes, and the effect of antisymmetric loading is analogous. The total result is found by superposition of both responses.

When there is a force or a moment applied just *at* the plane of symmetry, we divide it into two halves, which we treat as being applied on each side of the plane, infinitely close to it. The same treatment may be used with respect to a mass cut in half by the plane of symmetry—we think of it as two halves, one on each side of the plane, immediately adjacent to it.

Double symmetry and multiple symmetry are handled in essentially the same way.

SOLVED PROBLEMS

10-1 Consider the same system used in Prob. 9-3, but with viscous dampers added. The viscous and the elastic stiffness matrices are proportional, $\mathbf{c} = b\mathbf{k}$. What value should constant b have so that the damping ratio of the first mode is $\zeta_1 = 0.04$? What is then the value of ζ_2? The undamped frequencies are known from the reference problem.

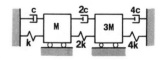

Equation 10.14 answers both questions. For the first mode where $\omega_1 = 49.83$ rad/s, $\zeta_1 = \frac{1}{2}b\omega_1$ gives $b = 1/622.88$ s/rad. Similarly

$$\zeta_2 = \frac{1}{2} \times \frac{1}{622.88} \times 86.70 = 0.0696$$

10-2 The system is similar to that in Fig. 10.4 except that all dampers are attached to the ground. Develop the damping matrix **c** and a formula for a modal ratio ζ_r when there is a proportionality between the damping and the mass constants, $c_i = \beta M_i$ (β is the same for all degrees of freedom).

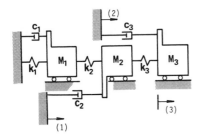

The equation of motion of mass M_2 will have the form

$$M_2 \ddot{u}_2 = -c_2 \dot{u}_2 - \cdots$$

in which the dots stand for the stiffness terms. The remaining equations will be analogous, which shows that the damping matrix is diagonal,

$$\mathbf{c} = \lceil c_1 \quad c_2 \quad c_3 \rfloor$$

Taking advantage of proportionality,

$$\mathbf{c} = \beta \mathbf{M}$$

From Eq. 10.6 we see that the proportionality also exists in the normal coordinates,

$$\bar{\mathbf{c}} = \beta \overline{\mathbf{M}}$$

The first two terms of Eq. 10.7 become

$$\overline{\mathbf{M}}\ddot{\mathbf{s}} + \beta \overline{\mathbf{M}}\dot{\mathbf{s}}$$

which, for the rth normal direction, gives

$$\ddot{s}_r + \beta \dot{s}_r$$

Comparing this with Eq. 10.9, we have $\zeta_r = \beta/(2\omega_r)$.

10-3 Write an expression for the response of this system to the force $P = P_0 \cos \Omega t$ applied to mass M_2. Use the matrices developed in Prob. 9-3. The damping ratio $\zeta = 0.02$ is the same for both modes.

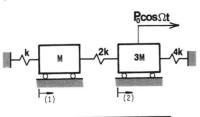

By the notation introduced in Sect. 10F, the amplitudes of the applied forces are $P_{01} = 0$ and $P_{02} = P_0$. The modal matrix determined earlier is

$$\boldsymbol{\Phi} = \begin{bmatrix} 1.0 & 1.0 \\ 0.8792 & -0.3792 \end{bmatrix}$$

Equation 10.17 gives us the modal forces:

$$Q_{01} = 1.0 \times 0 + 0.8792 P_0 = 0.8792 P_0$$

$$Q_{02} = 1.0 \times 0 - 0.3792 P_0 = -0.3792 P_0$$

Magnification factors are calculated from Eqs. 10.19 and phase angles from Eq. 10.20.

We need the modal stiffnesses $\bar{k}_r$ to calculate the response. Because the mass matrix is diagonal and the stiffness matrix is not, it is easier to calculate modal masses $\overline{M}_r$ from Eq. 9.21a and then to convert them to modal stiffnesses.

$$\overline{M}_1 = M + 3M \times 0.8792^2 = 3.319M$$

$$\overline{M}_2 = M + 3M \times 0.3792^2 = 1.4314M$$

The squares of natural frequencies were found in Prob. 9-3 as

$$\omega_1^2 = 1.2417 \frac{k}{M} \quad \text{and} \quad \omega_2^2 = 3.7584 \frac{k}{M}.$$

The modal stiffnesses are

$$\bar{k}_1 = \omega_1^2 \overline{M}_1 = 1.2417 \times 3.3190k = 4.1212k$$

$$\bar{k}_2 = 3.7584 \times 1.4314k = 5.3798k$$

The modal responses may now be found from Eq. 10.18:

$$\frac{Q_{01}}{\bar{k}_1} = \frac{0.8792 P_0}{4.1212k} = \frac{0.2133 P_0}{k}$$

$$\frac{Q_{02}}{\bar{k}_2} = \frac{-0.3792 P_0}{5.3798k} = \frac{-0.0705 P_0}{k}$$

and

$$s_1 = 0.2133 \mu_1 \frac{P_0}{k} \cos(\Omega t - \theta_1)$$

$$s_2 = -0.0705 \mu_2 \frac{P_0}{k} \cos(\Omega t - \theta_2)$$

We can return to the physical coordinates with the aid of Eq. 9.14a:

$$\begin{Bmatrix} u_1 \\ u_2 \end{Bmatrix} = \begin{Bmatrix} 0.2133 \\ 0.1875 \end{Bmatrix} \mu_1 \frac{P_0}{k} \cos(\Omega t - \theta_1)$$

$$- \begin{Bmatrix} 0.0705 \\ -0.0267 \end{Bmatrix} \mu_2 \frac{P_0}{k} \cos(\Omega t - \theta_2)$$

One can see that if the magnification factors are the same, the amplitudes of the first mode are higher than those of the second mode. This is because the static deflection under P_0 is similar to the first mode, not the second.

10-4 If we put $k=1000$ Lb/in. and $M=0.5$ Lb-s^2/in. in Prob. 10-3, we obtain $\omega_1=49.83$ and $\omega_2=86.70$ rad/s. Calculate the amplitudes and phase angles of the displacement response for the following values of the forcing frequency: (1) $\Omega=1.03\omega_1$; (2) $\Omega=0.5(\omega_1+\omega_2)$, and (3) $\Omega=0.97\omega_2$.

Following the development of solution in Prob. 10-3 we have:

1. $\Omega=1.03\omega_1=51.32$ rad/s
   $\mu_1=13.60$
   $\mu_2=1.538$
   $\tan\theta_1=-0.6765 \qquad \theta_1=2.5468$ rad
   $\tan\theta_2=0.0364 \qquad \theta_2=0.0364$ rad

$$\begin{Bmatrix} u_1 \\ u_2 \end{Bmatrix} = \begin{Bmatrix} 2.9008 \\ 2.5500 \end{Bmatrix} \frac{P_0}{k}\cos(\Omega t-2.5468)$$
$$-\begin{Bmatrix} 0.1084 \\ -0.0411 \end{Bmatrix} \frac{P_0}{k}\cos(\Omega t-0.0364)$$

2. $\Omega=0.5(\omega_1+\omega_2)=68.265$ rad/s
   $\mu_1=1.1383$
   $\mu_2=2.6223$
   $\tan\theta_1=-0.0625 \qquad \theta_1=3.0792$ rad
   $\tan\theta_2=0.0829 \qquad \theta_2=0.0827$ rad

$$\begin{Bmatrix} u_1 \\ u_2 \end{Bmatrix} = \begin{Bmatrix} 0.2428 \\ 0.2134 \end{Bmatrix} \frac{P_0}{k}\cos(\Omega t-3.0792)$$
$$-\begin{Bmatrix} 0.1849 \\ -0.0700 \end{Bmatrix} \frac{P_0}{k}\cos(\Omega t-0.0827)$$

3. $\Omega=0.97\omega_2=84.1$ rad/s
   $\mu_1=0.5406$
   $\mu_2=14.145$
   $\tan\theta_1=-0.0365 \qquad \theta_1=3.1051$ rad
   $\tan\theta_2=0.6565 \qquad \theta_2=0.5809$ rad

$$\begin{Bmatrix} u_1 \\ u_2 \end{Bmatrix} = \begin{Bmatrix} 0.1153 \\ 0.1014 \end{Bmatrix} \frac{P_0}{k}\cos(\Omega t-3.1051)$$
$$-\begin{Bmatrix} 0.9972 \\ -0.3777 \end{Bmatrix} \frac{P_0}{k}\cos(\Omega t-0.5809)$$

It may be clearly seen that when the forcing frequency is near the natural frequency of a particular mode, the amplitudes of this mode become the major components of the system deflections. By watching the flow of calculations for the case $\Omega=0.5(\omega_1+\omega_2)$ one can notice that damping has little effect on the magnification factors. It is another illustration of a well-known fact, namely, that damping is important only in the vicinity of resonance.

10-5 Calculate the spring forces in Prob. 10-3 using the numerical data and displacement solution of Prob. 10-4. Perform the computation for this instant of time when the response of the first mode shape reaches its first peak. The forcing frequency is $\Omega=68.265$ rad/s.

The function of time in the first modal component is $\cos(\Omega t-3.0792)$, which reaches its first peak value when

$$\Omega t_1-3.0792=0 \qquad \text{or} \qquad t_1=0.04511 \text{ s}$$

At this instant the second mode function has the value

$$\cos(\Omega t_1-0.0827)=-0.9895$$

The displacements are

$$u_1=[0.2428\times1.0-0.1849(-0.9895)]\frac{P_0}{k}=0.4258\frac{P_0}{k}$$

$$u_2=[0.2134\times1.0-(-0.07)(-0.9895)]\frac{P_0}{k}=0.1441\frac{P_0}{k}$$

The spring forces are calculated in sequence going from left to right in Fig. 10-3:

$$N_{01}=ku_1=0.4258P_0$$

$$N_{12}=2k(u_2-u_1)=-0.5634P_0$$

$$N_{23}=4k(-u_2)=-0.5764P_0$$

10-6 Calculate the upper bounds of displacements and spring forces in Prob. 10-3 using the numerical data and the solution of Prob. 10-4. The forcing frequency is $\Omega=68.265$ rad/s.

Summing the absolute values of modal displacement components, we have

$$\text{ub}(u_1)=(0.2428+0.1849)\frac{P_0}{k}=0.4277\frac{P_0}{k}$$

$$\text{ub}(u_2)=(0.2134+0.07)\frac{P_0}{k}=0.2834\frac{P_0}{k}$$

The modal spring forces going from left to right

1. $\begin{cases} N_{01}=ku_1=0.2428P_0 \\ N_{12}=2k(u_2-u_1)=-0.0588P_0 \\ N_{20}=4k(-u_2)=-0.8536P_0 \end{cases}$

2. $\begin{cases} N_{01}=-0.1849P_0 \\ N_{12}=0.5098P_0 \\ N_{20}=-0.28P_0 \end{cases}$

The upper bounds of spring forces are

$$\text{ub}(N_{01})=0.4277P_0$$

$$\text{ub}(N_{12})=0.5686P_0$$

$$\text{ub}(N_{23})=1.1336P_0$$

Note that the force in spring 0-1 calculated at an arbitrary time point in Prob. 10-5 as $0.4258P_0$ was not much different from its upper bound for this forcing frequency. The same is true of spring 1-2.

10-7 Prob. 10-3 gives the displacement solution for the system subjected to a harmonic force. That solution is presented as a sum of two normal modes. Determine the amplitudes of physical displacements when the forcing frequency is $\Omega = 68.265$ rad/s. Put $k = 1000$ Lb/in. and $M = 0.5$ Lb-s^2/in.

Taking advantage of the detailed calculations for this Ω in Prob. 10-4, we write the first displacement as

$$u_1 = 0.2428 \frac{P_0}{k} \cos(\Omega t - 3.0792) - 0.1849 \frac{P_0}{k} \cos(\Omega t - 0.0827)$$

The figure depicts the modal components of u_1 as rotating vectors with amplitudes $A_1 = 0.2428 P_0/k$ and $A_2 = 0.1849 P_0/k$. The vectors are shown at $\Omega t = 0$. (Note that the second component has its phase angle increased by π to treat A_2 as a positive number.) The angle between vectors A_1 and A_2 is

$$\phi = 2\pi - (\pi - 0.0827) - 3.0792 = 0.1451$$

From the law of cosines we have

$$A^2 = A_1^2 + A_2^2 + 2 A_1 A_2 \cos \phi$$

$$\therefore A = 0.4266 \frac{P_0}{k}$$

Repeating the operation for the amplitude of u_2 (with $A_1 = 0.2134 P_0/k$ and $A_2 = 0.07 P_0/k$, per Prob. 10-4),

$$\phi = 3.0792 - 0.0827 = 2.9965$$

$$A^2 = (0.2134^2 + 0.07^2 + 2 \times 0.2134 \times 0.07 \cos 2.9965) \frac{P_0^2}{k^2}$$

$$\therefore A = 0.1445 \frac{P_0}{k}$$

Note that u_1 and u_2 do not attain their respective extrema at the same time, because the phase angles of their amplitude vectors are different.

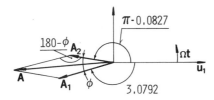

10-8 The system is subjected to the harmonic force $P_3 = P_0 \cos \Omega t$. The undamped frequencies are $\omega_1 = 15.445$, $\omega_2 = 72.526$, and $\omega_3 = 92.206$ rad/s for $k = 1000$ Lb/in. and $M = 0.5$ Lb-s^2/in. The modal matrix is

$$\Phi = \begin{bmatrix} 1.0 & 1.0 & 1.0 \\ 1.4404 & 0.185 & -0.6254 \\ 1.5317 & -0.5874 & 0.5557 \end{bmatrix}$$

If the forcing frequency is $\Omega = \omega_2$, compare the amplitudes of steady-state responses in all three modes. Damping ratio $\zeta_2 = 0.02$, while the other modal ratios are to be calculated from Eq. 10.14.

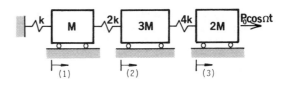

To find an amplitude of modal response in the rth mode, one must have μ_r, Q_{0r}, and $\bar{k}_r$ for that mode according to Eq. 10.18. Equation 10.14 implies that damping ratios are proportional to frequencies, therefore:

$$\zeta_1 = 0.0043 \qquad \zeta_2 = 0.02 \qquad \zeta_3 = 0.0254$$

The magnification factors are calculated from Eq. 10.19:

$$\mu_1 = 0.0475 \qquad \mu_2 = 25 \qquad \mu_3 = 2.608$$

where μ_1 is very small because the forcing frequency is several times larger than that of mode 1; μ_2 is largest because the forcing frequency is at resonance with mode 2. The modal forces, after Eq. 10.17,

$$Q_{01} = [1.0 \quad 1.4404 \quad 1.5317] \begin{Bmatrix} 0 \\ 0 \\ P_0 \end{Bmatrix} = 1.5317 P_0$$

Similarly, $Q_{02} = -0.5874 P_0$ and $Q_{03} = 0.5557 P_0$. Modal masses and stiffnesses:

$$\overline{M}_1 = 11.916 M \qquad \overline{M}_2 = 1.7928 M \qquad \overline{M}_3 = 2.791 M$$

from which

$$\bar{k}_1 = 15.445^2 \times 11.916 \times 0.5 = 1421.3 \text{ Lb/in.}$$

$$\bar{k}_2 = 72.526^2 \times 1.7928 \times 0.5 = 4715.1 \text{ Lb/in.}$$

$$\bar{k}_3 = 92.206^2 \times 2.791 \times 0.5 = 11,864 \text{ Lb/in.}$$

The successive modal amplitudes are, in accordance with Eq. 10.18,

$$\overline{A}_1 = \frac{\mu_1 Q_{01}}{\bar{k}_1} = 0.0475 \times \frac{1.5317 P_0}{1421.3} = 51.19 \times 10^{-6} P_0$$

$$\overline{A}_2 = -3.1145 \times 10^{-3} P_0 \qquad \overline{A}_3 = 0.1222 \times 10^{-3} P_0$$

The amplitude of the second mode is clearly dominating, even though the static deflection under P_0 has a shape very similar to that of the first mode.

10-9 In Prob. 10-8 the modal displacement amplitudes were found. Calculate the upper bounds of physical displacement components and the upper bounds of spring forces.

The absolute values of the modal amplitudes are:

$$\bar{A}_1 = 0.0512 \times 10^{-3} P_0 \qquad \bar{A}_2 = 3.1145 \times 10^{-3} P_0$$

$$\bar{A}_3 = 0.1222 \times 10^{-3} P_0$$

The equation $\vec{u} = \Phi \vec{s}$, which is used to go from modal to physical displacements, is used here in slightly altered form. Keeping in mind that our purpose is to obtain the combination of the absolute values of modal components, the negative signs in matrix Φ are dropped. Thus we have

$$\text{ub}\begin{Bmatrix} A_1 \\ A_2 \\ A_3 \end{Bmatrix} = \begin{Bmatrix} 3.2879 \\ 0.7264 \\ 1.9758 \end{Bmatrix} 10^{-3} P_0$$

These are the upper bounds of displacements expressed as a multiple of the forcing amplitude P_0. The upper bounds of spring forces are obtained by combining the displacements in the most unfavorable way. Going from left to right in Fig. 10-8, we get

$$\text{ub}(N_{01}) = \text{ub}\, k(A_1) = 3.2879 P_0$$

$$\text{ub}(N_{12}) = \text{ub}\, 2k(A_2 - A_1) = 8.0286 P_0$$

$$\text{ub}(N_{23}) = \text{ub}\, 4k(A_3 - A_2) = 10.8088 P_0$$

Note that the forces above are computed in a very conservative manner. A more realistic procedure is to find the forces induced by individual modes and then sum the absolute values.

10-10 Calculate the displacement amplitudes A_1 and A_2 of this system as a result of force $P_0 \sin \Omega t$ applied to mass 1. Sketch the nondimensional amplitudes as a function of Ω.

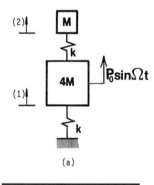

(a)

The stiffness and mass matrices are, respectively,

$$\mathbf{k} = \begin{bmatrix} 9k & -k \\ -k & k \end{bmatrix} \qquad \text{and} \qquad \mathbf{M} = \begin{bmatrix} 4M & 0 \\ 0 & M \end{bmatrix}$$

The frequency equation is

$$\begin{vmatrix} (9 - 4\tilde{\omega}^2) & -1 \\ -1 & (1 - \tilde{\omega}^2) \end{vmatrix} = 0$$

where $\tilde{\omega}^2 = \omega^2 M/k$. This is satisfied by

$$\tilde{\omega}_1^2 = 0.8246 \qquad \text{and} \qquad \tilde{\omega}_2^2 = 2.4254$$

or

$$\omega_1 = 0.9081 \left(\frac{k}{M} \right)^{1/2} \qquad \text{and} \qquad \omega_2 = 1.5574 \left(\frac{k}{M} \right)^{1/2}$$

From Eq. 9.12 we get

$$\Phi = \begin{bmatrix} 1.0 & 1.0 \\ 5.7016 & -0.7016 \end{bmatrix}$$

The external force vector is $\begin{Bmatrix} P_0 \\ 0 \end{Bmatrix}$

The modal forces are

$$Q_{01} = P_0 \qquad \text{and} \qquad Q_{02} = P_0$$

The modal masses are

$$\overline{M}_1 = 36.508 M \qquad \text{and} \qquad \overline{M}_2 = 4.4922 M$$

The modal spring constants:

$$\bar{k}_1 = \omega_1^2 \overline{M}_1 = 0.8246 \frac{k}{M} \times 36.508 M = 30.104 k$$

$$\bar{k}_2 = 10.895 k$$

Modal responses (Eq. 10.18):

$$s_1 = \mu_1 \frac{P_0}{30.104 k} \sin \Omega t$$

$$s_2 = \mu_2 \frac{P_0}{10.895 k} \sin \Omega t$$

Our physical displacements are found from Eq. 9.14a,

$$u_1 = \left(\frac{\mu_1}{30.104} + \frac{\mu_2}{10.895} \right) \frac{P_0}{k} \sin \Omega t \equiv A_1 \sin \Omega t$$

$$u_2 = \left(\frac{\mu_1}{5.28} - \frac{\mu_2}{15.529} \right) \frac{P_0}{k} \sin \Omega t \equiv A_2 \sin \Omega t$$

The amplitudes are plotted in (b). We note that displacement of mass M_1 is out of phase with forcing when Ω is slightly larger than ω_1 and is in phase again when Ω is slightly less than ω_2. Somewhere in between there is a value of Ω for which $A_1 = 0$. This is somewhat curious phenomenon, because this means that M_1, which we are shaking, is stationary at this value of Ω, while M_2, which we are not forcing directly, is vibrating.

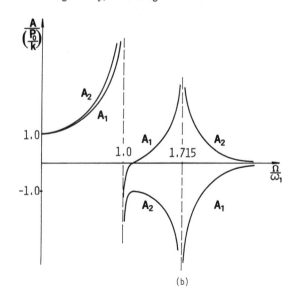

(b)

10-11 In Prob. 10-10 it was determined that at a certain forcing frequency denoted here as Ω_0, the mass to which the force was a applied remained motionless. Find a condition from which Ω_0 may be calculated for a general two-DOF system in which forcing $P_1 \sin \Omega_0 t$ applied along u_1 gives $u_1 = 0$. Use physical coordinates u_1 and u_2 to solve this problem.

Applying Eq. 10.21 to our system, we have

$$\begin{bmatrix} (k_{11} - \Omega^2 M_{11}) & (k_{12} - \Omega^2 M_{12}) \\ (k_{21} - \Omega^2 M_{21}) & (k_{22} - \Omega^2 M_{22}) \end{bmatrix} \begin{Bmatrix} A_1 \\ A_2 \end{Bmatrix} = \begin{Bmatrix} P_1 \\ 0 \end{Bmatrix}$$

Using Cramer's rule, we write

$$A_1 = \frac{1}{D} \begin{vmatrix} P_1 & (k_{12} - \Omega^2 M_{12}) \\ 0 & (k_{22} - \Omega^2 M_{22}) \end{vmatrix}$$

in which D is the determinant of the square matrix. The forcing frequency Ω_0 is obtained by setting $A_1 = 0$:

$$P_1 (k_{22} - \Omega_0^2 M_{22}) = 0 \qquad \therefore \ \Omega_0^2 = \frac{k_{22}}{M_{22}}$$

When $\Omega = \Omega_0$, displacement u_1 is equal to zero. We note that Ω_0 is the natural frequency of mass M_{22} attached to the ground with spring k_{22}. In the reference problem $k_{22} = k$ and $M_{22} = M$.

Usually there is a reverse problem. When we deal with a system subjected to a forcing frequency Ω, that system being mass $4M$ supported by the spring $8k$, we may want to add to it a *dynamic absorber* for the purpose of reducing the amplitude of motion. The natural frequency of that absorber, when treated separately, must then be equal to the prescribed forcing frequency.

10-12 The weight in Prob. 9-4 is subjected to a harmonically varying moment with an amplitude of 850 Lb-in. and a frequency of 75 Hz. Using the normal mode approach, estimate the upper bound of the bending moment in the beam. The modal damping is $\zeta = 0.02$.

From the reference problem:

$$M_{11} = 0.2591 \quad \text{and} \quad M_{22} = 4.8575$$

$$\Phi = \begin{bmatrix} 1.0 & 1.0 \\ 0.05077 & -1.051 \end{bmatrix} \quad \begin{aligned} \omega_1 &= 127.9 \\ \omega_2 &= 1072.4 \end{aligned}$$

The applied force is $P_2 = \mathfrak{M}_0 \sin \Omega t$ with $\Omega = 2\pi \times 75 = 471.2$ rad/s. Modal force amplitudes:

$$Q_1 = 0.05077 \mathfrak{M}_0 \qquad Q_2 = -1.051 \mathfrak{M}_0$$

Magnification factors:

$$\frac{1}{\mu_1^2} = \left(1 - \frac{471.2^2}{127.9^2}\right)^2 + \left(2 \times 0.02 \times \frac{471.2}{127.9}\right)^2$$

$$\therefore \ \mu_1 = 0.0795 \qquad \mu_2 = 1.239$$

Modal masses from the reference problem,

$$\overline{M}_1 = 0.2716 \qquad \overline{M}_2 = 5.6247$$

Using Eq. 10.18, the amplitude of a modal acceleration may be expressed as

$$\ddot{s}_r = -\mu_r \Omega^2 \frac{Q_{0r}}{k_r} = -\mu_r \left(\frac{\Omega}{\omega_r}\right)^2 \frac{Q_{0r}}{\overline{M}_r}$$

Ignoring the signs, we have

$$\ddot{s}_1 = 0.0795 \left(\frac{471.2}{127.9}\right)^2 \times \frac{0.05077}{0.2716} \times 850 = 171.4$$

$$\ddot{s}_2 = 38.0$$

The physical acceleration amplitudes are calculated from Eq. 9.14c, which we write here as

$$\begin{Bmatrix} \ddot{u}_1 \\ \ddot{u}_2 \end{Bmatrix} = \begin{Bmatrix} 171.4 \\ 8.702 \end{Bmatrix} \pm \begin{Bmatrix} 38.0 \\ -39.94 \end{Bmatrix}$$

When the accelerations of the first mode are multiplied by the respective masses, the inertia forces associated with the first mode are like those shown in the left sketch of (a). The right-hand sketch shows the bending moment distribution in the first mode. The second mode is handled in the same manner. The upper bounds of the bending moments are

$$1374 + 101 = 1475 \text{ Lb-in.} \qquad \text{at base}$$

$$42 + 194 = 236 \text{ Lb-in.} \qquad \text{at tip}$$

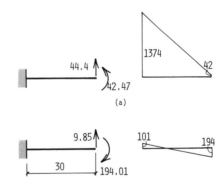

(a)

(b)

10-13 Calculate the deflection amplitudes at the nodal points due to the tip force $P = P_0 \cos \Omega t$, where $\Omega = 15$ rad/s and $P_0 = 1200$ Lb. Use all the necessary information from Prob. 9-26.

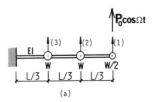

(a)

As there is no damping involved, one can use physical coordinates according to Eq. 10.21. We have

$$\frac{W}{g}\Omega^2 = \frac{20,000}{386} \times 15^2 = 11,658$$

and

$$\begin{bmatrix} -457.2 & -12,265 & 9,170 \\ -12,265 & 22,048 & -35,181 \\ 9,170 & -35,181 & 49,585 \end{bmatrix} \begin{Bmatrix} A_1 \\ A_2 \\ A_3 \end{Bmatrix} = \begin{Bmatrix} P_0 \\ 0 \\ 0 \end{Bmatrix}$$

$$\therefore \begin{Bmatrix} A_1 \\ A_2 \\ A_3 \end{Bmatrix} = \begin{Bmatrix} 0.1299 \\ -0.2569 \\ -0.2063 \end{Bmatrix}$$

When a force is applied at the end of beam, one usually expects the deflected shape to be close to the first mode. In our case, however, the deflection resembles the second mode, as illustrated by (b), because of proximity of the forcing frequency to ω_2.

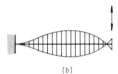

(b)

10-14 Calculate the distribution of bending moment and shear force when the beam in Prob. 10-13 reaches its extremum deflection.

The magnitude of the inertia force in the harmonic motion, taking place in the reference problem, is $M_n A_n \Omega^2$. This extreme value is reached at the same time as the deflection u_n, both quantities having the same sense. Simple multiplication gives us the inertia forces depicted in the figure. The amplitude A_1 is shown as positive in the reference problem, which means that it is in phase with the driving force P. Using only statics now, we can compute and tabulate the values of the shear force and the bending moment.

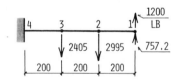

Stage	Shear (Lb)	Moment (Lb-in.)
1		0
	−1957.2	
2		0.3914×10^6
	1037.8	
3		0.1839×10^6
	3442.8	
4		-0.5047×10^6

10-15 The stiffness matrix for the system of cables and pulleys was derived in Prob. 8-13, assuming no slipping or slackening of cables. A twisting moment applied to the lower pulley is $M_t = M_{t0}\sin\Omega t$. Calculate the rotation amplitudes of the pulleys using the following data: $J = 0.7$ Lb-s^2-in. (moment of inertia of each pulley), $L_1 = 40$ in., $L_2 = 30$ in., $r = 6$ in., $EA = 14,000$ Lb, $M_{t0} = 1800$ Lb-in., and $\Omega = 66$ rad/s.

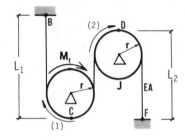

In the absence of damping, we can employ the method of Sect. 10F.

$$J\Omega^2 = 0.7 \times 66^2 = 3049.2$$

$$\mathbf{M}\Omega^2 = \begin{bmatrix} 3049.2 & 0 \\ 0 & 3049.2 \end{bmatrix}$$

The stiffness matrix from the reference problem:

$$EAr^2 \begin{bmatrix} \left(\dfrac{1}{L_1}+\dfrac{1}{L_2}\right) & \dfrac{1}{L_2} \\ \dfrac{1}{L_2} & \dfrac{2}{L_2} \end{bmatrix} = \begin{bmatrix} 29,400 & 16,800 \\ 16,800 & 33,600 \end{bmatrix}$$

From Eq. 10.21:

$$\begin{bmatrix} 26,351 & 16,800 \\ 16,800 & 30,551 \end{bmatrix} \begin{Bmatrix} \alpha_1 \\ \alpha_2 \end{Bmatrix} = \begin{Bmatrix} 1800 \\ 0 \end{Bmatrix}$$

$$\therefore \alpha_1 = 0.10519 \qquad \alpha_2 = -0.05784$$

While pulley 1 is turning clockwise, pulley 2 is going counterclockwise.

10-16 The double arrows on the left of the model of a two-story building symbolize centrifugal shakers that are capable of imposing a harmonic forcing on the structure. Select the frequencies and force amplitudes of shakers to obtain the first mode shape. Repeat the process for the second mode. Assume that there is no damping in the system. The natural frequencies and the mode shapes were previously given in the answer to Prob. 9-22.

$$\omega_1 = 7.1 \text{ rad/s} \qquad \mathbf{\Phi} = \begin{bmatrix} 1.0 & 1.0 \\ 1.7711 & -0.6211 \end{bmatrix}$$
$$\omega_2 = 17.385$$

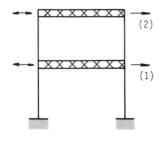

Since there is no damping and the forcing is harmonic, we can use the results of Sect. 10F. If some arbitrary exciting frequency Ω (not coinciding with ω_1 or ω_2) is selected, we can replace the vector of amplitudes $\vec{A}$ in Eq. 10.21 by the first mode shape,

$$(\mathbf{k}-\mathbf{M}\Omega^2)\vec{\Phi}_1=\vec{P}_0 \qquad (*)$$

and by a simple matrix multiplication determine the vector of force amplitudes $\vec{P}_0$. Let us choose $\Omega=10$ rad/s and calculate the coefficient matrix in this equation using the data from the reference problem.

$$\mathbf{M}\Omega^2 = \begin{bmatrix} 142,487 & 129,534 \end{bmatrix}$$

$$\mathbf{k}-\mathbf{M}\Omega^2 = \begin{bmatrix} 195,013 & -150,000 \\ -150,000 & 20,466 \end{bmatrix}$$

Carrying out the multiplication in ($*$) and doing the same for the second mode shape, the amplitudes of shaker forces are obtained:

$$\begin{Bmatrix} P_{01} \\ P_{02} \end{Bmatrix} = \begin{Bmatrix} -70,652 \\ -113,753 \end{Bmatrix} \qquad \begin{Bmatrix} P_{01} \\ P_{02} \end{Bmatrix}_2 = \begin{Bmatrix} 288,178 \\ -162,711 \end{Bmatrix}$$

(first mode) (second mode)

(The signs of forces have little meaning except for indicating that both components are in phase in the first mode and out of phase in the second.) These forces have excessively large magnitudes for any practical applications. We can obviously scale them down to the level where the displacement response can still be measured. A more effective way of reducing the required magnitude of forcing is to bring Ω closer to ω_1 or ω_2, whenever a particular mode is excited.

Note that if damping is involved, the problem becomes much more complicated because the phase angle is then another quantity to be determined.

10-17 Two masses are connected by an elastic link of stiffness $k=EA/L$. A step load $P=P_0H(t)$ is applied to mass M_1. Find the displacement response of this unconstrained system and the maximum force in the link. Use the free vibration results from Prob. 9-30.

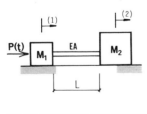

The modal matrix of the system is

$$\Phi = \begin{bmatrix} 1 & 1 \\ 1 & -\dfrac{M_1}{M_2} \end{bmatrix}$$

The modal masses:

$$\overline{M}_1 = M_1 + M_2 \qquad \text{and} \qquad \overline{M}_2 = M_1\left(1+\frac{M_1}{M_2}\right)$$

Equations of motion in modal coordinates are written as

$$\ddot{s}_r + \omega_r^2 s_r = \frac{1}{\overline{M}_r} Q_r(t)$$

The modal forces:

$$Q_1 = Q_2 = P_0 H(t)$$

The frequency of the first mode is zero, so the equation of motion reduces to

$$\ddot{s}_1 = \frac{1}{\overline{M}_1} P_0 H(t)$$

This is the equation of motion of a mass $\overline{M}_1$ subjected to a constant force P_0. From elementary mechanics we know that

$$s_1 = \frac{1}{2}\frac{P_0}{\overline{M}_1} t^2$$

when the system is at rest prior to the application of the force. The equation of the second mode is

$$\ddot{s}_2 + \omega_2^2 s_2 = \frac{1}{\overline{M}_2} P_0 H(t)$$

in which $\qquad \omega_2^2 = k(1/M_1 + 1/M_2)$.

The solution given in Prob. 2-12 is rewritten here to agree with this notation,

$$s_2 = \frac{P_0}{\omega_2^2 \overline{M}_2}(1-\cos\omega_2 t)$$

Now we can switch back to the physical coordinates

$$u_1 = s_1 + s_2$$

$$u_2 = s_1 - \frac{M_1}{M_2} s_2$$

Using a new notation

$$M = M_1 + M_2 \qquad \text{and} \qquad \frac{1}{M^*} = \frac{1}{M_1} + \frac{1}{M_2}$$

we get $\quad \overline{M}_1 = M \quad$ and $\quad \overline{M}_2 = M_1^2/M^*$.

The resultant displacements are

$$u_1 = \frac{P_0 t^2}{2M} + \frac{P_0}{k}\left(\frac{M^*}{M_1}\right)^2(1-\cos\omega_2 t)$$

$$u_2 = \frac{P_0 t^2}{2M} - \frac{P_0}{k}\frac{(M^*)^2}{M_1 M_2}(1-\cos\omega_2 t)$$

Each displacement has a rigid-body component and a vibratory component. The first one is the constant acceleration of the assembly. The compressive force in the link is

$$N = k(u_1 - u_2) = k\left(1 + \frac{M_1}{M_2}\right)s_2$$

or

$$N = \frac{M^*}{M_1}P_0(1 - \cos\omega_2 t)$$

Note that the maximum value is

$$N_{\max} = \frac{M^*}{M_1}\mu P_0 = \frac{M_2}{M_1 + M_2}\mu P_0$$

in which μ is the dynamic magnification factor equal to 2.0 in this case.

EXERCISES

10-18 The damping matrix of a system is a linear combination of stiffness and mass matrices, $\mathbf{c} = b\mathbf{k} + \beta\mathbf{M}$, in which b and β are constants. Develop a formula for the modal ratio ζ_r.

10-19 Find the damped natural frequencies of the two-mass system with an elastic link and a damper. The undamped version was analyzed in Prob. 9-30, where it was determined that

$$\mathbf{k} = \begin{bmatrix} k & -k \\ -k & k \end{bmatrix} \qquad \mathbf{\Phi} = \begin{bmatrix} 1 & 1 \\ 1 & -\dfrac{M_1}{M_2} \end{bmatrix}$$

$$\omega_1 = 0 \quad \text{and} \quad \omega_2^2 = \frac{EA}{L}\left(\frac{1}{M_1} + \frac{1}{M_2}\right) \equiv \frac{k}{M^*}$$

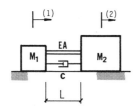

10-20 It was found in Prob. 10-8 that the second mode of vibration was predominant. Ignoring the effect of the first and the third mode, find the displacement response and the largest values of the spring forces.

10-21 The system analyzed in Prob. 10-8 is now subjected to two forces defined by the following

vector:

$$\overrightarrow{\mathbf{P(t)}} = \left\{ \begin{array}{c} P_0\cos\Omega t \\ 3P_0\cos\left(\Omega t - \dfrac{\pi}{2}\right) \\ 0 \end{array} \right\} = \left\{ \begin{array}{c} P_0 f_1(t) \\ 3P_0 f_2(t) \\ 0 \end{array} \right\}$$

Develop compact expressions for modal displacements when

$$\Omega = \omega_2 = 72.526 \text{ rad/s.}$$

10-22 Find the forcing frequency Ω_0 at which mass M_2 in Prob. 10-3 has zero displacement. Assume the system to be undamped.

10-23 The system is the same as in Prob. 10-10, except for the added damper. The damping is clearly nonproportional, since only one damper is present. Determine the damping matrix in generalized coordinates and equivalent modal damping, where $M = 1.2$ Lb-s^2/in., $k = 360$ Lb/in., and $c = 2.9$ Lb-s/in.

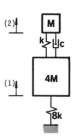

10-24 Using the stiffness matrix from Prob. 9-26 and the deflected shape determined in Prob. 10-13, calculate the joint forces and check whether they are indeed the same as those in Prob. 10-14.

10-25 The stiffness matrix for the system of three identical pulleys connected with a cable was derived in Prob. 8-28. Calculate the rotation amplitudes caused by the applied torques. Use $M_t = M_{t0}\sin\Omega t$ and put $L = 85$ in., $r = 7.5$ in., $EA = 20,000$ Lb, $J = 1.2$ Lb-s^2-in., $M_{t0} = 2400$ Lb-in., and $\Omega = 55$ rad/s.

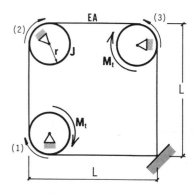

10-26 What are the amplitudes of cable forces induced by the loading in Prob. 10-15? What is the minimum cable preload N_0 that will prevent the cable from slackening? How big can the amplitude of the driving torque M_t become if the allowable cable load is $N_{all} = 550$ Lb?

10-27 The natural frequencies (squared) of the beam for the symmetric and antisymmetric modes are, respectively,

$$\omega_1^2 = \frac{162}{5}\frac{EI}{ML^3} \qquad \omega_2^2 = 486\frac{EI}{ML^3}$$

(See the answer to Prob. 9-25.) The applied forces are

$$P_1(t) = P_0\cos\Omega t \qquad P_2(t) = 2P_0\cos(2\Omega t)$$

Using the upper-bound approach, evaluate the bending moment for the following data: $M = 0.6$ Lb-s^2/in., $L = 120$ in., $EI = 4.8 \times 10^6$ Lb-in.^{2}, $P_0 = 95$ Lb, and $\Omega = 38$ rad/s.

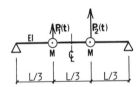

10-28 Compute the dynamic magnification factor for the peak bending moment in Prob. 10-27.

10-29 Develop an expression for the displacement solution for a system with n degrees of freedom, subjected to a load vector

$$\vec{\mathbf{P}}^T = [P_1\, P_2\, \cdots\, P_n]H(t)$$

which means that all load components are applied as step functions in time. Only the upper bound of the solution is of interest.

10-30 The system analyzed in Prob. 10-8 is subjected to two equal and opposite step loads. Calculate the upper bound of displacement response.

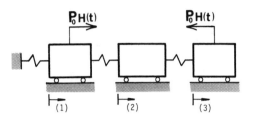

10-31 Calculate the amplitudes of the lateral harmonic forces that must be applied to the beam in Prob. 9-26 to induce the third deflected mode shape. The largest lateral deflection is to be 1.0 in., while the forcing frequency $\Omega = 1.2\omega_3$. Use all necessary information from the answer to Prob. 9-26.

10-32 Three masses are connected by elastic links, each of stiffness k. A step load $P = P_0 H(t)$ is applied to the left-end mass. Find the compressive forces in the links of this unconstrained system, taking advantage of structural symmetry. Designate the symmetric and the antisymmetric motion frequency by ω_1 and ω_2, respectively.

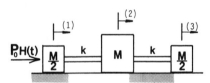

10-33 Solve Prob. 10-32 again; but instead of using symmetry principles, employ the normal-mode method. For free-vibration data refer to Prob. 9-17.

CHAPTER

11

Approximate Methods

The *Rayleigh method* allows us to determine an upper bound of a vibratory frequency, while the *Dunkerley formula* yields a lower bound. The *Holzer method* of successive iterations permits us to find not only frequencies, but also the associated mode shapes. The *static deflection method* provides us with a quick estimate of the fundamental frequency when the displacements under static loads are known. Quite often a *single-mode approximation* of the vibratory motion provides most of the information we need from dynamic analysis. Whether we use an exact or an assumed mode shape in this event, we are in fact reducing the structure to an *equivalent basic oscillator*. (The process of such a reduction is defined in this chapter.) The energy expressions for a system with a continuously distributed mass are also provided, but their use is limited to a single deflected shape. A rigorous approach to distributed-mass elements is presented later, in Chapter 15.

11A. Rayleigh's Method is an extension of the energy method, as presented in Chapter 1, to MDOF systems. Consider a body vibrating in a natural mode, every lumped mass harmonically oscillating with the same frequency ω. Observing that all masses reach their stationary points at the same time, we can write the energy balance equation (Eq. 1.8) for the whole system

$$\Pi_{\max} = T_{\max} \qquad (1.8)$$

from which the unknown frequency ω is calculated. As an example consider a beam with two lumped weights (Fig. 11.1). Let us assume that the deflected pattern during vibratory motion can be approximated by applying a certain set of static forces to the beam. We can choose them to be of the same magnitude as the forces of gravity. The strain energy of the deflected position is then

$$\Pi_{\max} = \tfrac{1}{2}(W_1 u_1 + W_2 u_2)$$

Figure 11.1 Deflections under forces proportional to gravity.

(As we know from Sect. 8B, the strain energy can be expressed in several ways. This equation happens to be the most convenient one in this case.) At the neutral position

$$T_{\max} = \frac{1}{2}\left(\frac{W_1}{g}v_1^2 + \frac{W_2}{g}v_2^2\right) = \frac{\omega^2}{2g}\left(W_1 u_1^2 + W_2 u_2^2\right)$$

by virtue of Eq. 2.6a. Using Eq. 1.8 we get the unknown frequency:

$$\omega^2 = g\frac{W_1 u_1 + W_2 u_2}{W_1 u_1^2 + W_2 u_2^2}$$

It is the task of an analyst to choose the set of static forces that gives a deflected pattern similar to the vibratory mode of interest. Of course there is more than one possibility in any particular case. In the example discussed one might apply a unit lateral force to the tip of the beam, which would result in a different ω value than one obtained with the help of gravity load. When several answers are obtained as a result of trials, the smallest one should be chosen as the best approximation. This is because this method overestimates the natural frequency. For some other forces, say P_1 and P_2 used in our example, we would get

$$\omega^2 = \frac{P_1 u_1 + P_2 u_2}{M_1 u_1^2 + M_2 u_2^2}$$

and the values of u_1 and u_2 would be different from those previously used. When not two but n masses are involved, more terms will appear in the frequency expressions;

$$\omega^2 = g\frac{W_1 u_1 + W_2 u_2 + \cdots + W_n u_n}{W_1 u_1^2 + W_2 u_2^2 + \cdots + W_n u_n^2} \quad (11.1)$$

when gravity load is applied, and

$$\omega^2 = \frac{P_1 u_1 + P_2 u_2 + \cdots + P_n u_n}{M_1 u_1^2 + M_2 u_2^2 + \cdots + M_n u_n^2} \quad (11.2)$$

when some other load pattern is used. In Eq. 11.1 summation is extended on all joints with masses. The same is true of the denominator of Eq. 11.2. The numerator of the latter may have as little as one term, because the applied load system may consist of a single force.

11B. Dunkerley's Method. Consider the same example used in the previous section. The first step in using this method is to calculate the displacements along the directions of interest, as shown in Fig. 11.2.

(By Sect. 8A, this is the calculation of entries a_{ii} on the main diagonal of the flexibility matrix.) After that the fictitious frequencies ω_{11} and ω_{22} are found from

$$\frac{1}{\omega_{11}^2} = a_{11}M_1 \quad \text{and} \quad \frac{1}{\omega_{22}^2} = a_{22}M_2$$

In the expressions above, ω_{11} would be the true natural frequency if only mass M_1 existed. Similarly, ω_{22} would be the frequency if the only nonzero mass were M_2. The fundamental frequency ω may now be approximated by

$$\frac{1}{\omega^2} = \frac{1}{\omega_{11}^2} + \frac{1}{\omega_{22}^2}$$

If there are n coupled degrees of freedom, the natural frequency ω is approximated by

$$\frac{1}{\omega^2} = \frac{1}{\omega_{11}^2} + \frac{1}{\omega_{22}^2} + \cdots + \frac{1}{\omega_n^2} \quad (11.3)$$

Each of the fictitious frequencies ω_{ii} is calculated ignoring all masses of the system except M_i. The fundamental frequency ω computed in this manner is always less than the true value.

When comparing Rayleigh's and Dunkerley's methods, note that the former has somewhat broader application because it can be used to find the frequency of any mode whose shape is approximately known. Dunkerley's method, which is applicable only to the first mode, does not require any knowledge of that deflected shape.

11C. Holzer's Method is very useful in calculation of natural frequencies and mode shapes of some unconstrained systems. Consider, for example, the chain of elements in Fig. 11.3a. When it is vibrating, both elastic and inertia forces are involved. The equation of motion for, say, mass M_2, is

$$M_2 \ddot{u}_2 = P_{23} - P_{12}$$

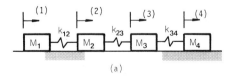

(a)

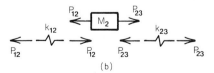

(b)

Figure 11.3 Holzer method: (a) notation; (b) equilibrium.

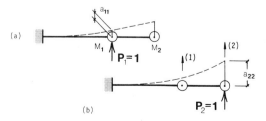

Figure 11.2 Deflections under unit forces.

If the system is oscillating in a natural mode, we have $\ddot{u}_2 = -\omega^2 u_2$, hence

$$P_{23} = P_{12} - \omega^2 M_2 u_2 \qquad (11.4)$$

The force in spring k_{23} is

$$P_{23} = k_{23}(u_3 - u_2)$$

$$\therefore u_3 = u_2 + \frac{P_{23}}{k_{23}} \qquad (11.5)$$

In the expression for M_1, which is analogous to Eq. 11.4, we must set $P_{01} = 0$, since there is no spring 01,

$$P_{12} = -\omega^2 M_1 u_1$$

We choose $u_1 = 1.0$, since we are allowed to do so in a free-vibration problem. Next, we write the equations for all springs and the remaining masses going from left to right. When some value is assumed for ω, we are then able to solve the equations one after another. If the assumed value is the same as a natural frequency, then the force P_{45} (in our case) computed from the last equation is equal to zero. (This is indeed the case, since there is no spring 45). A nonzero result for P_{45} indicates that a new ω value must be chosen and the computation repeated. Figure 11.4 shows the type of plot that is obtained from a number of iteration cycles. The intersection points of P_{45} curve with the abscissa are the natural frequencies. There is always a zero frequency present because the system is unconstrained. As a by-product of the solution, we can calculate the mode shapes. The system of equations is quite easy to solve, in the manner described, using a programmable calculator.

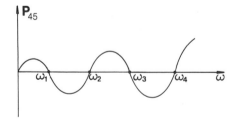

Figure 11.4   Force at free end as function of assumed ω.

11D. Static Deflection Method   is developed by extrapolating a certain SDOF system approach to an MDOF structure. The natural frequency ω of a weight W restrained by a spring k is found from

$$\omega^2 = \frac{gk}{W} = \frac{g}{(W/k)} = \frac{g}{u_{st}} \qquad (11.6)$$

where the quotient W/k is the static deflection u_{st} of the weight W under the force of gravity.

If a static deflection pattern of a particular MDOF system subjected to gravity is known, one can use this formula to calculate an approximate value of the natural frequency. It is assumed, of course, that the vibratory mode is similar to that deflected shape. The static deflection u_{st} should be the absolute value of the maximum deflection.

This method is less accurate than the previous two but often more convenient. Some static analysis usually precedes the dynamic calculations in any engineering project, and the maximum static deflection may be readily available. In computer-aided analysis it is also good practice to perform a static check first. By the use of Eq. 11.6 one may then get a quick estimate of ω, thereby making it easier to plan the scope of the dynamics work.

For very simple problems, such as those typically used in this chapter, the method usually underestimates the frequency. In large systems, however, this may not necessarily be the case, because it is often difficult to decide which deflection component to use as u_{st}.

11E. Calculation of Response by Single-Mode Approximation can be used when that particular mode is sufficient to describe the dynamic behavior of a structure. The development in Sect. 10A can then be repeated, giving the equation of motion

$$\overline{M}\ddot{s} + \overline{k}s = Q(t) \qquad (11.7)$$

If a three-DOF system is used as an example, the modal vector $\vec{\Phi}$ is defined by

$$\vec{\Phi}^T = [x_1 \quad x_2 \quad x_3]$$

and the generalized force Q is

$$Q(t) = x_1 P_1(t) + x_2 P_2(t) + x_3 P_3(t) \qquad (11.8)$$

where $P_n(t)$ is a force applied in the nth direction. The modal mass $\overline{M}$ is found from

$$\overline{M} = \vec{\Phi}^T \mathbf{M} \vec{\Phi} \qquad (11.9)$$

which is a simplified version of Eq. 9.21. When the mass matrix is diagonal, we have

$$\overline{M} = M_1 x_1^2 + M_2 x_2^2 + M_3 x_3^2 \qquad (11.10)$$

for our three-DOF system, while $\overline{k} = \omega^2 \overline{M}$.

Upon solving Eq. 11.7 we obtain the expression for the modal displacement $s(t)$, from which the physical displacements and accelerations are calculated:

$$\begin{Bmatrix} u_1 \\ u_2 \\ u_3 \end{Bmatrix} = \begin{Bmatrix} x_1 \\ x_2 \\ x_3 \end{Bmatrix} s(t) \qquad \begin{Bmatrix} \ddot{u}_1 \\ \ddot{u}_2 \\ \ddot{u}_3 \end{Bmatrix} = \begin{Bmatrix} x_1 \\ x_2 \\ x_3 \end{Bmatrix} \ddot{s}(t) \quad (11.11)$$

When damping is present, Eq. 10.9 may be used to calculate the modal response

$$\ddot{s} + 2\omega\zeta\dot{s} + \omega^2 s = \frac{Q(t)}{\overline{M}} \qquad (11.12)$$

If the forcing function changes harmonically with time, the steady-state solution is obtained by suppressing index r in the equations of Sect. 10F.

11F. Equivalent Oscillator Concept.

In the preceding section the motion of a system was expressed by one of the normal modes. Although it was not explicitly stated, the procedure was equivalent to reducing the system to an SDOF oscillator having modal mass $\overline{M}$ and stiffness $\overline{k}$, and acted on by a force $Q(t)$. Employing the actual mode shape is practical only when the modal characteristics of a structure are known, that is, after a good deal of computation. A slightly different approach is to use an assumed deflected shape, which, although usually less accurate, is more sensible in preliminary computations. The generation and handling of an equivalent oscillator is explained below.

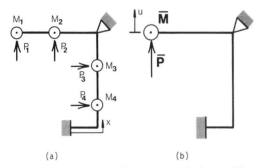

(a)   (b)

Figure 11.5 (a) Original frame; (b) equivalent oscillator.

Figure 11.5a shows a frame with four degrees of freedom, while in Fig. 11.5b the same structure is reduced to an SDOF oscillator. The deflection characterizing the latter device was selected to coincide with u_1. The deflected shape can be described by

$$u(x) = X(x) \cdot f(t) \qquad (11.13)$$

where $X(x)$ is a function of length x (measured along the frame axis) and f depends on time. Velocities are obtained by differentiation:

$$\dot{u}(x) = X(x) \cdot \dot{f}(t) \qquad (11.14)$$

Mass $\overline{M}$ is found from the condition that it have the same kinetic energy as the actual system,

$$\tfrac{1}{2}\overline{M}v^2 = \tfrac{1}{2}M_1 v_1^2 + \tfrac{1}{2}M_2 v_2^2 + \cdots + \tfrac{1}{2}M_n v_n^2 \quad (11.15)$$

where v is the velocity in the reference direction, $v = \dot{u}$. From Eq. 11.14 we have

$$v_r = X(x_r) \cdot \dot{f}(t) = X_r \cdot \dot{f}(t)$$

where X_r is the value of the shape function at $x = x_r$. Dividing both sides of Eq. 11.15 by v^2 and noting that

$$\frac{v_r^2}{v^2} = \frac{X_r^2}{X^2}$$

in which X stands for the value of $X(x)$ at the reference point, we get

$$\overline{M} = M_1 \left(\frac{X_1}{X}\right)^2 + M_2 \left(\frac{X_2}{X}\right)^2 + \cdots + M_n \left(\frac{X_n}{X}\right)^2 \qquad (11.16)$$

(In our example $X = X_1$.) The stiffness $\overline{k}$ of the equivalent oscillator is defined like that for any SDOF system once the reference deflection has been selected. When the parameters $\overline{k}$ and $\overline{M}$ are known, the natural frequency is found from $\omega^2 = \overline{k}/\overline{M}$.

To define the equivalent load $\overline{Q}$ in Fig. 11.5b, we equate its work with that of the original forces in Fig. 11.5a:

$$\overline{Q}X = P_1 X_1 + P_2 X_2 + \cdots + P_n X_n$$

which gives us

$$\overline{Q} = P_1\left(\frac{X_1}{X}\right) + P_2\left(\frac{X_2}{X}\right) + \cdots + P_n\left(\frac{X_n}{X}\right) \qquad (11.17)$$

We have been using discrete forces thus far. If a distributed load $q(x)$ is present, we may convert it to a set of discrete forces, but this is not always necessary. Instead, we can employ the formula

$$\overline{Q}X = \int_L X(x) q \, dx \qquad (11.18)$$

in which the integral on the right represents the work of the distributed load over the entire length L. (The symbol X standing alone has the same meaning as before.) It is only in this last equation that $X(x)$ must be a continuous function of x; otherwise it may be specified as a set of numbers.

When employing some form of a single-mode approximation, we automatically ignore the influence of higher modes. When a calculation is carried out using a more precise method, we can expect a higher level of response, at least in some portions of a structure. To make our simplified computation of response more realistic, we may multiply the results by a factor larger than unity.

11G. Energy Expressions for Continuous Systems.
Various expressions for the strain energy of discrete systems were presented in Sect. 8B. When we deal with a structure for which the internal forces are known, we may express the strain energy in terms of those forces. For a system consisting of beamlike elements with the internal forces designated by N, $\mathfrak{M}$, V, and M_t,

$$\Pi = \int \frac{N^2\,dx}{2EA} + \int \frac{\mathfrak{M}^2\,dx}{2EI} + \int \frac{V^2\,dx}{2GA_s} + \int \frac{M_t^2\,dx}{2GC}$$

(11.19)

The integration is extended over the whole system. If bending and shear take place in two planes, two additional integrals will appear.

The kinetic energy of a beam element of length dx is given by

$$dT = \tfrac{1}{2}(m\,dx)\dot{u}^2$$

where m is the mass per unit length and the velocity $\dot{u}$ is dependent on time and location. Using the deflected shape approximation given by Eqs. 11.13 and 11.14 we obtain

$$T = \tfrac{1}{2}\int m\dot{u}^2\,dx = \tfrac{1}{2}\dot{f}^2 \int mX^2(x)\,dx \quad (11.20)$$

in which the integration is extended over the whole structure. This expression is the counterpart of Eq. 11.15 for a system with lumped masses. In the particular case of a harmonic motion when $f(t) = \sin \omega t$, the maximum value of the kinetic energy is

$$T_{\max} = \frac{\omega^2}{2} \int mX^2(x)\,dx \qquad (11.21)$$

For a structure vibrating in a natural mode, we can equate this expression with the maximum strain energy from Eq. 11.19. This is an adaptation of the

Rayleigh method for structures with a continuously distributed mass. The natural frequency obtained from this procedure is higher than the true value.

SOLVED PROBLEMS

11-1 Find the fundamental frequency of the system described in Prob. 9-4 using Rayleigh's method. Use $E = 10 \times 10^6$ psi, $I = 4$ in.^{4}, $L = 30$ in., $W = 100$ Lb, and the mass moment of inertia at the tip $J = 4.8575$ Lb-s^2-in.

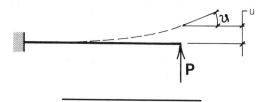

Assume that the deflected shape will be close to that of a cantilever under a tip load $P = 10,000$ Lb. From statics:

$$u = \frac{PL^3}{3EI} = \frac{10,000 \times 30^3}{3 \times 10^7 \times 4} = 2.25 \text{ in.}$$

$$\vartheta = \frac{PL^2}{2EI} = 0.1125 \text{ rad}$$

A form of Eq. 11.2 suitable for this problem is

$$\omega^2 = \frac{Pu + \mathfrak{M}_0\vartheta}{Wu^2/g + J\vartheta^2}$$

The moment applied to the tip, $\mathfrak{M}_0$, is zero by our choice,

$$\omega^2 = \frac{10,000 \times 2.25}{100 \times 2.25^2/386 + 4.8575 \times 0.1125^2}$$

$$\therefore \omega = 128.0 \text{ rad/s} \quad \text{or} \quad f = 20.37 \text{ Hz}$$

The frequency calculated by the exact method in Prob. 9-4 was 20.36 Hz, so the error of approximation is insignificant.

11-2 Calculate the fundamental frequency of the torsional system consisting of four discs. The angular stiffness K and mass moments of inertia J are tabulated below.

Node	$10^{-6}K$ (Lb-in./rad)	J (Lb-s^2-in.)
1		0.68
	35.87	
2		1.015
	71.89	
3		1.015
	82.35	
4		0.623
	0.8277	
5		—

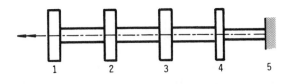

Subject the system to a unit angular acceleration so that at each disc there is a torque applied, equal to a mass moment of inertia of that disc. The calculation of angular deflections is shown in the table.

Node	M_{td} (Lb-in.)	M_t (Lb-in.)	$10^6 \Delta\alpha$ (rad)	$10^6 \alpha$ (rad)
1	0.68			4.1023
		0.68	0.0190	
2	1.015			4.0833
		1.695	0.0236	
3	1.015			4.0597
		2.71	0.0329	
4	0.623			4.0268
		3.333	4.0268	
5	—			0

where M_{td} stands for a torque applied to a particular disc while M_t is a resultant torque for a segment between discs. We have

$$\Delta\alpha = \frac{M_t}{K}$$

where K is taken from the first table. The increments $\Delta\alpha$ are summed going from the fixed end to obtain the angles of rotation α. When Eq. 11.2 is modified to represent torsional vibration, we get

$$\omega^2 = \frac{\sum M_{td}\alpha_d}{\sum J\alpha_d^2}$$

and the summation is extended to all discs. It follows that

$$\omega = 495.7 \text{ rad/s} \quad \text{or} \quad f = 78.89 \text{ Hz}$$

11-3 A cantilever beam of length $L = 26.3$ in. has a rotational restraint at the left end equivalent to $K = 9.234 \times 10^6$ Lb-in./rad. The beam stiffness parameter is $EI = 32.05 \times 10^6$ Lb-in.^{2} The lumped center mass is $M = 0.0108$ Lb-s^2/in. Calculate the fundamental frequency using the Rayleigh method.

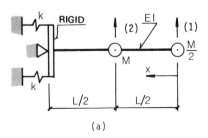

(a)

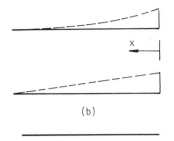

(b)

Let us assume that the tip force P gives a reasonable approximation of the first mode shape. Two deflection components due to beam and base flexibility, respectively, shown in (b), can be expressed as

$$u = \frac{PL^3}{6EI}\left(\frac{x^3}{L^3} - \frac{3x}{L} + 2\right) \quad \text{(beam flexibility)}$$

$$u = \frac{PL^2}{K}\left(1 - \frac{x}{L}\right) \quad \text{(base flexibility)}$$

Since both deflection components are additive, we have, at $x = 0$ and $x = L/2$, respectively,

$$u_1 = \frac{PL^3}{3EI} + \frac{PL^2}{K} \quad \text{and} \quad u_2 = \frac{5}{48}\frac{PL^3}{EI} + \frac{PL^2}{2K}$$

or

$$u_1 = \frac{3.5258PL^2}{K} \quad \text{and} \quad u_2 = \frac{1.2893PL^2}{K}$$

From Eq. 11.2:

$$\omega^2 = \frac{3.5258P + 1.2893 \times 0}{(0.0108/2) \times 3.5258^2 + 0.0108 \times 1.2893^2} \frac{K}{PL^2}$$

$$\therefore \omega = 743.79 \text{ rad/s} \quad \text{or} \quad f = 118.4 \text{ Hz}$$

11-4 Two discs are mounted on a shaft with fixed ends. Find the fundamental frequency of torsional vibrations using Dunkerley's method; J is the polar moment of inertia of each disc and K is the angular stiffness of each shaft segment.

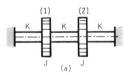

(a)

When the shaft is loaded through a disc as in (b), it may be treated as a combination of two angular springs working in parallel, one on each side of node 1. Their total stiffness is

$$K^* = K + 0.5K = 1.5K$$

in which $0.5K$ is the effective spring constant of the two segments effectively joined in series. The flexibility coefficients are

$$a_{11} = a_{22} = \frac{1}{1.5K}$$

The fictitious frequencies:

$$\frac{1}{\omega_{11}^2} = \frac{1}{\omega_{22}^2} = \frac{J}{1.5K}$$

Therefore

$$\frac{1}{\omega^2} = \frac{1}{\omega_{11}^2} + \frac{1}{\omega_{22}^2} = \frac{2J}{1.5K}$$

$$\therefore \omega = 0.866 \left(\frac{K}{J}\right)^{1/2}$$

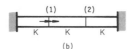

(b)

11-5 Dunkerley's method may be used to approximate a new natural frequency resulting from modifications in the inertial properties of a system. Suppose that in Prob. 9-20 the end mass is first increased from $2M$ to $3M$, then decreased to M. How will these changes affect the fundamental frequency?

The fundamental frequency of the original system is determined by

$$\omega_1^2 = \frac{0.11927k}{M}$$

The modified system with $3M$ at the end may be treated as consisting of the original system plus the additional mass M. When using Eq. 11.3, we can put $\omega_{11}^2 = \omega_1^2$ and then calculate ω_{22} as if M were attached at the end of three springs in the row. The effective stiffness in the latter case is

$$\frac{1}{k^*} = \frac{1}{k} + \frac{1}{2k} + \frac{1}{4k} = \frac{7}{4k}$$

and

$$\frac{1}{\omega_{22}^2} = \frac{M}{k^*} = \frac{7M}{4k}$$

The modified natural frequency ω_1' is

$$\frac{1}{(\omega_1')^2} = \frac{M}{0.11927k} + \frac{7M}{4k} \qquad \therefore \omega_1' = 0.3141 \left(\frac{k}{M}\right)^{1/2}$$

In the second case the modified frequency ω_1'' is found by assuming

$$\frac{1}{\omega_1^2} = \frac{1}{(\omega_1'')^2} + \frac{M}{k^*}$$

That is, we look at the original system as consisting of two—the first having three masses (M, $3M$, and M) and the second having only M at the end. Thus,

$$\frac{1}{(\omega_1'')^2} = \frac{M}{0.11927k} - \frac{7M}{4k} \qquad \therefore \omega_1'' = 0.3883 \left(\frac{k}{M}\right)^{1/2}$$

To summarize, we have

$$f_1 = 2.4581 \qquad f_1' = 2.2356 \qquad f_1'' = 2.7638 \text{ Hz}$$

11-6 Find the first frequency of a cantilever whose distributed mass m (Lb-s^2/in.^{2}) is lumped as shown in the figure. Use the static deflection method and take advantage of the flexibility matrix derived in Prob. 8-22.

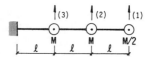

The deflections are found from the matrix equation $\vec{u} = a\vec{P}$

$$\begin{Bmatrix} u_1 \\ u_2 \\ u_3 \end{Bmatrix} = \frac{L^3}{EI} \begin{bmatrix} 0.33333 & 0.17284 & 0.04938 \\ 0.17284 & 0.09877 & 0.03086 \\ 0.04938 & 0.03086 & 0.01235 \end{bmatrix} \begin{Bmatrix} \dfrac{Mg}{2} \\ Mg \\ Mg \end{Bmatrix}$$

The deflections induced by the vector of gravity forces are

$$\begin{Bmatrix} u_1 \\ u_2 \\ u_3 \end{Bmatrix} = \begin{Bmatrix} 0.3889 \\ 0.2161 \\ 0.0679 \end{Bmatrix} \frac{MgL^3}{EI}$$

where u_1 is the desired value of u_{st} to be used in Eq. 11.6:

$$\omega^2 = \frac{g}{u_{st}} = \frac{EI}{0.3889ML^3}$$

But $M = mL/3$, so

$$\omega^2 = \frac{3EI}{0.3889mL^4} \qquad \text{and} \qquad \omega = 2.777 \left(\frac{EI}{mL^4}\right)^{1/2}$$

An exact method for a prismatic cantilever yields 3.516 as the value of the coefficient, which makes our error equal to 21%. Notice that this is the result of two separate inaccuracies:

1. Replacing a continuous cantilever by a lumped-mass model.
2. Performing an approximate calculation on that model.

11-7 A simply supported beam is subjected to its own gravity loading mg. Based on the static deflection method, estimate the natural frequency and compare with the actual value of

$$\omega = \frac{\pi^2}{L^2} \left(\frac{EI}{m}\right)^{1/2}$$

where m is the mass per unit length.

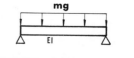

The center deflection, according to Ref. 7 is

$$u_{st} = \frac{5}{384} \frac{(mg)L^4}{EI}$$

From Eq. 11.6:

$$\omega^2 = \frac{g}{u_{st}} = \frac{384}{5} \frac{EI}{mL^4} \qquad \therefore \omega = \frac{8.7636}{L^2} \left(\frac{EI}{m} \right)^{1/2}$$

This is 11.2% less than the true value.

11-8   Generate an equivalent oscillator representing the fundamental mode of a three-story building in Prob. 9-9.

Assume the first mode to be reasonably well represented by a force P applied at the coordinate u_3 (i.e., at the top of the building). The deflections at the successive levels are:

$$u_1 = \frac{P}{1.6k} \qquad u_2 = u_1 + \frac{P}{1.4k} = 1.3393 \frac{P}{k}$$

$$u_3 = u_2 + \frac{P}{k} = 2.3393 \frac{P}{k}$$

Setting $u_1 = 1$ we get our assumed deflected shape:

$$\vec{X}^T = [1.0 \quad 2.1429 \quad 3.7429]$$

The equivalent weight is calculated from Eq. 11.16 by setting $X_3 \equiv X$ and using the weight data from the reference problem,

$$\overline{W} = W \left(\frac{1}{3.7429} \right)^2 + 0.67 W \left(\frac{2.1429}{3.7429} \right)^2 + W$$

$$= 1.291 W = 2.582 \times 10^6 \text{ Lb}$$

The stiffness along the reference direction:

$$\bar{k} = \frac{P}{u_3} = \frac{k}{2.3393} = 1.71 \times 10^6 \text{ Lb/in.}$$

The natural frequency of the oscillator:

$$\omega^2 = \frac{g\bar{k}}{W} \qquad \therefore \omega = 15.989 \text{ rad/s}$$

In spite of a fairly crude approximation, this is less than 3% above the exact solution in the reference problem.

11-9   The building analyzed in Prob. 9-9 is subjected to a pressure shock, which applies a uniformly distributed load along a wall, $q(t)$. The magnitude of this load grows instantaneously to $q_0 = 65$ Lb/in. and then decays exponentially according to $q = q_0 e^{-6.93t}$ (t in seconds). Assuming the single-mode response and using the equivalent oscillator developed in Prob. 11-8, calculate the maximum shear force at the base. The weights (Lb) lumped at the floors are $W_1 = 2 \times 10^6$, $W_2 = 1.34 \times 10^6$, and $W_3 = 2 \times 10^6$.

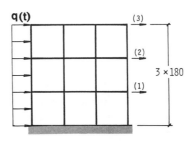

According to Prob. 11-8, the deflected shape is described by

$$\vec{X}^T = [1.0 \quad 2.1429 \quad 3.7429]$$

We cannot use Eq. 11.18 to determine the equivalent load $\overline{Q}$, because $X(x)$ is not given between the nodal points. We must first convert the loading to a set of concentrated forces based on the tributary length concept,

$$P_1 = P_2 = 180 \times 65 = 11,700 \text{ Lb}$$

$$P_3 = 90 \times 65 = 5850 \text{ Lb}$$

Since X_3 is the reference deflection, Eq. 11.17 yields

$$\overline{Q}_0 = 11,700 \left(\frac{1}{3.7429} \right) + 11,700 \left(\frac{2.1429}{3.7429} \right) + 5850 = 15,674 \text{ Lb}$$

This is the peak value of the equivalent load applied to the oscillator having

$$\overline{W} = 2.582 \times 10^6 \text{ Lb} \qquad \bar{k} = 1.71 \times 10^6 \text{ Lb/in.} \qquad \omega = 15.989 \text{ rad/s}$$

in accordance with Prob. 11-8. The solution for an oscillator acted on by the force varying according to

$$Q(t) = Q_0 e^{-at}$$

can be found in the answer to Prob. 2-41:

$$u(t) = \frac{Q_0}{M} \frac{1}{a^2 + \omega^2} \left(e^{-at} - \cos \omega t + \frac{a}{\omega} \sin \omega t \right)$$

After substituting the data, we find by numerical means that the function

$$u(t) = 7.7162 \times 10^{-3} (e^{-at} - \cos \omega t + 0.4334 \sin \omega t)$$

has a maximum 10.836×10^{-3} in. at $t = 0.1625$ s. From the definition of the deflected shape, we find the remaining two displacement components. In summary:

$$u_1 = 2.8951 \times 10^{-3} \quad u_2 = 6.2039 \times 10^{-3} \quad u_3 = 10.836 \times 10^{-3} \text{ in.}$$

To find the corresponding accelerations, one must differentiate the expression for $u(t)$ derived before. To make our task easier, we note that the exponential term is relatively small at $t = 0.1625$; therefore it appears reasonable to assume $a_{max}^2 = \omega^4 u_{max}^2$. The inertia forces are

$$R_1 = \frac{W_1}{g} \omega^2 u_1 = 3835 \text{ Lb} \qquad R_2 = 5506 \text{ Lb} \qquad R_3 = 14,353 \text{ Lb}$$

The base shear:

$$V = R_1 + R_2 + R_3 = 23,694 \text{ Lb}$$

11-10 Construct an equivalent oscillator for a portal frame that is to be loaded in the direction normal to its plane. The cross section is tubular, $A = 2.733$ in.^{2} and $I = 14.4$ in.^{4} The material is steel, $E = 29 \times 10^6$ psi, $\gamma = 0.284$ Lb/in.^{3}, $L = 40$ in.

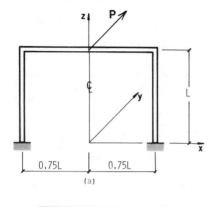

(a)

Looking down at the frame, vertically, we see the top beam as elastically restrained against translation and rotation in (b). The translational stiffness is associated with bending of the columns, each acting as a cantilever,

$$k = \frac{3EI}{L^3}$$

The angular restraints are caused by the torsional stiffness of the columns. Instead of computing K precisely and solving a redundant problem to find the center deflection, we will take a quicker and less accurate route. Suppressing the translational flexibility, we assume that the deflection due to P is an average of fixed-end and simply supported condition:

$$u_r \approx \frac{1}{2}\left(\frac{1}{48} + \frac{1}{192}\right)\frac{P(1.5L)^3}{EI} = \frac{1}{22.76}\frac{PL^3}{EI}$$

Denoting by u_c the deflection due to cantilever action of columns, we have

$$u_c = \frac{P}{2k} = \frac{PL^3}{6EI}$$

and

$$u = u_r + u_c = 0.2106\frac{PL^3}{EI}$$

The oscillator stiffness is therefore

$$\bar{k} = \frac{P}{u} = \frac{EI}{0.2106L^3} = 30,983 \text{ Lb/in.}$$

The mass per unit length of the frame is $m = \gamma A/g = 2.011 \times 10^{-3}$ Lb-s^2/in.^{2} The lumped masses will be placed at three locations. At

the center of beam we put

$$M_1 = \frac{1}{2} \times 1.5Lm = 60.33 \times 10^{-3} \text{ Lb-s}^2/\text{in.}$$

while at each corner we place

$$M_2 = \frac{1}{4} \times 1.5Lm + \frac{1}{2}Lm = 0.875Lm = 70.39 \times 10^{-3}$$

The previous displacement estimates tell us that if M_1 (at the reference point) deflects by 0.2106, the corner mass M_2 is displaced by $\frac{1}{6}$. The effective mass is found from Eq. 11.16,

$$\overline{M} = M_1 + 2M_2\frac{(1/6)^2}{0.2106^2} = 0.1485 \text{ Lb-s}^2/\text{in.}$$

The fundamental frequency of the out-of-plane motion may be estimated from

$$\omega = \left(\frac{30,983}{0.1485}\right)^{1/2} = 456.8 \text{ rad/s}$$

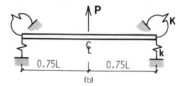

(b)

11-11 The load P applied to the portal frame in Prob. 11-10 is given by

$$P = P_0 \sin \Omega t \qquad \text{for} \quad \Omega t_0 \leqslant \pi$$

$$P = 0 \qquad \text{for} \quad \Omega t_0 > \pi$$

Determine the maximum bending moment at the point of load application and the maximum torque in the column if $P_0 = 1200$ Lb and $t_0 = 0.005$ s.

The natural period of the structure is

$$\tau = \frac{2\pi}{456.8} = 13.755 \times 10^{-3} \text{ s}$$

The duration of this sinusoidal pulse is only about $\tau/3$, therefore the maximum response is to be expected for $t > t_0$. From Prob. 2-33 we find the displacement amplitude of a simple oscillator due to a pulse of this type

$$u_{\max} = \frac{2(\Omega/\omega)}{|1 - \Omega^2/\omega^2|}\cos\left(\frac{\pi\omega}{2\Omega}\right)u_{\text{st}}$$

For $\Omega = \pi/0.005 = 628.3$ rad/s and $\omega = 456.8$ rad/s, we have

$$u_{\max} = 1.282 u_{\text{st}}$$

Once the dynamic effect of the load is known, we essentially deal with a static problem, except that we multiply P by the magnification factor equal to 1.282. Proceeding as in Prob. 11-10, we estimate the bending moment at the beam center to be

$$\mathcal{M}_{max} \approx \frac{1}{2}\left(\frac{1}{4} + \frac{1}{8}\right)1.282P(1.5L) = 17{,}307 \text{ Lb-in.}$$

which is the average of simply supported and fixed-end conditions. The figure shows one-half of the frame loaded with a force

$$\frac{1.282P}{2} = 769.2 \text{ Lb}$$

The torque in the column is

$$M_t = 769.2 \times 30 - 17{,}307 = 5769 \text{ Lb-in.}$$

Before the internal dynamic forces so derived are used for a strength check, they should be multiplied by some factor larger than unity, say 1.2. This is to account for an effect of higher modes, not included in the calculation.

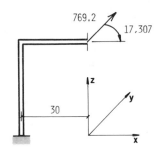

11-12 The system is identical with that of Prob. 9-17. Find the two nonzero frequencies by the Holzer method. Put $k = 100$ Lb/in. and $M = 1.0$ Lb-s^2/in.

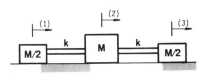

The equations of dynamic equilibrium, 11.4 and 11.5, become here

$$P_{12} = -0.5\omega^2$$

$$u_2 = 1 + \frac{P_{12}}{100}$$

$$P_{23} = P_{12} - \omega^2 u_2$$

$$u_3 = u_2 + \frac{P_{23}}{100}$$

$$P_{34} = P_{23} - 0.5\omega^2 u_3$$

upon substituting the figures and setting $u_1 = 1$. The force P_{34} is a reaction on mass M_3 from a nonexistent link 3-4 and therefore it is zero. If the assumed value of ω^2, however, does not coincide with its true value, P_{34} will be nonzero. Let us choose $\omega_1 = 10$ as the trial value and calculate displacements and forces:

$$P_{12} = -50 \quad u_2 = 0.5 \quad P_{23} = -100 \quad u_3 = -0.5 \quad P_{34} = -75$$

In the next trial $\omega_1 = 15$ gives us $P_{34} = 24.6$, which indicates that ω_1 is somewhere between the two values. Several more trials and we obtain

$$\omega_1 = 14.14 \text{ rad/s}$$

After repeating the search, we find $\omega_2 = 20$ rad/s. These answers agree with the reference problem. One may also easily check that the mode shapes calculated from the equations developed here are the same as those found previously.

EXERCISES

11-13 Find the fundamental frequency of the three-mass system using Rayleigh's method; $k = 1000$ Lb/in., $M = 0.5$ Lb-s^2/in. Force P is assumed to induce the first mode.

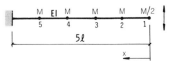

11-14 Find the first natural frequency of a uniform cantilever beam with length $L = 5\ell$ in. and bending stiffness EI. The distributed mass m(Lb-s^2/in.^{2}) is lumped as shown in the figure. Use the deflected shape due to the static tip load.

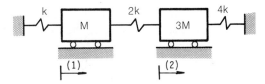

11-15 Using the Rayleigh method calculate the fundamental frequency of the two-mass system. Assume the corresponding mode shape to be induced by a constant acceleration. Put $M = 0.5$ Lb-s^2/in. and $k = 1000$ Lb/in. The flexibility matrix may be found in Prob. 8-2.

11-16 In the outline of a small turbine with a gear at one end and two wheels at the other shown in (a), the rotating part is supported by two flexible bearings nested in the rigid, stationary shell (hatched section). The inertial properties of the rotating assembly are listed below in accordance with the node numbers marked in (a). J' designates mass moment of inertia about the axis normal to the paper. The lumped mass model appears in (b).

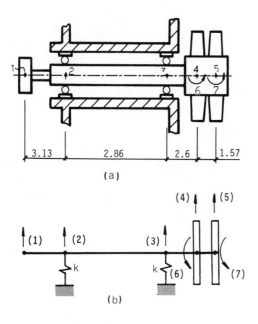

(a)

(b)

$M_1 = 1.00 \times 10^{-3}$ Lb-s²/in.

$M_2 = 3.73 \times 10^{-3}$ Lb-s²/in. $J_6' = 9.63 \times 10^{-3}$ Lb-s²-in.

$M_3 = 3.94 \times 10^{-3}$ Lb-s²/in. $J_7' = 9.04 \times 10^{-3}$ Lb-s²-in.

$M_4 = 10.76 \times 10^{-3}$ Lb-s²/in.

$M_5 = 8.97 \times 10^{-3}$ Lb-s²/in.

The effective I values for the shaft segments are shown below; $E = 28 \times 10^6$ psi. Each bearing stiffness is $k = 28{,}670$ Lb/in.

Segment:	1-2	2-3	3-4	4-5
I (in.⁴):	0.03	0.10	0.238	1.272

Using Rayleigh's method, find the fundamental frequency of the assembly. Assume the mode shape to be induced by gravity.

11-17   Using the Rayleigh method, determine the first natural frequency of a tapered cantilever with a continuously distributed mass. The ratio $H_1/H_3 = 5$ and the width normal to the paper is $B = $ constant. The material properties are E and ρ. Write the frequency expression in terms of the base section properties A_1 and I_1. *Hint.* Assume the deflected shape to be

$$X(x) = 1 - \cos \frac{\pi x}{2L}$$

and use Simpson's formula to integrate the energy expressions.

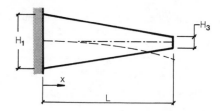

11-18   Find the fundamental frequency of a three-mass system from Prob. 9-20 by Dunkerley's method.

11-19   Calculate the first frequency of a shaft with four discs from Prob. 11-2 using Dunkerley's method.

11-20   Six discs are mounted on a shaft with fixed ends. Find the fundamental frequency of torsional vibration using Dunkerley's method. The polar moment of inertia of each disc is J, and K is the angular stiffness of each shaft segment.

11-21   Find the fundamental frequency of the system described in Prob. 9-4 using Dunkerley's method. Put $E = 10 \times 10^6$ psi, $I = 4$ in.⁴, $L = 30$ in., $W = 100$ Lb, and the mass moment of inertia at the tip $J = 4.8575$ Lb-s²-in.

11-22   Using Dunkerley's method, calculate the natural frequency of bending vibrations of the two-disc rotor. The shaft is weightless, while the masses and the diametral moments of inertia, respectively, are

$$M_1 = 0.8 \text{ Lb-s}^2/\text{in.} \qquad M_2 = 2.0 \text{ Lb-s}^2/\text{in.}$$

$$J_3' = 18.5 \text{ Lb-s}^2\text{-in.} \qquad J_4' = 82.0 \text{ Lb-s}^2\text{-in.}$$

The flexibility matrix for the set of coordinates 1 and 2 is

$$\mathbf{a} = \begin{bmatrix} 1.2585 & 0.9510 \\ 0.9510 & 0.9045 \end{bmatrix} 10^{-6}$$

as found in Prob. 8-15 using $L = 60$ in., $I_1 = 100$ in.⁴, $I_2 = 125$ in.⁴, and $E = 29 \times 10^6$ psi. Make a gross estimate of the angular flexibility terms.

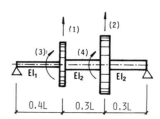

11-23 Calculate the natural frequency of the system in Prob. 11-16 by the static deflection method.

11-24 A ring is loaded with its own weight mg (per unit length of circumference) and is held at the lower point of the vertical diameter. From Ref. 7 we can find that the load causes shrinking of the vertical diameter by

$$0.4674 \frac{(mg)R^4}{EI}$$

Using the static deflection method, estimate the natural frequency.

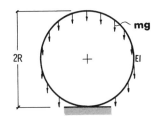

11-25 Find the formula for the first frequency of the tapered cantilever from Prob. 11-17 using the static deflection method. Compare the results with Prob. 11-17.

11-26 The shaft is the same as in Prob. 9-32. Calculate the two nonzero frequencies using the Holzer method. Put $K=1000$ Lb-in./rad and $J=10$ Lb-s^2-in.

11-27 Determine the fundamental frequency and mode shape using the Holzer method. Put $k=8$ Lb/in. and $M=0.5$ Lb-s^2/in.

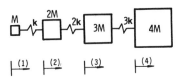

11-28 Calculate the response of the two-mass system in Prob. 11-15 to force $P_2=P_0\cos\Omega t$, using a single-mode approximation, setting $\Omega=0.9\omega$ and $\zeta=0.02$. Compare the result with the upper bound of solution in Prob. 10-3.

11-29 Consider again the dolly structure in Prob. 9-10. When the lateral force is applied to the CG in Fig. 9-10c, both the rotational and the translational springs deflect. Using the idea of combining the fictitious partial frequencies from Prob. 7-1, calculate the approximate first frequency of the system.

11-30 After the design of the dolly structure in Prob. 9-10 was completed, it was determined that the analysis was inadequate, because the flexibility of rails on which the dolly stands was not accounted for. What would be the change in the calculated fundamental frequency $\omega_1=76.2$ rad/s if the spring constant of rails were $k_r=7.6\times10^6$ Lb/in.? *Hint.* Use an equation developed in Prob. 7-1, treating ω_1 as a fictitious frequency rather than as an actual frequency.

CHAPTER

12

Response of Structure to Motion of Its Base

This chapter may be regarded as an extension of Chapter 5 to MDOF systems. In the direct method the usual first step is to replace the effect of base motion with the set of externally applied joint forces. A *response spectrum* approach is used when that motion is very irregular and complicated. Seismic ground motion, which falls into the latter category, is also analyzed in this chapter by some less sophisticated methods, such as one that is presented in the Uniform Building Code.

12A. Equations of Motion for Kinematic Forcing.
To set up equations of motion for kinematic forcing, we use reasoning similar to that in Sect. 5B, where a simple mass-spring system was investigated. Consider a frame with a few lumped masses in Fig. 12.1. The rectangular blocks represent the base elements, which are capable of deflecting and imposing forces on the frame via the base springs k_{b1}, k_{b2}, and k_{b3}.

When the base blocks rest undeflected at the reference level, the matrix equation of free vibrations of the undamped system can be written as

$$\mathbf{M\ddot{\vec{u}} + k\vec{u} = 0}$$

If the equation pertaining to the coordinate u_2 is isolated from this array, we notice it contains, among

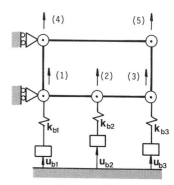

Figure 12.1 Displacements of base points.

others, the following terms:

$$\cdots + M_{22}\ddot{u}_2 + \cdots = \cdots - k_{b2}u_2 - \cdots$$

(k_{b2} is a component of the diagonal term k_{22}. Refer to Sect. 9A.) When the base points are moving, the equation is modified,

$$\cdots + M_{22}\ddot{u}_2 + \cdots = \cdots + k_{b2}(u_{b2} - u_2) - \cdots$$

This shows us that the displacement u_{b2} of the lower end of the base spring adds the term $k_{b2}u_2$ on the right side of the equation. There is a similar influence

146

at the other two locations, which modifies our matrix equation to the following:

$$\mathbf{M\ddot{\vec{u}}} + \mathbf{k\vec{u}} = \vec{\mathbf{P}}_{\mathbf{b}}(\mathbf{t}) = \begin{Bmatrix} k_{b1}u_{b1} \\ k_{b2}u_{b2} \\ k_{b3}u_{b3} \\ 0 \\ 0 \end{Bmatrix} \qquad (12.1)$$

The last two entries of the forcing vector are zero because the appropriate masses are not directly connected to the base. The displacements u_{bi} are some prescribed functions of time. When treated in this manner, the equation of motion of a structure that is excited through its base is a special case of Eq. 10.1.

When the base point motions are harmonic functions of time, all of them proportional to $\sin \Omega t$, our forcing may be written as Eq. 10.15 and the response amplitudes calculated in terms of physical displacements from Eq. 10.21.

12B. Concept of Response Spectrum. The system in Fig. 12.2a consists of a base with a row of SDOF oscillators, each with a different natural frequency ω_r. When the base performs its motion, each oscillator will respond differently.

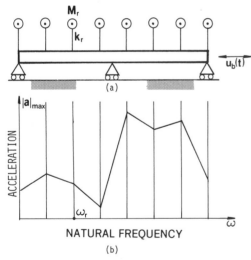

Figure 12.2 Generation of response spectrum: (a) test oscillators; (b) plot of peak responses.

Suppose that we record the peak acceleration $|a|_{max}$ attained during the test by every oscillator and plot this quantity in Fig. 12.2b. This figure is referred to as the *acceleration response spectrum*. When the motion of the base is random, that is, quite irregular

(e.g., during a seismic event), the response spectrum becomes a useful tool in the analysis of structures that are to be subjected to that particular loading. Although the damping elements are not explicitly shown in Fig. 12.2, they are assumed to exist and to represent some definite level of damping. Two identical time histories of base movement [i.e., two identical functions $u_b(t)$] will give different response spectra if used with different damping levels.

Note that a response spectrum gives us only the amplitudes and that phase relationships with respect to the base motion are lost.

12C. Application of Response Spectrum. In structural analysis this is based on the idea that any MDOF system may be treated as an assembly of SDOF oscillators. One the natural frequencies and the mode shapes have been computed, the response in each mode can be found from a prescribed spectrum. To be more specific, let us consider a simple oscillator whose base is undergoing an acceleration $\ddot{u}_b(t)$. As was shown in Sect. 5C, we can find the displacement $\bar{u}(t)$ of mass M with respect to the base if we treat the base as stationary and apply the force $-M\ddot{u}_b$ to the mass. The equation of motion relative to the base may be written as

$$\ddot{\bar{u}} + 2\omega\zeta\dot{\bar{u}} + \omega^2\bar{u} = -\ddot{u}_b$$

from which $\bar{u}$ can be found when $\ddot{u}_b$ is given. As shown in Prob. 12-6, the motion of an MDOF system vibrating in the rth mode is

$$\ddot{s}_r + 2\omega_r\zeta_r\dot{s}_r + \omega_r^2 s_r = -\Gamma_r\ddot{u}_b$$

The only meaningful difference between the two systems is the *participation factor* Γ_r, defined by

$$\Gamma_r = \frac{\vec{\Phi}_r^T \mathbf{M}\vec{\mathbf{j}}}{\overline{M}_r} \qquad (12.2)$$

in which $\vec{\Phi}_r^T$ is the transpose of the rth mode shape, $\overline{M}_r$ is the modal mass, and $\vec{\mathbf{j}}$ is the vector of directional coefficients showing how the masses are affected by the base movement. (The product $\mathbf{M}\vec{\mathbf{j}}$ results in the vector of inertia forces.) To reconcile this development with the general formulation given in Sect. 10C, the reader may compare the two equations above with Eq. 10.9.

Suppose now that the response spectrum representing the effect of base motion shows there is a relative acceleration response Z_{ar} for a given frequency ω_r. This means that the magnitude of the displacement response is Z_{ar}/ω_r^2 for a simple oscillator, while for our rth mode this quantity must be multiplied by Γ_r. The physical displacement components (relative to the base) are calculated from Eq. 9.14a written for an individual mode:

$$\vec{\mathbf{u}}_r = \frac{Z_{ar}}{\omega_r^2}\Gamma_r\vec{\Phi}_r \qquad (12.3)$$

To find the modal forces inducing this deformed pattern, we use Eq. 9.2:

$$\vec{\mathbf{P}}_r = \mathbf{k}\vec{\mathbf{u}}_r \qquad (12.4)$$

The internal forces applied by the rth mode may now be computed from static equilibrium. One may notice that when all degrees of freedom are of translational type and when they are parallel to the direction of base movement, each of the entries in the vector $\vec{\mathbf{j}}$ is a unity. The expression for a participation factor simplifies in this case to

$$\Gamma_r = \frac{M_1 x_1 + M_2 x_2 + \cdots + M_n x_n}{M_1 x_1^2 + M_2 x_2^2 + \cdots + M_n x_n^2} \qquad (12.5)$$

provided the mass matrix is diagonal. (Refer to Eq. 9.21a.) The numbers $x_1, x_2, \ldots, x_n$ are the entries of the rth mode.

A nodal force (or joint force) is a product of mass and the absolute acceleration. It is therefore simpler to obtain those forces directly than to go through the displacements and the stiffness matrix. Unfortunately, the development of theory based on the absolute acceleration is also more difficult, as one may guess from reading Sect. 5C. The following approximation may be used to determine the acceleration of the rth vibratory mode:

$$\ddot{s}_r = \Gamma_r Z_{ar}$$

The physical acceleration components can then be calculated from Eq. 9.14c,

$$\ddot{\vec{\mathbf{u}}}_r = \Gamma_r Z_{ar}\vec{\Phi}_r \qquad (12.6)$$

Multiplying these by the appropriate masses gives us a set of forces corresponding to the rth mode. The response spectrum Z_{ar} is typically given in terms of the absolute acceleration. For small damping, however, there is not much difference between the absolute and the relative acceleration in the rth mode and the distinction between the two has little practical value.

As mentioned in the previous section, no phase angle can be obtained from the response spectrum and all the modal responses are in terms of the maximum absolute values. A number of methods exists for combining those modal responses into resultants. Two of those will be presented here and, to make their description more tangible, let us look for the maximum value of bending moment $\mathfrak{M}_{max}$ at some point of a structure having n natural modes of vibration. The *sum-of-absolute-values method* gives us

$$\mathfrak{M}_{max} = |\mathfrak{M}_1| + |\mathfrak{M}_2| + \cdots + |\mathfrak{M}_n| \qquad (12.7)$$

in which the terms on the right are the bending moments computed for the individual modes. We see that $\mathfrak{M}_{max}$ calculated in this manner is the true upper bound of the response as defined in Sect. 10G, and that makes the method too conservative for most applications. A more popular approach is the *root-sum-square (RSS) method*, according to which

$$\mathfrak{M}_{max} = \left(\mathfrak{M}_1^2 + \mathfrak{M}_2^2 + \cdots + \mathfrak{M}_n^2\right)^{1/2} \qquad (12.8)$$

This method of calculating resultants is typically more realistic than the previous one, although it may be unconservative in some situations. For example, a combination of two vibratory components may approach an absolute sum if the natural frequencies involved are close enough. (That may be true even if the total time of a dynamic event is not too long.) In most cases the true results are somewhere between those given by Eqs. 12.7 and 12.8.

Spectrum response methods are widely used in the area of seismic analysis, which attempts to determine the structure response caused by earthquakes.

12D. Seismic Analysis per Building Code. The response spectrum method discussed in the preceding section is a popular approach to the earthquake response problem, but it is not the only one. Another frequently used procedure is to follow a simplified method of a building code for a particular geographic location. The Uniform Building Code, Ref. 1, will serve as a good example.

Figure 12.3 depicts a rigid stick with lumped masses, pivoting about a fixed point. If an angular acceleration ε is applied, the inertia force at h_r from the pivot point is $(W_r/g)\varepsilon h_r$. The total intertia shear force above the pivot point is

$$V' = \frac{\varepsilon}{g}\sum_r W_r h_r$$

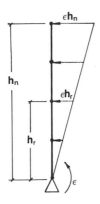

Figure 12.3 Kinematics of pivoting stick.

where summation is extended over all lumped masses. If V' is known, the angular acceleration can be computed from

$$\frac{\varepsilon}{g} = \frac{V'}{\sum\limits_{r} W_r h_r} \qquad (12.9)$$

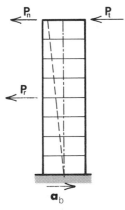

Figure 12.4 Inertia forces applied to a building.

The reason for discussing the kinetics of a rigid stick is that it serves as a model for calculation of seismic loads applied to a tall building. On the basis of observation, it is known that an axis of such a building remains nearly straight, while rotating about the base point, as shown in Fig. 12.4. At any floor level h_r at which a weight is lumped, there is an inertia force P_r associated with that fundamental mode. In addition, there is a force P_t, applied at the top of the building, which reflects the influence of higher modes. The sum of all the inertia forces is equal to the maximum shear V at the base of building. The amplitude of this force is given by the code

as

$$V = \tilde{Z} \tilde{K} \tilde{C} W \qquad (12.10)$$

where $\tilde{Z}$ is the seismic zone factor; $\tilde{Z} = 1.0$ for the areas known to exhibit the strongest earthquakes in the continental United States, and $\tilde{K}$ depends on how big a deformation a structure can withstand prior to failure; $\tilde{K} = 1.33$ for buildings in which the inertial load is resisted by shear walls and $\tilde{K} = 0.67$ when the load is resisted by bending of columns. The value of $\tilde{C}$ is calculated from

$$\tilde{C} = \frac{0.05}{T^{1/3}} \qquad (12.11)$$

where T is the fundamental period of the structure (in seconds). The upper limit of $\tilde{C}$ is 0.1 and this value should also be used when T is not known. The weight of the entire structure is W.

The next step after calculating V is find the inertia force at the top,

$$P_t = 0.004 V \left(\frac{h_n}{D}\right)^2 \qquad (12.12)$$

in which h_n is the height of building and D is its horizontal dimension at the base of structure. Since P_t is associated with the higher modes, the base shear of the "rigid-stick" mode is

$$V' = V - P_t \qquad (12.13)$$

The force V' is distributed along the height as discussed previously. The inertia force at the rth level is

$$P_r = \frac{\varepsilon}{g} W_r h_r \qquad (12.14)$$

where the coefficient ε/g is found from Eq. 12.9. After all the external inertia forces have been calculated, one can proceed to the internal forces and stress levels in members.

There are two main reasons for an axis of a tall building to appear more like a rigid stick than a flexural cantilever when a horizontal acceleration is applied. The first is that the ground acts like a rotational spring giving the axis a slope at the base. The second, which is especially pronounced in framed structures, is the presence of shear deformation.

12E. Approximate Methods for Base Motion. The response spectrum method described in Sect. 12C requires a good deal of labor and may often be approximated by using a single-mode response or an equivalent oscillator concept as described in Chapter 11. Omitting higher modes is justifiable as long as the fundamental mode predominates. To compensate for this omission, we may multiply our single-mode results by some factor larger than unity. In my experience, the factor of 1.25 is sufficient in most situations.

If an acceleration response spectrum is prescribed, but no dynamic properties of a structure are available, one can make a crude estimate by applying the peak spectrum acceleration to the masses of the structure. The inertia forces obtained in this manner are treated as any other static load. This is sometimes called a *uniform acceleration method*.

SOLVED PROBLEMS

12-1 The ground support at the right end of the structure is suddenly moved by the distance δ to the right. Calculate the undamped displacement response using the modal matrix developed in Prob. 9-3,

$$\mathbf{\Phi} = \begin{bmatrix} 1.0 & 1.0 \\ 0.8792 & -0.3792 \end{bmatrix}$$

The modal spring constants (Prob. 10-3) are

$$\bar{k}_1 = 4.1212k \quad \text{and} \quad \bar{k}_2 = 5.3798k$$

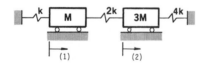

The effect of the movement of the base point is the application of the force $4k\delta$ to mass M_2. Noting that the direction is positive, our vector of base forces in Eq. 12.1 becomes

$$\vec{P}_b = \begin{Bmatrix} 0 \\ 4k\delta \end{Bmatrix}$$

with the understanding that the displacement δ is applied suddenly, as so-called step loading. We can now calculate the modal forces

$$Q_1 = 0.8792 \times 4k\delta = 3.5168k\delta$$

$$Q_2 = -0.3792 \times 4k\delta = -1.5168k\delta$$

The equations of motion in normal coordinates, after Eq. 10.3b, are

$$\overline{M}_1 \ddot{s}_1 + \bar{k}_1 s_1 = 3.5168k\delta$$

$$\overline{M}_2 \ddot{s}_2 + \bar{k}_2 s_2 = -1.5168k\delta$$

If we set $3.5168k\delta = P_0$ in the first equation we can, upon changing notation, use the results of Prob. 2-12:

$$s_1 = \frac{3.5168k\delta}{\bar{k}_1} (1 - \cos \omega_1 t)$$

and similarly for the second equation,

$$s_2 = \frac{-1.5168k\delta}{\bar{k}_2} (1 - \cos \omega_2 t)$$

Employing Eq. 9.14a we revert to physical coordinates,

$$\begin{Bmatrix} u_1 \\ u_2 \end{Bmatrix} = \begin{Bmatrix} 0.5714 - 0.8533 \cos \omega_1 t + 0.2819 \cos \omega_2 t \\ 0.8571 - 0.7502 \cos \omega_1 t - 0.1069 \cos \omega_2 t \end{Bmatrix} \delta$$

The first, constant term in each row is the translation taking place when δ is applied in a static manner.

12-2 The system is the same as in Prob. 9-20. Calculate the maximum inertia forces due to harmonic base motion $u_b = 0.1827 \sin 65t$ (u_b is in inches if t is in seconds). The matrices for the fixed-base conditions are

$$\mathbf{M} = \begin{bmatrix} 0.5 & 1.5 & 1.0 \end{bmatrix}$$

$$\mathbf{k} = \begin{bmatrix} 3000 & -2000 & 0 \\ -2000 & 6000 & -4000 \\ 0 & -4000 & 4000 \end{bmatrix}$$

The base spring stiffness is $k = 1000$ Lb/in.

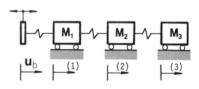

The only nonzero entry in the forcing vector, Eq. 12.1, is $ku_b = 182.7 \sin 65t$, along direction 1. Using Eq. 10.21 we have

$$\mathbf{M}\Omega^2 = \begin{bmatrix} 2112.5 & 6337.5 & 4225.0 \end{bmatrix}$$

and

$$\begin{bmatrix} 887.5 & -2000 & 0 \\ -2000 & -337.5 & -4000 \\ 0 & -4000 & -225 \end{bmatrix} \begin{Bmatrix} A_1 \\ A_2 \\ A_3 \end{Bmatrix} = \begin{Bmatrix} 182.7 \\ 0 \\ 0 \end{Bmatrix}$$

$$\therefore \begin{Bmatrix} A_1 \\ A_2 \\ A_3 \end{Bmatrix} = \begin{Bmatrix} 0.2199 \\ 0.0062 \\ -0.1105 \end{Bmatrix} \quad \text{and} \quad \Omega^2 \begin{Bmatrix} A_1 M_1 \\ A_2 M_2 \\ A_3 M_3 \end{Bmatrix} = \begin{Bmatrix} 464.5 \\ 39.3 \\ -466.9 \end{Bmatrix}$$

The last column gives us the amplitudes of the inertia forces. These forces are proportional to $\sin 65t$.

12-3 Three beams have their ends clamped in the walls of an enclosure. Each beam has a lumped mass at the midheight of the box. There is a clearance $b=0.08$ in. between the masses. Every beam has the same bending stiffness $EI=21,000$ Lb-in². The enclosure is under acceleration $a=45\sin\Omega t$ (a is in in./s² if t is in seconds) with slowly varying Ω. What frequency range must be avoided so that the masses do not impact one another? Solve for $L=8.4$ in. and $M=0.015$ Lb-s²/in. How would light damping affect the answer?

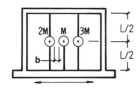

The most significant response will occur in the vicinity of a natural frequency. If a beam has length L and concentrated mass M, we have

$$\omega^2 = 3\frac{EI}{ML^3} \qquad \text{(cantilever)}$$

$$\omega^2 = 192\frac{EI}{ML^3} \qquad \text{(ends fixed)}$$

Inserting our mass and stiffness data,

$$\frac{EI}{ML^3} = \frac{21,000}{0.015\times 8.4^3} = 2362.06$$

$$\omega_1^2 = 192\frac{EI}{2ML^3} = \frac{192}{2}\times 2362.06 = 226,758 \qquad \text{(left side)}$$

$$\omega_2^2 = 3\frac{EI}{M(L/2)^3} = 24\times 2362.06 = 56,689 \qquad \text{(center)}$$

$$\omega_3^2 = 192\frac{EI}{3ML^3} = 151,172 \qquad \text{(right side)}$$

This already sets the initial limits to our answer. We know that to avoid a resonance, we must have

$$\Omega < \omega_2 = 238.1 \text{ rad/s}$$

$$\Omega > \omega_1 = 476.2 \text{ rad/s}$$

Considering the actual amplitudes of displacements will further broaden the undesirable range of Ω.

The relative motion of each mass is described by Eq. 5.3. The amplitude of the inertia force applied, for example, to the left mass is $45\times 2M$. The amplitude of relative displacement response is

$$\bar{A}_1 = \frac{1}{1-\Omega^2/\omega_1^2}\frac{45\times 2M}{2M\omega_1^2} = \frac{0.1984\times 10^{-3}}{1-\Omega^2/\omega_1^2}$$

(Note that the identity $k=M\omega^2$ was used in the denominator.) To

be on the safe side, let us assume that the adjacent masses move in opposite directions. The sum of their amplitudes must then be less than the gap b. We may notice from this equation that the resonance conditions must be very closely approached to induce such big deflections. For example, the condition

$$\bar{A}_1 = \frac{b}{2} = 0.04 \text{ in.}$$

requires $\Omega/\omega_1 = 0.9975$. The conclusion is that the previously established limits are approximately valid, and to avoid interference, the forcing frequency should satisfy

$$\Omega < 226 \text{ rad/s}$$

$$\Omega > 500 \text{ rad/s}$$

(The original limits were adjusted by about 5% to provide some measure of safety.) A small damping, say $\zeta < 0.05$, will not influence these limits to any noticeable degree.

12-4 By means of a single-mode approximation, find the peak bending moment response in the beam. The forcing is in the form of base acceleration of magnitude a_0, which is a step function of time. It is known from Prob. 9-26 that $\omega = 2.982$ rad/s and

$$\vec{\Phi}_1^T = [1.0 \quad 0.5401 \quad 0.1618]$$

while $W = 20,000$ Lb. Use $a_0 = 3.5g$, $\zeta = 0.12$, and $L = 600$ in.

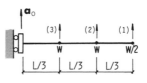

Let us ignore damping at first and introduce it later, in the final stage of calculation. The inertia forces induced by the base movement are

$$P_1 = -\frac{1}{2}\frac{W}{g}a_0 \qquad P_2 = -\frac{W}{g}a_0 \qquad P_3 = -\frac{W}{g}a_0$$

The modal force and mass, respectively, are

$$Q = (1.0\times 0.5 + 0.5401\times 1.0 + 0.1618\times 1.0)\left(-\frac{W}{g}a_0\right)$$

$$= -1.2019\frac{W}{g}a_0$$

$$\bar{M}_1 = (0.5\times 1.0^2 + 1.0\times 0.5401^2 + 1.0\times 0.1618^2)\frac{W}{g} = 0.8179\frac{W}{g}$$

Equation 11.12, without the damping term, becomes

$$\ddot{s} + \omega^2 s = -\frac{1.2019}{0.8179}a_0 = -1.4695a_0$$

If the acceleration a_0 was statically applied (i.e., growing very slowly to the a_0 level), we could speak of a static force

$$Q_0 = -1.2019\frac{W}{g}a_0$$

and the static displacement

$$s_{st} = \frac{Q_0}{k} = \frac{Q_0}{M\omega^2}$$

This was done to take advantage of the solution to Prob. 2-12, in which the displacement response of the basic oscillator under similar conditions was given as

$$u = u_{st}(1-\cos\omega t)$$

therefore, its acceleration was

$$u = \omega^2 u_{st}\cos\omega t$$

Considering the sign of loading in both problems, we have

$$\ddot{s} = -\omega^2 s_{st}\cos\omega t = -\frac{Q_0}{M}\cos\omega t = -1.4695 a_0\cos\omega t$$

These are the accelerations relative to the base. From Eq. 11.11 the physical components are calculated as

$$\begin{Bmatrix}\ddot{u}_1\\\ddot{u}_2\\\ddot{u}_3\end{Bmatrix} = \begin{Bmatrix}1.4695\\0.7937\\0.2378\end{Bmatrix}(-a_0\cos\omega t)$$

The resultant acceleration is obtained by adding the base acceleration a_0 to each of these components. At $t = \pi/\omega$ we have the following accelerations and inertial forces:

$$\begin{Bmatrix}\ddot{u}_1\\\ddot{u}_2\\\ddot{u}_3\end{Bmatrix} = \begin{Bmatrix}2.4695\\1.7937\\1.2378\end{Bmatrix}a_0 \qquad \begin{Bmatrix}P_1\\P_2\\P_3\end{Bmatrix} = \begin{Bmatrix}4.3216\\6.2780\\4.3323\end{Bmatrix}W$$

Using simple statics, we can now find that the peak moment is at the base and that it has the magnitude of $\mathfrak{M} = 0.1194\times10^9$ Lb-in. From Sect. 6C it is known that the effect of damper is to reduce the impact force of the basic oscillator by the factor

$$\left(\frac{1}{1-\zeta^2}\right)^{1/2}\exp\left[\left(2\zeta-\frac{\pi}{2}\right)\zeta\right]$$

Taking advantage of the analogy between these phenomena, we apply this factor to our maximum bending moment. For $\zeta = 0.12$ we have

$$\mathfrak{M}' = 0.8586\times0.1194\times10^9 = 0.1025\times10^9 \text{ Lb-in.}$$

A computer-aided analysis involving all three modes of vibrations shows

$$\mathfrak{M}' = 0.1015\times10^9 \text{ Lb-in.}$$

12-5 The system is like that shown in Fig. 12.2a, except there is a damper between each mass M_r and the base. There are 20 oscillating cantilevers with natural frequencies varying from 1.0 to 20.0 Hz. The

damping ratio for each cantilever is $\zeta_r = 0.05$. The base performs sinusoidal motion with frequency $\Omega = 30$ rad/s and amplitude $a_b = 0.25g$. Calculate and plot the acceleration response spectrum Z_a.

Our objective is to find the absolute acceleration of a mass attached to the base for which we can use the transmissibility Eq. 5.9. Denoting by Z_a the maximum magnitude of acceleration response, we have

$$\frac{Z_a^2}{a_b^2} = \frac{1+(2\zeta\Omega/\omega)^2}{(1-\Omega^2/\omega^2)^2+(2\zeta\Omega/\omega)^2} = \frac{1+(0.1f/f_n)^2}{(1-f^2/f_n^2)^2+(0.1f/f_n)^2}$$

The magnitude of base acceleration is $a_b = 0.25g$. The strongest excitation will be applied to the cantilever with $f_n = 5$ Hz, because its f_n is closest to the forcing frequency $f = 30/2\pi = 4.775$ Hz. The table below shows the values of Z_a for some selected natural frequencies.

f_n(Hz)	$Z_a(g)$
3	0.412
4	0.810
5	1.766
6	0.425
8	0.140
10	0.075

We can easily notice that increasing the number of cantilevers, which sense the base motion, gives better accuracy in determining the spectrum.

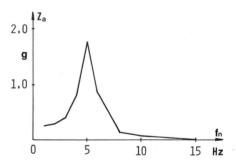

12-6 Properties of an MDOF system are described by a set of natural frequencies ω_r and the corresponding mode shapes $\vec{\Phi}_r$. A response spectrum Z_a, which gives acceleration levels relative to the moving base, is available. Develop the equations for the contribu-

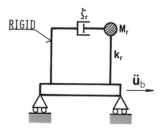

tion of the rth mode to relative displacements and for the joint forces in the system.

If the base displacement is denoted by u_b, the relative displacement $\bar{u}$ of the mass of a simple oscillator can be found from the equation of motion

$$M\ddot{\bar{u}} + c\dot{\bar{u}} + k\bar{u} = -M\ddot{u}_b \qquad (5.6)$$

This approach, described in Sect. 5C, allows us to treat our system as stationary and subjected to a known inertia force. A similar method is used for our MDOF system, except that we have to multiply the masses by some coefficients that show the extent to which those masses are affected by the base movement. (For example, if translational excitation is involved, the angular inertias of the system will not be directly excited.) For a two-DOF system, this may be put as

$$\begin{bmatrix} M_1 & 0 \\ 0 & M_2 \end{bmatrix} \begin{Bmatrix} j_1 \\ j_2 \end{Bmatrix} = \begin{Bmatrix} M_1 j_1 \\ M_2 j_2 \end{Bmatrix} \equiv \mathbf{M\vec{j}}$$

where j_r are called the directional coefficients. A special form of the equation of motion (10.5) applicable to this case is

$$\mathbf{M\ddot{\bar{u}}} + \mathbf{c\dot{\bar{u}}} + \mathbf{k\bar{u}} = -\mathbf{M\vec{j}}\ddot{u}_b$$

where $\ddot{u}_b$ is a function of time. Proceeding as in Sect. 10C, we find an equation of motion in the rth vibratory mode:

$$\ddot{s}_r + 2\omega_r\zeta_r\dot{s}_r + \omega_r^2 s_r = -\vec{\Phi}_r^T \mathbf{M\vec{j}}\frac{\ddot{u}_b}{\overline{M}_r} \equiv -\Gamma_r\ddot{u}_b$$

$\vec{\Phi}_r^T$ is the transpose of the rth mode shape and the coefficient

$$\Gamma_r = \frac{\vec{\Phi}_r^T \mathbf{M\vec{j}}}{\overline{M}_r}$$

is called the rth participation factor. If an actual oscillator, as shown in the figure, were subjected to the base motion, its differential equation would be the same as the one above, except that the right-hand side would be equal to $-\ddot{u}_b$. This means that the forcing of the rth mode is Γ_r times the forcing of an actual oscillator with the same characteristics. When the response spectrum shows that for a given frequency ω_r there is an acceleration response Z_{ar}, the magnitude of the displacement response, also relative to the base, is Z_{ar}/ω_r^2. This leads us directly to Eqs. 12.3 and 12.4.

12-7 The model of a two-story building in (a) (previously considered in Prob. 9-22) is subjected to an

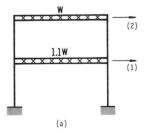

(a)

earthquake whose horizontal acceleration spectrum is constant, $Z_a = 1.6g$. Calculate the joint forces and the resultant shear forces using RSS summation of modal components for $W = 500,000$ Lb.

Copying the answer to the reference problem, we have

$$\begin{aligned} \omega_1 &= 7.1 \text{ rad/s} \\ \omega_2 &= 17.385 \text{ rad/s} \end{aligned} \qquad \Phi = \begin{bmatrix} 1.0 & 1.0 \\ 1.7711 & -0.6211 \end{bmatrix}$$

The lumped masses are

$$M_1 = 1.1 \times \frac{500,000}{386} = 1424.9 \quad \text{and} \quad M_2 = 1295.3$$

The modal masses are calculated from Eq. 9.21a,

$$\overline{M}_1 = M_1 + 1.7711^2 M_2 = 5488$$

$$\overline{M}_2 = M_1 + 0.6211^2 M_2 = 1924.6 \text{ Lb-s}^2/\text{in.}$$

Both masses are directly subjected to the inertia forces due to base movement; therefore the entries of the $\vec{j}$ vector in Eq. 12.2 are unities. Consequently,

$$\mathbf{M\vec{j}} = \begin{Bmatrix} M_1 \\ M_2 \end{Bmatrix}$$

and the participation factors are calculated as follows.

$$\vec{\Phi}_1^T \mathbf{M\vec{j}} = \begin{bmatrix} 1.0 & 1.7711 \end{bmatrix} \begin{Bmatrix} M_1 \\ M_2 \end{Bmatrix} = 3719$$

$$\vec{\Phi}_2^T \mathbf{M\vec{j}} = 620.4$$

$$\Gamma_1 = \frac{3719}{5488} = 0.6777 \qquad \Gamma_2 = 0.3224$$

The acceleration response is the same for each mode,

$$Z_{ar} = 1.6g = 617.6 \text{ in./s}^2$$

We can now use Eq. 12.3 to calculate the modal displacements:

$$\vec{u}_1 = \frac{Z_{a1}}{\omega_1^2}\Gamma_1\vec{\Phi}_1 = \frac{617.6}{7.1^2} \times 0.6777 \begin{Bmatrix} 1.0 \\ 1.7711 \end{Bmatrix} = \begin{Bmatrix} 8.303 \\ 14.705 \end{Bmatrix}$$

$$\vec{u}_2 = \begin{Bmatrix} 0.6588 \\ -0.4092 \end{Bmatrix}$$

The joint forces are:

$$\begin{Bmatrix} P_1 \\ P_2 \end{Bmatrix}_1 = \begin{bmatrix} 337,500 & -150,000 \\ -150,000 & 150,000 \end{bmatrix} \begin{Bmatrix} 8.303 \\ 14.705 \end{Bmatrix} = \begin{Bmatrix} 596,513 \\ 960,300 \end{Bmatrix}$$

$$\begin{Bmatrix} P_1 \\ P_2 \end{Bmatrix}_2 = \begin{Bmatrix} 283,725 \\ -160,200 \end{Bmatrix}$$

The RSS summation yields

$$P_1 = \left(596{,}513^2 + 283{,}725^2\right)^{1/2} = 660{,}551$$

$$P_2 = \left[960{,}300^2 + (-160{,}200)^2\right]^{1/2} = 973{,}571$$

The modal components and the resultant nodal forces are shown in (b). A similar illustration of shear forces is in Fig. (c). In both figures the model of the building is presented as a single column.

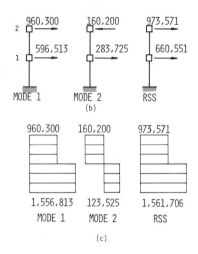

(b)

(c)

12-8 The model of the dolly whose free vibrations were analyzed in Prob. 9-10 is subjected to an earthquake with an acceleration spectrum shown in (a). Compute the dynamic forces applied to the lumped masses in the reference problem using RSS summation.

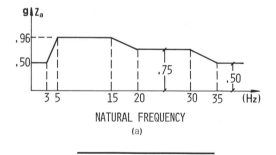

NATURAL FREQUENCY

(a)

The nodal masses, after the reduction carried out in the reference problem, are:

$$M_1 = M = \frac{151{,}000}{386} = 391.19 \text{ Lb-s}^2/\text{in.}$$

$$M_2 = J = 1.418 \times 10^6 \text{ Lb-s}^2\text{-in.}$$

The modal matrix is:

$$\boldsymbol{\Phi} = \begin{bmatrix} 1.0 & 1.0 \\ 0.01 & -0.02757 \end{bmatrix}$$

Since only the translational mass is excited by the base movement,

Eq. 12.5 defining the participation factors has but one nonzero term in the numerator,

$$\Gamma_1 = \frac{391.19 \times 1.0}{391.19 \times 1.0^2 + 1.418 \times 10^6 \times 0.01^2} = 0.734$$

$$\Gamma_2 = \frac{391.19 \times 1.0}{391.19 \times 1.0^2 + 1.418 \times 10^6 \times 0.02757^2} = 0.2663$$

The natural frequencies are

$$\omega_1 = 76.2 \text{ rad/s} \quad \text{or} \quad f_{n1} = 12.128 \text{ Hz}$$

$$\omega_2 = 173.65 \text{ rad/s} \quad \text{or} \quad f_{n2} = 27.637 \text{ Hz}$$

The response spectrum tells us that the corresponding accelerations are $0.96g$ and $0.75g$, respectively. The modal components, after Eq. 12.6:

$$\begin{Bmatrix} \ddot{u}_1 \\ \ddot{u}_2 \end{Bmatrix}_1 = 0.734 \times 0.96g \begin{Bmatrix} 1.0 \\ 0.01 \end{Bmatrix} = \begin{Bmatrix} 272 \\ 2.72 \end{Bmatrix}_1$$

$$\begin{Bmatrix} \ddot{u}_1 \\ \ddot{u}_2 \end{Bmatrix}_2 = 0.2663 \times 0.75g \begin{Bmatrix} 1.0 \\ -0.02757 \end{Bmatrix} = \begin{Bmatrix} 77.09 \\ -2.125 \end{Bmatrix}_2$$

Note that $\ddot{u}_2$ is measured in rad/s^2 and one must multiply it by the distance from the CG to obtain the translational component applied to a particular mass. The distances and the contributions of rotation $\ddot{u}_2$ in both vibratory modes are shown in (b). The masses calculated from the weights in the reference problem are:

$$M_1' = 103.63 \qquad M_2' = 33.68 \qquad M_3' = 54.92 \qquad M_4' = 198.96$$

These masses are multiplied by the resultant accelerations in each mode as shown in the table. The distinction between positive and negative sense of applied joint force is made within each mode, but not for the RSS sums.

	Mode 1		Mode 2		RSS	
Station	a (in/s^2)	P (Lb)	a (in./s^2)	P (Lb)	a (in./s^2)	P (Lb)
1	509.4	52,789	−108.4	−11,233	520.8	53,971
2	400.6	13,492	−23.4	−788	401.3	13,516
3	264.6	14,532	82.9	4553	277.3	15,229
4	128.6	25,586	189.1	37,623	228.7	45,502

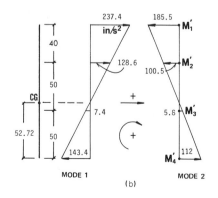

MODE 1   MODE 2

(b)

12-9 When the structure considered in Probs. 9-10 and 12-8 is under seismic excitation, the inertial forces tend to overturn it. Before that takes place, however, the wheels on one side must "lift off" or lose contact with the rail. Calculate the net lifting-off moment resulting from the prescribed earthquake.

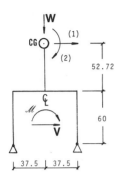

A schematic representation of the base of dolly in Fig. 9-10a is shown in (a), where $W = 151,000$ Lb is the weight of the system, while V and $\mathcal{M}$ represent the resultant seismic load reduced to the level of bottom of wheels. The following are the modal components applied at the CG, in accordance with Prob. 12-8:

Mode 1. $P = 151,000 \times \dfrac{272}{386} = 106,404$ Lb

$\mathcal{M}_c = 1.418 \times 10^6 \times 2.72 = 3.857 \times 10^6$ Lb-in.

Mode 2. $P = 151,000 \times \dfrac{77.09}{386} = 30,157$ Lb

$\mathcal{M}_c = -1.418 \times 10^6 \times 2.125 = -3.013 \times 10^6$ Lb-in.

At the bottom level we have

Mode 1. $\mathcal{M} = 106,404 \times 112.72 + 3.857 \times 10^6 = 15.851 \times 10^6$
Mode 2. $\mathcal{M} = 386,297$ Lb-in.

The RSS sum of the last two figures is

$$\mathcal{M} = 15.856 \times 10^6 \text{ Lb-in.}$$

The net lift-off moment is the difference between $\mathcal{M}$ and the stabilizing effect of weight

$$\mathcal{M}' = 15.856 \times 10^6 - 151,000 \times 37.5 = 10.19 \times 10^6 \text{ Lb-in.}$$

Since the weight of the dolly does not nearly compensate for the seismic forces, there will be a strong tendency for the wheels to lift off the rails. To prevent this, special restraining devices must be installed.

12-10 A tower with platforms at three levels is excited by a rocking motion of the base about its center B. To determine the displacements relative to the base, what expression for the forcing function must be put on the right side of Eq. 10.5?

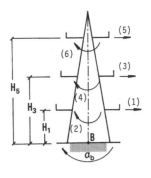

The figure clearly implies the existence of the angular inertia terms in the model of the tower. When the acceleration $\ddot{\alpha}_b$ is imposed, we have the following inertia forces at the second level at H_3 above ground:

$$P_3 = -M_3 H_3 \ddot{\alpha}_b \qquad \text{and} \qquad P_4 = -J_4 \ddot{\alpha}_b$$

in which P_4 is the moment applied along coordinate 4. The forcing function in the equations of motion is

$$-\mathbf{M}\vec{\mathbf{j}}\ddot{\alpha}_b = -\begin{Bmatrix} M_1 H_1 \\ J_2 \\ M_3 H_3 \\ J_4 \\ M_5 H_5 \\ J_6 \end{Bmatrix} \ddot{\alpha}_b$$

in which $\mathbf{M}$ is the mass matrix,

$$\mathbf{M} = \begin{bmatrix} M_1 & J_2 & M_3 & J_4 & M_5 & J_6 \end{bmatrix}$$

and $\vec{\mathbf{j}}$ is the vector of displacement coefficients,

$$\vec{\mathbf{j}}^T = \begin{bmatrix} H_1 & 1 & H_3 & 1 & H_5 & 1 \end{bmatrix}$$

The forcing function herein developed is quite similar to what was used for a translational base movement.

12-11 Calculate the lateral seismic loads applied to the building in Prob. 9-9 based on the acceleration response spectrum used in Prob. 12-8. Use two independent methods:

1.   Uniform acceleration along the height.
2.   Equivalent oscillator developed for this structure in Prob. 11-8.

Explain the differences in the results of both methods.

The fundamental frequency is $\omega = 15.587$ rad/s. The weights, going from the bottom up, are

$$W_1 = W \qquad W_2 = 0.67W \qquad W_3 = W$$

where $W = 2 \times 10^6$ Lb. The same response spectrum as used in

Prob. 12-8 gives us $Z_a = 0.5g$ for $f_n = 15.587/2\pi = 2.481$ Hz. When uniformly applied, this acceleration yields the following inertia forces:

$$P_1 = 0.5W \qquad P_2 = 0.335W \qquad P_3 = 0.5W,$$

which completes part 1 of the problem.

The equivalent oscillator in Prob. 11-8 has natural frequency $\omega = 15.989$ rad/s, weight $\overline{W} = 1.291W$ and the deflected shape

$$\vec{X}^T = [1.0 \quad 2.1429 \quad 3.7429]$$

To correctly use this model, we must first find out what the base acceleration $\ddot{u}_b$ really means in this case. The actual structure experiences the forces of magnitude

$$P_1 = \frac{W\ddot{u}_b}{g} \qquad P_2 = \frac{0.67W\ddot{u}_b}{g} \qquad P_3 = \frac{W\ddot{u}_b}{g}$$

Since the reference displacement is at the coordinate 3 (top), the generalized force is given by Eq. 11.17 as

$$Q = (1.0 \times 1.0 + 0.67 \times 2.1429 + 1.0 \times 3.7429) \frac{W\ddot{u}_b}{3.7429g}$$

$$= 1.6508 \frac{W\ddot{u}_b}{g}$$

The acceleration experienced by the mass of oscillator is

$$\frac{1.6508}{1.291}\ddot{u}_b = 1.2787\ddot{u}_b$$

The new natural frequency is slightly above the true value, but it still gives $Z_a = 0.5g$ from the spectrum in Prob. 12-8. Our adjusted response acceleration is

$$a = 1.2787 \times 0.5g = 0.6394g$$

This is for the reference coordinate 3. Using the mode shape to scale the accelerations in the remaining directions, we have

$$\begin{Bmatrix} a_1 \\ a_2 \\ a_3 \end{Bmatrix} = \begin{Bmatrix} 0.1708g \\ 0.3661g \\ 0.6394g \end{Bmatrix} \quad \text{and} \quad \begin{Bmatrix} P_1 \\ P_2 \\ P_3 \end{Bmatrix} = \begin{Bmatrix} 0.1708W \\ 0.2453W \\ 0.6394W \end{Bmatrix}$$

Comparing the results from both methods, we note not only that the resultant base shear is different ($1.335W$ vs. $1.055W$), but that there also is a significant difference in the distribution of seismic load along the height. Using an equivalent oscillator typically gives a smaller total base shear than a uniform acceleration method. This is because the former is a single-mode approximation (using an assumed or the actual deflected shape in that mode) and therefore it uses only a part of the total forcing. When employed for design, the results of a single-mode approximation should be applied by a correction factor larger than 1.0 to account for the presence of higher modes.

12-12 Perform a seismic analysis of the two-story building described in Prob. 9-22 in accordance with the Uniform Building Code. Use the dimensions in the figure. The natural frequency of the first mode $f_n = 1.13$ Hz; $\tilde{Z} = 1.0$, and $\tilde{K}\tilde{C}$ must not be less than 0.1. Calculate the maximum shears and moments in the columns.

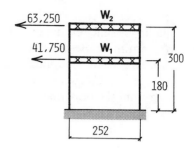

From the reference problem, $W_1 = 550,000$ Lb and $W_2 = 500,000$ Lb. Following Sect. 12D we have

$$\tau = \frac{1}{1.13} = 0.885 \text{ s} \qquad \text{and} \qquad \tilde{C} = \frac{0.05}{0.885^{1/3}} = 0.052$$

For this type of structure $\tilde{K} = 0.67$ and $\tilde{K}\tilde{C} = 0.0349$. Using the minimum value from the problem statement,

$$V = 1.0 \times 0.1W = 0.1(W_1 + W_2) = 105,000 \text{ Lb}$$

The force to be applied to the top is

$$P_t = 0.004V \left(\frac{300}{252}\right)^2 = 0.0057V$$

We will assume $P_t = 0$, since this is permitted by the code for $(h/D) \leqslant 3.0$. Thus

$$V' = V - P_t \approx V$$

$$\frac{\varepsilon}{g} = \frac{V'}{W_1 h_1 + W_2 h_2} = \frac{1}{2371.4}$$

The inertial forces applied to the lumped masses are calculated from Eq. 12.14 and shown in the figure. The shear forces per column are

$$\frac{63,250}{2} = 31,625 \text{ Lb} \qquad \text{(upper)}$$

$$\frac{105,000}{2} = 52,500 \text{ Lb} \qquad \text{(lower)}$$

The columns are rigidly connected to the floors and consequently there is zero-moment point at midheight of each level. This allows us to calculate the bending moment in each column:

$$\frac{31,625(300 - 180)}{2} = 1.8975 \times 10^6 \text{ Lb-in.}$$

$$52,500 \times \frac{180}{2} = 4.725 \times 10^6 \text{ Lb-in.}$$

EXERCISES

12-13 Find the acceleration response of the system in Prob. 12-1 when the forcing is a constant acceleration a_0 of the left base point, oriented toward the left.

12-14 The base of the system vibrates according to $u_b = u_0 \cos \Omega t$. Determine the forcing frequency Ω_0 at which M_1 is motionless. Also solve a reverse problem, namely, select k_d and M_2 so that at a designated forcing frequency M_1 remains motionless.

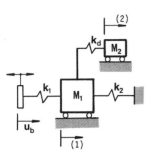

12-15 Rework Prob. 12-7, obtaining the nodal forces directly from spectral accelerations. Compare the results with the reference problem.

12-16 Rework Prob. 12-8 using displacements and stiffness matrices instead of accelerations and compare the results. *Hint.* Notice that the moment about the CG may be converted to horizontal accelerations.

12-17 The figure shows the effective load-carrying area near the base of cylinder analyzed in Probs. 9-10 and 12-8. Using the modal loads and RSS summation, compute the resultant bending moment $\mathcal{M}$ and shear force V at the base section of the cylinder. Also calculate the dynamic stress due to bending and shear.

12-18 The frame is excited by a horizontal ground motion. The translational masses are associated with coordinates 1, 3, 5, 7, and 8 while the rotational mass exists at coordinate 2. (Coordinates 4 and 6 are needed for definition of deflected pattern, but no inertia is associated with them.) Develop the vector of directional coefficients $\vec{\mathbf{j}}$ to be used in the equations of motion.

12-19 Calculate the lateral seismic loads applied to the building in Prob. 9-9. Use the code formulas with $\tilde{Z} = 1$ and $\tilde{K}\tilde{C} = 0.1$. Determine bending and shear of a single column at the bottom level, assuming all columns to be the same.

12-20 A water tank is to be constructed in a seismic zone with $\tilde{Z} = 1.0$. The maximum value of the factor $\tilde{K}\tilde{C}$ for this type of structure is 0.25. The specific weight of water is 62.4 Lb/ft³ and the tank is assumed to be full. The shell dimensions in the figure pertain to the interior; therefore, when calculating weights, allow an additional 5% for steel. Perform the seismic analysis according to the code and calculate the required shell thickness at the base if the allowable direct stress is 22,000 psi. (The weight may be lumped at two points, as implied in the drawing.) Put $P_t = 0$.

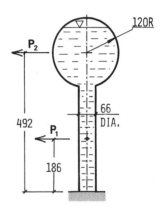

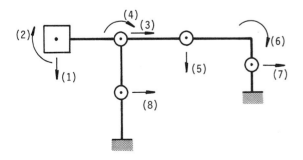

12-21 A storage tank on four legs may be subjected to a seismic load along any horizontal direction. Which is worse as far as the axial and the bending load in the legs are concerned, direction 1 or direction 2?

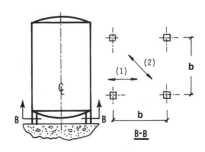

CHAPTER

13

Shock, Impact, and Collision

This is a continuation of Chapter 6 into problems involving several degrees of freedom. The word *collision* used in the title implies a sudden contact of two bodies with masses of the same order of magnitude.

The impact and the shock are applied to either constrained or unconstrained systems, and the response is usually computed by some approximate method. Both slender and compact bodies are featured, with either linear or nonlinear contact characteristics.

13A. Central Collision of Bodies takes place when the contact point of the colliding objects is located on the same line along which the centers of gravity are moving. Like a similar, but simpler phenomenon described in Sect. 6A, the collision is separated into four distinct stages illustrated in Fig. 13.1.

At the beginning we have ball M_1 approaching ball M_2, $v_1 > v_2$. Stage 1 lasts until the balls touch each other, and at this moment they still have their original velocities. During Stage 2 the first ball tries to accelerate the second one by means of its inertia, and the resulting contact force deforms both bodies. Stage 2 ends at that instant when both balls travel with the same velocity v_0. Stage 3 is the recovery process during which the balls return to their original shape (if the collision is elastic) and their relative velocities change sign, so that at the end of this stage they barely touch each other. At Stage 4 the balls continue with their new velocities, V_1 and V_2. The time of the

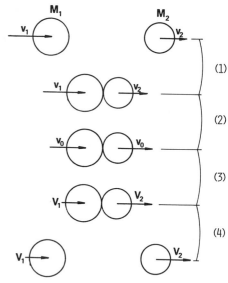

Figure 13.1 Four stages of collision.

collision itself (Stages 2 and 3) is very short in comparison with entire process described.

Figure 13.2 illustrates another way of looking at the collision of two bodies. If an observer is moving with velocity v_0, he sees the balls approaching each other, then going apart following the collision. In fact, he can think of each ball rebounding from a rigid wall that appears stationary to him. This interpretation helps us to treat collision as a form of impact against a rigid barrier that is moving with the previously defined velocity v_0.

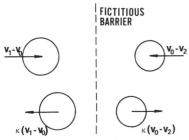

Figure 13.2 Collision treated as rebound from moving barrier.

The common velocity v_0, which is in effect at that instant when the balls are stationary with respect to one another, may be found from the principle of preservation of momentum: $M_1v_1 + M_2v_2 = (M_1 + M_2)v_0$:

$$v_0 = \frac{M_1v_1 + M_2v_2}{M_1 + M_2} \qquad (13.1)$$

To find the velocities after the collision we return for a moment to Fig. 6.1, which shows us that the final velocity V with respect to the barrier is $-\kappa v$, where κ is the restitution coefficient and v is the impact velocity. The analogous change of sign and magnitude is shown in Fig. 13.2. Taking into account that the entire system is moving with the velocity v_0, we have

$$\begin{aligned} V_1 &= v_0 - \kappa(v_1 - v_0) \\ V_2 &= v_0 + \kappa(v_0 - v_2) \end{aligned} \qquad (13.2)$$

Positive velocity is directed to the right. By subtracting both sides of these equations, we can eliminate v_0 and obtain the relationship between the relative velocities:

$$V_2 - V_1 = -\kappa(v_2 - v_1) \quad \text{or} \quad V_r = -\kappa v_r \quad (13.3)$$

The energy loss during collision is expressed by

$$\Delta T = \tfrac{1}{2}M_1(v_1^2 - V_1^2) + \tfrac{1}{2}M_2(v_2^2 - V_2^2)$$

which can be written in a manner similar to Eq. 6.4:

$$\Delta T = \tfrac{1}{2}M^* v_r^2(1 - \kappa^2) \qquad (13.4)$$

Where M^* is an equivalent mass calculated from

$$\frac{1}{M^*} = \frac{1}{M_1} + \frac{1}{M_2} \qquad (13.5)$$

while v_r is the relative velocity of both bodies, which was used in Eq. 13.3. The collision may be perfectly elastic ($\kappa = 1.0$), inelastic ($0 < \kappa < 1.0$), or perfectly plastic ($\kappa = 0$). The meaning of these terms is the same as described in Sec. 6A.

The developments above were carried out using the techniques of the rigid-body mechanics. The deformability was accounted for in a general way, by means of the restitution coefficient. To establish such important parameters as the duration of contact and the magnitude of contact force, some information is needed with regard to the resistance-deflection properties of the colliding bodies.

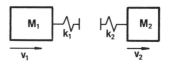

Figure 13.3 Collision of bodies with linear characteristics.

13B. Parameters of Collision with Linear Characteristic. Two objects that are to come in contact are shown in Fig. 13.3. The midpoint of impact (with respect to time) is characterized by the common velocity v_0 of both bodies. At this moment the compression of springs reaches its maximum. A comparison of kinetic plus strain energy at the beginning and the end of the loading phase of contact gives

$$\tfrac{1}{2}M_1v_1^2 + \tfrac{1}{2}M_2v_2^2 = \tfrac{1}{2}(M_1 + M_2)v_0^2 + \tfrac{1}{2}k_1\delta_1^2 + \tfrac{1}{2}k_2\delta_2^2$$

in which δ_1 and δ_2 are the maximum spring deflections. The maximum contact force resulting from this energy balance is

$$R_m = v_r(M^*k^*)^{1/2} \qquad (13.6)$$

and v_r is the relative velocity before the contact, as used in Eq. 13.3. In addition, M^* is the effective mass and k^* is the effective stiffness of the series connection of springs:

$$\frac{1}{k^*} = \frac{1}{k_1} + \frac{1}{k_2}$$

Note that Eq. 13.6 is a generalization of Eq. 6.7.

Deflection of any of the two springs is found from

$$R_m = k_1\delta_1 = k_2\delta_2 \qquad (13.7)$$

Let t_1 be the loading phase of contact, which is depicted as Stage 2 in Fig. 13.1. This interval may be identified with one-quarter of the natural period of a system consisting of two masses and a spring of stiffness k^* joining them. The natural frequency of such an arrangement was found in Prob. 9-30 to be

$$\omega = \left(\frac{k^*}{M^*}\right)^{1/2}$$

using the notation of this section. We thus get

$$t_1 = \frac{\pi}{2}\left(\frac{M^*}{k^*}\right)^{1/2} \qquad (13.8)$$

which is analogous to Eq. 6.10 pertaining to the loading phase of impact. If the collision is indeed perfectly elastic, the entire time of contact is $2t_1$. Otherwise, the time of the unloading phase t_2 may be calculated using the unloading stiffness of the springs.

13.C.   Parameters of Collision with Nonlinear Characteristic. One may treat the colliding bodies as rigid after putting their entire flexibility into a "bumper" spring separating them. A general form of a force-deflection relationship for such a spring is written as

$$R(\delta) = k_0 f(\delta) \qquad (13.9)$$

where δ is the relative movement of the bodies and k_0 is a constant coefficient, which only sometimes has a meaning of initial stiffness. If the properties are given separately for each body as $R_1(\delta)$ and $R_2(\delta)$, those two springs must be joined in series so that we can think of one bumper spring as in Fig. 13.4, whose deformability is given by Eq. 13.9. The kinetic energy lost in the loading phase of collision changes into strain energy Π,

$$\tfrac{1}{2}M_1 v_1^2 + \tfrac{1}{2}M_2 v_2^2 - \tfrac{1}{2}(M_1 + M_2)v_0^2 = \Pi$$

or

$$\Pi = \tfrac{1}{2}M^* v_r^2 \qquad (13.10)$$

Having the value of Π and knowing the resistance·deflection curve, we are able to calculate the maximum reaction R_m and the associated deflection δ_m.

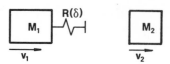

Figure 13.4 Collision of bodies with nonlinear characteristics of contact.

The time of the loading phase t_1 is the same as a quarter of the period of free vibration. The natural frequency for this nonlinear system was determined in Prob. 3-10, and we can conclude that

$$t_1 = \frac{\pi}{2}\left(\frac{M^*}{k_0 F}\right)^{1/2} \qquad (13.11)$$

where F is the function of amplitude δ_m,

$$F = F(\delta_m) = \frac{1}{3\delta_m}(f_0 + 2.8284 f_c) \qquad (13.12)$$

with $f_0 = f(\delta_m)$ and $f_c = f(0.7071\delta_m)$. This approximation is valid only when $R(\delta)$ has a smoothly varying slope.

13.D.   Eccentric Impact Against Frictionless Barrier takes place when the velocity of the CG is not aligned with the normal to the surface at the impact point.

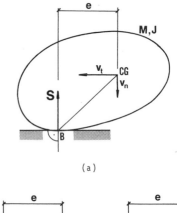

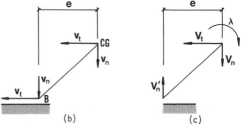

Figure 13.5 Eccentric impact, frictionless barriers: (a) impacting body; (b) velocities before impact; (c) velocities after impact.

Figure 13.5a shows a body with the initial velocity components v_n and v_t impacting the surface at point B. The impulse S of the interaction force is directed along the normal at point B. Figures 13.5b and 13.5c show the kinematic conditions just prior to impact and just after it ends. In the first case we have a translational motion defined by v_n and v_t, while in the second all three components of CG motion are involved: V_n, V_t, and λ. The velocity of rebound at the contact point V_n', is the geometric sum of the velocity of the CG and of the relative velocity with respect to the CG,

$$V_n' = \lambda e - V_n \qquad (13.13)$$

The approach to this type of impact is similar to that used in Sect. 6A. The coefficient of restitution κ again determines the ratio of velocities after and before the time of contact, $V_n' = \kappa v_n$ or

$$\lambda e - V_n = \kappa v_n \qquad (13.14)$$

Two impulses are applied to the body, S and κS, both in the same direction. The change of linear and angular momentum, respectively,

$$M(v_n - V_n) = (1 + \kappa)S \qquad (13.15)$$

$$J\lambda = (1 + \kappa)Se \qquad (13.16)$$

The tangential velocity remains unchanged, $v_t = V_t$, because there was no friction. Equations 13.14, 13.15, and 13.16 form a system with three unknowns: V_n, λ, and S. The solution is given by

$$V_n = \frac{1 - \kappa(i/e)^2}{1 + (i/e)^2} v_n \qquad (13.17)$$

$$S = \frac{J}{J + Me^2} Mv_n \qquad (13.18)$$

in which $i^2 = J/M$ and J is the mass moment of inertia about the axis through the CG. The angular velocity λ may be calculated form Eq. 13.14. It is also worth remembering that the positive signs are those shown in Figs. 13.5b and 13.5c.

A barrier is considered *rough* when the friction force is so large that no slip along the contact area is possible. In physical world this is likely to happen when a sharp corner presses on the surface, as Fig. 13.6 shows. During the loading phase of impact, segment p rotates about the contact point.

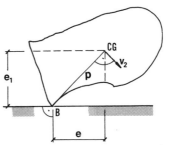

Figure 13.6 Impact against rough barrier.

13E. Impact Against Body Free to Rotate About Fixed Axis. A particular type of this impact is considered, when the direction of the contact force is perpendicular to the line determined by the pivot point and the center of gravity (Fig. 13.7).

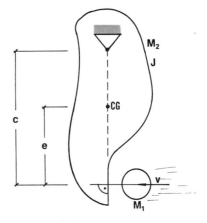

Figure 13.7 Impact against body free to rotate about fixed axis.

Similarly to what was done in Sect. 13A, we isolate an instant when the impacting object has velocity v_0 that is the same as the velocity of the contact point of the rotating body. Since the angular momentum of the system is preserved, we can write

$$M_1 vc = M_1 v_0 c + \frac{J_0 v_0}{c}$$

$$\therefore v_0 = \frac{M_1 v}{M_1 + J_0/c^2} \qquad (13.19)$$

where J_0 is the mass moment of inertia of the body about the axis of rotation,

$$J_0 = J + M_2(c - e)^2$$

in which J is about the center of gravity. Note that if we put

$$M_r = \frac{J_0}{c^2} \qquad (13.20)$$

then Eq. 13.19 becomes a particular case of Eq. 13.1 for the impacted mass M_r in the state of initial rest, $v_2 = 0$. For this reason we can use the formulas of Sect. 13A if we replace the body with the rotational degree of freedom by an object free to translate and having the *reduced mass M_r*. In particular, the velocities after the impact are, by Eqs. 13.2:

$$V = v_0 - \kappa(v - v_0) \qquad (13.21)$$

$$\lambda = \frac{v_0}{c}(1 + \kappa) \qquad (13.22)$$

in which V is the post-impact velocity of M_1, positive in the same direction as v, while λ is the angular velocity of the constrained body.

13F. Contact Force and Contact Time of Eccentric Impacts. The analysis of an eccentric impact against a frictionless surface shows that the impulse applied to the impacting body during the loading phase is expressed by Eq. 13.18. It may also be written as $S = M_r v_n$ in which

$$M_r = \frac{J}{J + Me^2} M \qquad (13.23)$$

is the mass of the body reduced to the point of impact. In what follows M_r is treated as the mass impacting the surface along the normal direction with the impact velocity v_n.

The corresponding value for a rough surface impact is

$$M_r = \frac{Me_1^2 + J}{Mp^2 + J} M \qquad (13.24)$$

As Fig. 13.6 shows, p is the distance between the impact point and the CG while e_1 is the height of CG above the impacted surface.

To calculate the time of the loading phase of impact, we need the spring stiffness k, which characterizes the deformability of the impacting body. If this figure is available, we can write on the basis of Sect. 6C

$$t_1 = \frac{\pi}{2}\left(\frac{M_r}{k}\right)^{1/2} \qquad (13.25)$$

This is good only for a linear spring. In the particular case of a contact that proceeds according to the Hertz theory, Eq. 6.15 holds with M_r replacing M. For a general nonlinear characteristic we can apply the method of finding t_1 described in Sect. 6F.

The impact force may be calculated by using a modified form of Eq. 6.7,

$$R_m = v_n(M_r k)^{1/2} \qquad (13.26)$$

in which v_n is the velocity normal to the surface at the contact point. When the characteristic is nonlinear, one can follow the procedure of Sect. 13C with v_n and M_r replacing v_r and M^*, respectively.

Note that Eq. 13.26 does not include the effect of gravity on the impacting body. If R_m calculated from that equation is not many times larger than the weight of body, a more general work-energy relation should be set up.

An impact against a body free to rotate about a fixed axis is a special case of a collision of two bodies. For a linear characteristic of springs representing a combined flexibility of both bodies, one can obtain the impact force and time from Eqs. 13.5, to 13.8 by putting

$$M_2 = M_r = \frac{J_0}{c^2} \qquad (13.27)$$

as defined in Sect. 13E. The approach in Sect. 13C should be used when a nonlinear characteristic is involved.

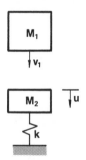

Figure 13.8 Impact against spring-supported mass.

13G. Impact Against Spring-Supported Mass. Figure 13.8 shows mass M_1 moving against a stationary mass M_2 that is supported by spring of stiffness k. One can find, on the basis of Eqs. 13.1 and 13.2, that the velocity of M_2 is

$$V_2 = \frac{M_1 v_1(1 + \kappa)}{M_1 + M_2} \qquad (13.28)$$

following the impact. In a particular case of plastic contact $\kappa=0$ and both masses travel with V_2. In this event and when the force of gravity is present, the maximum spring deflection caused by impact is

$$u_m=u_{st}+\left(u_{st}^2+\frac{M_1}{M_1+M_2}\frac{2T_1}{k}\right)^{1/2} \quad (13.29)$$

in which

$$u_{st}=\frac{M_1g}{k} \quad \text{and} \quad T_1=\tfrac{1}{2}M_1v_1^2$$

while the total spring deflection, measured from unstrained position is $M_2g/k+u_m$. When the force of gravity is not involved, we set $u_{st}=0$ and obtain

$$u_m=\frac{M_1v_1}{[k(M_1+M_2)]^{1/2}} \quad (13.30)$$

13H. Impact Against Constrained Beam. The engineering theory of beam impact is an attempt to reduce the problem to that considered in the previous section using a single-mode approximation. The deflected shape itself is usually assumed, and the calculation is carried out according to Sect. 11F. When the moving mass M collides with one of the lumped masses of the beam, as in Fig. 13.9, the entire beam is assumed to acquire the initial velocity at that instant.

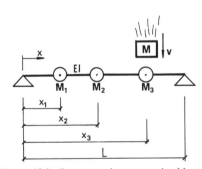

Figure 13.9 Impact against constrained beam.

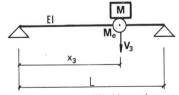

Figure 13.10 Simplified beam impact.

The contact is usually treated as perfectly plastic; that is, there is no rebound at least until the maximum deflection is reached. The effective mass M_e is located at the impact point and calculated from

$$M_e=M_1\left(\frac{X_1}{X}\right)^2+M_2\left(\frac{X_2}{X}\right)^2+\cdots+M_n\left(\frac{X_n}{X}\right)^2 \quad (13.31)$$

in which X_r and X are the values of the shape function at $x=x_r$ and at the location of M_e, respectively. A deflection at point x_r is

$$u_r=X_rf(t)$$

in which $f(t)$ is a time function. The stiffness k of the equivalent oscillator, which replaces our beam, is measured at the impact point, $x=x_3$ in Fig. 13.9 or 13.10. Putting k in Eq. 13.29 or 13.30 allows us to calculate the maximum deflection u_m at the impact point. The maximum spring reaction in Fig. 13.8 is

$$R_m=ku_m$$

In terms of the model in Fig. 13.10, R_m is the resultant support reaction or the force exerted by M_e+M on the massless beam. This statement allows us to find the maximum dynamic bending moment and shear force.

At this point a more elaborate and also more realistic procedure may be used to determine the maximum bending moment. The inertia force applied to the ith mass is

$$P_i=-M_i\ddot{u}_i=\omega^2M_iX_if(t)$$

We can sum the maximum inertia forces and equate them with a previously calculated reaction

$$R_m=\omega^2f\sum M_iX_i \quad (13.32)$$

in which f stands for the extreme value of $f(t)$. Computing the coefficient ω^2f from this equation allows us to determine the values of the individual inertia forces and thereby obtain a more realistic load distribution.

The engineering theory of beam impact explained above gives a reasonable approximation of the status of the system some time after the contact has begun. At the outset the deflected pattern is quite irregular because of superposition of the elastic waves originating where the force was suddenly applied. Only after about $T_1/4$, where T_1 is the fundamental period, the deflected shape is close to that of the fundamental mode or to the static deflection under the contact force.

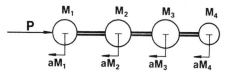

Figure 13.11 Dynamic load applied to unconstrained body.

13J. Shock Load on Unconstrained Bodies — Dynamic Equilibrium. Using d'Alembert's principle, we can look at an accelerating body as being in a special state of equilibrium where the active, applied loads are balanced by the inertia forces. An example in Fig. 13.11 shows an axial bar with lumped masses, subjected to a force P at one end. When we think of the bar as a rigid body, the magnitude of the inertia force applied to mass M_i is

$$R_i = aM_i$$

where $a = P/(M_1 + M_2 + M_3 + M_4)$ is the acceleration of the body. The load P is a function of time, and the inertia forces are also time dependent. The rigid-body type of distribution just presented is only one possibility, but an important one, since it provides a reference level for calculation of internal forces. Consider the simplest possible problem of this type illustrated in Fig. 13.12*a*. The exact analysis (Prob. 10-17) shows that if a load of magnitude P_0 is suddenly applied to M_1 (step load), the force in the elastic link may be calculated in two operations. In the first we obtain its value corresponding to the dynamic equilibrium in rigid-body motion, Fig. 13.12*b*

$$N = aM_2 = \frac{P_0}{M_1 + M_2} M_2$$

Figure 13.12 (*a*) Load applied to unconstrained body; (*b*) compressive component of that load.

This compression would be valid for the accelerating load, which very slowly attains its P_0 level. The second operation is therefore to account for the sudden growth of load by using the magnification factor equal to 2.0 in this case. Thus,

$$N_{max} = \frac{2P_0}{M_1 + M_2} M$$

When this procedure is extended to larger systems, like that in Fig. 13.11, the results should be treated as a gross approximation, which may, at places, considerably deviate from the true values.

13K. Impact Against Unconstrained Beam. The engineering theory of beam impact presented in Sect. 13H is valid only when the beam is supported in a manner preventing any rigid-body component of motion. When dealing with an underconstrained beam or an unconstrained beam, the concept of the dynamic equilibrium developed in Sect. 13J is particularly helpful.

The procedure consists of calculating flexibility relative to the dynamic equilibrium condition. Once that flexibility is known and visualized as an external spring, the contact load and the impact duration may be found using the equations in Sect. 13F.

The reader ought to be aware that the accuracy of the method outlined in this section is very limited. It must be regarded as merely the first approximation to a solution of a complex problem. It has value in that it gives an analyst the order of magnitude of the quantities desired, which is a good thing to have even if a more sophisticated analysis is to be carried out later.

SOLVED PROBLEMS

13-1 A rail car weighing $W_1 = 20,000$ Lb is rolling with the velocity $v_1 = 10$ mph when it strikes a standing car weighing $W_2 = 8000$ Lb. The stiffness of bumpers on each car is $k = 12,000$ Lb/in. What will be the velocities following the impact, and what will be the overload factor to which an occupant of the lighter car would be exposed? Assume $\kappa = 1.0$.

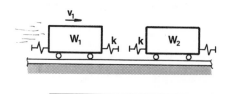

$$v_0 = \frac{W_1 v_1}{W_1 + W_2} = \frac{20{,}000 \times 10}{20{,}000 + 8000} = 7.143 \text{ mph} \qquad \text{(Eq. 13.1)}$$

$$V_1 = 2v_0 - v_1 = 2 \times 7.143 - 10 = 4.286 \text{ mph} \qquad \text{(Eq. 13.2)}$$

$$V_2 = 2v_0 - v_2 = 2 \times 7.143 - 0 = 14.286 \text{ mph} \qquad \text{(Eq. 13.2)}$$

The effective mass, Eq. 13.5, is:

$$\frac{1}{M^*} = \frac{386}{20{,}000} + \frac{386}{8000} \qquad \therefore M^* = 14.804 \text{ Lb-s}^2/\text{in.}$$

The relative velocity prior to impact is:

$$|v_r| = 10 - 0 = 10 \text{ mph} = 10 \times 5280 \times \frac{12}{3600} = 176 \text{ in./s}$$

The effective stiffness is:

$$\frac{1}{k^*} = \frac{1}{12{,}000} + \frac{1}{12{,}000} \qquad \therefore k^* = 6000 \text{ Lb/in.}$$

Equation 13.6 yields the impact force:

$$R_m = v_r (M^* k^*)^{1/2} = 176(14.804 \times 6000)^{1/2} = 52{,}454 \text{ Lb}$$

The overload factor to which an occupant of the lighter car would be subjected is $n_2 = R_m / W_2 = 6.557$. This means that if the entire weight W_2 is rigid, the acceleration to which it is subjected is 6.557 that of gravity. Including the flexibility of the structure itself in the calculation would reduce this number.

13-2 Two balls with velocities $v_1 = 80$ in./s and $v_2 = 30$ in./s, radii $r_1 = 1.0$ in. and $r_2 = 1.5$ in., respectively, collide. The material is steel, $E = 29 \times 10^6$ psi, $\nu = 0.3$, and $\gamma = 0.282$ Lb/in.^{3}. Based on the Hertz theory, calculate the maximum contact force R_m and the maximum local deflection δ_m. The collision is concentric.

If the Hertz theory is applied, the contact area characteristic is given by Eq. 6.14 as

$$R = k_0 \delta^{3/2}$$

and does not matter whether rigid-body motion of the impacted object is possible. Following the reasoning in Sect. 6E and replacing M by M^* and v by v_r in Eq. 6.15, we get

$$\delta_m = \left(\frac{5 M^* v_r^2}{4 k_0} \right)^{0.4}$$

Inserting the numbers,

$$M_1 = \frac{\gamma}{g} \times \frac{4}{3} \pi r_1^3 = \frac{0.282}{386} \times \frac{4}{3} \pi \times 1.0^3 = 3.06 \times 10^{-3} \text{ Lb-s}^2/\text{in.}$$

$$M_2 = 10.33 \times 10^{-3} \text{ Lb-s}^2/\text{in.}$$

$$v_r = v_1 - v_2 = 80 - 30 = 50 \text{ in./s}$$

When both bodies have the same material, Eq. 6.12 becomes

$$\delta = \left\{ \left[\frac{3}{2} \left(\frac{1 - \nu^2}{E} \right) \right]^2 \left(\frac{1}{r_1} + \frac{1}{r_2} \right) \right\}^{1/3} R^{2/3} = 15.456 \times 10^{-6} R^{2/3}$$

or

$$R = 16.456 \times 10^6 \delta^{3/2} \qquad \therefore k_0 = 16.456 \times 10^6$$

The effective mass is:

$$\frac{1}{M^*} = \frac{1}{M_1} + \frac{1}{M_2}$$

$$\therefore M^* = 2.361 \times 10^{-3} \text{ Lb-s}^2/\text{in.}$$

The maximum relative deflection is:

$$\delta_m = \left(\frac{5}{4} \frac{2.361 \times 10^{-3} \times 50^2}{16.456 \times 10^6} \right)^{0.4} = 2.888 \times 10^{-3} \text{ in.}$$

and maximum contact force is:

$$R_m = 16.456 \times 10^6 (2.888 \times 10^{-3})^{3/2} = 2554 \text{ Lb}$$

13-3 Develop an equation for the maximum collision force of two masses. The shock isolator attached to one of them consists of spring k and damper c in parallel.

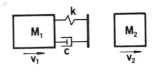

If a single mass M with this type isolator impacts a barrier, the maximum force is, according to Eq. 6.11,

$$R_m = v \left(\frac{Mk}{1 - \zeta^2} \right)^{1/2} \exp \left[\left(2\zeta - \frac{\pi}{2} \right) \zeta \right]$$

A similar relationship may be written for the contact force between the two colliding masses, based on the observation that during the loading phase of collision the system behaves like that described in Prob. 10-19. Here we have

$$R_m = v_r \left(\frac{M^* k}{1 - \zeta^2} \right)^{1/2} \exp \left[\left(2\zeta - \frac{\pi}{2} \right) \zeta \right]$$

in which

$$v_r = v_1 - v_2, \qquad \frac{1}{M^*} = \frac{1}{M_1} + \frac{1}{M_2}, \qquad \zeta = \frac{c}{2(kM^*)^{1/2}}$$

Note that this relationship is the extension of Eq. 13.6 in the same sense as Eq. 6.11 is the extension of Eq. 6.7.

13-4 A punch press, schematically shown in (a), is reduced to a simple model in (b) in the following manner. The slide with the upper die is depicted as a free-falling mass M_0 (driving mechanism not shown). The bed with the lower die, the columns with guides, and the crown are lumped into one mass M_1. The foundation M_2 is supported by spring k_{20} representing the ground stiffness. The isolators between the bed and the foundation are modeled by spring k_{12}. During stamping operation M_0 impacts M_1 with velocity v and rebounds with a coefficient κ. Using the energy method, calculate the maximum dynamic deflection of the foundation, u_{2m}, for the following parameters: $gM_0 = 330$ Lb, $gM_1 = 5500$ Lb, $gM_2 = 20,000$ Lb, $k_{12} = 35 \times 10^6$ Lb/in., $k_{20} = 5.5 \times 10^6$ Lb/in., $v = 280$ in./s, $\kappa = 0.5$.

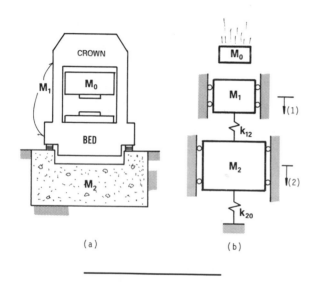

(a)　　　　　　(b)

Using Eqs. 13.1 and 13.2 with adjusted notation, we evaluate velocity V_1 of the impacted mass M_1:

$$v_0 = \frac{M_0 v}{M_0 + M_1} \qquad V_1 = v_0 + \kappa v_0$$

Substituting, we find $V_1 = 23.77$ in./s.

From this moment on, the striking mass M_0 is out of the picture. The kinetic energy of M_1 is converted into strain energy of the two-mass system when it reaches its stationary position or the end of the loading phase. It is further assumed that both masses are subjected to the same acceleration a at the stationary position giving rise to the following spring forces:

$$N_{12} = aM_1 \quad \text{and} \quad N_{20} = a(M_1 + M_2)$$

Strain energy in springs:

$$\Pi = \frac{1}{2k_{12}}(aM_1)^2 + \frac{1}{2k_{20}}(aM_1 + aM_2)^2$$

$$= \frac{a^2}{2}\left[\frac{M_1^2}{k_{12}} + \frac{(M_1 + M_2)^2}{k_{20}}\right] = 0.3996 \times 10^{-3} a^2$$

Equating Π with the kinetic energy:

$$T = \tfrac{1}{2}M_1 V_1^2 = 4025.4$$

we obtain $a = 3173.7$ in./s^2. The dynamic deflection of M_2 is

$$u_{2m} = \frac{a}{k_{20}}(M_1 + M_2) = 0.03812 \text{ in.}$$

13-5 A rod of length $L = 20$ in. and weight $W = 5$ Lb is falling down with a velocity $v = 150$ in./s when it hits a stop. The eccentricity of impact is $e = 3$ in., while the coefficient of restitution is $\kappa = 0.6$. Calculate the distribution of velocity along the axis of the rod just after the impact.

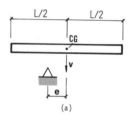

(a)

Moment of inertia about CG is $ML^2/12$, therefore

$$i^2 = J/M = L^2/12; \text{ then}$$

$$\frac{i^2}{e^2} = \frac{L^2}{12e^2} = 3.704$$

With $v_n = v$ and $V_n = V$ we find from Eq. 13.17

$$V = \frac{1 - \kappa(i/e)^2}{1 + (i/e)^2}v = \frac{1 - 0.6 \times 3.704}{1 + 3.704} \times 150 = -38.98 \text{ in./s}$$

According to the sign convention in Fig. 13.5, the negative sign of $V = V_n$ means that the CG is moving up. The λ is calculated from Eq. 13.14:

$$\lambda = \frac{1}{e}(\kappa v + V) = \frac{1}{3}(0.6 \times 150 - 38.98) = 17.01 \text{ rad/s}$$

This angular velocity results in

$$v_e = \lambda \frac{L}{2} = 17.01 \times 10 = 170.1 \text{ in./s at each end}$$

Superposing this pure rotation with a translation determined by V, we obtain the velocity plot as in (b).

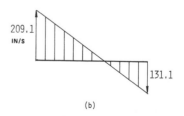

209.1
IN/S

131.1

(b)

13-6 A cylindrical container weighing 150,000 Lb is dropped from a height of 360 in. The spring shown at the impacting corner represents cask deformability, $k = 36.63 \times 10^6$ Lb/in. The surface on which the cask lands is to be considered rough; that is, no slipping is possible. Treating the cask as a homogeneous cylinder find the maximum impact force for $\alpha = 0°$ and $\alpha = 15°$, where α is the angle that the diagonal makes with the vertical, measured clockwise.

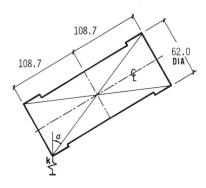

Moment of inertia about the CG:

$$J = \frac{W}{12g}(h^2 + 0.75D^2) = \frac{150,000}{12 \times 386}(217.4^2 + 0.75 \times 62^2)$$

$$= 1.6239 \times 10^6 \text{ Lb-s}^2\text{-in.}$$

The eccentricity parameters, according to Fig. 13.6,

$$p = (108.7^2 + 31^2)^{1/2} = 113.03 \text{ in.}$$

$$e_1 = p\cos 15° = 109.18 \text{ in.}$$

(Notice that these are needed only for $\alpha = 15°$, since the impact becomes central for $\alpha = 0$.) The mass is $M = 388.6$ Lb-s^2/in. The reduced mass, from Eq. 13.24, is

$$M_r = \frac{Me_1^2 + J}{Mp^2 + J}M = 369.0 \text{ Lb-s}^2\text{/in.}$$

The velocity of free drop from $H = 360$ in. is

$$v = (2gH)^{1/2} = 527.18 \text{ in./s}$$

The maximum vertical reaction component is calculated from Eq. 13.26,

$$R_m = v(M_r k)^{1/2} = 61.29 \times 10^6 \text{ Lb} \quad \text{for} \quad \alpha = 15°$$

When $\alpha = 0$, which means that the diagonal is vertical during impact, we simply put $M_r = M$ and obtain

$$R_m = 62.90 \times 10^6 \text{ Lb}$$

The reason for R_m to be larger during the impact along the

diagonal is that the entire kinetic energy is converted into strain energy at the end of the loading phase. The same cannot be said about the position corresponding to $\alpha = 15°$, because there is some angular velocity at the beginning of rebound. The maximum deflection for $\alpha = 0$ is

$$\delta_m = \frac{R_m}{k} = \frac{62.90 \times 10^6}{36.63 \times 10^6} = 1.7172 \text{ in.}$$

This is negligibly small in comparison with the drop height of 360 in., which makes the use of Eq. 13.26 justified.

13-7 An impulse of magnitude S is applied with an eccentricity e to a rod in (a). Calculate the reduced mass M_r of the rod defined as follows. M_r is the mass that, when subjected to S, acquires the same velocity as the impact point of the rod. In (b) two rods are moving against each other. What will be their respective velocities at the contact point after impact with $\kappa = 0.5$?

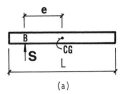

(a)

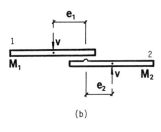

(b)

The off-center impulse is equivalent to impulse S applied at the CG plus the moment of impulse Se about the CG. The velocity of translation acquired is $V = S/M$ and the angular velocity is $\lambda = Se/J$, where J is the mass moment of inertia about the center. The resultant velocity at B:

$$V_B = \frac{S}{M} + \frac{Se^2}{J} = \frac{S}{M}\frac{J + Me^2}{J}$$

On the other hand, mass M_r subjected to S acquires the velocity $V = S/M_r$. By requesting that $V = V_B$ we get the expression for the mass of the rod reduced to point B

$$M_r = M\frac{J}{J + Me^2}$$

in which the denominator is the mass moment of inertia of the rod about point B. Introducing this concept allows us to replace the rod with a point mass to facilitate the calculation of post-impact velocities. In the problem shown in (b) we have

$$M_{r1} = \frac{M_1 J_1}{J_1 + M_1 e_1^2} \qquad M_{r2} = \frac{M_2 J_2}{J_2 + M_2 e_2^2}$$

The velocities are calculated with the aid of Eqs. 13.1 and 13.2:

$$v_0 = \frac{M_{r1} - M_{r2}}{M_{r1} + M_{r2}} v \qquad V_1 = 1.5v_0 - 0.5v$$

$$V_2 = 1.5v_0 + 0.5v \qquad \text{as} \quad v_1 = v, \text{ and } v_2 = -v$$

13-8 A simply supported beam has seven lumped masses, each of magnitude M_s, evenly spaced along the length. (The masses over the supports are not to be included.) Mass M impacts the beam at the center with a velocity v. What is the effective mass M_e that must be placed at the impact point to represent the inertia of the beam?

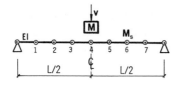

We can write Eq. 13.31 as

$$M_e = M_s \sum_{i=1}^{7} \left(\frac{X_i}{X}\right)^2$$

The static deflection curve for a prismatic beam loaded at the center is readily available, but to simplify the task somewhat we will use

$$X(x) = \sin\frac{\pi x}{L}$$

as the shape approximation. At the center $X(x) = 1$, therefore

$$\left(\frac{X_i}{X}\right)^2 = \left(\sin\frac{\pi x_i}{L}\right)^2$$

It follows that

$$M_e = M_s \sum_{i=1}^{7} \left(\sin\frac{\pi x_i}{L}\right)^2$$

$$= M_s\left[\left(\sin\frac{\pi}{8}\right)^2 + \left(\sin\frac{2\pi}{8}\right)^2 + \cdots + \left(\sin\frac{7\pi}{8}\right)^2\right]$$

$$= 4.0 M_s$$

If the figure is meant to show a lumped model of a beam with a uniformly distributed mass, then the entire mass of the beam is

$$M_b = 0.5 M_s + 7 M_s + 0.5 M_s = 8 M_s$$

after allowing for the two end masses. The result tells us that $0.5 M_b$ is to be placed at the center.

13-9 Calculate the impact force applied to the beam in Prob. 13-8. Find the maximum dynamic bending moment based on (1) the effective mass M_e at the center of beam and (2) actual mass distribution. Use $M = 10$ Lb-s^2/in., $M_s = 1$ Lb-s^2/in., $v = 100$ in./s, $L = 40$ in., and $EI = 20 \times 10^6$ Lb-in.^{2} Assume plastic impact.

The effective mass of the beam is

$$M_e = 4.0 M_s = 4.0 \text{ Lb-s}^2/\text{in.}$$

The stiffness of the simply supported beam, loaded at the center, is:

$$k = \frac{48EI}{L^3} = 48 \times 20 \times \frac{10^6}{40^3} = 15,000 \text{ Lb/in.}$$

Referring now to the model in Fig. 13.9 we have

$$T_1 = \tfrac{1}{2} M v_1^2 = \tfrac{1}{2} \times 10 \times 100^2 = 50,000 \text{ Lb-in.}$$

$$u_{st} = \frac{Mg}{k} = 10 \times 386/15,000 = 0.2573 \text{ in.}$$

The maximum deflection, Eq. 13.29, with $M_1 = M$,

$$u_m = 0.2573 + \left(0.2573^2 + \frac{10}{10+4}\frac{2 \times 50,000}{15,000}\right)^{1/2} = 2.4546 \text{ in.}$$

The dynamic reaction,

$$R_m = ku_m = 15,000 \times 2.4546 = 36,819 \text{ Lb}$$

The bending moment distribution caused by impact in part 1 of the problem is the same as from the static loading with the force of magnitude R_m applied to the center,

$$\mathfrak{M}_{max} = \tfrac{1}{4} R_m L = \tfrac{1}{4} \times 36,819 \times 40 = 368,190 \text{ Lb-in.}$$

The impacting mass M is included in the summation of inertial forces according to Eq. 13.32,

$$\frac{R_m}{\omega^2 f} = 2 M_s\left[\sin\left(\frac{\pi}{8}\right) + \sin\left(\frac{2\pi}{8}\right) + \sin\left(\frac{3\pi}{8}\right)\right]$$

$$+ (M_s + M)\sin\left(\frac{4\pi}{8}\right) = 15.027$$

On the other hand,

$$R_m = 36,819 \text{ Lb}, \qquad \therefore \ \omega^2 f = 2450.1$$

After calculating the inertia forces in accordance with Sect. 13H, we find the maximum bending moment at the center to be 325,480 Lb-in. This is about 12% less than the result found using the simplified method.

13-10 A weight $W = 100$ Lb is dropped freely from the height $h = 10$ in. onto a flange of a column consisting of three distinct segments. Find the distribution of axial forces due to landing of the weight, assuming the impact to be concentric. Use $L = 15$ in., $A = 8$ in.^{2}, and $E = 10 \times 10^6$ psi. The weight is lumped into $W_1 = 84.6$ Lb; $W_2 = 50.8$ Lb, and $W_3 = 17.6$ Lb.

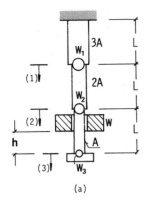

(a)

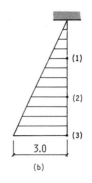

(b)

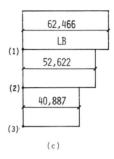

(c)

Assume the deflection curve $X(x)$ to be the straight line as shown in (b), which gives

$$X_1 = 1 \qquad X_2 = 2 \qquad X_3 = 3$$

The effective weight W_e located at node 3 is found from Eq. 13.31.

$$W_e = 84.6 \left(\tfrac{1}{3}\right)^2 + 50.8 \left(\tfrac{2}{3}\right)^2 + 17.6 = 49.58 \text{ Lb}$$

The deflection of the column under the load P applied to the end is:

$$u = \frac{PL}{3EA} + \frac{PL}{2EA} + \frac{PL}{EA} = 1.8333 \frac{PL}{EA} = \frac{P}{2.909 \times 10^6}$$

stiffness $k = \dfrac{P}{u} = 2.909 \times 10^6$ Lb/in.

We now have all the parameters of the model in Fig. 13.10. The maximum deflection is calculated from Eq. 13.29.

$$u_{st} = \frac{W}{k} = \frac{100}{2.909 \times 10^6} = 34.376 \times 10^{-6} \text{ in.}$$

The kinetic energy is the same as Wh,

$$T_1 = Wh = 100 \times 10 = 1000 \text{ Lb-in.}$$

$$u_m = u_{st} + \left(u_{st}^2 + \frac{100}{100 + 49.58} \times \frac{2 \times 1000}{2.909 \times 10^6} \right)^{1/2}$$

$$= 21.473 \times 10^{-3} \text{ in.}$$

The impact load is therefore

$$R_m = ku_m = 62{,}466 \text{ Lb}$$

The inertia forces are calculated using Eq. 13.32,

$$P_i = \omega^2 M_i X_i f$$

$$\frac{R_m}{\omega^2 f} = \frac{84.6 \times 1 + 50.8 \times 2 + 17.6 \times 3 + 100 \times 3}{386} = 1.3964$$

which gives $\omega^2 f = 44{,}734$ and

$$P_1 = 9804 \text{ Lb} \qquad P_2 = 11{,}775 \text{ Lb} \qquad P_3 = 40{,}887 \text{ Lb}$$

The distribution of axial forces is shown in (c).

13-11 The axial bar with three lumped masses is acted on by the unbalanced force $P = 200$ Lb. Show the dynamic equilibrium under the active and the inertia forces, where M_1, M_2, and M_3 are 10, 2 and 8 Lb-s^2/in., respectively.

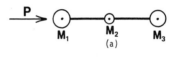

(a)

The rigid-body acceleration is:

$$a = \frac{P}{M} = \frac{200}{10 + 2 + 8} = 10 \text{ in./s}^2$$

The inertia forces are:

$$R_1 = M_1 a = 10 \times 10 = 100 \text{ Lb}$$

$$R_2 = 2 \times 10 = 20 \text{ Lb}$$

$$R_3 = 8 \times 10 = 80 \text{ Lb}$$

In the free-body diagram (b), note that $P - R_1 = 200 - 100 = 100$ Lb is the resultant effectively applied to M_1.

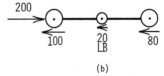

(b)

13-12 A two-segment beam can be impacted either at the center (1) or at the end (2). Using the concept of dynamic equilibrium, derive a spring constant representing beam flexibility for each of the impacts. The mass is M and the bending stiffness is EI.

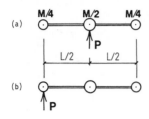

1. The acceleration is $a = P/M$ (see (a)). The system of active and inertia forces is such that from the static stiffness viewpoint we have a beam loaded at the center and supported at the ends. The center load is now designated by Q, since its magnitude has no effect on stiffness, which is

$$k = \frac{Q}{\delta} = \frac{48EI}{L^3}$$

2. The linear acceleration is $a = P/M$, while the angular

$$\varepsilon = \frac{PL}{2J} \quad (\text{see } (b)).$$

$$J = 2\left(\frac{L}{2}\right)^2 \frac{M}{4} = \frac{1}{8}ML^2 \qquad \therefore \varepsilon = \frac{4P}{ML}$$

The net effect of both the translational and angular actions (including the active and the inertia forces) is such that we can treat the beam as loaded at the left end and supported at the center and right end.

The tip deflection may be determined by thinking of the two-span beam as of a cantilever with an angular spring of stiffness K at the right end,

$$K = \frac{3EI}{L/2} = \frac{6EI}{L}$$

The moment $\mathfrak{M} = QL/2$ applied at the base of the cantilever gives the slope

$$\vartheta = \frac{\mathfrak{M}}{K} = \frac{QL}{2}\frac{L}{6EI} = \frac{QL^2}{12EI}$$

The total tip deflection is

$$\delta = Q\frac{(L/2)^3}{3EI} + \frac{L}{2}\frac{QL^2}{12EI} = \frac{QL^3}{12EI}$$

The flexural stiffness with respect to the tip load is thus

$$k = \frac{Q}{\delta} = \frac{12EI}{L^3}$$

13-13 A beam with a square, 4×4 in. cross section and length $L = 50$ in. is falling with the velocity $v = 200$ in/s. when its end hits a stationary wedge. Calculate the maximum bending moment in the beam resulting from that impact. Use the three-mass model

together with the results of Prob. 13-12. The material is steel, $E = 29 \times 10^6$ psi and $\gamma = 0.282$ Lb/in.^{3}

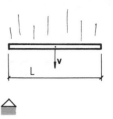

The first step is to find the reduced mass from Eq. 13.23

$$M_r = \frac{J}{J + Me^2}M$$

The moment of inertia about the CG is $J = \frac{1}{12}ML^2$, while $e = L/2$,

$$\therefore M_r = \frac{L^2}{L^2 + 12e^2}M = \frac{M}{4}$$

The second step is a determination of beam stiffness, when impacted at the end. This was done in Prob. 13-12, using a three-mass model of the beam,

$$k = \frac{12EI}{L^3}$$

The impact force may now be found

$$R_m = v(M_r k)^{1/2} = v\left(\frac{M}{4}\frac{12EI}{L^3}\right)^{1/2} = v\left(\frac{3MEI}{L^3}\right)^{1/2}$$

The mass is:

$$M = 4 \times 4 \times 50 \times \frac{0.282}{386} = 0.5845 \text{ Lb-s}^2/\text{in.}$$

The area moment of inertia is:

$$I = \frac{4^4}{12} = 21.333 \text{ in.}^4$$

and the impact force is:

$$R_m = 200\left(3 \times 0.5845 \times 29 \times 10^6 \frac{21.333}{50^3}\right)^{1/2} = 18,632 \text{ Lb}$$

To find the maximum bending moment, we must again return to the solution of Prob. 13-12. The load P at the end of the beam is reduced by the inertia force of the lumped mass at this location. The net resultant is $P/4$. Our maximum bending moment is therefore

$$\mathfrak{M} = \frac{R_m}{4}\frac{L}{2} = \frac{1}{8} \times 18,632 \times 50 = 116,450 \text{ Lb-in.}$$

13-14 A beam with properties described in Prob. 13-13 impacts a frictionless surface at an angle of $60°$ with the horizontal. The velocity of drop, $v_0 = 200$

in./s, is directed vertically. Calculate the impact force taking the axial and bending stiffness of the bar, respectively, to be

$$k_a = \frac{EA}{L} \quad \text{and} \quad k_b = \frac{12EI}{L^3}$$

Assume that the effect of local deformation of the bar itself and the surface on which it drops is to double the flexibility computed on the basis of k_a and k_b alone.

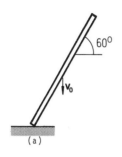

(a)

The moment of inertia about the CG is $J = ML^2/12$. To calculate the reduced mass M_r from Eq. 13.23 we put

$$e^2 = \left(\frac{L}{2}\cos 60°\right)^2 = \frac{L^2}{16}$$

and

$$M_r = \frac{L^2/12}{(L^2/12)+(L^2/16)}M = 0.5714M$$

It is shown in (b) that the resultant surface reaction has the bending component R_{mb} and the axial component R_{ma}. It follows that the combined bar flexibility along R_m is

$$\frac{1}{k^*} = \frac{\sin^2\alpha}{k_a} + \frac{\cos^2\alpha}{k_b}$$

With our data, this gives

$$\frac{1}{k^*} = \frac{L}{EA}\times 0.75 + \frac{L^3}{12EI}\times 0.25 = 80.82\times 10^{-9} + 4.209\times 10^{-6}$$

$$\therefore k^* = 233{,}090 \text{ Lb/in.}$$

By including the effect of the other deformation components, as given in the problem statement, the stiffness is decreased to

$$k_{ef} = \frac{k}{2} = 116{,}545 \text{ Lb/in.}$$

The mass of the bar is $M = 0.5845$ Lb-s^2/in., from Prob. 13-13. Equation 13.26 is used to determine the impact force:

$$R_m = v_0(M_r k_{ef})^{1/2} = 200(0.5714\times 0.5845\times 116{,}545)^{1/2}$$

$$= 39{,}460 \text{ Lb}$$

This is considerably more than was calculated in the reference problem for the same impact velocity, but different orientation. Note that the offset due to finite depth of the beam was ignored in the calculation of e.

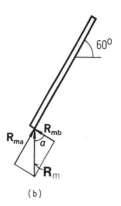

(b)

EXERCISES

13-15 The first body is moving with velocity v_1 and the second with v_2. If their masses are the same, what will their velocities be after an elastic ($\kappa = 1$) collision?

13-16 Two balls rolling without friction with velocities $v_1 = 10$ in./s and $v_2 = -5$ in./s collide and as a result the first ball stops. The restitution coefficient is $\kappa = 0.9$. What is the mass ratio of the balls?

13-17 The first ball is suspended on a thread, while the second, also restrained by a thread, is released from a horizontal position with no initial velocity. Calculate the angle α that determines the peak point of the first ball after it is struck by the second one. The coefficient of restitution is $\kappa = 0.85$.

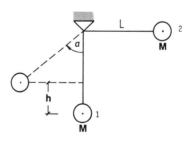

13-18 Consider two rail cars as described in Prob. 13-1. The light one is standing at a distance $L = 80$ in. from the rigid barrier. After the first collision, one expects that car to rebound from the barrier and a second collision to follow. Is there enough time for the lighter car to completely rebound before the second collision? If so, calculate the maximum force of the second contact between the cars. Assume all rebounds to be elastic, $\kappa = 1.0$.

13-19 From Sect. 127 of Ref. 3 it is known that the time of the loading phase of collision between two bodies deforming according to the Hertz theory is

$$t_1 = \frac{1.47\delta_m}{v_r}$$

where δ_m is the maximum relative deflection and v_r is the relative velocity prior to contact. Using the results obtained in Prob. 13-2, calculate t_1 by this formula and then perform the approximate calculation according to Sect. 13C.

13-20 The figure shows two identical bars moving against each other. Each has velocity $v = 300$ in./s with respect to a fixed frame, cross section $A = 2$ in.^{2} and length $L = 16$ in. The material is steel, $E = 29 \times 10^6$ psi, $\gamma = 0.282$ Lb-s^2/in.^{4} The impact is concentric and the faces of bars are flat. Lump the mass of each bar at the end away from the impact point. Does the impact stress exceed the yield stress of material, $\sigma_y = 50,000$ psi?

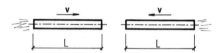

13-21 A bar of length $L_1 = 20$ in., circular cross section, and a spherical end with $r_1 = 1.0$ in., moving with the velocity $v = 50$ in./s, collides with a similar bar at rest, $L_2 = 30$ in. and $r_2 = 1.5$ in. The material is steel, $E = 29 \times 10^6$ psi, $\gamma = 0.282$ Lb/in.^{3} The local characteristic of two spheres of such properties is calculated in Prob. 13-2 as $R = k_0 \delta^{3/2}$ with $k_0 = 16.456 \times 10^6$. Find the maximum force of collision R_m using (1) local flexibility only and (2) both general flexibility and local flexibility.

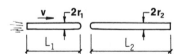

13-22 Develop an equation for the maximum collision force of the two masses, each equipped with a shock isolator.

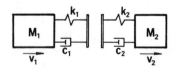

13-23 Solve Prob. 13-4 using the normal-mode method to calculate the response to the initial condi-

tions. Find only the upper bound of the dynamic deflection rather than the actual maximum.

13-24 Consider again Prob. 13-1 with two impacting cars. What is the overload factor if a damper, $c = 310$ Lb-s/in., is attached parallel to each spring?

13-25 A more sophisticated model of the punch press in Prob. 13-4 is employed here. The crown and part of columns are now treated as a separate mass, so that $gM_3 = 1800$ Lb, and $gM_1 = 3700$ Lb, while $M_2 = 20,000$ Lb as before. The additional spring is $k_{13} = 24 \times 10^6$ Lb/in. The overall damping ratio is $\zeta = 0.02$. Keeping the rest of parameters as in the reference problem, find the dynamic deflection of M_2 using the energy method. (Assume an undamped system at first and introduce the damping correction at the end.)

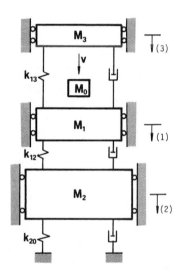

13-26 A sphere with mass M and velocity v is approaching a surface at an angle α with the normal. The surface is frictionless and the coefficient of restitution is κ. Calculate the angle of rebound β (with the normal) and the rebound velocity V.

13-27 The force-deflection characteristic of a cask in Prob. 13-6 was assumed to be a straight line. The actual relationship is presented as a set of points, tabulated below. Calculate the maximum contact force and the associated displacement. Assume central impact.

Point	δ (in.)	R (10^6 Lb)
1	0.02869	2.182
2	0.04126	3.118
3	0.10348	6.235
4	0.3404	12.47
5	1.7752	31.18
6	5.297	62.35

13-28 A prismatic rod of length $L=20$ in. is impacted at $c=15$ in. below its pivot point. Calculate the velocities immediately following the elastic impact. What should be the coefficient of restitution κ if the impacting ball is to drop vertically under its own weight following impact? Use $W_1=2$ Lb, $W_2=5$ Lb, and $v=30$ in./s.

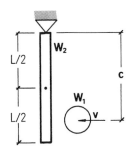

13-29 What should the distance c in Prob. 13-28 be to avoid a reaction at the pivot point?

13-30 A plate in the shape of an ellipse is free to rotate about an axis located at a distance d from the center. An impact load is applied at point B, normal to the plate. What should be the value of d to ensure that there is no reaction in the bearings holding the axis when an impact takes place? The moment of inertia of the plate about the x-axis is $J_x=\frac{1}{4}Ma^2$.

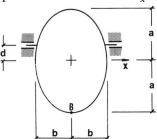

13-31 Calculate the impact force as a result of the mass of magnitude $1.5M$ striking the tip of the cantilever beam with velocity $v=250$ in./s. Also find the maximum bending moment based on the effective mass at the impacted tip. Use the parabolic approximation of the deflected shape, $X=(x/L)^2$ and the following data: $M=0.8$ Lb-s^2/in., $L=60$ in., and $EI=50\times10^6$ Lb-in.^{2}

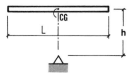

13-32 A beam with length $L=480$ in. and cross section having $I=15{,}375$ in.^{4} is dropped from a height $h=48$ in. onto a stationary wedge. Calculate the maximum bending moment in the beam resulting from that impact. Use the three-mass model together with the results of Prob. 13-12. The material is steel, $E=29\times10^6$ psi and $\gamma=0.282$ Lb/in.^{3} The beam weighs 25.8 Lb per inch of length.

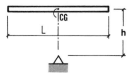

13-33 A shaft with three discs is acted on by an unbalanced torque $M_t=5000$ Lb-in. applied to the center disc. Show the dynamic equilibrium under the active and inertia torques for $J=25$ Lb-s^2-in.

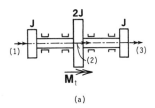

(a)

13-34 An unbalanced force $P=5000$ Lb is applied to the left end of a three-mass beam. What is the dynamic equilibrium under the active and inertia forces? Use $M=20$ Lb-s^2/in. and $a=15$ in.

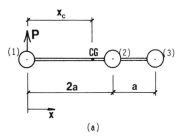

(a)

13-35 Force $P=1200$ Lb is suddenly applied to the right end of a beam that can pivot about its left end. Treating the beam as rigid, calculate the reaction at the pivot point. Use $M=6$ Lb-s^2/in. and $b=8$ in.

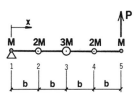

13-36 An unbalanced force $P = 10,000$ Lb is applied to the right end of a beam. Calculate the support reaction, assuming the dynamic equilibrium condition. Use $M = 5$ Lb-s^2/in. and $c = 10$ in.

13-37 The material of the beam in Prob. 13-14 begins to yield when the maximum dynamic stress exceeds 36,000 psi. Is the described impact strong enough to cause yielding?

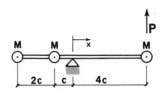

CHAPTER

14

Miscellaneous Problems

One of the distinct types of problems considered in this chapter is the dynamic behavior of a rigid block attached to an elastic half-space, the practical application being that of a structure resting on soil. The second topic is the concept of a shear beam and a shear plate. The remaining problems are associated with the ideas developed in previous chapters.

faces xy are presumed to be attached to an elastic half-space. As illustrated in Fig. 14.2a, the half-space is characterized by constants E, G, and ν, related by a well-known equation,

$$G = \frac{E}{2(1+\nu)}$$

From the point of view of the elastic resistance, the half-space may be replaced by a set of springs attached at the base of a block, Fig. 14.2b. The spring constants for the round block are, after Ref. 18,

$$k_z = \frac{4Gr_0}{1-\nu} \qquad k_y = \frac{32(1-\nu)Gr_0}{7-8\nu}$$

$$K_\psi = \frac{8Gr_0^3}{3(1-\nu)} \qquad K_\phi = \frac{16}{3}Gr_0^3 \qquad (14.1)$$

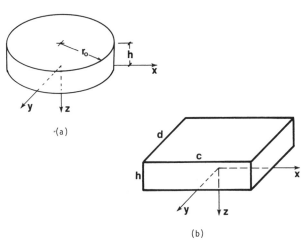

(a)

(b)

Figure 14.1 Rigid blocks under consideration: (a) round; (b) rectangular.

14A. Rigid Block on Elastic Half-Space. Figure 14.1 shows two types of rigid block, the motion of which will be the subject of our investigation. Their

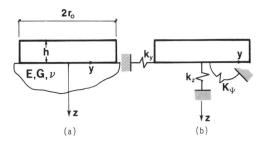

(a)   (b)

Figure 14.2 (a) Block on elastic half-space; (b) block on equivalent springs.

The last constant, K_ϕ, pertains to twisting deformation about the z-axis. When the block is rectangular, as in Fig. 14.1b, the following modifications must be made:

$$r_0 = \left(\frac{cd}{\pi}\right)^{1/2} \quad \text{in} \quad k_z \text{ and } k_y \qquad (14.2a)$$

$$r_0' = \left[\frac{cd}{6\pi}(c^2 + d^2)\right]^{1/4} \quad \text{in} \quad K_\phi \text{ (instead of } r_0) \qquad (14.2b)$$

$$K_\psi = \frac{Gcd^2}{1-\nu}\left(0.3923 + \frac{d}{10.59c}\right) \qquad (14.2c)$$

The last equation is valid for rocking about the x-axis. When this motion takes place about the y-axis, the symbols c and d must be interchanged.

14B.   Shear Beam.   The lateral deflection of a two-dimensional beam has two components—bending and shear. The first is due to curving of the axis, while the second results from distortion of elements normal to the axis. When only the latter is accounted for, we speak of a *shear beam* as depicted in Fig. 14.3. This

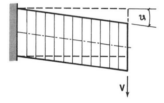

Figure 14.3   Shear beam.

model can be thought of as consisting of thin layers capable of shear distortion, but infinitely rigid in the direction normal to their thickness. The shear angle ϑ is defined by

$$\vartheta = \frac{V}{GA_s}$$

where A_s is the shear area. When the lateral force is constant along the length as in Fig. 14.3, V is the same for every vertical slice of the beam. The tip deflection is then

$$u = \vartheta L = \frac{VL}{GA_s} \qquad (14.3)$$

Although some structures behave nearly like ideal shear beams, the main usefulness of this concept is in

visualizing the shear component of deflection of actual beams. The stiffness matrix, which includes shear deformation, is given by Eq. 8.14. When the lateral force is continously changing along the length and the shear area of the cross section is also varying, one can use the third integral in Eq. 8.20 to calculate this deflection component. The influence of shear is likely to be substantial when one or more of the following conditions take place:

1.   The beam is short, say a cantilever whose length is only twice its depth.
2.   The cross section is hollow or branched (as opposed to a compact section like a solid circle or rectangle).
3.   The beam is of sandwich type with a lightweight core.

By definition, the elastic rotation of a cross section cannot take place in a shear beam. From this viewpoint it makes no difference whether the end section is completely fixed or merely restrained against translation. The fixity in Fig. 14.3 is needed for rigid-body equilibrium.

14C.   Shear Plate.   The relation between shear and flexural deflections for plates is similar to that previously discussed for beams. The shear component of deflection may be visualized by introducing a concept of a *shear plate*. Consider a circular plate in Fig. 14.4.

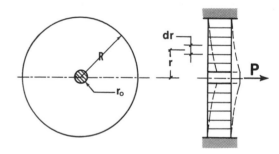

Figure 14.4   Shear plate.

The picture of shear deformation in a radial section from center to edge is similar to that in Fig. 14.3. The plate model can be thought of as consisting of thin rings capable of shear distortion, but infinitely rigid in the radial direction. The only major difference is that now the load P is applied through a rigid insert with a small, but finite radius r_0.

The shear angle ϑ is defined similarly as for a beam. The shear area is now referred to a circumferential ring section:

$$A_s = 2\pi r h_s$$

where h_s is a shear thickness. (For a solid plate with actual thickness h we have $h/h_s = 1.2$, because for a beam with a rectangular section $A/A_s = 1.2$.) The increment of center deflection due to a ring with a radial width dr is

$$du = \vartheta \, dr = \frac{P \, dr}{2\pi r h_s G}$$

integrating from r_0 to R gives us

$$u_c = \frac{P}{2\pi h_s G} \ln \frac{R}{r_0} \qquad (14.4)$$

The conditions under which this shear deflection is significant in comparison with the flexural component are analogous to those previously spelled out for beams.

For this shear plate model there is no difference between laterally supported and completely fixed edges. If the loading is applied not at a center point, but on a radius R_0, Eq. 14.4 remains valid with R_0 replacing r_0. The boundary conditions inside R_0 will not matter then; it can be the elastic plate, or the rigid insert, or the hole with radius R_0.

SOLVED PROBLEMS

14-1 A rigid block rests on soil with the following properties: $G = 7500$ psi and $\nu = 0.35$. Calculate the constants of six springs associated with the six degrees of freedom of the block.

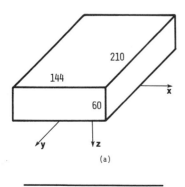

(a)

Using Eqs. 14.1 and 14.2 we have

$$r_0 = \left(144 \times \frac{210}{\pi}\right)^{1/2} = 98.11 \text{ in.}$$

$$k_z = \frac{4Gr_0}{1-\nu} = \frac{4 \times 7500 \times 98.11}{1 - 0.35} = 4.528 \times 10^6 \text{ Lb/in.}$$

$$k_x = k_y = \frac{32(1-\nu)Gr_0}{7-8\nu} = \frac{32(1-0.35) \times 7500 \times 98.11}{7 - 8 \times 0.35}$$

$$= 3.644 \times 10^6 \text{ Lb-in.}$$

$$r_0' = \left[\frac{cd}{6\pi}(c^2 + d^2)\right]^{1/4} = \left[\frac{144 \times 210}{6\pi}(144^2 + 210^2)\right]^{1/4} = 101 \text{ in.}$$

$$K_\phi = \frac{16}{3} Gr_0'^3 = \frac{16}{3} \times 7500 \times 101^3 = 41.212 \times 10^9 \text{ Lb-in./rad}$$

$$K_{\psi x} = \frac{Gcd^2}{1-\nu}\left(0.3923 + \frac{d}{10.59c}\right) = 38.836 \times 10^9 \text{ Lb-in./rad}$$

Similarly

$$K_{\psi y} = \frac{7500 \times 210 \times 144^2}{1 - 0.35}\left(0.3923 + \frac{144}{10.59 \times 210}\right)$$

$$= 22.964 \times 10^9 \text{ Lb-in./rad}$$

Some of these constants are illustrated in (b) and Fig. 14.2b.

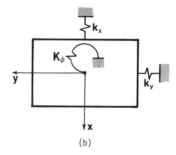

(b)

14-2 Calculate the center deflection of the shear plate in Fig. 14.4 due to uniformly distributed load q (psi).

The total shear force acting on a central portion of the plate with radius r is

$$V = \pi r^2 q$$

which one can easily find from the equilibrium condition. This force acting on the shear area

$$A_s = 2\pi r h_s$$

gives the distortion angle

$$\vartheta = \frac{V}{GA_s} = \frac{qr}{2Gh_s}$$

Following the procedure in Sect. 14C, the center deflection is computed as

$$u_c = \int_0^R \vartheta\, dr = \frac{q}{2Gh_s} \int_0^R r\, dr = \frac{qR^2}{4Gh_s}$$

Notice that there was no need to introduce a rigid insert, since the loading did not have a singularity at the center.

14-3 A body of mass M and moment of inertia J with respect to the CG is restrained by a translational spring and a rotational spring, having constants k and K, respectively. Considering the horizontal translation of the CG as well as rotation about it, establish the frequency equation for the system.

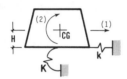

Sketching the displaced pattern as well as the forces associated with it, first for u_1 and then for u_2, we obtain the stiffness matrix:

$$\mathbf{k} = \begin{bmatrix} k & -kH \\ -kH & (K+kH^2) \end{bmatrix}$$

while the mass matrix is diagonal. The frequency equation (9.13) becomes

$$\begin{vmatrix} (k-\omega^2 M) & -kH \\ -kH & (K+kH^2-\omega^2 J) \end{vmatrix} = 0$$

It is evident that the translation u_1 and the rotation u_2 are coupled.

14-4 In Prob. 14-1 we found the six elastic spring constants characterizing the half-space under a rectangular block. Assuming the weight density of the concrete block to be 150 Lb/ft^3, compute the natural frequencies of the system.

The block weight W is $V\gamma = 144 \times 210 \times 60 \times 150/12^3 = 157{,}500$ Lb, and mass $M = 408.03$ Lb-s^2/in. The moments of inertia about the axes through the CG are:

$$J_x = \frac{M}{12}(d^2+h^2) = \frac{408.03}{12}(210^2+60^2) = 1.622 \times 10^6 \text{ Lb-s}^2\text{-in.}$$

$$J_y = 0.8275 \times 10^6 \qquad J_z = 2.205 \times 10^6 \text{ Lb-s}^2\text{-in.}$$

The only uncoupled vibratory components are vertical translation and torsion. Their respective frequencies are:

$$\omega_z^2 = \frac{k_z}{M} = \frac{4.528 \times 10^6}{408.03} \qquad \therefore \omega_z = 105.3 \text{ rad/s}$$

$$\omega_\phi^2 = \frac{K_\phi}{J_z} = \frac{41.212 \times 10^9}{2.205 \times 10^6} \qquad \therefore \omega_\phi = 136.7 \text{ rad/s}$$

Figure 14-1b shows which planes of reference are associated with which stiffness parameters. The frequency equation for the yz-plane is obtained from Prob. 14-3, after an appropriate adjustment in notation.

$$\begin{vmatrix} k_y - \omega^2 M & -k_y H \\ -k_y H & (K_{\psi x} + k_y H^2 - \omega^2 J_x) \end{vmatrix} = 0$$

But

$$k_y H = 3.644 \times 10^6 \times 30 = 109.32 \times 10^6$$

and

$$K_{\psi x} + k_y H^2 = 42.116 \times 10^9$$

Therefore

$$\begin{vmatrix} (3.644 \times 10^6 - 408.03\omega^2) & -109.32 \times 10^6 \\ -109.32 \times 10^6 & (42.116 \times 10^9 - 1.622 \times 10^6 \omega^2) \end{vmatrix} = 0$$

$$\therefore \quad \omega_{yz1} = 89.05 \quad \text{and} \quad \omega_{yz2} = 164.22 \text{ rad/s}$$

A similar frequency equation will hold for the xz-plane, where we have

$$\omega_{xz1} = 86.44 \quad \text{and} \quad \omega_{xz2} = 182.13 \text{ rad/s}$$

14-5 A simply supported segment of a pipe with $D = 24$ in. and wall thickness $t = 0.968$ in. has a distributed mass $m = 0.05141$ Lb-s^2/in.^{2} Using a lumped-mass model with M at the center point, calculate the natural frequencies for $L/D = 2$, 5, and 10. The cross-sectional and material data are:

$$A = 70.11 \text{ in.}^2 \qquad I = 4650 \text{ in.}^4$$

$$E = 25.7 \times 10^6 \text{ psi} \qquad G = 9.9 \times 10^6 \text{ psi}$$

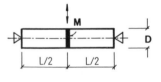

The tributary length approach explained in Sect. 9G implies one-half of the total mass lumped at the center with the remainder at the support points. Thus

$$M = \tfrac{1}{2}mL$$

The flexural and shear deflections, when P is applied at the center are, respectively:

$$u_b = \frac{PL^3}{48EI} \qquad \text{and} \qquad u_s = \frac{PL}{4GA_s}$$

The shear area of a thin annular section is approximately $A_s = 0.5A$.

The resultant deflection is a sum of the bending and the shear components, which gives us the effective stiffness k^* as

$$\frac{1}{k^*} = \frac{L^3}{48EI} + \frac{L}{2GA}$$

The natural frequency is found from

$$\frac{1}{\omega^2} = \frac{M}{k^*} = L^2\left(\frac{m}{4}\right)\left(\frac{L^2}{24EI} + \frac{1}{GA}\right)$$

Substituting $L=48$, 120, and 240 in., in turn, we obtain:

$$\omega = 3879 \qquad \omega = 914.5 \qquad \omega = 250.5 \text{ rad/s}$$

The details of calculation indicate that for $L/D=2$ the influence of shear deflection is predominant, while for $L/D=10$ it is negligible.

Since the flexural and shear beam models have additive deflections, we can think of them as of springs connected in series and use the appropriate equation developed in Prob. 7-1. This is useful when the natural frequencies of both models are separately computed at first.

14-6 Calculate an approximate natural frequency of the first symmetric mode for a pipe segment from Prob. 14-5, using a three-segment model. Take into account not only both types of flexibility, but also both types of inertia. Perform the calculation for $L/D=2$, 5, and 10.

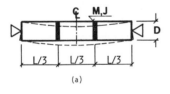

(a)

The natural frequency may be approximated by

$$\frac{1}{\omega^2} = \left(\frac{1}{\omega_{11}^2} + \frac{1}{\omega_{22}^2}\right) + \left(\frac{1}{\omega_{33}^2} + \frac{1}{\omega_{44}^2}\right)$$

To explain and derive this formula, let us first ignore the rotatory inertia and define the following component frequencies:

ω_{11} natural frequency when only bending deflections are taken into account

ω_{22} natural frequency when only shear deflections are considered

Since the lateral deflections due to bending and shear are additive, we can refer to Prob. 7-1 to see that

$$\frac{1}{\omega_M^2} = \frac{1}{\omega_{11}^2} + \frac{1}{\omega_{22}^2}$$

defines the approximate frequency (series connection analogy). In the next step we ignore the translational masses and concern ourselves only with the rotatory inertias J. Defining ω_{33} and ω_{44}

just like ω_{11} and ω_{22} before, we again have

$$\frac{1}{\omega_J^2} = \frac{1}{\omega_{33}^2} + \frac{1}{\omega_{44}^2}$$

Recalling the Dunkerley method (Eq. 11.3), the formulas defining ω_M and ω_J are combined into the equation given at the outset. The lumped inertias in our problem are

$$M = \tfrac{1}{3}mL \qquad \text{and} \qquad J = \tfrac{1}{3}\rho IL = \tfrac{1}{3}mL\frac{I}{A}$$

The component frequency ω_{11} can be found from the answer to Prob. 9-25,

$$\omega_{11}^2 = \frac{162}{5}\frac{EI}{ML^3} = 97.2\frac{EI}{mL^4}$$

From symmetry principles we know that there is zero shear at the center segment. When we think of shear deformations only (i.e., when no section rotation is allowed), each end segment is a cantilever with a flexibility and frequency, respectively, of

$$\frac{u}{P} = \frac{L}{3GA_s} = \frac{L}{1.5GA} \qquad \text{and} \qquad \frac{1}{\omega_{22}^2} = \frac{mL}{3}\frac{L}{1.5GA} = \frac{mL^2}{4.5GA}$$

The effect of the rotatory inertia on the beam in a symmetric mode is shown in (b). The action of two equal and opposite bending moments influences only the center portion of the beam. Each inertia may be thought of as being placed at the end of a cantilever of length $L/6$. The natural frequency is found from

$$\frac{1}{\omega_{33}^2} = J\frac{L}{6EI} = \frac{mL^2}{18EA}$$

There is no shear deflections associated with the rotatory inertia load, and we may put

$$\frac{1}{\omega_{44}^2} = 0$$

Combining all expressions,

$$\frac{1}{\omega^2} = mL^2\left(\frac{L^2}{97.2EI} + \frac{1}{4.5GA} + \frac{1}{18EA}\right)$$

The second term in parentheses, associated with shear deflection, is about 10 times as large as the third, representing the influence of rotatory inertia. Substituting the numbers,

$$\frac{1}{\omega^2} = mL^2\left(\frac{L^2}{11.616} + 351\right)10^{-12}$$

for $L=48$, 120, and 240 in., we obtain, respectively,

$$\omega = 3920 \text{ rad/s} \qquad \omega = 921.52 \text{ rad/s} \qquad \omega = 252.19 \text{ rad/s}$$

Compared with the results of Prob. 14-5, the differences are quite small.

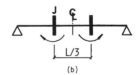

(b)

14-7 A shaft with length L and bending stiffness EI has n equally spaced discs, each with a moment of inertia J about the diametral axis normal to the paper. Assuming that the true deflected shape of the first bending mode is given by

$$X(x) = \sin\frac{\pi x}{L}$$

find the corresponding natural frequency using the Rayleigh method. Assume the discs to be so closely spaced that we can use a distributed rotatory inertia $j = nJ/L$.

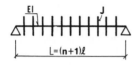

According to the formulation of Sect. 11F, the deflection and slope are given, respectively, by

$$u = X(x) \cdot f(t) \qquad \text{and} \qquad u'(x) = X'(x) \cdot f(t)$$

The strain energy may be calculated by using the second integral in Eq. 11.19, since the bending moment is determined by the inertia forces. However, it is more convenient to employ a different expression here. It is known from the strength of materials that

$$\frac{\mathcal{M}}{EI} = \frac{1}{\rho} = u''$$

where ρ is beam curvature. The strain energy is therefore

$$\Pi = \int_0^L \frac{1}{2EI}(EIu'')^2\,dx = \frac{1}{2}\int_0^L EI(u'')^2\,dx$$

It is permissible to use this expression only when $X(x)$ does not deviate much from the actual deflected shape, otherwise the error in the second derivative may become very large. Here we have

$$u''(x) = -\left(\frac{\pi}{L}\right)^2 f\sin\frac{\pi x}{L} \qquad \text{and} \qquad \Pi = \frac{1}{2}EI\left(\frac{\pi}{L}\right)^4 f^2\frac{L}{2}$$

Taking advantage of the fact that the discs are densely spaced, we may write the kinetic energy expression similarly to Eq. 11.20,

$$T = \frac{1}{2}\dot{f}^2\int_0^L j(X')^2\,dx$$

in which the distributed mass m is replaced by distributed rotatory inertia $j = nJ/L$, and instead of translation $X(x)$ there is a rotation $X'(x)$. Integrating, we have

$$T = \frac{1}{2}\dot{f}^2 j\int_0^L \left(\frac{\pi}{L}\right)^2\cos^2\frac{\pi x}{L}\,dx = \frac{1}{2}\dot{f}^2 j\left(\frac{\pi}{L}\right)^2\frac{L}{2}$$

In harmonic motion, we have $\dot{f}^2 = \omega^2 f$. Equating peak values of Π and T, we get

$$\omega^2 = \frac{\pi^2 EI}{jL^2} = \frac{\pi^2 EI}{nLJ}$$

14-8 Calculate the stiffness of the sandwich beam under the tip load as well as the natural frequency. The width, normal to paper, is 1.0 in. The facing sheets are aluminum, $E = 10.6 \times 10^6$ psi, while the core is a hexagonal honeycomb with $G = 68,000$ psi. The tip weight is 1.2 Lb, including a portion of the beam itself. Compare the results with those obtained from flexural properties alone.

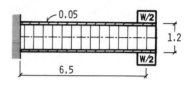

The bending properties of the section are characterized by the outside depth $H_2 = 1.2$ in. and inside depth $H_1 = 1.2 - 2 \times 0.05 = 1.1$ in. The second area moment is:

$$I = \frac{1}{12}(1.2^3 - 1.1^3) = 0.03308 \text{ in.}^4$$

Bending flexibility is:

$$\frac{u_b}{P} = \frac{L^3}{3EI} = \frac{6.5^3}{3 \times 10.6 \times 10^6 \times 0.03308} = \frac{261.1}{10^6}$$

The facing sheets are built into the wall rather than being hinged from it. This will make it difficult for the beam to have the shear angle $\vartheta = P/(GA_s)$ in the vicinity of the wall like the beam in Fig. 14.3. To allow for this end effect we can reduce the shear length of the beam by one depth of the cross section,

$$L' = 6.5 - 1.2 = 5.3$$

Shear flexibility is:

$$\frac{u_s}{P} = \frac{L'}{GA_s} = \frac{5.3}{68,000 \times 1.0 \times 1.1} = \frac{70.86}{10^6}$$

where A_s is the section area of the honeycomb. The resultant stiffness is

$$\frac{1}{k} = \frac{261.1}{10^6} + \frac{70.86}{10^6} \qquad \therefore\ k = 3012.4 \text{ Lb/in.}$$

Natural frequency:

$$\omega^2 = \frac{kg}{W} = \frac{3012.4 \times 386}{1.2} \qquad \therefore\ \omega = 984.4 \text{ rad/s}$$

One may check that a computation based on the bending stiffness only would give

$$k = 3830 \text{ Lb/in.} \qquad \text{and} \qquad \omega = 1110 \text{ rad/s}$$

This shows that although the beam is not very short (length/depth = 5.4), the influence of shear is quite noticeable.

14-9 A shear beam vibrates in a natural mode as illustrated. The extreme displacement of every vibrating mass M_0 is the same. What is the value of the participation factor with respect to the lateral base motion? Generalize the result to a shear beam that has $2n$ segments and is vibrating in a mode that has n moving masses.

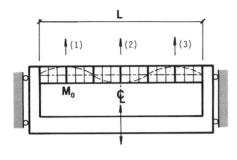

The beam consists of identical segments. Since the extreme displacements are the same for each half-wave, we are clearly dealing with a repeating pattern along the length. The mode shape $\vec{\Phi}$ may be written as

$$\vec{\Phi}^T = \begin{bmatrix} 1 & -1 & 1 \end{bmatrix}$$

and the mass matrix as

$$\mathbf{M} = \begin{bmatrix} M_0 & M_0 & M_0 \end{bmatrix}$$

According to Eq. 12.5, the participation factor is

$$\Gamma = \frac{M_0(1) + M_0(-1) + M_0(1)}{M_0 + M_0 + M_0} = \frac{1}{3}$$

(Notice that we could have used a more general form by including also the stationary masses, but that would not change the expression for Γ.) If we now increase the number of half-waves to 4, Γ will become 0. This is true for any mode with an even number of half-waves and could be also deduced from symmetry. (Symmetric loading cannot excite an antisymmetric shape.) When the number n of vibrating masses is odd, we find

$$\Gamma = \frac{1}{n}$$

14-10 The base of a shear beam in Prob. 14-9 is subjected to harmonic acceleration $a_b = a_0 \sin \Omega t$. Determine the relative displacement and the shear force response in the nth vibratory mode in terms of the mode number n. This mode has n moving masses and $2n$ segments. Each moving mass M_n is determined from $mL = 2nM_n$, where m is the mass of beam per unit length. The damping ratio is ζ.

According to Sect. 12C, the equation of relative motion in the nth natural mode is

$$\ddot{s}_n + 2\omega_n \zeta_n \dot{s}_n + \omega_n^2 s_n = -\Gamma_n a_b$$

where Γ_n is the nth participation factor. This factor was calculated in Prob. 14-9, which gives us

$$-\Gamma_n a_b = -\frac{1}{n} a_0 \sin \Omega t$$

To use the solution for a simple oscillator given by Eq. 2.15, we must put

$$\frac{P_0}{M_n} = -\frac{a_0}{n} \quad \text{or} \quad P_0 = -\frac{M_n}{n} a_0$$

where M_n is the modal mass. The amplitude of the steady-state response is then

$$s_n = \mu_n |P_0| \frac{1}{k_n} = \mu_n \frac{M_n}{n} a_0 \left(\frac{1}{M_n \omega_n^2} \right) = \frac{\mu_n a_0}{n \omega_n^2}$$

Since the modal vector in Prob. 14-9 consists only of unit entries, the value of s shown above is also the amplitude of physical displacement, relative to the base,

$$u_n = \mu_n \frac{a_0}{n \omega_n^2}$$

The shear force associated with this displacement is

$$V_n = k_n u_n = M_n \omega_n^2 u_n = \mu_n M_n \frac{a_0}{n}$$

The magnification factor is found from

$$\frac{1}{\mu_n^2} = \left(1 - \frac{\Omega^2}{\omega_n^2} \right)^2 + \left(2\zeta \frac{\Omega}{\omega_n} \right)^2$$

The modal mass is $M_n = mL/(2n)$. The modal stiffness k_n is determined by the center deflection of a simply supported shear beam with length $l = L/n$:

$$\frac{1}{k_n} = \frac{u}{P} = \frac{l}{4GA_s} \qquad \therefore k_n = \frac{4nGA_s}{L}$$

The natural frequency is:

$$\omega_n^2 = \frac{k_n}{M_n} = \frac{8n^2 GA_s}{mL^2}$$

The response expressions can now be rewritten,

$$u_n = \mu_n \frac{a_0 mL^2}{8n^3 GA_s} \qquad V_n = \mu_n \frac{mLa_0}{2n^2}$$

14-11 A flexural beam with hinged ends is subjected to base acceleration $a_b = a_0 \sin \Omega t$. When the forcing frequency $\Omega = \omega_3$, where

$$\omega_3^2 = \frac{48EI}{M_0 L'^3} \qquad \text{with} \qquad L' = \frac{L}{3}$$

the beam is at resonance and has the deflected shape as illustrated (all masses shown as undeflected are moving the same as the base). The system damping is small, $\zeta = 0.05$. What is the relative displacement amplitude of vibration? What is the maximum inertia force applied to a vibrating mass?

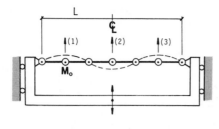

Notice that because of similarity in the deformed pattern, the approach to a flexural beam is essentially the same as that used for the shear beam in Probs. 14-9 and 14-10. The equation of motion in this mode becomes

$$\ddot{s}_3 + 2\omega_3\zeta\dot{s}_3 + \omega_3^2 s_3 = -\tfrac{1}{3}a_b$$

Following Prob. 14-10, we find the modal displacement as

$$s_3 = -\frac{\mu a_0}{3\omega_3^2} = -\frac{1}{2\zeta}\frac{a_0}{3\omega_3^2} = -\frac{10}{3}\frac{a_0}{\omega_3^2}$$

The physical displacement amplitudes are

$$\vec{u} = \vec{\Phi}_3 s_3 \qquad \text{or} \qquad u_1 = -u_2 = u_3 = -\frac{10}{3}\frac{a_0}{\omega_3^2}$$

We can write the frequency as

$$\omega_3^2 = \frac{k_3}{M_0}$$

where k_3 is the stiffness, in accordance with the equation in the problem statement. The magnitude of each individual joint force is

$$P_i = k_3 u_i = \frac{10}{3}\frac{a_0 M_0}{k_3}k_3 = \frac{10}{3}a_0 M_0$$

At the extreme deflection there is no viscous force and P_i is equal to the inertia force.

14-12   Develop a generalized solution to Prob. 14-11. Specifically, the forcing frequency $\Omega = \omega_n$, where

$$\omega_n^2 = \frac{48EI}{M_0 L'^3} \qquad \text{with} \quad L' = \frac{L}{n}$$

Each length L' may be treated as a simply supported beam with M_0 at the center. Determine the maximum inertia force applied to a vibrating mass and the associated bending moment and shear force. If this is a model of a continuous beam such that $mL = 2nM_0$, what form would the solution take? There is the same, small damping ζ in each mode.

First of all, let us notice that if we want to treat the problem as a function of n, we are speaking of a different beam model for each value of n. Repeating the procedure in Prob. 14-9, we find $\Gamma_n = 0$

when n is even and $\Gamma_n = 1/n$ when n is odd. The equation of motion in an odd-numbered mode is

$$\ddot{s}_n + 2\omega_n\zeta\dot{s}_n + \omega_n^2 s_n = -\frac{1}{n}a_b$$

The amplitude of modal response is

$$s_n = \frac{1}{2\zeta}\left(\frac{1}{\omega_n^2}\right)\left(-\frac{a_0}{n}\right)$$

The physical displacements and joint forces, respectively, are

$$|u_n| = \frac{1}{2\zeta}\frac{M_0}{k_n}\frac{a_0}{n} \qquad \text{and} \qquad P_0 = k_n|u_n| = \frac{M_0}{2\zeta}\frac{a_0}{n}$$

Each piece of length L' is treated as a simply supported beam, which makes the computation of shear and bending quite straightforward,

$$V = \frac{P_0}{2} \qquad \mathfrak{M} = \frac{P_0 L'}{4}$$

If the model is a representation of a beam with a continuously distributed mass m, our expressions become, for $n = 1, 3, 5, \ldots$,

$$P_0 = \frac{mLa_0}{4\zeta n^2} \qquad V_0 = \frac{mLa_0}{8\zeta n^2} \qquad \mathfrak{M} = \frac{mL^2 a_0}{16\zeta n^3}$$

The higher the mode number excited, the smaller the internal forces become. The nth mode frequency can be written as a function of the mode number by putting

$$M_0 = \frac{mL}{2n} \qquad \text{and} \qquad L' = \frac{L}{n}$$

into the opening equation for ω_n:

$$\omega_n^2 = \frac{96EIn^4}{mL^4}$$

14-13   Two mating gears are capable of freely rotating about their respective centers. The flexibility of contacting teeth is symbolically shown as a short beam with stiffness k. If a torque $M_{t0} = 1500$ Lb-in. is applied to the larger gear as a step load in time, what is the maximum circumferential contact force between the teeth resulting from this shock load? Use $r_1 = 5$ in., $r_2 = 3$ in., $J_1 = 0.72$ Lb-s^2-in., and $J_2 = 0.09$ Lb-s^2-in.

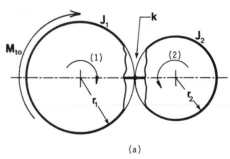

(a)

Let us first calculate the accelerations ε_1 and ε_2 resulting from rigid-body rotation of the system under M_{t0}. When the smaller gear is rotating with acceleration ε_2, the torque applied to it must be $J_2\varepsilon_2$. This gives a tangential force as $J_2\varepsilon_2/r_2$ and the torque, which the large gear must overcome, is

$$J_2\varepsilon_2\frac{r_1}{r_2}=J_2\varepsilon_1\left(\frac{r_1}{r_2}\right)^2$$

since $\varepsilon_1 r_1=\varepsilon_2 r_2$ from kinematics. The total inertial resistance to the turning of the larger wheel is therefore

$$J_1\varepsilon_1+J_2\varepsilon_1\left(\frac{r_1}{r_2}\right)^2=0.97\varepsilon_1$$

This is equal to the external torque M_{t0},

$$\therefore\quad\varepsilon_1=\frac{1500}{0.97}=1546.4\text{ rad/s}^2$$

Having this reference value computed, we can now apply the method outlined in Sect. 13J. The actual loading is resolved into two subcases. The first one, in (b), drives both wheels in a rigid-body manner, without any interaction between them:

$$M'_{t1}=J_1\varepsilon_1=1113.4$$

$$M'_{t2}=J_2\varepsilon_2=232.0\text{ Lb-in.}$$

The second subcase is the external load that causes the elastic interaction (c),

$$M''_{t1}=M_{t0}-M'_{t1}=1500-1113.4=386.6$$

$$M''_{t2}=M'_{t2}=232\text{ Lb-in.}$$

(Only the absolute values of the torques are given, their signs are shown in the figures.) It is easy to check that the system is in equilibrium under the loading in (c).

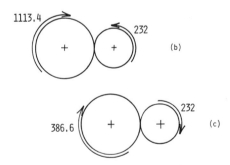

All the quantities established thus far refer to a pseudostatic condition of the system. To account for the dynamic nature of the loading, we must multiply the interaction force by the appropriate magnification factor, which is 2.0 in this case,

$$P=\mu\frac{M''_{t2}}{r_2}=2.0\times\frac{232}{3}=154.7\text{ Lb}$$

14-14 During normal operation of the propulsion system (Prob. 8-10) the torque applied to the rotor of the turbine, $M_{t3}=472,700$ Lb-in., is balanced by the resisting torques, $M_{t2}=649,500$ and $M_{t1}=681,340$ Lb-in. Occasionally, however, a sudden decrease of resistance at both pump stages reduces M_{t2} and M_{t3} to 25% of their operating values. Assuming this condition to occur instantaneously, while the turbine torque remains unchanged, estimate the peak twisting moments in shafts K_1, K_2, and K_3. Moments of inertia of discs are

$$J_1=1780\text{ Lb-s}^2\text{-in.}\qquad J_2=423.5\text{ Lb-s}^2\text{-in.}$$

$$J_3=722.5\text{ Lb-s}^2\text{-in.}$$

and their gear ratios are

$$g_1=\frac{r'_1}{r''_1}=0.2242\qquad g_2=\frac{r'_2}{r''_2}=0.4926$$

The decrease of the resisting torques may be treated as a dynamic load expressed by

$$\Delta M_{t1}=-0.75\times681,340=-511,005\text{ Lb-in.}$$

$$\Delta M_{t2}=-487,125\text{ Lb-in.}$$

and suddenly applied to the system. This dynamic load will now be resolved into two subcases, one pertaining to rigid-body motion and the other to the elastic interaction. The kinematics of the first subcase is illustrated in (a). If the acceleration of the turbine is ε_3, the torque applied to it must be $J_3\varepsilon_3$. Similarly, the external torques applied to the remaining discs have to be $J_1\varepsilon_1$ and $J_2\varepsilon_2$, respectively. If we think of disc J_3 as being driven, the torque that must be applied to it to overcome the inertia of J_1 is $(J_1\varepsilon_1)g_1$, while to overcome the inertia of J_2 we need $(J_2\varepsilon_2)g_2$. The total torque driving J_3 will be then

$$J_3\varepsilon_3+J_2\varepsilon_2 g_2+J_1\varepsilon_1 g_1$$

The total torque applied to J_3 is actually

$$\Delta M_{t1}g_1+\Delta M_{t2}g_2=354,525\text{ Lb-in.}$$

Noting that $\varepsilon_2=\varepsilon_3 g_2$ and $\varepsilon_1=\varepsilon_3 g_1$ in a rigid-body motion, we equate both expression for the total torque on J_3:

$$(722.5+423.5\times0.4926^2+1780\times0.2242^2)\varepsilon_3=354,525$$

$$\therefore\varepsilon_3=387.57\text{ rad/s}^2$$

This enables us to calculate the external torques applied to the discs in the rigid-body subcase as presented in (b). For example,

$$M'_{t1}=J_1\varepsilon_1=J_1\varepsilon_3 g_1=154,670\text{ Lb-in.}$$

The second subcase is the difference between what is actually applied and what is required by the rigid-body pattern:

$$M''_{t1}=M_{t1}-M'_{t1}=511,005-154,670=356,335\text{ Lb-in.}$$

$$M''_{t2}=406,272\text{ Lb-in.}$$

$$M''_{t3}=280,020\text{ Lb-in.}$$

with the actual signs in (*c*), which shows a self-balanced system with elastic interaction between members. The torques depicted as applied to the disc are actually the shaft torques. Allowing for the dynamic action of the load, the maximum shock torques are

$$K_1: -712,670 \text{ Lb-in.} \qquad K_2: -812,544 \text{ Lb-in.}$$

$$K_3: 560,040 \text{ Lb-in.}$$

(The signs were selected according to the following convention: a torque is positive if the vector of torque appears to be stretching the shaft.) The dynamic components so derived oscillate between zero and the values shown above. Writing the operating torques with the same convention, we have

$$K_1: 681,340 \text{ Lb-in.} \qquad K_2: 649,500 \text{ Lb-in.}$$

$$K_3: -472,700 \text{ Lb-in.}$$

When a computer program utilizing all three modes was employed, the following shock loads were found acting on the shaft:

$$K_1: -1,059,715 \text{ Lb-in.} \qquad K_2: -798,157 \text{ Lb-in.}$$

$$K_3: 558,045 \text{ Lb-in.}$$

The errors of our approximation are negligible except for K_1, where we underestimate the true result by 32.7%.

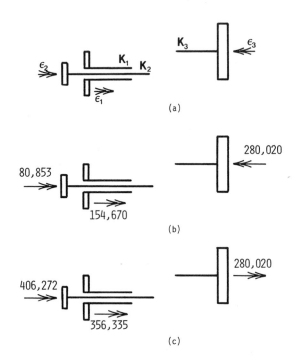

In conclusion we may say that (when signs are taken into account) the dynamic torques resulting from sudden unloading of the system do not cause an increase in the resultant torques, when superposed on the operating values. (This is true regardless of whether the approximate or the accurate shock torques are employed.)

EXERCISES

14-15 In Prob. 8-6 a flexibility matrix of this beam was derived using only bending properties:

$$\mathbf{a}_b = \begin{bmatrix} 1 & \dfrac{5}{16} \\[2mm] \dfrac{5}{16} & \dfrac{1}{8} \end{bmatrix} \dfrac{L^3}{3EI}$$

Develop the flexibility matrix for the corresponding shear beam and calculate the entries of the resultant flexibility matrix if the cross section is of the wide-flange type and the following are given:

$$A = 38.21 \text{ in.}^2 \qquad I = 4009 \text{ in.}^4$$
$$h = 24.25 \text{ in.} \quad L = 97 \text{ in.} \quad t = 0.565 \text{ in.}$$
$$E = 29 \times 10^6 \text{ psi} \qquad G = 11 \times 10^6 \text{ psi}$$

14-16 A building is located on soil with elastic properties $G = 5940$ psi and $\nu = 0.45$. The width of building is 900 in., its weight $W = 26.5 \times 10^6$ Lb, and the moment of inertia about the CG is $J = 16.7 \times 10^9$ Lb-s²-in. Calculate the natural frequencies of motion in the *yz*-plane, treating the building as a rigid body.

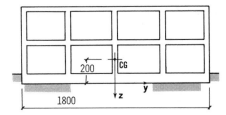

14-17 Estimate the natural frequencies associated with the deflected shapes in (*b*) and (*c*) for the doubly symmetric frame. *Hint.* Refer to Prob. 7-12.

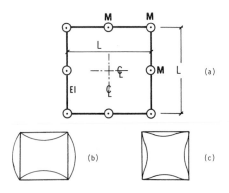

14-18 The horizontal force $P = P_0 \sin \Omega t$ applied to a block foundation has $P_0 = 11,000$ Lb and $\Omega = 42$ rad/s and is located above the surface of the block. The weight of the block is $W = 340,000$ Lb and the moment of inertia about the CG axis, $J = 27.3 \times 10^6$ Lb-s²-in. The elastic constants are $k_y = 3.36 \times 10^6$ Lb-in. and $K_\psi = 25.3 \times 10^9$ Lb-in./rad. Calculate the amplitudes of horizontal and angular displacements and the maximum shear force acting on the ground. Use physical displacement coordinates, as described in Sect. 10F.

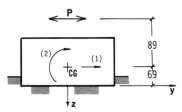

14-19 The figure shows an outline of the foundation block analyzed in Probs. 14-1 and 14-4. Vertical force $P_z = P_0 \cos \Omega t$ is applied at the point near the upper right corner, as seen in this view. Determine the upper bound of u_z displacement of that corner for $P_0 = 5000$ Lb and $\Omega = 75$ rad/s. Ignore the coupling between the rotational and horizontal translation of the CG.

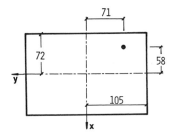

14-20 The weightless beam with mass M at the tip has a force $P(t)$ applied at midlength. Describe how to find the tip response by means of reducing the problem to an SDOF system. The flexibility matrix, as calculated in Prob. 8-6, is

$$\mathbf{a} = \frac{L^3}{3EI} \begin{bmatrix} 1 & \dfrac{5}{16} \\ \dfrac{5}{16} & \dfrac{1}{8} \end{bmatrix}$$

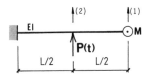

14-21 The rotor has the same properties as the one in Prob. 1-11; $K_1 = 69.98 \times 10^6$ Lb-in./rad (thicker segment), $K_2 = 34.56 \times 10^6$ Lb-in./rad (thinner segment), and $J = 20$ Lb-s²-in. At the transition plane of the shaft (angular coordinate 2) there is a torque applied, $M_t = 1500t$ (torque in pound-inches if t in seconds). No angular inertia is associated with coordinate 2. What is the equation for the disc rotation $\alpha_1(t)$? *Hint.* Refer to Prob. 2-31.

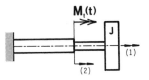

14-22 The weight $W = 8$ Lb impacts a plate with velocity $v_0 = 120$ in./s. The plate is built into another part so that its ends may be treated as midway between fixed and simply supported. The moving weight has a short tube at its end and may be considered to be nondeformable. The plate material data are $E = 10.6 \times 10^6$ psi, $G = 4.0 \times 10^6$ psi, and $\nu = 0.325$. Treating the plate as weightless, find the impact force.

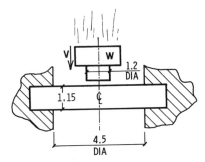

PART III

SPECIAL TOPICS

CHAPTER

15

Bodies with Continuous Mass Distribution

In the preceding chapters we dealt with systems having masses lumped at discrete points. This chapter describes the motion that occurs when the mass is continuously distributed and solutions of applicable equations of motion. The dynamic properties of second-order elements (bar, shaft, cable, and shear beam) and flexural beams are established. The responses to initial conditions as well as to the external loads are determined using the normal-mode method. The response to harmonic forcing is treated in detail. Although only very simple systems may be analyzed by the methods presented in this chapter, the results are nevertheless quite valuable, because they provide, among other things, important clues for finite-element modeling.

15A. Equations of Motion—Bar and Shaft. A segment of an axial bar shown in Fig. 15.1 is in the state of dynamic equilibrium. The external load of intensity q (Lb/in.) is balanced by the increment of the internal force N and by the inertia force of magnitude $m\ddot{u}\,dx$, where $m = A\rho$ is the mass per unit length (Lb-s^2/in.^{2}). The rate of change of the stretching force N is $\partial N/\partial x$, therefore its value changes by $(\partial N/\partial x)\,dx$ over the segment dx. Projecting all forces on the x-axis, we obtain

$$N + \frac{\partial N}{\partial x}\,dx - N + q\,dx - m\ddot{u}\,dx = 0$$

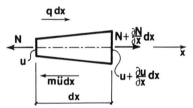

Figure 15.1 Dynamic equilibrium of bar element.

The elongation of the segment is, by an elementary relation,

$$\frac{\partial u}{\partial x} = \frac{N}{EA} \qquad \text{or} \qquad N = EAu' \qquad (15.1)$$

If a bar is of variable cross section, A is a function of x. The derivative of N is, in this case,

$$\frac{\partial N}{\partial x} = \frac{\partial}{\partial x}\left(\frac{\partial u}{\partial x}EA\right) \qquad (15.2)$$

Our equation of motion becomes now

$$m\ddot{u} - (EAu')' = q \qquad (15.3)$$

in which the prime denotes differentiation with respect to x. For a constant axial stiffness, $EA=$

189

constant, the equation simplifies to

$$\ddot{u} - c^2 u'' = \frac{q}{m} \qquad (15.4)$$

where

$$c = \left(\frac{EA}{m}\right)^{1/2} = \left(\frac{E}{\rho}\right)^{1/2} \qquad (15.5)$$

can be shown to be the speed of propagation of a longitudinal elastic disturbance along the bar.

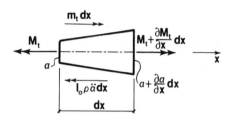

Figure 15.2   Dynamic equilibrium of shaft element.

Figure 15.2 shows a segment of a shaft in dynamic equilibrium; M_t is the twisting moment at a certain location, while the external distributed load has the intensity m_t Lb-in./in. The moment of inertia about the shaft axis is $I_0 \rho\, dx$ in which I_0 is the polar moment of inertia of the cross section. The independent variable is the angle of rotation α. There is a complete analogy between a bar and a shaft. Projecting the vectors pertaining to the angular quantities on the x-axis, we obtain

$$\frac{\partial M_t}{\partial x} + m_t - I_0 \rho \ddot{\alpha} = 0$$

If the elementary equation for the torsional deformation is applied to a shaft segment of length dx, we get

$$d\alpha = \frac{M_t\, dx}{GC}$$

in which $d\alpha$ is the angle of twist and C is the torsional constant of the cross section. This may also be written as

$$\frac{\partial \alpha}{\partial x} = \frac{M_t}{GC} \qquad \text{or} \qquad M_t = GC\alpha' \qquad (15.6)$$

Limiting ourselves to shafts with a constant section we obtain

$$\frac{\partial M_t}{\partial x} = GC \frac{\partial^2 \alpha}{\partial x^2} \qquad (15.7)$$

Our equation of motion becomes now

$$\ddot{\alpha} - c_1^2 \alpha'' = \frac{m_t}{\rho I_0} \qquad (15.8)$$

where

$$c_1 = \left(\frac{GC}{\rho I_0}\right)^{1/2} \qquad (15.9)$$

and c_1 is the speed of propagation of angular disturbances. For circular cross sections $C = I_0$.

Although the material to follow refers to a bar, everything stated is also valid for a shaft if the notation is suitably changed. To solve Eq. 15.4 we must know the initial conditions in the form

$$u(x,0) = u_0(x) \qquad \text{and} \qquad \dot{u}(x,0) = v_0(x)$$

that is, the initial position and velocity of any point of the bar axis at time equal to zero. Also *boundary conditions*, which are displacements or forces (or combinations thereof) at the ends must be given. Our concern at this point is limited to solving Eq. 15.4 in its homogeneous form, that is, when the right side is zero. A general solution is assumed as

$$u(x,t) = X(x) \cdot f(t) \qquad (15.10)$$

where X is a function of x only and it is referred to as a mode shape, while f depends only on time. The detailed expressions for X and f are

$$X(x) = D_1 \cos\frac{\omega x}{c} + D_2 \sin\frac{\omega x}{c} \qquad (15.11)$$

$$f(t) = B_1 \cos \omega t + B_2 \sin \omega t \qquad (15.12)$$

There are infinitely many natural frequencies ω and the like number of natural modes given by Eq. 15.11. The complete solution is the sum of those individual components:

$$u(x,t) = u_1 + u_2 + u_3 + \cdots = \sum_{n=1}^{\infty} X_n(x) \cdot f_n(t)$$

$$(15.13)$$

Substitution of boundary conditions into Eq. 15.10 gives us a frequency equation for a particular system, with which we can establish the set of natural frequencies ω_i. (Note that Eq. 9.13 is the corresponding expression for discrete systems.) In the process we also find the relationship between the constants D_1

and D_2. Suppose, for example, that we have found $D_2 = 0.5D_1$. This allows us to put

$$X(x) = D_1\left(\cos\frac{\omega x}{c} + 0.5\sin\frac{\omega x}{c}\right)$$

We can now set $D_1 = 1$, since a scale factor of a mode shape has no meaning. The compete solution, from Eq. 15.13, is

$$u(x,t)$$
$$= \sum_{n=1}^{\infty}\left(\cos\frac{\omega_n x}{c} + 0.5\sin\frac{\omega_n x}{c}\right)(B_{1n}\cos\omega_n t + B_{2n}\sin\omega_n t)$$

The series of constants B_{1n} and B_{2n} are determined by the initial conditions or by the forcing function.

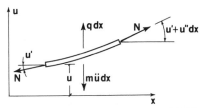

Figure 15.3 Dynamic equilibrium of cable element.

15B. Equations of Motion—Cable and Shear Beam. The segment of cable (string) in Fig. 15.3 is subjected to a laterally distributed load q (Lb/in.), to a stretching force N, which is assumed constant, and to the distributed inertia force of magnitude $m\ddot{u}\,dx$. The lateral displacement u is assumed to be very small in comparison with the length of cable. By projecting all forces on the vertical axis we obtain

$$\ddot{u} - c_2^2 u'' = \frac{q}{m} \qquad (15.14)$$

which is analogous to Eqs. 15.4 and 15.8. The constant c_2 is, as before, the speed of propagation of displacements. In this case,

$$c_2 = \left(\frac{N}{m}\right)^{1/2}$$

The solution of Eq. 15.14 when the right-hand side equals zero is analogous to that for bars and shafts.

The *shear beam* is the fourth member of the group called *second-order elements*. Static deformation properties were explained in Sect. 14B, while the equation of motion is analogous to those of a bar, a shaft, and

a cable (Prob. 15-15). The speed of propagation is

$$c_3 = \left(\frac{GA_s}{m}\right)^{1/2}$$

The equation of motion for this group is of the second order with respect to the x-variable; hence the name.

15C. Equations of Motion—Flexural Beam. The beam element in Fig. 15.4a is in the state of dynamic equilibrium. The bending stiffness EI (Lb-in.^{2}) is assumed constant and so is the mass per unit length $m = A\rho$. The equation of motion is of the fourth order with respect to x:

$$m\ddot{u} + EIu'''' = q \qquad (15.15)$$

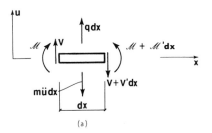

(a)

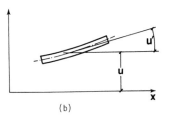

(b)

Figure 15.4 Flexural beam element: (*a*) dynamic equilibrium; (*b*) displacement and slope.

The other important equations for the beam segment are

$$EIu'' = \mathfrak{M} \qquad (15.16)$$

$$EIu''' = V \qquad (15.17)$$

The first one is the relation between the bending moment and the curvature of the beam axis. The second may be obtained from the first by differentiation and by noting that $M' = V$ if the higher order terms are ignored in the angular equilibrium of beam element. Equation 15.15 may be obtained by differentiation of 15.17 and by writing the equation of vertical equilibrium for Fig. 15.4a.

The boundary conditions involve deflection and slope (Fig. 15.4b) as well as shear forces and bending moments. The general form of solution of free vibration ($q=0$) is the same as in Eq. 15.10, while the functions of position and time are, respectively,

$$X(x) = D_1 \sin(px) + D_2 \cos(px)$$
$$+ D_3 \sinh(px) + D_4 \cosh(px) \quad (15.18)$$

$$f(t) = B_1 \cos(\omega t) + B_2 \sin(\omega t) \quad (15.19)$$

The parameter p and the natural frequency ω are related by

$$p^4 = \frac{\omega^2 m}{EI} = \omega^2 \frac{A\rho}{EI} \quad (15.20)$$

It is useful to list the derivatives of X, since some of them always occur in a statement of boundary conditions:

$$X' = p\big[D_1 \cos(px) - D_2 \sin(px)$$
$$+ D_3 \cosh(px) + D_4 \sinh(px)\big]$$

$$X'' = p^2\big[-D_1 \sin(px) - D_2 \cos(px)$$
$$+ D_3 \sinh(px) + D_4 \cosh(px)\big] \quad (15.21)$$

$$X''' = p^3\big[-D_1 \cos(px) + D_2 \sin(px)$$
$$+ D_3 \cosh(px) + D_4 \sinh(px)\big]$$

15D. Response to Initial Conditions. The axial displacement of a freely vibrating bar may be written as

$$u(x, t) = X_1(x) \cdot f_1(t) + X_2(x) \cdot f_2(t)$$
$$+ X_3(x) \cdot f_3(t) + \cdots \quad (15.22)$$

according to Eq. 15.13. To obtain the velocity, we differentiate this expression with respect to t. At $t=0$ both the displacement and the velocity are known, prescribed functions of x, as stated in Sect. 15A. With the help of Eq. 15.12, our initial conditions become

$$u_0(x) = B_{11}X_1 + B_{12}X_2 + B_{13}X_3 + \cdots \quad (15.23)$$

$$v_0(x) = \omega_1 B_{21}X_1 + \omega_2 B_{22}X_2 + \omega_3 B_{23}X_3 + \cdots \quad (15.24)$$

The mode shapes X_n, which are calculated from the boundary conditions, are treated as known functions in this expression. (A constant that appears after solving a boundary value problem is set equal to unity.) To calculate the coefficient B_{1n}, let us multiply both sides of Eq. 15.23 by X_n and integrate from zero to L. We first note that

$$\int_0^L X_m X_n \, dx = 0 \quad (15.25)$$

whenever $m \neq n$. This is known as the *orthogonality property* of the natural modes. In consequence we have

$$\int_0^L X_n u_0(x) \, dx = B_{1n} \int_0^L X_n^2 \, dx \quad (15.26)$$

which, upon integration, gives us the value of B_{1n}. The same operation on Eq. 15.24 yields

$$\int_0^L X_n v_0(x) \, dx = \omega_n B_{2n} \int_0^L X_n^2 \, dx \quad (15.27)$$

from which B_{2n} can be found.

The orthogonality property expressed by Eq. 15.25 holds true for all boundary conditions except when lumped masses are present. In the latter case this property is expressed differently and the response calculation becomes more complicated. What was said about the response of bars is also valid for the remaining second-order elements.

The beam equation of motion is different from that of the bar and the difference in the solution is visible by comparing Eqs. 15.11 and 15.18. Yet, the described procedure of calculating the response to the initial conditions is also valid for beams, provided there are no lumped inertia elements.

This method is analagous to the normal-mode approach for discrete systems. The initial displacement and/or velocity is first presented as a linear combination of natural modes and this serves to establish the magnitude of response in each mode. (Some modes may predominate, others may not be excited.) The solution of the equation of motion is a linear combination of those modal responses.

15E. Response to Applied Load. When the forcing function on the right side of Eq. 15.4 is not equal to zero, we have the case of forced vibration. The first step toward obtaining the solution is to represent the forcing function $q(x, t)$ in terms of the natural modes:

$$q(x, t) = \sum_{n=1}^{\infty} X_n(x) \cdot q_n(t) \quad (15.28)$$

This is accomplished by using the same technique employed in handling the initial conditions $u_0(x)$ and $v_0(x)$ in Sect. 15D. Here we have

$$\int_0^L X_n(x) \cdot q(x,t)\, dx = q_n(t) \int_0^L X_n^2\, dx \quad (15.29)$$

from which $q_n(t)$ can be found. We again present $u(x,t)$ as a sum of its modal components, Eq. 15.13,

$$u = \sum_{n=1}^{\infty} X_n f_n$$

The time function $f(t)$ is determined from

$$\ddot{f}_n + \omega_n^2 f_n = \frac{1}{m} q_n \quad (15.30)$$

This is an equation of an undamped oscillator having mass m and natural frequency ω_n. Solving for f_n gives us the magnitude of response in the nth vibratory mode. For this reason q_n may be called the *modal load*. The solution of Eq. 15.30 can be found either by predicting its general form (see Sect. 2C) or by employing Laplace transforms as outlined in Sect. 2J. (Other, less practical methods also exist.) The modal solutions are summed according to Eq. 15.13.

When a concentrated load is applied at some point of a bar, it may be treated as a special case of a distributed load of infinitely large magnitude applied over an infinitely short segment of length. Suppose that the magnitude of a concentrated load applied at $x = x_0$ is P. Employing a unit delta function defined in Appendix III we may write

$$q(x,t) = P\delta(x - x_0)$$

Owing to the properties of the delta function, the computation of modal forces $q_n(t)$ is simplified, because instead of Eq. 15.29 we have

$$PX_n(x_0) = q_n(t) \int_0^L X_n^2\, dx \quad (15.31)$$

(Of course P may be time dependent.)

The calculation of a shaft response is mathematically the same as that of the bar, except that notation should be changed as follows:

$$q(x,t) \rightarrow m_t(x,t) \qquad q_n(t) \rightarrow m_{tn}(t)$$

$$m \rightarrow \rho I_0 \qquad\qquad P \rightarrow M_t$$

Similar changes are needed for a cable and a shear beam.

Determination of a flexural beam response is quite similar. We seek the series solution in the form of Eq. 15.13 for our equation of motion

$$m\ddot{u} + EIu'''' = q(x,t)$$

The time functions are again calculated from Eq. 15.30. When a concentrated lateral force P is applied, the modal forces are calculated from Eq. 15.31.

When computing modal loads according to this method it is implicitly assumed that the orthogonality condition as given by Eq. 15.25 holds true. This is not so when lumped inertia elements are present, and this approach is not applicable to such cases.

15F. Forcing Function Varying Harmonically with Time. Suppose that the distributed load q applied to a bar is expressed as

$$q(x,t) = q(x) \cdot e^{i\Omega t} \quad (15.32)$$

which, according to Chapter 7, is equivalent to considering two separate cases, namely, $q(x) \cdot \cos \Omega t$ and $q(x) \cdot \sin \Omega t$. The steady-state solution is then assumed to be

$$u(x,t) = u(x) e^{i\Omega t} \quad (15.33)$$

That is, it is assumed to vary harmonically with time at the same frequency Ω. Upon differentiation and substitution into Eq. 15.4, we get

$$u'' + \frac{\Omega^2}{c^2} u = -\frac{q(x)}{mc^2} \quad (15.34)$$

This is an ordinary differential equation, and u is a function of x only. It is similar to the equation of motion of an undamped oscillator (2.7) except that differentiation is performed with respect to x, not t. By analogy with Eq. 2.8 we have

$$u(x) = D_1 \cos \frac{\Omega x}{c} + D_2 \sin \frac{\Omega x}{c} + u_p(x) \quad (15.35)$$

where $u_p(x)$ is a particular solution of Eq. 15.34.

Employing the method described here leads to a closed-form solution for $u(x,t)$. Acting similarly, it is also possible to obtain a closed-form solution of the beam equation (15.15). Since there are four constants to be determined now, the series solution resulting from the application of the normal-mode method is often an easier alternative.

15G. Approximate Methods. Even the simplest systems may pose serious computational difficulties when handled by means of differential equations. For this reason numerous simplifications are used in practical engineering work. All methods of Chapter 11 (except that of Holzer, designed specifically for discrete systems) can be employed here, provided the reader knows that they are applicable to a particular situation.

SOLVED PROBLEMS

15-1 An elastic bar of length L performs free axial vibrations. Find the natural frequencies and the associated mode shapes. Develop a formula for the general solution $u(x, t)$.

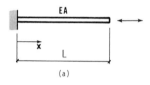

(a)

At the left end $u = 0$ or $X = 0$, where X is the shape function defined by Eq. 15.11. A substitution yields $D_1 = 0$. At the right end the axial force is zero, or by Eq. 15.1, $u' = 0$. Equivalently,

$$\frac{\partial X}{\partial x} = D_2 \frac{\omega}{c} \cos \frac{\omega x}{c} = 0 \quad \text{for} \quad x = L$$

Clearly D_2 may not be zero, otherwise u would vanish; therefore

$$\cos \frac{\omega L}{c} = 0 \quad \therefore \quad \omega = \frac{\pi c}{2L}, 3 \frac{\pi c}{2L}, 5 \frac{\pi c}{2L}, \ldots$$

The nth mode shape is thus

$$X_n = \sin n \frac{\pi x}{2L} \qquad n = 1, 3, 5, \ldots .$$

upon setting D_2 (a scale factor) equal to unity. With the knowledge of the mode shapes, a solution component may be written as $u_n = (B_{1n} \cos \omega_n t + B_{2n} \sin \omega_n t) \sin n(\pi x/2L)$ for $n = 1, 3, 5, \ldots$, and the complete solution is

$$u(x, t) = \sum_{n=1,3,\ldots}^{\infty} u_n(x, t)$$

The first two mode shapes are shown in (b).

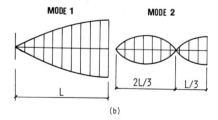

MODE 1 MODE 2

(b)

15-2 A disc with moment of inertia J_D about its rotary axis is attached to a circular shaft having mass moment of inertia $J_S = I_0 \rho L$. Establish the frequency equation for this system and find the fundamental frequency for (1) $J_D/J_S = 1$, (2) $J_D/J_S = 0$, and (3) $J_D/J_S = \infty$. (The last two are the limiting cases.)

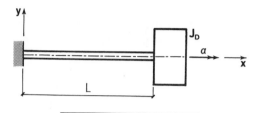

At the left end $\alpha(0, t) = 0$ and from Eq. 15.11, adapted for torsion case, $D_1 = 0$.

$$\therefore \quad X(\alpha) = \sin \frac{\omega x}{c_1}$$

When the disc is moving with an acceleration $\ddot{\alpha}$, it means that there is a torque of magnitude $J_D \ddot{\alpha}$ applied to it. There also is an equal and opposite torque applied to the shaft. Noting the orientation of this latter quantity, we can write, with the aid of Eq. 15.6,

$$GC \frac{\partial \alpha}{\partial x} = -J_D \ddot{\alpha} \quad \text{at} \quad x = L$$

From Eq. 15.10:

$$\alpha = (B_1 \cos \omega t + B_2 \sin \omega t) \sin \frac{\omega x}{c_1}$$

for one of the modes. Differentiating and substituting the second boundary condition gives

$$GC \frac{\omega}{c_1} (B_1 \cos \omega t + B_2 \sin \omega t) \cos \frac{\omega x}{c_1}$$

$$= J_D \omega^2 (B_1 \cos \omega t + B_2 \sin \omega t) \sin \frac{\omega x}{c_1} \quad \text{at} \quad x = L$$

or

$$\tan \frac{\omega L}{c_1} = \frac{GC\omega}{J_D \omega^2 c_1} = \frac{c_1}{\omega L} \cdot \frac{J_S}{J_D}$$

We can now think of $\omega L/c_1$ as an independent variable z and plot the functions $\tan z$ and $(J_S/J_D)/z$. For $J_S/J_D = 1$ we get the first intersection between $z = 0$ and $\pi/2$ and by a numerical search we find the first answer:

1. $\omega_1 = 0.8603 \dfrac{c_1}{L}$

   When $J_D/J_S = 0$, the right side of the frequency equation becomes infinitely large. The smallest value of the argument for which it happens is $\pi/2$, which gives us the second answer:

2. $\omega_1 = \dfrac{\pi c_1}{2L}$

   Similarly, the last answer is

3. $\omega_1 = \dfrac{\pi c_1}{L}$

Notice that case 2 corresponds to the free end, while case 3 is equivalent to the end being fixed.

15-3 A shaft is fixed at one end and elastically restrained at the other. The angular stiffness of the restraint is K (Lb-in./rad). Formulate the frequency equation of the system. Find the first frequency when the restraint is twice as stiff as the shaft itself.

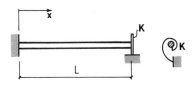

The boundary conditions are:

$$\alpha=0 \quad \text{at} \quad x=0 \quad \text{and} \quad GC\frac{\partial\alpha}{\partial x}=-K\alpha \quad \text{at} \quad x=L$$

The second one is written on the basis of Eq. 15.6, and by noting that if the right end turns by α, there is a torque $-K\alpha$ applied to it by the restraint. Using Eqs. 15.10 to 15.12 the first condition gives us $D_1=0$, so that

$$X(x)=D_2\sin\frac{\omega x}{c_1}$$

while from the second boundary condition,

$$GCD_2\frac{\omega}{c_1}\cos\frac{\omega L}{c_1}=-KD_2\sin\frac{\omega L}{c_1}$$

or

$$\tan\frac{\omega L}{c_1}=-\frac{GC}{KL}\left(\frac{\omega L}{c_1}\right)$$

In our case $GC/KL=\frac{1}{2}$ and we can briefly write our equation as $\tan z=-z/2$. By plotting both sides of the equation, we find our solution to be between $\pi/2$ and $3\pi/2$. A numerical search gives $z=2.2889$, from which we conclude that

$$\omega_1=2.2889\frac{c_1}{L}$$

15-4 The fundamental frequency of a bar fixed at one end and free at the other is

$$\omega_1=\frac{\pi c}{2L}$$

Suppose that we want to use a simplified model in which the bar is a weightless, elastic element with a lumped mass M at the end. What fraction of the total bar mass mL should M be for the frequencies of both systems to be the same?

Since

$$\frac{c}{L}=\frac{1}{L}\left(\frac{E}{\rho}\right)^{1/2}=\left(\frac{k}{mL}\right)^{1/2}$$

in which $k=EA/L$, we can write

$$\omega_1=\frac{\pi}{2}\left(\frac{k}{mL}\right)^{1/2}$$

For a bar with a lumped mass we have

$$\omega_1=\left(\frac{k}{M}\right)^{1/2}$$

Equating the right-hand sides of both relations we get

$$M=0.4053\,mL$$

15-5 Find the natural frequencies and the mode shapes of a simply supported beam.

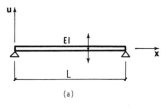

(a)

At the left end the deflection and the bending moment are zero, which we write as $u=0$ and $u''=0$ at $x=0$.

From Eqs. 15.18 and 15.21 we obtain

$$D_2+D_4=0 \quad \text{and} \quad -D_2+D_4=0$$

which results in $D_2=D_4=0$. The mode shape formula reduces to

$$X(x)=D_1\sin px+D_3\sinh px$$

At the right end we also have $u=0$ and $u''=0$, which gives

$$D_1\sin pL+D_3\sinh pL=0$$

$$-D_1\sin pL+D_3\sinh pL=0$$

Adding these equations yields $D_3\sinh pL=0$. We therefore set $D_3=0$, as $\sinh pL=0$ only for $p=0$. This leads to

$$D_1\sin pL=0$$

We cannot have $D_1=0$, because all constants would then vanish and so would our solution. The alternative is to put $\sin pL=0$ or $p_n=\pi/L, 2(\pi/L), 3(\pi/L),\ldots$.

From Eq. 15.20:

$$\omega_n^2=p_n^4\frac{EI}{m}=\left(\frac{n\pi}{L}\right)^4\frac{EI}{A\rho} \quad \text{for} \quad n=1,2,3,\ldots$$

The mode shape equation is

$$X_n(x)=\sin\frac{n\pi x}{L}$$

The first three mode shapes are plotted in (b).

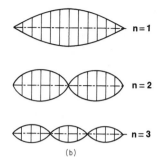

(b)

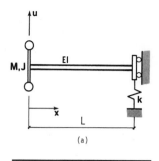

(a)

15-6   The left end of the beam cannot translate and is restrained by an angular spring of stiffness K. At the right end, which is supported by a spring k, there is a lumped mass M. Formulate the boundary conditions.

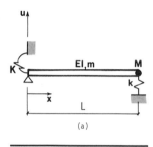

(a)

Whenever there is an elastic constraint, an elastic reaction will appear. Also, a presence of inertia will result in the application of an inertia reaction to the beam. To visualize those effects, assume the rotation of the left end as well as translation and acceleration of the right end, all quantities being positive. The associated reactions appear in (b). Figure 15.4a shows that positive moment bends an infinitesimal beam element so that the compressed fibers are on top, while positive shear rotates that element clockwise (provided, of course, the coordinate system u–x is the same). Now that we can establish the signs of the end forces applied to the beam, we have, from Eqs. 15.16 and 15.17:

$$EIu''=Ku' \quad \text{at} \quad x=0$$

$$EIu'''=M\ddot{u}+ku \quad \text{at} \quad x=L$$

The remaining two conditions are zero translation at the left end and zero moment at the right:

$$u=0 \quad \text{at} \quad x=0$$

$$u''=0 \quad \text{at} \quad x=L$$

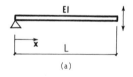

(b)

15-7   At the left end of beam there is a body of mass M and moment of inertia J attached. The right end is prevented from rotating and has a translational elastic support. Formulate the boundary conditions.

The elastic and inertial reactions are shown in (b). Using Eqs. 15.16 and 15.17 and keeping in mind the sign convention from Fig. 15.4a, we have

$$EIu''=J\ddot{u}' \quad \text{and} \quad EIu'''=-M\ddot{u} \quad \text{at} \quad x=0$$

$$u'=0 \quad \text{and} \quad EIu'''=ku \quad \text{at} \quad x=L$$

The symbol $\ddot{u}'$ used above stands for the angular acceleration:

$$\ddot{u}'\equiv\frac{\partial^2}{\partial t^2}\left(\frac{\partial u}{\partial x}\right)$$

15-8   An underconstrained beam of length L performs free bending vibrations. Develop a frequency equation and sketch a plot indicating where to find the first few frequencies.

EI

(a)

The boundary conditions are:

$$X=X''=0 \quad \text{at} \quad x=0$$

$$X''=X'''=0 \quad \text{at} \quad x=L$$

Using Eqs. 15.18 and 15.21 we obtain:

$$D_2+D_4=0 \quad \therefore D_2=D_4=0$$

$$-D_2+D_4=0$$

$$-D_1\sin pL+D_3\sinh pL=0$$

$$-D_1\cos pL+D_3\cosh pL=0$$

The determinant of the coefficients is now set equal to zero so that D_1 and D_3 may have nonzero values:

$$\begin{vmatrix} -\sin pL & \sinh pL \\ -\cos pL & \cosh pL \end{vmatrix}=0$$

which is equivalent to $\tan pL=\tanh pL$.

The intersection points of the two functions constituting the frequency equation are shown in *b*. The zero frequency must also be considered, since it corresponds to free rotation.

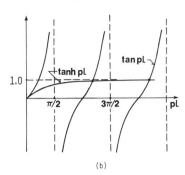

(b)

15-9 A circular shaft that is rotating with a steady angular speed λ (rad/s) is suddenly stopped at the cross section $x=0$. Determine the subsequent displacement response of the end segment, $0 \leqslant x \leqslant L$.

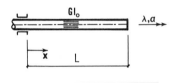

The angular speed λ is treated as the initial velocity. The angle of rotation α is measured from the position occupied at the instant the shaft is stopped. Thus, our initial conditions are.

$$\alpha = 0 \quad \text{at} \quad t=0$$

$$\dot{\alpha} = \lambda \quad \text{at} \quad t=0$$

The mode shapes and frequencies are analogous to those in Prob. 15-1:

$$X_n = \sin \frac{n\pi x}{2L} \quad \text{and} \quad \omega_n = \frac{n\pi c_1}{2L} \quad \text{for} \quad n=1,3,5,\ldots.$$

The integrals on both sides of Eq. 15.27 are now calculated.

$$\int_0^L X_n^2 \, dx = \int_0^L \sin^2 \frac{n\pi x}{2L} \, dx = \frac{L}{2}$$

(See Appendix I.)

$$\int_0^L \lambda \sin \frac{n\pi x}{2L} \, dx = \frac{2L}{n\pi} \lambda$$

Substituting in Eq. 15.27 gives us

$$\frac{2L}{n\pi} \lambda = \frac{n\pi c_1}{2L} \frac{L}{2} B_{2n}$$

$$\therefore B_{2n} = \frac{8L\lambda}{n^2 \pi^2 c_1} \quad \text{for} \quad n=1,3,5,\ldots.$$

Obviously $B_{1n}=0$ because $u_0(0)=0$. The time function is thus

$$f_n(t) = \frac{8L\lambda}{n^2 \pi^2 c_1} \sin \omega_n t$$

and the solution may be briefly written as

$$\alpha(x,t) = \sum_{n=1,3,5,\ldots}^{\infty} X_n f_n$$

15-10 Investigate how the twisting moment changes with time at the base of the shaft in Prob. 15-9.

From Eq. 15.6:

$$M_t = GC\alpha' = GI_0 \sum_{n=1,3,5,\ldots}^{\infty} X_n' f_n$$

But

$$X_n' = \frac{n\pi}{2L} \cos \frac{n\pi x}{2L} \quad \text{and} \quad X_n'(0) = \frac{n\pi}{2L}$$

$$\therefore M_t(0,t) = \frac{4\lambda}{\pi c_1} GI_0 \sum_{n=1,3,5,\ldots}^{\infty} \frac{1}{n} \sin n\omega_1 t$$

where $\omega_1 = \pi c_1/(2L)$. We first note that $M_t=0$ for $t=0$ and $t=\pi/\omega_1$. At $t=\pi/(2\omega_1)$ we get

$$M_t = \frac{4\lambda}{\pi c_1} GI_0 \left(1 - \frac{1}{3} + \frac{1}{5} - \ldots\right) = \frac{\lambda}{c_1} GI_0$$

One can find by differentiation of $M_t(0,t)$ that at $t=\pi/(2\omega_1)$ the value of the twisting moment reaches its maximum. This similarity to an SDOF system having natural frequency ω_1 comes from the fact that all system frequencies are multiples of ω_1. Let us now determine M_t at some intermediate time point, say $t=\pi/(4\omega_1)$:

$$M_t = \frac{4\lambda}{\pi c_1} GI_0 \frac{1}{\sqrt{2}} \left(1 + \frac{1}{3} - \frac{1}{5} - \frac{1}{7} + \frac{1}{9} + \ldots\right)$$

The sum of the series in parentheses is slightly larger than unity. For an SDOF system we would have

$$M_t = \frac{4\lambda}{\pi c_1} GI_0 \frac{1}{\sqrt{2}}$$

at this instant.

15-11 The mode shapes and the natural frequencies of a bar fixed at both ends are, respectively:

$$X_n = \sin \frac{n\pi x}{L} \quad \text{and} \quad \omega_n = \frac{n\pi c}{L}$$

for $n=1,2,3,\ldots$

The intensity of the distributed axial load is

$$q = q_0 \sin \frac{\pi x}{L}$$

Assuming homogeneous initial conditions, find the response due to a sudden application of this load.

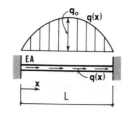

Calculate the value of integrals in Eq. 15.29:

$$\int_0^L X_n(x) \cdot q(x,t)\, dx = q_0 H(t) \int_0^L \sin\frac{n\pi x}{L} \sin\frac{\pi x}{L}\, dx$$

$$= H(t) q_0 \frac{L}{2} \qquad (=0 \text{ for } n \neq 1, \text{ see Eq. A1.6})$$

[Symbol $H(t)$ is used to indicate that the load is the step function in time.]

$$\int_0^L X_n^2\, dx = \frac{L}{2}$$

After substitution we obtain

$$q_1(t) = H(t) q_0, \qquad q_n(t) = 0 \quad \text{for} \quad n \neq 1$$

In this particular case our modal expansion of $q(x,t)$ consists of only one term. Equation 15.30 becomes

$$\ddot{f}_1 + \omega_1^2 f_1 = \frac{1}{m} q_0 H(t)$$

$$\therefore f_1(t) = \frac{q_0}{m\omega_1^2}(1 - \cos\omega_1 t) = \frac{q_0}{m}\frac{L^2}{\pi^2 c^2}(1 - \cos\omega_1 t)$$

The function $H(t)$ was not shown after the differential equation was solved. Our complete solution

$$u(x,t) = X_1 f_1 = \frac{q_0 L^2}{\pi^2 EA} \sin\frac{\pi x}{L}(1 - \cos\omega_1 t)$$

consists of only one term. This is because the shape of the forcing function is the same as that of the first vibratory mode and in effect only this mode is excited.

15-12 Determine the response of a simply supported beam subjected to a concentrated load

$$P(t) = P_0 H(t)$$

applied at a distance a from the left end.

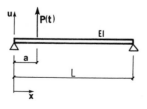

With the expressions for the mode shapes and frequencies available from Prob. 15-5 we have

$$X_n(a) = \sin\frac{n\pi a}{L}$$

which we substitute into Eq. 15.31:

$$P_0 H(t) \sin\frac{n\pi a}{L} = q_n(t) \int_0^L \sin^2\frac{n\pi x}{L}\, dx$$

$$\therefore q_n(t) = \frac{2}{L} P_0 \sin\frac{n\pi a}{L} H(t)$$

Equation 15.30 gives us

$$\ddot{f}_n + \omega_n^2 f_n = \frac{2}{mL} P_0 \sin\frac{n\pi a}{L} H(t)$$

We can now easily find that

$$f_n(t) = \frac{2}{mL\omega_n^2} P_0 \sin\frac{n\pi a}{L}(1 - \cos\omega_n t)$$

and

$$u(x,t) = \sum_{n=1}^{\infty} X_n f_n$$

where

$$X_n = \sin\frac{n\pi x}{L}$$

15-13 A simply supported beam is subjected to a load concentrated at the midspan, which is a step function in time. Discuss how the deflection changes with time and sketch the deflected shape for $t = \tau_1/50$, where $\tau_1 = 2\pi/\omega_1$ is the fundamental period. Use the results of Prob. 15-12.

With $a = L/2$ the time function is

$$f_n(t) = \frac{2P_0}{mL\omega_n^2} \sin\frac{n\pi}{2}(1 - \cos\omega_n t)$$

Noting that $\sin(n\pi/2) = (-1)^{(n-1)/2}$ for $n = 1,3,5,\ldots$ and $mL\omega_n^2 = n^4\pi^4 EI/L^3$, we get

$$f_n(t) = \frac{2P_0 L^3}{n^4\pi^4 EI}(-1)^{(n-1)/2}(1 - \cos\omega_n t)$$

Using the following relations:

$$\omega_n = n^2\omega_1 = n^2\left(\frac{2\pi}{\tau_1}\right)^2$$

$$1 - \cos\omega_n t = 2\sin^2\frac{\omega_n t}{2}$$

and observing that $u_{st} = P_0 L^3/48EI$ is the static deflection under P_0 at the middle, we have

$$f_n(t) = \frac{192}{\pi^4 n^4} u_{st}(-1)^{(n-1)/2} \sin^2\frac{n^2\pi t}{\tau_1} \qquad (*)$$

And finally

$$u(x,t) = \sum_{n=1,3,5,\dots}^{\infty} \sin\frac{n\pi x}{L} f_n(t) \qquad (**)$$

since all even-numbered terms vanish. First we check on the displacement at the midspan,

$$u\left(\frac{L}{2},t\right) = \frac{192}{\pi^4} u_{st} \sum_{n=1,3,5,\dots}^{\infty} \frac{1}{n^4} \sin^2\frac{n^2\pi t}{\tau_1}$$

At $t=0$, $u=0$. For $t=\tau_1/2$ we have

$$\sum = 1 + \frac{1}{3^4} + \frac{1}{5^4} + \cdots = \frac{\pi^4}{96}$$

$$\therefore u\left(\frac{L}{2},\frac{\tau_1}{2}\right) = 2u_{st}$$

It can also be shown that

$$u\left(\frac{L}{2},\frac{\tau_1}{4}\right) = u_{st}$$

which indicates that the motion of the midpoint is similar to that of an oscillator with a natural period τ_1 (See Prob. 2-12) This is possible only because each higher frequency is the multiple of the first one. Finding the deflected shape for $t=\tau_1/50$ is a little more involved. Let us first determine the values of $f_n(\tau_1/50)$ from Eq. (*):

n:	1	3	5	7	9	11
$\frac{1}{u_{st}}f_n\!\left(\frac{\tau_1}{50}\right)$:	$\frac{1}{128.68}$	$\frac{-1}{143.13}$	$\frac{1}{317.09}$	≈ 0	$\frac{1}{3850.4}$	$\frac{-1}{7917.6}$

It is apparent that in general the coefficients decrease rapidly with growing n, although a few successive ones may sometimes have nearly the same absolute value. The table that follows shows the computed values of nondimensional deflection u/u_{st} using Eq. (**) and including only as many terms as shown in the table above.

$\dfrac{x}{L}$	$\dfrac{u}{u_{st}}$
0.05	0.00041
0.10	0.00002
0.15	−0.00126
0.20	−0.00230
0.25	−0.00158
0.30	0.00129
0.35	0.00573
0.40	0.01113
0.45	0.01615
0.50	0.01830

It is interesting to notice that the deflected shape in the figure is quite different from the static curve. A plot made for $t<\tau_1/50$ would show an even wavier line.

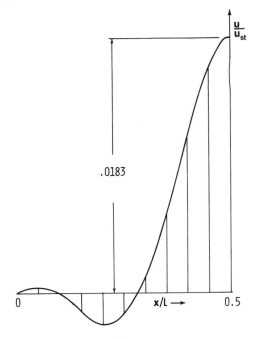

15-14 The fundamental frequency of a square, simply supported plate is

$$\omega = \frac{5.7}{a^2}\left(\frac{Et^2}{\rho(1-\nu^2)}\right)^{1/2}$$

where E, t, ρ, and ν are, respectively, Young's modulus, the thickness, the mass density, and Poisson's ratio. Suppose that we want to develop a model in which the plate is a weightless element having mass M lumped at its center. What should be the ratio of M to the actual plate mass to preserve the natural frequency ω?

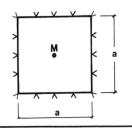

According to Ref. 7, the center deflection of a simply supported, rectangular plate is

$$u = 0.1391(1-\nu^2)\frac{Pa^2}{Et^3}$$

where P is the load at the center. The natural frequency ω of the model is given by

$$\omega^2 = \frac{k}{M} = \frac{Et^3}{0.1391(1-\nu^2)a^2 M}$$

Equating this with the actual plate frequency yields

$$\frac{Et^3}{0.1391(1-\nu^2)a^2M} = \frac{5.7^2 Et^3}{a^2 a^2 \rho t(1-\nu^2)}$$

We note the plate mass is $a^2\rho t$ and therefore

$$M = 0.2213(a^2\rho t)$$

15-15 A general beam deflection has two components—bending and shear:

$$u = u_b + u_s$$

The first one is due to the action of bending moment (Fig. 15.4b and Eq. 15.16), while the second results from the action of shear forces, as shown in the figure and defined by

$$\vartheta = \frac{V}{GA_s}$$

where A_s is the shear area. When shear distortion is ignored, we have *flexural beam*, and when flexure is neglected (or prevented) we deal with a *shear beam*. Develop the equation of motion of a shear beam. (*Note.* In this chapter we limit ourselves to flexural beams unless a statement is made to the contrary.)

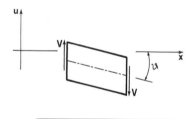

On the basis of equilibrium in Fig. 15.4a,

$$q - m\ddot{u} - \frac{\partial V}{\partial x} = 0$$

The relationship between a force and a deformation is, according to the problem statement,

$$V = GA_s\vartheta = -GA_s u'$$

Differentiating and substituting gives us

$$\ddot{u} - c_3^2 u'' = \frac{q}{m}$$

in which

$$c_3^2 = \frac{GA_s}{m} = \frac{GA_s}{A\rho}$$

The equation is analogous to that of a bar, a shaft, or a cable.

15-16 Use the Dunkerley method for the determination of the natural frequency of a shaft with a lumped inertia at the end (Prob. 15-2) with $J_D/J_S = 1$.

The shaft stiffness is $K = GC/L$. When the distributed mass is ignored, we get

$$\omega_{11}^2 = \frac{K}{J_D} = \frac{GC}{J_D L} = \frac{\rho I_0 L}{J_D}\frac{c_1^2}{L^2}$$

When the lumped mass is ignored, we have a simple shaft with the fundamental frequency

$$\omega_{22} = \frac{\pi c_1}{2L}$$

(This is from Prob. 15-1 after replacing c with c_1.) This is our other fictitious frequency to be used with the Dunkerley formula. Noting that in our problem $J_D/(\rho I_0 L) = 1$, the frequency is

$$\frac{1}{\omega^2} = \frac{1}{\omega_{11}^2} + \frac{1}{\omega_{22}^2} = \frac{L^2}{c_1^2}\left(1 + \frac{4}{\pi^2}\right)$$

$$\therefore \omega = \frac{0.8436 c_1}{L}$$

This is about 2% less than calculated in the reference problem.

EXERCISES

15-17 A bar of length $L = 200$ in. lies on a frictionless horizontal plane. Calculate the first three natural frequencies when the material properties of the bar are $E = 29 \times 10^6$ psi and $\gamma = 0.282$ Lb/in.^{3}

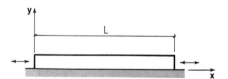

15-18 Formulate the boundary conditions for the bar with a mass and a spring at each end. *Hint.* As a preparatory step, assume positive displacement and accelerations at each end and draw the forces applied to the ends.

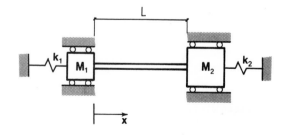

15-19 Determine the frequency equation for a bar that has a mass attached at each end. Find the first vibratory frequency when $M_1 = M$; $M_2 = 2M$, and $M_0 = AL\rho = M$. Compare the answer with that of Prob. 9-30, where the mass of the bar was ignored.

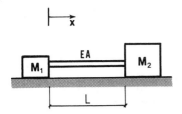

15-20 A disc with a mass moment of inertia J is attached to the end of a circular shaft, whose other end is fixed. The moment of inertia of the shaft itself is also J. The fundamental frequency of the system is

$$\omega_1 = \frac{0.8603 c_1}{L}$$

from Prob. 15-2. If we want to construct a simplified model, in which the shaft is a weightless, elastic element and the inertia is lumped at the end, how big should that inertia $\bar{J}$ be for the frequencies of both systems to be the same?

15-21 A bar of length $2L$ is restrained at the center by a set of axial springs having an effective stiffness k. Formulate the boundary conditions and derive the frequency equation. Write the expression for the mode shape over the left half of the bar. *Hint.* Formulate a separate equation for each half of the bar. Besides three displacement conditions, there is a static equation, which may be deduced from (b).

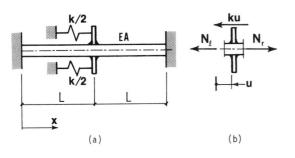

(a)   (b)

15-22 A cantilever beam of length L performs free bending vibrations. Develop a frequency equation and a formula for the fundamental frequency. Sketch the first mode shape.

15-23 An unconstrained beam of length L performs free bending vibrations. Develop a frequency equa-

tion and a formula for the fundamental vibratory frequency. Sketch the corresponding mode shape.

15-24 List the boundary conditions for the beam restrained by two angular springs and one linear spring.

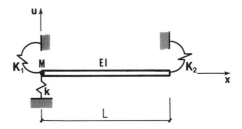

15-25 Formulate a frequency equation for a beam with completely fixed ends.

15-26 Develop a frequency equation for the cantilever beam with an additional mass M lumped at the moving end.

15-27 Determine the first frequency when the ratio of the lumped mass to the distributed mass in Prob. 15-26 is (1) $M/(mL) = 0.5$ and (2) $M/(mL) = 1.5$. *Hint.* Notice that the fundamental frequency of a beam with a lumped mass is always less than the frequency of the corresponding beam without such a mass.

15-28 The fundamental frequency of a cantilever beam with a uniformly distributed mass m is

$$\omega_1 = \frac{1.875^2}{L^2} \left(\frac{EI}{m} \right)^{1/2}$$

Consider another cantilever beam, which is itself weightless but has a lumped mass M at the end. What fraction of mL should M be, if the frequencies of both cantilevers are to be the same?

15-29 A cable having the axial preload N is subjected to a lateral displacement δ and then suddenly released from the displaced position. After finding the natural frequency and the mode shapes, determine the motion of the cable after release.

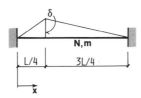

15-30 The bar, initially at rest, is subject to a suddenly applied distributed load

$$q = \frac{q_0 x}{L} H(t)$$

in which $H(t)$ is the unit step function. Determine the subsequent motion of the bar.

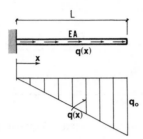

15-31 Let $u(L)$ denote the maximum dynamic displacement of the free end in Prob. 15-30 and let u_{st} stand for the static deflection at the same place. Calculate the maximum value of the dynamic magnification factor $\mu = u(L)/u_{\text{st}}$. *Hint.* Notice that static deflection due to a prescribed load pattern may be obtained by integration of Eq. 15.1.

15-32 The force applied at the midpoint of the bar is $P = P_0(t/t_0)$. Calculate the displacement response.

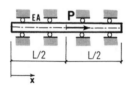

15-33 The mode shapes and natural frequencies of a simply supported beam are, respectively,

$$X_n = \sin\frac{n\pi x}{L} \quad \omega_n = \left(\frac{n\pi}{L}\right)^2 \left(\frac{EI}{A\rho}\right)^{1/2} \quad n = 1, 2, 3, \ldots$$

Calculate the response to a load distributed according to a sinusoid:

$$q(x, t) = q_0 H(t) \sin\frac{2\pi x}{L}$$

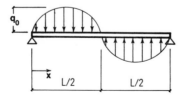

15-34 A cable fixed at both ends is subjected to a lateral load, constant along the length and varying according to $q = q_0 \sin \Omega t$. Find a closed-form solution for displacement response.

15-35 Solve Prob. 15-34 again using a general normal-mode approach discussed in Sect. 15E. The natural frequencies and the mode shapes are, respectively,

$$\omega_n = \frac{n\pi c_2}{L} \quad \text{and} \quad X_n = \sin\frac{n\pi x}{L} \quad \text{for} \quad n = 1, 2, 3, \ldots$$

15-36 A bar, fixed at one end and free at the other has a force $P = P_0 \cos \Omega t$ applied to it at the free end. Calculate the displacement response using the external force as a boundary condition.

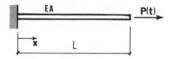

15-37 Solve Prob. 15-12 when the load applied to the beam is harmonic with respect to time:

$$P(t) = P_0 e^{i\Omega t}$$

The natural frequencies and mode shapes are, respectively:

$$\omega_n = \left(\frac{n\pi}{L}\right)^2 \left(\frac{EI}{A\rho}\right)^{1/2} \quad X_n = \sin\frac{n\pi x}{L} \quad n = 1, 2, 3, \ldots$$

Find only the steady-state response.

15-38 A simply supported beam is loaded at midspan by a harmonic force

$$P = P_0 e^{i\Omega t}$$

Find the expression for the bending moment at this location, using the result of Prob. 15-37.

15-39 A multistory frame consists of two flexible columns and rigid horizontal elements. If the axial deformation of the columns is small compared to the flexure and if the number of stories is large, the framework may be treated as a continuous shear beam. Shear stiffness k of each segment is defined as

$$k = \frac{V}{u_{i+1} - u_i} = \frac{V}{\Delta u}$$

Determine the steady-state displacement response to a force $P = P_0 \cos \Omega t$ applied at the top level.

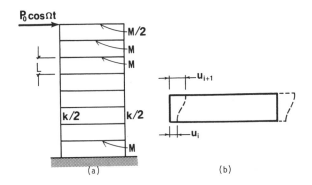

(a)   (b)

15-40 Determine the first frequency when the ratio of the lumped mass to the distributed mass in Prob. 15-26 is (1) $M/(mL)=0.5$ and (2) $M/(mL)=1.5$. Use the Dunkerley method and compare the answer with that for Prob. 15-27.

15-41 A heavy weight, $W=100$ Lb and $J=4.8575$ Lb-s^2-in., is attached to the end of the cantilever with $L=30$ in. The natural frequencies of the system with the beam itself treated as weightless were determined in Prob. 9-4 and the fundamental frequency was found to be $f=20.36$ Hz. What is the actual value of f, if the distributed mass, previously disregarded, is such that

$$mgL=\frac{W}{3}$$

Use the Dunkerley method and put $E=10\times10^6$ psi and $I=4$ in.^{4}

15-42 The base of a simply supported beam is driven with a lateral acceleration $a_b=a_0\sin\Omega t$. Calculate the maximum bending moment and the shear force induced by the first mode response. The forcing frequency $\Omega=\omega_1$, and the presence of small damping causes the magnification factor to be equal to μ_1.

CHAPTER

16

Geometric Stiffening

The basic idea of geometric stiffening and its application for axial members was introduced in Chapter 1. Those concepts are extended here to other element types and to MDOF systems. In formulating the dynamic problem, the axial preload modifies the stiffness matrix of the system. When the modification is significant enough to render the matrix singular, the computed fundamental frequency becomes zero. The axial load capable of inducing this condition is the same as the buckling load of the structure. (The latter may also be found from purely static reasoning.) Calculation of natural frequencies altered by the application of axial loads is one of the main problem topics in this chapter.

16A. Behavior of Rigid, Pin-Ended Bar is the simplest example of the effects investigated in this chapter. When the horizontal force N shown in Fig. 16.1 is zero, the angular stiffness of the system is K.

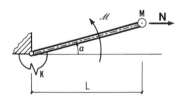

Figure 16.1 Rigid, pin-ended bar.

The presence of N makes it more difficult to rotate the bar because of the additional moment $NL\alpha$ op-

posing rotation. The total apparent stiffness is thus

$$\tilde{K} = K + NL = K + K_g \qquad (16.1)$$

where K_g is called the geometric stiffness. When the axial force is directed opposite so that it puts the bar in compression, the apparent stiffness $\tilde{K}$ is less than the elastic stiffness K. If the magnitude of the compressive force is sufficient, $\tilde{K}$ may become zero, which means that no resistance is offered to rotation. We then say that the system becomes unstable and write

$$K - N_{cr}L = 0 \qquad \text{or} \qquad N_{cr} = \frac{K}{L} \qquad (16.2)$$

where N_{cr} is the *critical force* or the *buckling force*. Using this concept we can put the expression for the apparent stiffness in a different form:

$$\tilde{K} = \left(1 + \frac{N}{N_{cr}}\right) K \qquad (16.3)$$

in which positive N means tension. When J is the mass-moment of inertia of the bar with respect to the pivot point, the natural frequency in absence of N is calculated from $\omega^2 = K/J$. Taking the axial force into account merely modifies the stiffness, therefore

$$\tilde{\omega}^2 = \omega^2 \left(1 + \frac{N}{N_{cr}}\right) \qquad (16.4)$$

where $\tilde{\omega}$ is the natural frequency calculated in the presence of the axial force N, positive when tensile.

The equation of motion of the system in Fig. 16.1 is

$$J\ddot{\alpha} = -c'\dot{\alpha} - K\alpha - NL\alpha + \mathfrak{M}$$

This is essentially the same as Eq. 2.1 for the translational motion except for the added axial force term. When written as

$$J\ddot{\alpha} + c'\dot{\alpha} + \tilde{K}\alpha = \mathfrak{M}(t) \qquad (16.4a)$$

with $\tilde{K}$ defined by Eq. 16.1, the similarity is even more pronounced. (Note that a damper is not shown in Fig. 16.1, the damping term was added to the equation for the sake of completeness.) The conclusion is that once we replace the elastic stiffness K by its apparent value $\tilde{K}$, we can ignore the effect of axial force on the motion of an SDOF system.

16B. Equation of Motion of Beam-Column.

A beam subjected to an axial force in addition to lateral loading is usually referred to as a beam-column. Figure 15.3, although drawn mainly for a cable, shows well how the force N influences the equilibrium of a deformed beam element. The net effect of N is the same as if a distributed load of intensity Nu'' were applied to the beam. Equation 15.15, which describes the lateral motion, can now be written as

$$m\ddot{u} + EIu'''' - Nu'' = q \qquad (16.5)$$

The solution is again given by Eq. 15.10. The time function retains its form, Eq. 15.12, except that ω is replaced by $\tilde{\omega}$ to indicate the influence of N. The shape function is given by

$$EIX'''' - NX'' - m\tilde{\omega}^2 X = 0 \qquad (16.6)$$

Solving the equation in an exact manner is rarely done because of mathematical difficulties.

16C. Beam Frequency Calculation by Ritz-Galerkin Method.

The simplest form of this method may be described as follows. Assume a mode shape $X(x)$ that satisfies at least the boundary conditions regarding deflection and slope. The corresponding natural frequency is found from

$$\int_0^L (EIX'''' - NX'' - m\tilde{\omega}^2 X) X\, dx = 0 \qquad (16.7)$$

when the prescribed integration is performed. (The term in parentheses is identical with the left-hand side of Eq. 16.6.) Apart from $\tilde{\omega}$ for a given value of N we may also find ω (i.e, the frequency of the same mode in the absence of N as well as the critical force N_{cr}). If the last two quantities are known beforehand, their newly calculated values provide a measure of how accurately $X(x)$ was assumed.

16D. Buckling Force, Critical Force, and Their Relation to Vibratory Frequency.

When a beam is under the influence of an axial force N, its static equilibrium is described by Eq. 15.16 in which $\mathfrak{M}$ is replaced by Nu:

$$EIu'' = Nu \qquad (16.8)$$

A solution of this equation for a beam with some particular boundary conditions gives us a series of forces

$$N_{cr1}, N_{cr2}, N_{cr3}, \cdots$$

and a series of the associated functions

$$u_1(x), u_2(x), u_3(x), \cdots$$

representing some deformed shapes of the beam. Those special values of the axial load are called *critical forces* and they are always negative, or compressive. The name stems from the fact that for $N = N_{cr}$ even the slightest lateral load applied to the beam will cause the deflections to become infinite. The smallest of all critical forces (in terms of absolute values), N_{1cr}, is called the *buckling force* because it is associated with the lateral instability, or buckling, during static compression test. A beam cannot carry a sustained compressive load larger than the buckling force, and for this reason the remaining critical values are of lesser importance.

A dynamic definition of a critical force may be obtained from Eq. 16.6. When a natural frequency $\tilde{\omega}$ treated as a function of N becomes zero, we say that the critical force associated with that $\tilde{\omega}$ has been reached. This implies

$$EIX'''' - NX'' = 0$$

as a condition for a critical state. We note that the same relation may be obtained from Eq. 16.8, which means that the axial force equal to N_{cr} not only causes static instability, but also reduces the natural frequency of vibrations to zero.

A natural frequency $\tilde{\omega}_r$ of a beam-column vibrating in the rth mode may be approximately expressed as

$$\tilde{\omega}_r^2 = \omega_r^2 \left(1 + \frac{N}{N_{crr}}\right) \quad \text{or} \quad \tilde{\omega}_r^2 = \omega_r^2 \left(1 - \frac{N}{N_{crr}}\right)$$

$$(16.9)$$

where ω_r is the natural frequency in absence of N and N_{crr} is the magnitude of critical force associated with the rth mode. The first equation is used for the tensile and the second for compressive N. (Writing two equations instead of one allows us to treat all symbols as positive quantities.) As shown in Prob. 16-3, these expressions are exactly true for a simply supported beam. For a given value of N, the first-mode frequency is most significantly affected, while the influence on the higher modes is typically much less important.

A convenient way of specifying N or N_{cr} is by expressing it as a multiple of *Euler force N_e*,

$$N_e = \frac{\pi^2 EI}{L^2} \qquad (16.10)$$

where N_e is the buckling force for a simply supported beam made of linearly elastic material.

For most structures, even those that look quite slender, the buckling load calculated on the elastic basis exceeds the true value determined with the use of actual material properties. From our viewpoint, however, this is of little concern, because as long as the structure is deforming within the elastic range, it is the elastic value of N_{cr} that influences stiffness and natural frequency.

16E. Effect of Axial Force on Beam Stiffness Matrix. The stiffness matrix of an axial bar in presence of an axial load was given in Chapter 8. It can be briefly put down as

$$\tilde{\mathbf{k}} = \mathbf{k} + \mathbf{k}_g \qquad (16.11)$$

where $\mathbf{k}$ is the elastic and $\mathbf{k}_g$ the geometric component of the effective or apparent stiffness matrix $\tilde{\mathbf{k}}$. The same applies to a beam element. The apparent stiffness matrix can be obtained from Eq. 16.5, (when $m = 0$ and $q = 0$) and then presented as a sum of two components. The exact expressions are rather involved, so only the first-order approximations are quoted here. For the beam-column in Fig. 16-2, the elastic component is given by Eq. 8.12 while the

geometric part is

$$\mathbf{k}_g = \begin{bmatrix} \dfrac{6}{5} & -\dfrac{L}{10} \\[2ex] -\dfrac{L}{10} & \dfrac{2L^2}{15} \end{bmatrix} \dfrac{N}{L} \qquad (16.12)$$

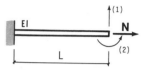

Figure 16.2 Reference directions for stiffness matrix.

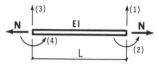

Figure 16.3 Reference directions for stiffness matrix.

When a more general case of a 4×4 beam matrix (Fig. 16.3) is considered, the geometric stiffness is

$$\mathbf{k}_g = \begin{bmatrix} \dfrac{6}{5} & -\dfrac{L}{10} & -\dfrac{6}{5} & -\dfrac{L}{10} \\[2ex] -\dfrac{L}{10} & \dfrac{2L^2}{15} & \dfrac{L}{10} & -\dfrac{L^2}{30} \\[2ex] -\dfrac{6}{5} & \dfrac{L}{10} & \dfrac{6}{5} & \dfrac{L}{10} \\[2ex] -\dfrac{L}{10} & -\dfrac{L^2}{30} & \dfrac{L}{10} & \dfrac{2L^2}{15} \end{bmatrix} \dfrac{N}{L}$$

$$\quad (1) \qquad (2) \qquad (3) \qquad (4)$$

$$(16.13)$$

The elastic part is given by Eq. 8.13 (see p. 390 of Ref. 6 for the derivation of the complete matrix). The relation between forces and displacements is now

$$\vec{\mathbf{P}} = \tilde{\mathbf{k}}\vec{\mathbf{u}} \qquad (16.14)$$

which is a generalization of Eq. 8.1. Once the axial force is "hidden" in the apparent stiffness matrix, we need not concern ourselves with it during the rest of the calculation. Equation 16.14 applies not only to a single beam but to any structural assembly that exhibits geometric stiffening.

When a beam is put in compression, we find that at some values of the axial force the determinant of the apparent stiffness matrix becomes zero:

$$|\tilde{\mathbf{k}}| = |\mathbf{k} + \mathbf{k}_g| = 0 \qquad (16.15)$$

Again, as previously, a force capable of inducing this condition is called a critical force, while the smallest of all critical forces is referred to as a buckling force.

An actual geometric stiffness matrix almost always contains nonlinear terms. The first-order approximations given here by Eqs. 16.12 and 16.13 are useful in computing response when the axial load is much smaller than the buckling load. They are not good, however, for calculating the buckling load itself, especially if the members of a structure are not equally influenced by the axial load. If we are capable to somehow estimate the elastic buckling load for even a complicated structure, we can expect relations of the type of Eq. 16.3, 16.4, and 16.9 to give realistic results, provided there is some similarity between the buckling pattern involved and the vibratory mode of interest.

SOLVED PROBLEMS

16-1 The elastically supported rigid beam has the following natural frequencies:

$$\omega_1^2 = \frac{2k_v}{M} \qquad \omega_2^2 = \frac{24K + 6k_v L^2}{ML^2} \qquad \omega_3^2 = \frac{2k_h}{M}$$

from Prob. 9-8. What should be the compressive force N_0 to make $\tilde{\omega}_1 = \tilde{\omega}_2$? What is the buckling force?

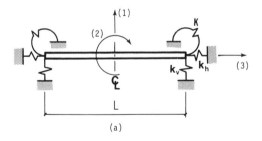

(a)

Apply a tensile force N to the beam. As long as there is no rotation, the effect of this force will not be felt, from which we conclude that ω_1 and ω_3 will not be influenced. That is,

$$\tilde{\omega}_1 = \omega_1 \qquad \text{and} \qquad \tilde{\omega}_3 = \omega_3$$

To calculate $\tilde{\omega}_2$ we look at the beam when rotated by an angle α, as in (b). The presence of N results in an additional restraining moment $NL\alpha$. The total angular stiffness is therefore

$$\tilde{k}_{22} = 2K + \tfrac{1}{2}k_v L^2 + NL$$

With $J = (1/12)ML^2$ we have

$$\tilde{\omega}_2^2 = \frac{24K + 6k_v L^2 + 12NL}{ML^2}$$

To obtain $\tilde{\omega}_1 = \tilde{\omega}_2$, we apply $N = -N_0$

$$2k_v = \frac{24K}{L^2} + 6k_v - \frac{12N_0}{L}$$

or

$$N_0 = \frac{1}{3}\left(k_v L + \frac{6K}{L}\right)$$

The applied force must, of course, be compressive. From the expression for $\tilde{\omega}_2^2$ we may also find the buckling force:

$$N_{cr} = \frac{1}{2}\left(k_v L + \frac{4K}{L}\right)$$

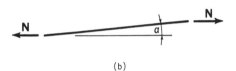

(b)

16-2 Derive the stiffness matrix for a double pendulum under the influence of gravity.

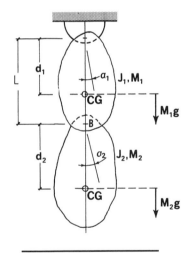

Let us think of gravity as of concentrated forces applied at the respective CGs. Equation 8.10 tells us that when we displace point B by a unit distance, the force of resistance, perpendicular to L, is N/L, provided N is a stretching force along L. This also means that the moment of resistance for a unit rotation is NL. Taking into account the actual distribution of stretching load, we have

$$k_{11} = M_1 g d_1 + M_2 g L$$

since from the viewpoint of the upper link, $M_2 g$ is applied at point B. A similar relation holds if only the lower link is rotated by unity:

$$k_{22} = M_2 g d_2$$

The coefficient $k_{12}=0$ because rotating one link does not impose any tendency to rotate on the other. Assembling the matrix, we have

$$\mathbf{k}=\begin{bmatrix} M_1d_1+M_2L & 0 \\ 0 & M_2d_2 \end{bmatrix}g$$

A more involved generation of this matrix starting with the dynamic equilibrium equations may be found, for example, in Ref. 2.

16-3 Find the natural frequencies of a simply supported beam-column subjected to stretching load with a magnitude

$$N=0.5N_e=\frac{0.5\pi^2EI}{L^2}$$

where N_e is the Euler buckling force. The natural frequencies and the mode shapes in the absence of N were determined in Prob. 15-5:

$$\omega_n^2=\left(\frac{n\pi}{L}\right)^4\frac{EI}{m} \quad \text{and} \quad X_n=\sin\frac{n\pi x}{L}$$

Function $X_n(x)$, which was developed for an ordinary beam, also happens to be a solution of Eq. 16.6. Upon substitution, we get the natural frequency:

$$\tilde{\omega}^2=\frac{EI}{m}\left(\frac{n\pi}{L}\right)^4+\frac{N}{m}\left(\frac{n\pi}{L}\right)^2$$

By inserting $n=1,2,\ldots$, a sequence of natural frequencies is obtained. The same reasoning used in Sect. 16A gives us the value of critical force

$$N_{crn}=\left(\frac{n\pi}{L}\right)^2EI=n^2N_e$$

and an equation for natural frequency,

$$\tilde{\omega}_n^2=\omega_n^2\left(1+\frac{N}{N_{crn}}\right)$$

In our particular case we have

$$\frac{N}{N_{crn}}=\frac{N}{n^2N_e}=\frac{1}{2n^2} \quad \text{and} \quad \tilde{\omega}_n^2=\omega_n^2\left(1+\frac{1}{2n^2}\right)$$

For the successive frequencies,

$$\tilde{\omega}_1=1.2247\omega_1 \quad \tilde{\omega}_2=1.0607\omega_2 \quad \tilde{\omega}_3=1.0274\omega_3$$

The influence of the axial force quickly diminishes with the growing mode number.

16-4 A beam having one end fixed and the other end free has the fundamental frequency ω. Because of

vibratory environment it is required that the beam be subjected to a sustained tensile load N_0. How much can ω increase due to this preload if the allowable tensile stress is $\sigma_0=60,000$ psi? Length $L=15$ in., the cross section is tubular, $D_0=4.00$ in., and $D_i=3.5$ in., $E=29\times10^6$ psi.

The maximum allowable tensile force is

$$N_{\max}=\frac{\pi}{4}\left(D_0^2-D_i^2\right)\sigma_0=\frac{\pi}{4}(4.0^2-3.5^2)60,000=176,715 \text{ Lb}$$

The buckling force for these end conditions is known to be

$$N_{cr}=\frac{N_e}{4}=\frac{\pi^2EI}{4L^2}=1.6538\times10^6 \text{ Lb}$$

From Eq. 16.9:

$$\frac{\tilde{\omega}}{\omega}=\left(1+\frac{N}{N_{cr}}\right)^{1/2}=\left(1+\frac{176,715}{1.6538\times10^6}\right)^{1/2}=1.052$$

As we see, the preload will increase the natural frequency by only about 5%. With the same allowable stress, the increase would be larger for a more slender beam.

16-5 Find the fundamental frequency of the beam attached to a disc rotating with angular velocity Ω. The beam axis is placed along the disc radius. To simplify the problem, assume that the beam is under a constant axial load N_{av} equal to the axial force at the midpoint of beam. Find the expression for the buckling velocity Ω_b.

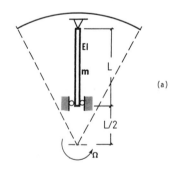

(a)

The centrifugal force applied to a beam segment of length dx, located at a distance r from the center of disc, is

$$dP=\Omega^2rm\,dx=q\,dx$$

in which q is the distributed load shown in (b),

$$q=\Omega^2m\left(\frac{L}{2}+x\right)$$

From the equilibrium of a beam segment having length x, also in

(b), we find the compressive force:

$$N_x = \int_0^x q\,dx = \int_0^x \Omega^2 m\left(\frac{L}{2}+x\right)dx = \frac{1}{2}\Omega^2 mx(L+x)$$

According to the simplification suggested in the problem statement, we assume the beam to be under a constant compressive load corresponding to the midpoint,

$$N_{av} = N_x\left(\frac{L}{2}\right) = \frac{1}{2}\Omega^2 m\frac{L}{2}\left(L+\frac{L}{2}\right) = \frac{3}{8}\Omega^2 mL^2$$

The fundamental frequency in the absence of axial load was given in Prob. 16-3, while the buckling load is known to be

$$N_{cr} = \frac{\pi^2 EI}{L^2}$$

Using Eq. 16.9, we have

$$\tilde{\omega} = \frac{\pi^2}{L^2}\left(\frac{EI}{m}\right)^{1/2}\left(1 - \frac{3\Omega^2 mL^4}{8\pi^2 EI}\right)^{1/2}$$

And the buckling velocity, at $\tilde{\omega}=0$,

$$\Omega_b^2 = \frac{8\pi^2 EI}{3mL^4}$$

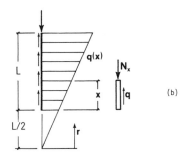

(b)

16-6 A tubular column of height $H=300$ in. supports a weight $W=1800$ Lb. To provide better restraint against overturning forces, two cables are attached, each preloaded to $N_0 = 7500$ Lb. Estimate the natural frequency of vibration in the horizontal direction before and after the addition of the cables. The outer diameter of the column is $D_0 = 10$ in. and the wall thickness is $t=0.65$ in. The effective section area of the cable is $A_c = 0.375$ in.[2] The material is steel, $E=29\times10^6$ psi and $\gamma=0.282$ Lb/in.[3] The natural frequency of a prismatic beam fixed at one end and supported at the other is

$$\omega = \frac{15.42}{L^2}\left(\frac{EI}{m}\right)^{1/2}$$

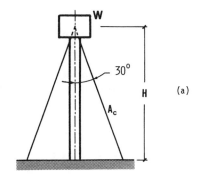

(a)

The second area moment of the tube section is $I=209.65$ in.[4] The stiffness of column, parallel to the ground is:

$$k_1 = \frac{3EI}{L^3} = 3\times29\times10^6\times\frac{209.65}{300^3} = 675.54 \text{ Lb/in.}$$

If we treat the column as a weightless cantilever with a weight at the tip, we have

$$\omega_1 = \left(\frac{k_1 g}{W}\right)^{1/2} = \left(\frac{675.54\times386}{1800}\right)^{1/2} = 12.04 \text{ rad/s}$$

To evaluate the contribution of a cable in (b), refer to Eq. 8.11,

$$k_{11} = c_x^2\frac{EA_c}{L} + \left(1-c_x^2\right)\frac{N_0}{L}$$

with $c_x = \cos 60° = \frac{1}{2}$ and $L=300/\cos30° = 346.41$ in., we get

$$k_{11} = 7848.4 + 16.2 = 7864.6 \text{ Lb/in.}$$

The lateral stiffness of both cables is thus

$$2k_{11} = 2\times7864.6 \approx 15{,}730 \text{ Lb/in.}$$

The cables are schematically represented by springs in (c). Their stiffness ($2k_{11}$) is considerably larger than that of the column, k_1. If the column is neglected, the natural frequency is

$$\omega' = \left(\frac{15{,}730\times386}{1800}\right)^{1/2} = 58.08 \text{ rad/s}$$

On the other hand, we may ignore W and treat the column as a beam, simply supported at the upper end. The distributed mass is

$$m = A\frac{\gamma}{g} = \frac{\pi}{4}(10.0^2 - 8.7^2)\times\frac{0.282}{386} = 0.01395 \text{ Lb-s}^2/\text{in.}^2$$

The problem statement gives the frequency formula for this mode,

$$\omega_2'' = \frac{15.42}{300^2}\left(\frac{29\times10^6\times209.65}{0.01395}\right)^{1/2} = 113.1 \text{ rad/s}$$

The buckling load of this column treated as a beam fixed at one end and pin supported at the other is

$$N_{cr} = \frac{2.045\pi^2 EI}{L^2} = 1.3635\times10^6 \text{ Lb}$$

The compressive load applied by cables:

$$N = 2N_0 \cos 30° = 12,990 \text{ Lb}$$

Using Eq. 16.9 we see that $\tilde{\omega}_2 \approx \omega_2$, which shows that this effect of the axial load is negligible. The Dunkerley formula, Eq. 11.3, may be used to estimate the actual system frequency:

$$\frac{1}{\omega_2^2} = \frac{1}{58.08^2} + \frac{1}{113.1^2} \qquad \therefore \omega_2 = 51.67 \text{ rad/s}$$

Installing the preloaded cables has changed the natural frequency from 12.04 to 51.67 rad/s.

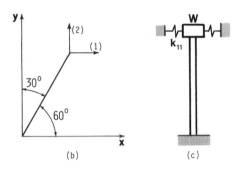

(b)   (c)

16-7 The structure is the same as in Prob. 9-13. Estimate the magnitude of the vertical forces P_0 that will make the system buckle. Consider the symmetric and the antisymmetric modes.

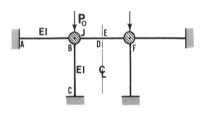

The use of the linearized geometric stiffness matrix, Eq. 16.12, is not suitable for the purpose of finding the critical load, as mentioned in Sect. 16E. A gross estimate will be performed instead. Let us first notice that according to the elementary theory, the buckling load of column BC is between

$$N_{cr1} = 2.045 N_e \qquad \text{and} \qquad N_{cr2} = 4.0 N_e$$

where $N_e = \pi^2 EI/L^2$. The lower limit corresponds to perfectly flexible beams AB and BD, while the upper limit is obtained when those beams are infinitely rigid. Let us assume the average

$$N_{cr}' = \tfrac{1}{2}(2.045 + 4.0) N_e = 3.023 N_e$$

as the buckling load of the symmetrical mode. Since P_0 is the same as the compressive force in column BC, the condition $P_0 = 3.023 N_e$ defines the symmetric buckling.

The rotational joint stiffness for the unloaded structure is

$$K_s = \frac{10EI}{L} \qquad \text{and} \qquad K_a = \frac{14EI}{L}$$

for the symmetric and antisymmetric modes, respectively (Prob. 9-13). One can find the antisymmetric buckling load N_{cr} by interpolating with respect to flexibility $1/K$, the interpolation being based on two points:

$$\text{at} \quad \frac{1}{K_s} = \frac{L}{10EI} \qquad N_{cr} = 3.023 N_e$$

$$\text{at} \quad \frac{1}{K} = 0 \qquad N_{cr} = 4.0 N_e$$

Consequently

$$\text{at} \quad \frac{1}{K_a} = \frac{L}{14EI} \approx 0.0714 \frac{L}{EI}$$

$$N_{cr}'' = \left(4 - \frac{4 - 3.023}{0.1} \times 0.0714\right) N_e = 3.302 N_e$$

which we can indirectly verify by sketching a plot of N_{cr} versus $1/K$. Since N_{cr}' is smaller, buckling takes place when $P_0 = N_{cr}' = 3.023 N_e$.

The exact solution of this problem in Ref. 11 is given as $N_{cr} = 3.1041 N_e$.

EXERCISES

16-8 A rigid bar as in Fig. 16.1 has a natural frequency $f = 100$ Hz when free from the axial load. Calculate the modified frequencies (1) when $N = 250$ Lb and (2) when $N = -250$ Lb is applied. The buckling load is $N_{cr} = 1000$ Lb. From the results, conclude whether the change due to tension or that due to compression is more significant in this case.

16-9 Assume that the beam subjected to the axial load N rotates by a small angle θ without any elastic deflection, while N remains parallel to its initial orientation. Upon expressing the displacement vector in Eq. 16.14 by L and θ, find the load vector necessary to enforce this rotation. Compare the results with what is known about a preloaded axial bar.

16-10 A weightless beam with mass M at the center is subjected to a compressive force $N_0 = 0.3(\pi^2 EI/L^2)$. Find the expression for the natural frequency.

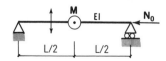

16-11 Assuming the first mode shape as

$$X = 1 - \cos\frac{2\pi x}{L}$$

find the natural frequency $\tilde{\omega}$ of a beam with both ends fixed. The compressive force N_0 has the magnitude $1.2 N_e$.

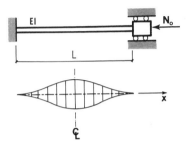

16-12 A beam is subjected to the simultaneous action of compressive force N_0 and end moments $\mathfrak{M}$. Using the apparent stiffness matrix $\tilde{\mathbf{k}}$, find the magnitude of N_0 for which the moment $\mathfrak{M}$ needed for some prescribed rotation θ becomes infinitely small.

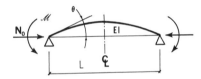

16-13 A beam fixed at the left end may have three possible conditions at the right end. Calculate the buckling forces by setting the determinant of the apparent stiffness matrix equal to zero. Discuss the accuracy of using the approximate geometric matrix given by Eq. 16.12 if the exact answers are: (a) $N_{cr} = N_e$, (b) $N_{cr} = 2.045 N_e$, and (c) $N_{cr} = 0.25 N_e$, where $N_e = \pi^2 EI / L^2$.

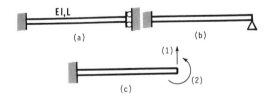

16-14 The system is the same as in Prob. 16-5 except that the supports are switched. Will the expression for the fundamental frequency be the same as before, except for the change of sign? If $\Omega^2 mL^2 = 1000$ Lb and $\pi^2 EI / L^2 = 2000$ Lb, how many times will the natural frequency increase as a result of the switching of supports?

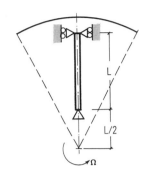

16-15 The weight of the platform (together with a portion of columns) is $W_1 = 18{,}500$ Lb and the additional load that may be placed on it is $W_2 = 8500$ Lb. The natural frequency of the lateral vibrations of the system without W_2 is $f_{n1} = 14.5$ Hz; with W_2 it is $f_{n2} = 9.8$ Hz. Determine the factor of safety against elastic buckling, relative to $W_1 + W_2$.

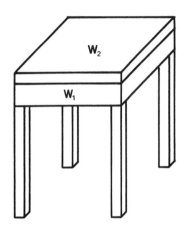

16-16 Find the expression for the displacement response of the beam-column with bending stiffness EI and distributed mass m. Use the approach outlined in Sect. 15E. Compare the result with a special case of the answer to Prob. 15-37.

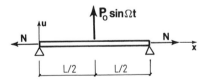

16-17 The frame in Prob. 16-7 is symmetrically loaded with $\mathfrak{M} = \mathfrak{M}_0 \sin \Omega t$ applied to each joint having angular inertia J. What is the amplitude of joint rotation if there is a constant force $P_0 = 8150$ Lb applied as in Fig. 16-7? Solve for $\mathfrak{M}_0 = 1800$ Lb-in., $\Omega = 32$ rad/s, $J = 47$ Lb-s^2-in., $EI = 1.32 \times 10^6$ Lb-in.^{2}, $L = 36$ in., and $\zeta = 0.03$ damping. Also use the frequency formula for the symmetric mode from Prob. 9-13:

$$\omega^2 = \frac{10EI}{JL}$$

16-18 A cable has only axial stiffness and when stretched with a preload N, it will resist the lateral forces owing to the presence of that preload. The natural frequencies of the cable are

$$\omega_n^2 = \left(\frac{n\pi}{L}\right)^2 \frac{N}{m} \qquad n = 1, 2, 3, \ldots$$

A flexural beam has only bending stiffness, which allows it to resist lateral loads. The natural frequencies of a simply supported beam were found in Prob. 15-5:

$$\omega_n^2 = \left(\frac{n\pi}{L}\right)^4 \frac{EI}{m} \qquad n = 1, 2, 3, \ldots$$

A beam-column may be viewed as having the combined stiffness of a cable and a flexural beam working in parallel. Taking advantage of a formula in Prob. 7-1, write the equation for the natural frequency of a simply supported beam-column.

CHAPTER

17

Rotating Machinery

A typical rotor consists of a disc (or discs) set on a shaft held in space by bearings. This system can be associated with a large number of distinct problems. If the axis of rotation and the principal axis of inertia are not aligned, the forces of *unbalance* act on the system. The flexibility of the shaft may give rise to *whirling*, in which the rotating motion of a disc about the shaft axis is superposed on the orbital motion of this axis itself. A rotor at rest and one spinning about its axis will have different natural frequencies of transverse motion. All these phenomena are analyzed in this chapter for both rigid and flexible rotors. The main effort is to reduce the effect of rotation to the action of oscillating lateral forces, so that a plane instead of a spatial problem may be analyzed. When the axis of a spinning and whirling rotor is projected on a plane fixed in space, it appears to be a beam vibrating in that plane. Under certain circumstances it is possible to determine *critical speeds* of rotors by calculating the natural frequencies of associated beams. Also, the unbalance response may be reduced to the harmonic forcing known from Chapter 10, provided some modifications are made to account for the spinning motion. The torsional vibration problems, which are dispersed in the preceding chapters, are not included here because they are not associated with the above-mentioned phenomena.

17A. Unbalance Forces in Rigid Rotors. To determine the nature of forces exerted by a rotor on the bearings that hold the shaft, it is sufficient to con-

sider a single disc and then superpose the effects from all discs.

A condition of a rotor known as *static unbalance* is illustrated in Fig. 17.1a. The axis of a disc mounted on the shaft does not coincide with the axis of rotation, but is shifted by a distance e from it. The angular velocity Ω (also called speed) induces the centrifugal force

$$P = \Omega^2 M e \qquad (17.1)$$

in which M is the mass of disc. This force is balanced by the bearing reaction R, which is resolved into R_1 and R_2 as Fig. 17.1b implies.

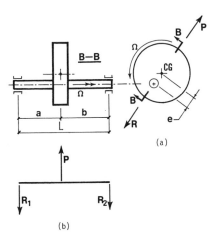

Figure 17.1 Static unbalance: (a) kinetics; (b) forces applied to shaft.

213

Figure 17.2a shows the *dynamic unbalance* in which the center of gravity is located on the axis of rotation, but the disc itself is mounted in such a way that its own axis makes a small angle ϕ with the rotation axis. By considering the relation between the external moment applied to the rotor (by the bearings) and the rate of change of the angular momentum of the rotor, we find the equilibrium of loads to be as shown in Fig. 17.2b. In that figure $\mathfrak{M}_g$ is the moment applied to the shaft by the disc. It represents the effect of disc inertia and is called a gyroscopic moment. We find that

$$R = \frac{\mathfrak{M}_g}{L}$$

where

$$\mathfrak{M}_g = -(J - J')\Omega^2 \sin\phi \cos\phi \approx -(J - J')\Omega^2\phi \quad (17.2)$$

in which J and J' are the mass moments of inertia:

$$J = \frac{1}{8}MD^2 \qquad \text{(about the disc axis)}$$

$$J' = \frac{M}{12}(l^2 + 0.75D^2)$$

(about the diametral axis through the CG)

(l is disc thickness measured along rotor axis). For a thin disc we have $J' \approx \frac{1}{2}J$ and

$$\mathfrak{M}_g = -\frac{1}{2}J\Omega^2\phi \qquad (17.3)$$

Figure 17.2b is meant to show the actual direction of $\mathfrak{M}_g$ for ϕ as marked in Fig. 17.2a, as long as J is larger than J'. The negative sign in Eqs. 17.2 and 17.3 indicates that $\mathfrak{M}_g$ acts opposite to ϕ. (The angle ϕ is usually very small.)

The gyroscopic moment is comparable to the centrifugal force appearing in the static unbalance because both are the inertia effects of rotating masses. Notice that while that force is trying to increase the magnitude of unbalance by pulling the CG away from the axis of rotation, the gyroscopic moment $\mathfrak{M}_g$ acts to reduce the dynamic unbalance measured by angle ϕ, provided J is larger than J'.

In practical applications there is always some static and dynamic unbalance at the same time. The planes in which e and ϕ are measured (Figs. 17.1 and 17.2, respectively) do not, in general, coincide, and the equations for the resultant reactions are obtained by superposition. The term "rigid" used in the title of this section means that the deformations of the rotor resulting from rotation are insignificant and need not be considered in an analysis.

The centrifugal force and the gyroscopic moment caused by the described deviations from perfect alignment are fixed in space with respect to the rotor. This means they rotate, relative to a fixed system, with the same angular velocity Ω as the rotor.

17B. Whirling of Rigid Rotor in Flexible Mounting. A rotor is said to be *whirling* when its axis moves about the initial, undeflected position. This component of motion is independent of the angular velocity Ω of the rotor about its own axis. To illustrate both motions, Fig. 17.3 shows various positions of a cross section of a whirling shaft. The speed of rotation Ω is shown by changing locations of a point C marked on the surface of shaft, while the *whirl speed* λ pertains to the travel of the axis.

The whirl is said to be *synchronous* when the whirl speed is the same as the rotational speed, $\lambda = \Omega$. In Fig. 17.3 we clearly have a nonsynchronous whirl, since $\lambda > \Omega$. The path of the shaft axis is called an *orbit*.

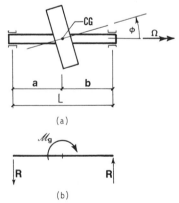

Figure 17.2 Dynamic unbalance: (a) geometry; (b) forces applied to shaft.

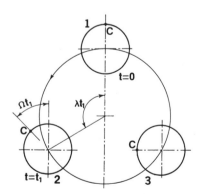

Figure 17.3 Speed of rotation Ω and whirl speed λ.

From Sect. 17.A we know that when the principal axis of the rotor is not parallel to the axis of rotation, the gyroscopic moment appears. Another reason, which concerns us here, is that the axis of rotation itself does not remain parallel to its undeflected position. With reference to Fig. 17.4, the gyroscopic moment is expressed by

$$\mathfrak{M}_g = -(J\Omega - J'\lambda)\lambda\alpha \qquad (17.4)$$

when the angle α between the deflected and undeflected axis is small. For $J\Omega > J'\lambda$ this moment acts to reduce α. The other effect of the inertia is the centrifugal force

$$P = \lambda^2 Mu \qquad (17.5)$$

due to the deflection u of the CG of the rotor. For a rotor in a synchronous whirl, we have

$$\mathfrak{M}_g = -(J - J')\lambda^2\alpha \qquad (17.6)$$

The gyroscopic moment that occurs as a result of a dynamic unbalance and was discussed in the previous section may be thought of as a special case of synchronous whirl.

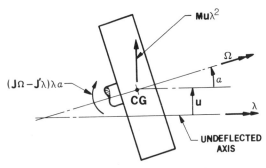

Figure 17.4 Kinetics of whirling disc.

The whirl speed λ is positive when it has the same sense as the projection of Ω on the undeflected axis. Otherwise we have *negative whirl*. Notice that when λ is negative then $\mathfrak{M}_g$ acts to increase α and the absolute value of this moment calculated from Eq. 17.4 is larger than for the same positive λ. The terms *forward precession* and *backward procession* are sometimes used in place of positive and negative whirl.

When the whirling shaft axis remains parallel to its undeflected position, we speak of a cylindrical whirl. The other distinct possibility is a conical whirl.

17C. Cylindrical Whirl of Rigid Rotor takes place in situations like that shown in Fig. 17.5. The mountings at both ends are identical and isotropic. The unbalance is located at the midspan, and the whole system is symmetric with respect to the midplane perpendicular to the rotor axis. The angular velocity Ω causes the rotor axis to oscillate about its undeflected position because of the effect of unbalance. Since the excitation is located at the midplane, there is no gyroscopic effect associated with tilting of the rotor axis. The external damping force acting on the rotor is assumed to be proportional to the velocity of translation of the rotor axis. The driving torque M_{t0} is needed to overcome the effect of damping.

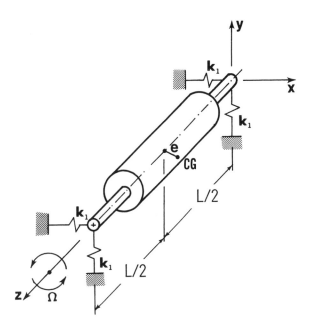

Figure 17.5 Elastically supported, rigid rotor.

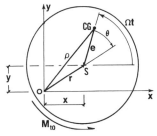

Figure 17.6 Synchronous, cylindrical whirl.

A cross-sectional view of the rotor in a steady-state motion with a constant velocity Ω is shown in Fig. 17.6. Point O is the trace of the undeflected rotor (or shaft axis) while point S marks its deflected position. The center of gravity of the rotor is marked CG. Rotation of segment e with respect to the system x–y induces the deflection of the shaft axis by r and, in effect, the whole triangle O–S–CG rotates with angular velocity Ω. Since the speed of vector r is by definition the whirl speed λ, we find this condition to be a synchronous whirl. The three vectors r, e, and ρ do not change their lengths, in time as long as Ω is constant. The constant radius of the orbit is the reason for referring to this motion as a cylindrical or circular whirl. The x- and y-coordinates of the shaft axis are given by

$$x = \mu \frac{\Omega^2 M e}{k} \cos(\Omega t - \theta) \qquad (17.7a)$$

$$y = \mu \frac{\Omega^2 M e}{k} \sin(\Omega t - \theta) \qquad (17.7b)$$

where μ and θ are defined by Eqs. 2.16 and 2.17, respectively. Constant $k = 2k_1$ designates the resultant radial stiffness of both mountings and ω the natural frequency of radial vibrations of the rotor.

By comparing Eqs. 17.7 with those in Sect. 2G we see the motion of the rotor axis as that of an elastic system subjected to harmonic forcing separately along the x- and y-axes. The amplitude of the applied load is $P_0 = \Omega^2 M e$ and it is equal to the centrifugal force that would be acting on the CG of an undeflected shaft. The damping ratio ζ has the same meaning as previously and is referred to the mass of the rotor moving with a velocity Ωr_0, where r_0 is the length of the rotating vector r.

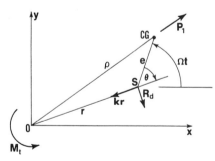

Figure 17.7 Forces acting on whirling shaft.

The forces acting on the shaft are shown in Fig. 17.7. Besides the elastic resistance of the mounting k and the centrifugal force

$$P_1 = \Omega^2 M \rho_0 \qquad (17.8)$$

there is a damping force

$$R_d = \Omega r_0 c = \Omega r_0 (2 M \omega \zeta) \qquad (17.9)$$

Length $r_0 = |r|$ is found from

$$r_0 = \mu \frac{M \Omega^2 e}{k} \qquad (17.10)$$

Once e, θ, and r_0 are known, $\rho_0 = |\rho|$ can be calculated from the triangle in Fig. 17.7. Notice that when damping is absent, we get $\theta = 0$ and vectors e, r, and ρ become aligned.

An important remark may be made with reference to the system depicted in Fig. 17.8. The mass shown there cannot rotate, but its center moves the same as point S in Fig. 17.6 owing to the presence of force P_0 rotating with velocity Ω. The action of this force replaces the combined effect of the rotation and unbalance of the original system.

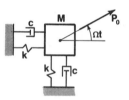

Figure 17.8 Simplified system under rotating force.

17D. Conical Whirl of Rigid Rotor occurs in its pure form when the rotor has only dynamic unbalance. The system must be symmetric with respect to its midplane, as described in the preceding section, and the principal axis must intersect the undeflected axis of rotation at the midlength (Prob. 17-18). The approach to the analysis of conical whirl is essentially the same as that for cylindrical whirl previously discussed. Both phenomena occur simultaneously in actual rotating machinery.

17E. Whirling of Disc on Flexible Shaft is analyzed as a special case of free vibrations. The discussion in Sect. 9A shows that the matrix expression

$$\mathbf{k\vec{u} = \vec{P}} \qquad (17.11)$$

may be treated as an equation of motion of an undamped system, provided $\vec{P}$ represents the inertia forces induced by that motion.

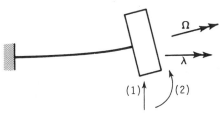

Figure 17.9 Whirling disc on flexible shaft.

Consider a single disc mounted on a massless shaft in Fig. 17.9. The speed of rotation is Ω and the whirl speed is λ, both measured with respect to a fixed coodinate system. The inertia force components consist of the centrifugal force and the gyroscopic moment, both shown in Fig. 17.4. Noting the signs of deflections in Fig. 17.9, the inertia forces may be written in the matrix form as

$$\begin{Bmatrix} P_1 \\ P_2 \end{Bmatrix} = \begin{bmatrix} M\lambda^2 & 0 \\ 0 & -(J\Omega - J'\lambda)\lambda \end{bmatrix} \begin{Bmatrix} u_1 \\ u_2 \end{Bmatrix} \quad (17.12)$$

Equation 17.11 becomes now

$$\begin{bmatrix} k_{11} & k_{12} \\ k_{21} & k_{22} \end{bmatrix} \begin{Bmatrix} u_1 \\ u_2 \end{Bmatrix} = \begin{bmatrix} M\lambda^2 & 0 \\ 0 & -(J\Omega - J'\lambda)\lambda \end{bmatrix} \begin{Bmatrix} u_1 \\ u_2 \end{Bmatrix}$$

or, using the speed ratio $h = \Omega/\lambda$,

$$\begin{bmatrix} (k_{11} - M\lambda^2) & k_{12} \\ k_{21} & k_{22} - (J' - Jh)\lambda^2 \end{bmatrix} \begin{Bmatrix} u_1 \\ u_2 \end{Bmatrix} = \begin{Bmatrix} 0 \\ 0 \end{Bmatrix}$$

$$(17.13)$$

The translation u_1 and the rotation u_2 are both measured in the plane, which revolves with the shaft at the angular velocity λ. Upon solving this system of equations, we obtain two *critical speeds* λ_1 and λ_2, and two associated mode shapes. This is done for a fixed value of the shaft speed Ω. Changing Ω will also change the corresponding critical speeds.

When we set $\Omega = 0$ and simultaneously replace the symbol λ by ω, Eqs. 17.13 become the equations of the lateral vibrations of a body with mass M and moment of inertia J', the latter about the axis normal to the plane of the paper. (The reader can refer to Prob. 9-4.) This observation opens the possibility of looking at Eqs. 17.13 in a different way. Instead of thinking of the disc as whirling in space about the undeflected axis with velocity λ, we can treat it as vibrating in plane with the circular frequency λ. (This is equivalent to saying that we choose to look at the

projected view of the disc.) Our equations represent the in-plane vibrations now, and the only effect of the rotation Ω is the presence of an additional moment showing up as an extra term

$$J\Omega\lambda = Jh\lambda^2$$

in the lower diagonal entry. Problem 17-5 illustrates one possible way of handling this term.

17F. Critical Speed of a rotor is the angular velocity at which the amplitude of whirl reaches its maximum. In an idealized case when no damping is present, this amplitude would become infinitely large, according to linear theory. When the principal rotor axis is kept parallel to its undeflected position, the gyroscopic effect disappears and the critical speed has the same value as the natural frequency of the transverse vibrations.

In Sects. 17B, C, and D we discussed whirling of a rigid rotor in flexible bearing. The critical speeds obtained from such a model are called the *rigid-body critical speeds*. This term is also applicable to a shaft, not necessarily rigid itself, whose mounting is flexible enough to allow this type of whirling. When bending of the rotor axis appears to be most significant component of motion, we speak of *bending critical speeds*.

When only shaft twisting is involved, the *torsional critical speeds* may be an important property of a system. Since no whirling takes place during twisting motion, finding torsional critical speeds does not differ from calculating natural frequencies of a torsional system and therefore nothing new is involved here.

Equations for finding critical speeds of an elastic shaft with a single rotor were given in Sect. 17E. They can easily be generalized for a shaft with an arbitrary number of discs (see Prob. 17-8). Free vibration methods may be employed in the calculation if some physical adjustments in a system are made to account for the gyroscopic effect.

17G. Unbalance Response. Consider again a rigid rotor with unbalance e as shown in Fig. 17-5. If we want to calculate the displacement of rotor axis projected on, say, the fixed yz-plane, we apply a force

$$P_y = \Omega^2 Me \sin\Omega t$$

to our system, provided Ωt is measured from the xz-plane. We can also say that we apply a projection of the centrifugal force $P_0 = \Omega^2 Me$ on the yz-plane. The same approach will give correct results when we have a rotor in the form of a disc on an elastic shaft. For any rotor that can be treated as an SDOF system in a particular reference plane, we can write

$$M\ddot{u} + ku = \Omega^2 M e \sin \Omega t \qquad (17.14)$$

when there is no damping involved and when angle Ωt is measured as in the example above.

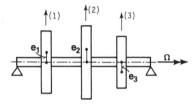

Figure 17.10   Static unbalances in three-disc rotor.

If there are several discs on a shaft, each of them is a subject to a similar forcing function. In the system shown in Fig. 17.10 all unbalance eccentricities are assumed to lie in the same plane passing through the undeflected shaft axis. The displacements of disc centers in the plane of the paper are found from the general equation of forced, undamped vibrations (10.1),

$$\mathbf{M\ddot{u}} + \mathbf{k\vec{u}} = \begin{Bmatrix} \Omega^2 M_1 e_1 \\ \Omega^2 M_2 e_2 \\ -\Omega^2 M_3 e_3 \end{Bmatrix} \sin \Omega t \qquad (17.15)$$

Following Sect. 10F we have

$$\vec{u} = \vec{A} \sin \Omega t$$

where the amplitudes A_r are found from Eq. 10.21,

$$\begin{bmatrix} \left(k_{11} - M_1\Omega^2\right) & k_{12} & k_{13} \\ k_{12} & \left(k_{22} - M_2\Omega^2\right) & k_{23} \\ k_{13} & k_{23} & \left(k_{33} - M_3\Omega^2\right) \end{bmatrix}$$

$$\times \begin{Bmatrix} A_1 \\ A_2 \\ A_3 \end{Bmatrix} = \begin{Bmatrix} \Omega^2 M_1 e_1 \\ \Omega^2 M_2 e_2 \\ -\Omega^2 M_3 e_3 \end{Bmatrix} \qquad (17.16)$$

To take account of rotatory inertia of the discs, one would have to add three angular coordinates in the plane of the paper, thus doubling the number of

equations. If dynamic unbalances (as defined in Sect. 17A) are present, the forcing terms will appear in these added equations.

The other missing item in Eqs. 17.16 is the gyroscopic effect. This may be included by adding an appropriate term to every equation associated with an angular coordinate. (Refer to Sect. 17E.) It is worthwhile to keep in mind that in positive whirl the effects of rotatory inertia and gyroscopic moment oppose each other, thereby reducing somewhat the cumulative error if both effects are ignored.

The reader should be aware that describing the displacement response of a rotor having several discs with the aid of only a single projection plane has limited use. When the unbalances are not located in the same axial plane, when there is damping in the system, and when the rotor mounting is not radially isotropic, the second reference plane may be necessary, which doubles the number of equations involved.

SOLVED PROBLEMS

17.1   A thin disc weighing $W = 300$ Lb is set on the shaft with eccentricity $e = 0.05$ in. The plane of the disc is tilted by $0.12°$ with respect to the axis of rotation. The angle between the axial planes of both unbalances is $45°$. Determine the forces in the x- and y-directions exerted by the rotor on the bearings when the speed is $n = 3000$ rpm; $r = 15$ in. and $L = 25$ in.

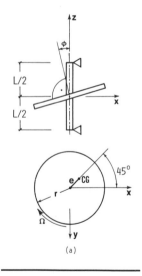

The angular velocity is

$$\Omega = \frac{\pi n}{30} = 100\pi = 314.16 \text{ rad/s}$$

From Eq. 17.1,

$$P = \Omega^2 e \frac{W}{g} = 314.16^2 \times 0.05 \times \frac{300}{386} = 3835.4 \text{ Lb}$$

which gives

$$P_1 = \frac{P}{2} = 1917.7 \text{ Lb}$$

per bearing. The polar moment of inertia of the disc

$$J = \frac{1}{2} \frac{W}{g} r^2 = \frac{1}{2} \times \frac{300}{386} \times 15^2 = 87.44 \text{ Lb-s}^2\text{-in.}$$

The gyroscopic moment, according to Eq. 17.3,

$$\mathfrak{M} = -\frac{1}{2} J \Omega^2 \phi = -\frac{1}{2} \times 87.44 \times 314.16^2 \left(0.12 \times \frac{\pi}{180} \right)$$

$$= -9037.3 \text{ Lb-in.}$$

The negative sign indicates that $\mathfrak{M}$ is directed opposite to ϕ. Each bearing is subjected to

$$\pm P_2 = \frac{\mathfrak{M}}{L} = \frac{9037.3}{25} = 361.5 \text{ Lb.}$$

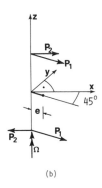

(b)

Figure 17-1*b* shows how the forces are applied to the bearings when the shaft is in the same position as in (*a*). It is seen that the resultant applied to the top bearing is larger than at the bottom. At this position the components are

$$P'_x = P_2 + P_1 \cos 45° = 1718 \text{ Lb}$$

$$P'_y = -P_1 \cos 45° = -1356 \text{ Lb}$$

These components are rotating with angular velocity Ω, and a stationary observer will measure the following:

$$P_x = P'_x \cos \Omega t - P'_y \sin \Omega t$$

$$P_y = P'_x \sin \Omega t + P'_y \cos \Omega t$$

17-2 An unbalance of a rigid rotor is represented by weights $W_a = 0.10$ Lb and $W_b = 0.06$ Lb placed on the outside surface with radius r, as in (*a*). The correction weights are to be placed in planes I and II,

on the radius equal to $0.75r$. Calculate the magnitude and the angular location of those weights.

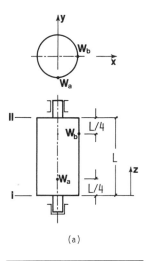

(a)

The centrifugal forces applied by the unbalance weights are:

$$P_a = \Omega^2 \frac{W_a}{g} r \qquad \text{and} \qquad P_b = \Omega^2 \frac{W_b}{g} r$$

The necessary balancing forces, from (*b*), are:

$$R_{x1} = 0.25 P_b \qquad R_{y1} = 0.75 P_a$$

$$R_{x2} = 0.75 P_b \qquad R_{y2} = 0.25 P_a$$

The resultants of the required balancing forces in each of the correction planes; from (*c*), are:

$$R_1 = \left(R_{x1}^2 + R_{y1}^2 \right)^{1/2} = \frac{\Omega^2 r}{g} \left[(0.25 \times 0.06)^2 + (0.75 \times 0.10)^2 \right]^{1/2}$$

$$= 0.07649 \frac{\Omega^2 r}{g}$$

$$R_2 = 0.051478 \frac{\Omega^2 r}{g}$$

A corrective weight W applies a force

$$R = 0.75 \Omega^2 \frac{W}{g} r$$

when placed at $0.75r$. This means

$$0.07649 = 0.75 W_1 \qquad \therefore W_1 = 0.102 \text{ Lb}$$

$$0.051478 = 0.75 W_2 \qquad \therefore W_2 = 0.06864 \text{ Lb}$$

For a system of coordinates tied with the rotor, the weights have to be placed at the following angles, measured as shown in (*c*):

$$\alpha_1 = \arctan \left(\frac{0.75 \times 0.10}{0.25 \times 0.06} \right) = 78.69°$$

$$\alpha_2 = \arctan \left(\frac{0.25 \times 0.10}{0.75 \times 0.06} \right) = 29.05°$$

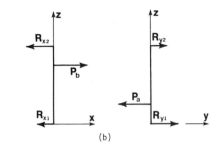

(b)

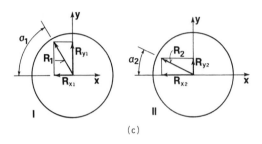

(c)

17-3 The figure shows two identical centrifugal vibrators attached to the base of a structure in the xy-plane. Each consists of two rotating masses, each mass exerting the force $\Omega^2 Me$ on the vibrator. The angular speed and eccentricity is the same for each mass. By selecting the sense of angular speed (clockwise or counterclockwise) and the phase angle we get different resultants acting on the structure. Write the expressions for angles α_1, α_2, β_1, and β_2 as functions of time so that the base of the structure is subjected to (1) a single pulsating force along the z-direction, and (2) a pulsating moment about the x-axis.

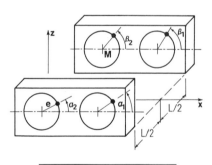

There are four rotating masses, all applying centrifugal force of the same magnitude. Imagine that the shafts of the front vibrator are turning opposite each other,

$$\alpha_1 = \Omega t \quad \text{and} \quad \alpha_2 = \pi - \Omega t$$

then the resultant for this machine is a vertical pulsating force. If we set also

$$\beta_1 = \Omega t \quad \text{and} \quad \beta_2 = \pi - \Omega t$$

we have the first answer. To obtain a moment about the x-axis, the

resultants for the front and the rear vibrators must act opposite each other. Leaving α_1 and α_2 unchanged, we put

$$\beta_1 = -\Omega t \quad \text{and} \quad \beta_2 = \Omega t - \pi$$

17-4 A rigid rotor (Fig. 17.5) is supported by an isotropic bearing at each end. The bearings have the same radial stiffness, $k_1 = 15,000$ Lb/in. each. The unbalance $e = 0.002$ in. is at the midlength of the rotor, which weighs $W = 50$ Lb. The damping ratio is $\zeta = 0.12$. Find:

1. The critical speed of cylindrical whirl.
2. The radius r_0 of the shaft orbit.
3. The radius ρ_0 of the CG.
4. The phase lag angle θ.

The amplitudes are to be calculated for $n = 6000$ rpm.

The effective stiffness is $k = 2k_1 = 30,000$ Lb/in. The critical speed $\omega = (kg/W)^{1/2} = 481.25$ rad/s. (This is the same as the critical whirl speed λ, since we are dealing with the synchronous whirl.) Angular velocity $\Omega = \pi n/30 = 628.32$ rad/s. The magnification factor μ and the angle θ are found from Eqs. 2.16 and 2.17:

$$\frac{1}{\mu} = \left[\left(1 - \frac{628.32^2}{481.25^2} \right)^2 + \left(2 \times 0.12 \frac{628.32}{481.25} \right)^2 \right]^{1/2} \quad \therefore \mu = 1.2968$$

$$\tan \theta = \frac{2 \times 0.12 \times (628.32/481.25)}{1 - (628.32/481.25)^2} = -0.4447 \quad \therefore \theta = 156.02°$$

since we know that for $\Omega > \omega$, θ is between 90° and 180°. Equation 17.10 gives us the whirl radius

$$r_0 = 1.2968 \left(\frac{50}{386} \times 628.32^2 \times \frac{0.002}{30,000} \right) = 0.004421 \text{ in.}$$

From Fig. 17.6 we can find the radius of the CG orbit ρ_0:

$$\rho_0^2 = r_0^2 + e^2 + 2r_0 e \cos \theta$$

$$= 0.004421^2 + 0.002^2 + 2 \times 0.004421 \times 0.002 \cos 156.02°$$

$$= 0.002718 \text{ in.}$$

17-5 Equations 17.13 describe whirling motion of a rotating shaft with a single disc. As mentioned in Sect. 17E, we can also think of those equations as representing the in-plane vibrations provided an adjustment is made for the shaft speed Ω. How should the physical properties of the model be altered to obtain such an adjustment?

The only difference in the description of whirling and in-plane motion is in the a_{22} entry of the coefficient matrix. Let us assume first that the whirl is positive, which means that λ is directed the same as the projection of Ω on the undeflected axis. The diagonal term in question can then be written as

$$(k_{22}+J\Omega\lambda)-J'\lambda^2=\bar{k}_{22}-J'\lambda^2$$

This means that the effect of Ω is to increase the angular stiffness k_{22} to $\bar{k}_{22}$, the increase being $J\Omega\lambda$. This difference may be visualized as an external angular spring restraining the disc. The vibration analysis of the system in (a) will give correct values of critical speeds of the original system.

If λ is a negative number (which indicates negative whirl, as previously defined) we can write the entry as

$$k_{22}-(J'-Jh)\lambda^2=k_{22}-\tilde{J}\lambda^2$$

This time the effect of Ω is to increase the angular inertia from J' to $\tilde{J}$, the increase being $Jh=J\Omega/\lambda$. To account for this change, we attach an additional angular inertia to the original system (see (b)).

To obtain a relationship between Ω and λ in positive whirl we assume the value of the product $\Omega\lambda$, attach a fictitious spring to the system, and obtain λ_1 and λ_2 as a result of vibration analysis. The divisions

$$\frac{\Omega\lambda}{\lambda_1}=\Omega_1 \qquad \text{and} \qquad \frac{\Omega\lambda}{\lambda_2}=\Omega_2$$

in which the numerators are the assumed values, give us the associated shaft speeds. Repeating the procedure, we get enough points to plot λ_1 and λ_2 as functions of Ω.

In negative whirl, it is simpler to assume the value of ratio $|\Omega/\lambda|$. The in-plane vibration analysis again gives us the natural frequencies and their associated shaft speeds.

It should be obvious to the reader that using the in-plane vibration analysis to obtain critical speeds is of little value for a shaft with a single disc, since we can handle Eq. 17.13 directly, without interpreting its physical aspects. It is only for larger systems that using the established, general methods to determine critical speeds may be beneficial.

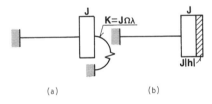

(a)　　　　(b)

17-6 When the axis of a whirling disc is projected on a selected, fixed reference plane passing through the undeflected axis, it appears that the shaft is undergoing flexural vibrations in that plane. By investigating the deformation of the shaft material, determine the difference between whirling and simple flexing of the shaft from that point of view. Consider both positive and negative whirl.

In (a) we have positive whirl with the shaft speed Ω equal to the whirl speed λ. In (b) there is $\lambda=-\Omega$, and in (c) and (d) the flexural vibrations are shown. Point C of the cross section is selected and the sign of bending stress is marked at several positions of the shaft on its orbit. In (a) point C is in tension at all times, while in (b) it changes from tension to compression and back to tension with two such cycles per revolution. In simple flexure (c) and (d) the stress at point C alternates between tension and compression.

As described in Sect. 17E, the whirling motion with the whirl speed λ may, under certain circumstances, be treated as flexural vibrations with frequency λ. This similarity, however, does not extend to the stressing of shaft material.

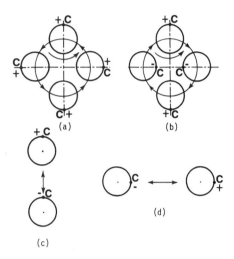

(a)　　　　(b)

(c)　　　　(d)

17-7 In Prob. 9-6 the analysis of bending vibrations of a rotor was performed. Taking gyroscopic effect into account, find the first critical whirl speed for

$$h=\frac{\Omega_1}{\lambda_1}=1 \qquad \text{and} \qquad h=-1$$

Calculate the desired speeds by finding the natural frequencies of a modified system. The polar mass moment of inertia of the disc is $J=9.72$ Lb-s^2-in. The first frequency of bending vibrations is 952.01 rad/s.

To obtain the whirl frequency equation from the bending frequency expression, the adjustments to the latter will be made in accordance with Sect. 17E.

$h=1$, *Positive Whirl.* The angular stiffness is increased by $J\Omega\lambda=J\lambda^2$. Since λ is still unknown, assume that it will exceed the natural frequency, as calculated in the reference problem, by 10%. Put $\Omega\lambda=(1.1\times952.01)^2=1.097\times10^6$.
The angular stiffness term becomes

$$156.67\times10^6+9.72\times1.097\times10^6=167.33\times10^6$$

and the new frequency equation is

$$\begin{vmatrix} (1.2715\times10^6-0.2627\lambda^2) & (-12.544\times10^6) \\ (-12.544\times10^6) & (167.33\times10^6-4.86\lambda^2) \end{vmatrix}=0$$

$$\therefore \lambda_1=1066.8 \text{ rad/s} \qquad \therefore \lambda_2=6175.2 \text{ rad/s}$$

The shaft speed corresponding to λ_1 is

$$\Omega_1 = \frac{1.097 \times 10^6}{1066.8} = 1028.3 \text{ rad/s}$$

This means the true value of h is

$$h = \frac{1028.3}{1066.8} = 0.9639$$

Treating λ as a linear function of h (based on points $h=0$ and $h=0.9639$), we get for $h=1.0$:

$$\lambda_1 = \Omega_1 = 1071.1 \text{ rad/s}$$

$h=-1$, *Negative Whirl*. Use the modified mass moment of inertia:

$$J' + J|h| = 4.86 + 9.72 = 14.58 \text{ Lb-s}^2\text{-in.}$$

The frequency equation is now

$$\begin{vmatrix} (1.2715 \times 10^6 - 0.2627\lambda^2) & (-12.544 \times 10^6) \\ (-12.544 \times 10^6) & (156.67 \times 10^6 - 14.58\lambda^2) \end{vmatrix} = 0$$

$$\therefore -\lambda_1 = 857.83 \text{ rad/s} \qquad -\lambda_2 = 3853.5 \text{ rad/s}$$

(We obtain λ^2 as a solution and choose λ to be negative according to our convention.) The shaft speed at the first frequency is:

$$\Omega_1 = -\lambda_1 = 857.83 \text{ rad/s}$$

17-8 Assuming the matrices of mass and stiffness of the rotor in the figure to be known, construct the equations of motion from which the critical speeds can be found. Describe how to generalize the equations for a system with n discs.

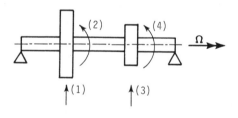

The expressions for inertia forces are similar for each disc; we can therefore use Eqs. 17.12 twice, with a suitable change of notation:

$$\begin{Bmatrix} P_1 \\ P_2 \\ \hline P_3 \\ P_4 \end{Bmatrix}$$

$$= \begin{bmatrix} M_1\lambda^2 & 0 & & \\ 0 & -(J_2\Omega - J_2'\lambda)\lambda & & \\ \hline & & M_3\lambda^2 & 0 \\ & & 0 & -(J_4\Omega - J_4'\lambda)\lambda \end{bmatrix} \begin{Bmatrix} u_1 \\ u_2 \\ \hline u_3 \\ u_4 \end{Bmatrix}$$

Inserting this into Eq. 17.11 and using speed ratio $h = \Omega/\lambda$, we obtain

$$\begin{bmatrix} (k_{11} - M_1\lambda^2) & k_{12} \\ k_{12} & [k_{22} - (J_2' - J_2 h)\lambda^2] \\ k_{13} & k_{23} \\ k_{14} & k_{24} \end{bmatrix}$$

$$\begin{bmatrix} k_{13} & k_{14} \\ k_{23} & k_{24} \\ (k_{33} - M_3\lambda^2) & k_{34} \\ k_{34} & [k_{44} - (J_4' - J_4 h)\lambda^2] \end{bmatrix} \begin{Bmatrix} u_1 \\ u_2 \\ u_3 \\ u_4 \end{Bmatrix} = \begin{Bmatrix} 0 \\ 0 \\ 0 \\ 0 \end{Bmatrix}$$

Equating the determinant of the coefficient matrix to zero will yield the whirl speeds λ. They will have different values for each assumed shaft speed Ω. To modify the system so that the whirl speeds can be found from vibration analysis, we must modify the disc properties as in Prob. 17-5.

A system with n discs will be described by $2n$ equations of motion. According to our convention, the odd-numbered equations will pertain to the translational displacement components. It is quite easy to see from our results how the system of equations will change when the successive discs are added.

17-9 Problem 9-6 analyzed the bending vibrations of a rotor. Using the same numerical data, find the displacement response to the unbalance $e = 0.005$ in. at $\Omega = 3000$ rad/s. The polar mass moment of inertia of the disc is $J = 9.72$ Lb-s^2-in. Include the effect of rotatory inertia and gyroscopic stiffening for the case of synchronous whirl. Discuss how to generalize the approach when several discs are present.

Our starting point consists of the equations of motion (17.13), where we put $\lambda = \Omega$ (synchronous whirl) and introduce the translational forcing term $\Omega^2 Me$, from Sect. 17G:

$$\begin{bmatrix} (k_{11} - M\Omega^2) & k_{12} \\ k_{12} & [k_{22} - (J' - J)\Omega^2] \end{bmatrix} \begin{Bmatrix} A_1 \\ A_2 \end{Bmatrix} = \begin{Bmatrix} \Omega^2 Me \\ 0 \end{Bmatrix}$$

where A_1 and A_2 are the unknown amplitudes of translation and rotation, respectively. The frequency equation in the reference problem may be used to write the coefficient matrix in this expression after an adjustment for the presence of gyroscopic term. We replace J' by

$$J' - J = 4.86 - 9.72 = -4.86$$

And the diagonal coefficients become

$$1.2715 \times 10^6 - 0.2627\Omega^2 = -1.0928 \times 10^6$$

$$156.67 \times 10^6 + 4.86\Omega^2 = 200.41 \times 10^6$$

For $\Omega^2 Me = 3000^2 \times 0.2627 \times 0.005 = 11,822$ our equations become

$$\begin{bmatrix} -1.0928 \times 10^6 & -12.544 \times 10^6 \\ -12.544 \times 10^6 & 200.41 \times 10^6 \end{bmatrix} \begin{Bmatrix} A_1 \\ A_2 \end{Bmatrix} = \begin{Bmatrix} 11,822 \\ 0 \end{Bmatrix}$$

$$\therefore A_1 = -6.295 \times 10^{-3} \text{ in.} \qquad A_2 = -0.394 \times 10^{-3} \text{ rad}$$

To generalize our approach we may refer to Prob. 17-8, where the system of equations for a shaft with two discs was developed. If the centrifugal forces $\Omega^2 M_r e_r$ are placed at each rth translational (odd) equation, we can then solve for the unknown deflection amplitude. If all eccentricities have the same sense, all centrifugal forces have positive sign.

EXERCISES

17-10 The rotor consists of a disc of length l, diameter D, and a massless shaft. Due to a dynamic unbalance of magnitude ϕ, there is a gyroscopic moment $\mathfrak{M}_g$ applied by the disc to the shaft. Discuss how this moment changes as a function of l/D.

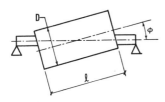

17-11 In Fig. 17.2 the dynamic unbalance of a disc with mass M and moments of inertia J and J' is shown as an angle of misalignment ϕ. An alternative way of depicting the same effect is by means of two eccentric masses, as shown. Calculate e and c as functions of J, J', and ϕ so that the gyroscopic moment is the same in both cases. Give an equation for a general case as well as for a thin disc.

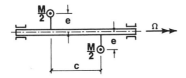

17-12 The rotor is the same as in Figs. 17.5 and 17.6 except that each spring constant is $k/2$ instead of k_1. The system of springs will thus represent a flexible, isotropic mounting with an effective stiffness k in either direction. Show that in a steady state of motion, $\Omega = $ constant, when the shaft-driving torque and the damping force are absent, the segments OS and e are aligned as in either (a) or (b). Develop an equation for the distance beween the CG and the undeflected axis O.

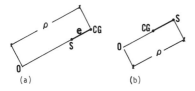

(a)   (b)

17-13 The system is the same as in Figs. 17.5 and 17.6. The rotating velocity Ω about the shaft axis S is constant. The damping force R_d is proportional to the velocity of point S and directed oppositely to it. The proportionality constant is c. The driving torque M_{t0} maintains the constant speed Ω. Calculate: (1) components x and y of the shaft axis and (2) whirl velocity λ. *Hint.* Use complex variable $r = x + iy$ to combine two equations of motion into one.

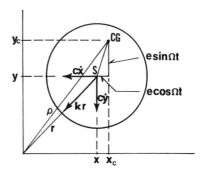

17-14 For a rotor described in Prob. 17-4, find the whirl radius r_0 and the CG radius ρ_0 at the critical speed of the cylindrical whirl.

17-15 Consider again the rotor in Prob. 17-12, which is in the synchronous whirl of velocity Ω. If we are interested in the projection of radial displacement r on the x-axis, it is more convenient to use a translational scheme as shown than the original whirling system. What should be the force $P(t)$, shown above, as a function of the parameters of the original system, to give us the correct projected displacement x?

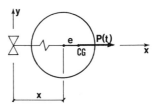

17-16 What is the resultant force transmitted to the bearings of a damped elastic rotor whirling with angular velocity Ω? The following data are given: rotor mass M, unbalance e, shaft stiffness k, and external damping ratio ζ.

17-17 A disc of mass M and unbalance e is attached to a shaft of stiffness k. The bow of the shaft at rest has the magnitude δ. Assuming that the maximum bow and the unbalance are collinear in the axial view, find the radius of the shaft center S when the system is rotating with the velocity Ω. Assume $e > \delta$.

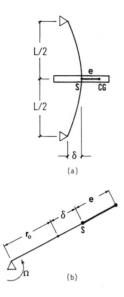

(a)

(b)

17-18 Consider the rigid, elastically supported rotor with the dynamic unbalance ϕ shown in (a). The two possibilities of conical whirl are shown in (b) and (c). Assuming a *long rotor*, for which $J' > J$ and treating J, J', k, L, and ϕ as known, develop the expression for the angular deflection α as a function of shaft speed Ω.

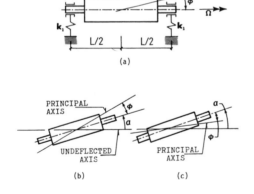

(a)

PRINCIPAL
AXIS

UNDEFLECTED
AXIS

(b)

PRINCIPAL
AXIS

(c)

17-19 In a short rotor the angle of dynamic unbalance is denoted by ϕ and the angle of elastic deflection by α. Find the relationship between α and the shaft speed Ω. At what Ω_1, in terms of system properties, will the angle between the deflected principal axis and the undeflected axis of rotation be equal to $\phi/2$?

17-20 The rotor is the same as one in Fig. 17.5, except that each vertical spring constant is k_y while each horizontal spring constant is k_x. The speed of rotation Ω is constant, and there is no damping. Determine the coordinates x and y of the deflected axis (point S in Fig. 17.6) as functions of time.

17-21 A disc of mass M and unbalance e is attached to a shaft of stiffness k. The ends of shaft are placed in nonisotropic bearings, each characterized by spring constants k_1 and k_2. Determine the coordinates of shaft center S during rotation with angular velocity Ω. The shaft and the bearings are to be considered massless.

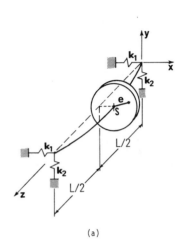

(a)

17-22 The first frequency of lateral vibration of a shaft in Prob. 11-22 was found using Dunkerley's method. Calculate the synchronous whirl speed ($h = \Omega/\lambda = 1$) using the same approach and simulating the gyroscopic effect by attaching an angular spring to each disc. The polar moments of inertia are $J_1 = 37$ Lb-s^2-in. and $J_2 = 164$ Lb-s^2-in.

17-23 The rotor discs have unbalances e_1 and e_2, respectively, both located in the same axial plane. Calculate the amplitudes of deflections u_1 and u_2 for $\Omega = 500$ rad/s. Also, masses, $M_1 = 0.8$ Lb-s^2/in., $M_2 = 2.0$ Lb-s^2/in., unbalances, $e_1 = 0.0020$ in., $e_2 = 0.0025$ in., natural frequencies, $\omega_1 = 611.3$ rad/s, $\omega_2 = 2673.9$ rad/s, and stiffness matrix (refer to Prob. 9-21):

$$\mathbf{k} = \begin{bmatrix} 3.8668 & -4.0656 \\ -4.0656 & 5.3802 \end{bmatrix} 10^6$$

Disregard the rotatory inertia and the gyroscopic effect.

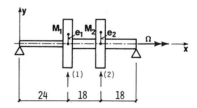

17-24 What are the magnitudes of the rotating forces of unbalance applied to the bearings in Prob. 17-23 when $\Omega = 500$ rad/s? *Hint.* Below the first natural frequency, the elastic deflections add with eccentricities.

17-25 A two-disc rotor is the same as in Prob. 17-23 except that no unbalance of the second disc exists, $e_2 = 0$. The rotor speed is gradually changed from $\Omega = 500$ to $\Omega = 3000$ rad/s. At what speed, if any, does the amplitude of lateral vibration become zero?

Random Vibration

When a forcing function applied to a structure has an irregular, nonperiodic shape, we speak of *random forcing*. This type of excitation, as well as the response that it induces, are described indirectly, by statistical means. Forcing is understood to be a continuous function of the excitation frequency (which, from a practical viewpoint means that a great number of excitation frequencies are involved) and is prescribed by a *power spectral density* (PSD). The most often encountered types of random loading are base motion and acoustic pressure. Although the main topic of this chapter is the SDOF system response, some methods of attacking MDOF systems are also presented. The subject of fatigue due to random vibration is also discussed.

18A. Deterministic Versus Probabilistic Approach to Vibrations. When one describes a forcing function acting on a system by making the following statement:

The applied force, varying between a minimum of 10 Lb and a maximum of 14 Lb, is a sine function of time.

He is speaking about a *deterministic* force, that is a force that is a prescribed function of time. On the other hand, a *probabilistic* force of similar order of magnitude may be described by saying

The applied force oscillates in an irregular fashion and

there is a 95% probability that the magnitude of the force is between 10 and 14 Lb.

Even if the second statement is made more informative by supplying some additional data, no attempt will be made to define the magnitude of forcing versus time. Using a probabilistic statement implies that the forcing is *random* and that statistical quantities will be used to indirectly characterize it. This is essentially different from the deterministic approach in which we think of forcing as having some particular value at a given point in time.

When the properties of a system are known, we can calculate its response to any given excitation. This means that a deterministic forcing is always associated with a deterministic response. Conversely, a random forcing implies random response.

18B. Some Global Properties of Random Functions. Figure 18.1 shows a randomly varying function of time, $u(t)$. The values of $u(t)$ were recorded during a

Figure 18.1 Random, time-dependent function u.

time interval t_m. Although no mathematical expression is available for the function itself, there are some important properties pertaining to the entire plot. The *mean value* of $u(t)$ is

$$\overline{u(t)} \equiv \bar{u} = \frac{1}{t_m} \int_0^{t_m} u(t)\, dt \qquad (18.1)$$

Figure 18.2 shows the same function squared. The *mean-square* of $u(t)$ is defined as

$$\overline{u^2(t)} \equiv \overline{u^2} = \frac{1}{t_m} \int_0^{t_m} u^2(t)\, dt \qquad (18.2)$$

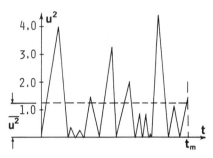

Figure 18.2 Random function squared, u^2.

The measure of how widely the function $u - \bar{u}$ oscillates is given by its *variance*,

$$\sigma^2 = \frac{1}{t_m} \int_0^{t_m} (u - \bar{u})^2\, dt \qquad (18.3)$$

When the expression behind the integral sign is expanded, we find that

$$\sigma^2 = \overline{u^2} - (\bar{u})^2 \qquad (18.4)$$

which means that the variance is equal to the mean-square, minus the square of the mean. Quite often the mean value is zero and then the variance and the mean-square are identical. The *root mean square* of $u(t)$ is defined as

$$\text{RMS}(u) = \left(\overline{u^2} \right)^{1/2} \qquad (18.5)$$

The *standard deviation* of $u(t)$ is the square root of the variance,

$$\sigma(u) = \sqrt{\sigma^2} \qquad (18.6)$$

When the mean value is zero, RMS coincides with σ.

If a random function plot is a record of a certain physical activity, we must make certain that the ob-

servation lasts long enough to make our derived global properties (e.g., the mean-square value) valid for the entire process to be described. It is also obvious that the character of the physical process may not undergo significant changes with the passage of time, so that those global properties do not change. The class of random functions satisfying these requirements are called *stationary* and *ergodic*, and our interest is limited to such functions.

18C. Distribution of Probability. Figure 18.3 shows a fragment of a record of a random function $u(t)$. If we wish to determine the probability of u being in the range of values (u_1, u_2), we may draw the lines corresponding to those levels and measure the corresponding time increments $(\Delta t)_i$. The ratio

$$P(u_1, u_2) = \frac{\Delta t_1 + \Delta t_2 + \cdots + \Delta t_n}{t_m}$$

calculated for the entire record length t_m will be the desired probability of u being between u_1 and u_2 at some selected instant. A slightly different question, involving the probability that $|u|$ will be larger than some given value u_3 may be answered by measuring the time increments along the lines at u_3 and at $-u_3$. The resulting probability is designated by $P(|u| > u_3)$.

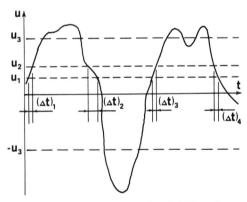

Figure 18.3 Determination of probability values.

Suppose now that our random function $u(t)$ is prescribed by its global properties, namely, the mean value $\bar{u}$ and the standard deviation σ. To be able to make statements regarding the magnitude of $u(t)$, we must know something about the *probability density function* $\mathfrak{p}(u)$ which is involved. One such function is shown in Fig. 18.4.

Having $\mathfrak{p}(u)$ prescribed, we define the probability P of u being in the range (u_1, u_2) at any selected

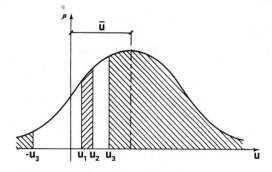

Figure 18.4 Gaussian probability density function.

point in time by means of the following integral

$$P(u_1, u_2) = \int_{u_1}^{u_2} \mathfrak{p}(u)\, du \qquad (18.7)$$

The probability that $|u| > u_3$ (which means that $u > u_3$ or $u < -u_3$) is similarly defined and shown as two rather than one shaded area in Fig. 18.4.

As u tends to infinity in either direction, $\mathfrak{p}$ asymptotically diminishes to zero. The area under the entire $\mathfrak{p}$–u curve is equal to 1.

The most commonly used probability function is the *normal distribution*, also referred to as the *Gaussian distribution* and expressed by

$$\mathfrak{p}(u) = \frac{1}{\sqrt{2\pi}} \frac{1}{\sigma} \exp\left[\frac{-(u - \bar{u})^2}{2\sigma^2} \right] \qquad (18.8)$$

The shape of this curve is the same as that shown in Fig. 18.4. It is easy to notice that the plot is symmetric about the mean value $\bar{u}$. The integral appearing in Eq. 18.7 is tabulated in many sources; see, for example, Ref. 9. Equation 18.8 tells us how probable it is that the u will be so-and-so big at some arbitrary instant, which is the same as saying that Eq. 18.8 allows us to predict an *instantaneous magnitude* on a statistical basis.

18D. Discrete Versus Continuous Spectrum of Forcing. Consider the oscillator shown in Fig. 18.5 subjected to a set of harmonic forces, each expressed by

$$P_r(t) = P_{0r} \sin \Omega_r t \qquad (18.9)$$

The amplitude of force P_{0r} may be converted to the amplitude of forcing displacement, $u_{0r} = P_{0r}/k$. (This is done to make our development more general, although in this particular case u_{0r} by itself has little meaning.) Let us now determine the mean-square excitation of this composite forcing. For a single

harmonic function $u = u_0 \sin \Omega t$, this quantity is, according to Eq. 18.2,

$$\overline{u^2(t)} = \frac{1}{t_m} \int_0^{t_m} u_0^2 \sin^2 \Omega t\, dt$$

If t_m is an exact multiple of the period $T = 2\pi/\Omega$, our result will be $u_0^2/2$. If it is not, the equation

$$\overline{u^2(t)} = \tfrac{1}{2} u_0^2$$

may still be regarded as a good approximation, because the time involved t_m is by assumption considerably larger than T. When there are a number of forcing functions, each defined by an amplitude u_{0r}, the resulting mean-square excitation is

$$\overline{u^2(t)} = \tfrac{1}{2}\left(u_{01}^2 + u_{02}^2 + \cdots + u_{0n}^2 \right)$$

The contributions of the individual components in Fig. 18.5 are depicted in Fig. 18.6. A plot of this type is called a *discrete forcing spectrum*.

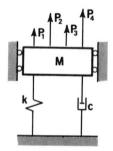

Figure 18.5 Oscillator with several harmonic forces applied.

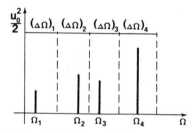

Figure 18.6 Discrete forcing spectrum.

When there are many exciting frequencies present in a discrete spectrum, it becomes more natural to think of the density of forcing as a continuous function of Ω. Let us define the *power spectral density* (PSD) of the displacement excitation, $S_u(\Omega)$, by means of the following

$$\Delta \overline{u^2(\Omega)} = S_u(\Omega) \frac{\Delta\Omega}{2\pi} \qquad (18.10a)$$

in which $\Delta\overline{u^2}(\Omega)$ is the contribution of the forcing in the interval $\Delta\Omega$ to the mean-square excitation. This is best explained by Fig. 18.7, which is a plot of S_u. The hatched area, when divided by 2π, gives the left side of Eq. 18.10a. Although the continuous spectrum is introduced as a limiting case of a dense discrete spectrum, the reverse approach is also possible. That means that starting out with a continuous plot of S_u we can convert it into a discrete equivalent.

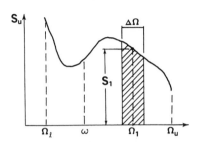

Figure 18.7 Power spectral density of displacement excitation.

When dealing with theory we employ rad/s as units of frequency, while in most practical problems the frequency is given in hertz. In the latter case we employ a slightly different definition of S_u, namely,

$$\Delta\overline{u^2(f)} = S_u(f)\cdot\Delta f \qquad (18.10b)$$

in which $f(\text{Hz})$ is the frequency. This duality allows us to use the same figure for $S_u(\Omega)$ measured in in.^{2} per rad/s as for $S_u(f)$ which is, by definition, in in.^{2}/Hz. (Refer to Prob. 18-3.)

Although only the PSD of displacement excitation has been defined thus far, we can similarly introduce the density of acceleration S_a, velocity S_v, or force S_P. The term *power spectral density* is often used with respect to the spectral forcing because the energy per unit of time (power) is proportional to the square of the displacement.

In our analysis forcing is treated as a random function, which means its time dependence is known only in terms of probabilistics. The forcing is typically defined by its mean value and the spectral density of the oscillating component. Since the latter is a continuous function, we can view such forcing as a superposition of infinitely many harmonic waves with various amplitudes and phase angles.

18E. Response of Oscillator to Spectral Forcing. When there is only one load applied to the oscillator in Fig. 18.5, the forcing function and the displacement response are, respectively,

$$P(t)=P_0\sin\Omega t \qquad \text{and} \qquad u(t)=\mu u_0\sin(\Omega t-\theta)$$

in which $u_0=P_0/k$. (Refer to Sect. 2G.) The mean-square response is

$$\overline{u^2} = \frac{1}{t_m}\int_0^{t_m}\mu^2 u_0^2\sin^2(\Omega t-\theta)\,dt = \tfrac{1}{2}\mu^2 u_0^2$$

(As mentioned before, this is a good approximation, because $t_m \gg T$.) The mean-square forcing is $\overline{P^2}=P_0^2/2$, and we conclude that

$$\overline{u^2} = \tfrac{1}{2}\mu^2 u_0^2 = \mu^2\frac{1}{k^2}\,\overline{P^2} \qquad (18.11)$$

This relationship demonstrates a proportionality between the mean-squares of excitation and response if only a single frequency is present. When there are n frequencies involved, the mean-square displacement response is

$$\overline{u^2(t)} = \tfrac{1}{2}\mu_1^2 u_{01}^2 + \tfrac{1}{2}\mu_2^2 u_{02}^2 + \cdots + \tfrac{1}{2}\mu_n^2 u_{0n}^2$$

in which $u_{0r}=P_{0r}/k$. Using the mean-squares of the individual forcing components from Eq. 18.11, we have

$$\overline{u^2(t)} = \mu_1^2\,\overline{u_1^2} + \mu_2^2\,\overline{u_2^2} + \cdots + \mu_n^2\,\overline{u_n^2} \quad (18.12)$$

The magnification factors μ_r are expressed by

$$\frac{1}{\mu_r^2} = \left(1-\frac{\Omega_r^2}{\omega^2}\right)^2 + \left(2\zeta\frac{\Omega_r}{\omega}\right)^2 \qquad (18.13)$$

The response to a continuous spectrum S_u is found as a limiting case of a discrete spectrum response. Taking an interval $(\Delta\Omega)_r$ about the frequency Ω_r, we have from Eq. 18.10a

$$\overline{u_r^2} = S_u(\Omega_r)\frac{(\Delta\Omega)_r}{2\pi} \qquad (18.14)$$

$(\Delta\overline{u_r^2}=\overline{u_r^2}$, because we replace, for a moment, our continuous spectrum by a discrete one.) Substituting into Eq. 18.12 yields

$$\overline{u^2(t)} = \frac{1}{2\pi}\Big[\mu_1^2 S_u(\Omega_1)\cdot(\Delta\Omega)_1$$
$$+ \cdots + \mu_n^2 S_u(\Omega_n)\cdot(\Delta\Omega)_n\Big] \quad (18.15)$$

In the limit, we put $(\Delta\Omega)_r=d\Omega$ and obtain an integral expression

$$\overline{u^2(t)} = \frac{1}{2\pi}\int_0^\infty\mu^2 S_u\,d\Omega \qquad (18.16)$$

keeping in mind that both μ^2 and S_u are function of Ω. It is easy to see that the density of forcing near the natural frequency ω will have a much larger influence on the mean-square response than the density far away from ω.

In the particular case of *white noise*, which is characterized by a constant PSD for the entire infinite spectrum of frequencies, the integration of Eq. 18.16 gives

$$\overline{u^2(t)} = \frac{\omega S_u}{8\zeta} = \frac{\pi f_n S_u}{4\zeta} \qquad (18.17)$$

Since we are referring to the situation in Fig. 18.5, it is equivalent to specify either an exciting force or a displacement. If a spectral density S_P of force P is prescribed instead of S_u, we put

$$S_u = \frac{S_P}{k^2} \qquad (18.18)$$

Many types of random forcing occurring in engineering practice are referred to as Gaussian, which means that an instantaneous magnitude has the probability distribution defined by Eq. 18.8. In this case the magnitude of response also obeys the same law.

18F. Acoustic Response is the response of structural elements subjected to sound waves. The pressure oscillations associated with the noise of aircraft or rocket engines, for example, can often be damaging to sensitive systems. The *sound pressure level* is defined in relative terms as

$$L = 10\log_{10}\frac{p^2}{p_{\text{ref}}^2} \qquad \text{(decibels)} \qquad (18.19)$$

where p is the pressure at some instant of time, while p_{ref} is the reference pressure, usually 2.9×10^{-9} psi.

The sound pressure is actually the difference between the instantaneous pressure and the pressure of equilibrium at a particular location. The total, true pressure is positive at any time, but the sound pressure p may have either sign.

The simplest type of sound is a *pure tone*, which contains only one frequency. In engineering applications, however, we are most often concerned with noises, that is, sounds containing so many adjacent frequencies that distinguishing between them is impractical. A characteristic of such a sound called the *sound spectrum* $G(f)$ is defined as follows. The mean-square pressure $\overline{p^2(f)}$ associated with frequency f is

obtained from

$$\overline{p^2(f)} = G(f) \cdot (\Delta f) \qquad (18.20)$$

The definition is quite similar to that of Eq. 18.10b. The spectral density G is measured in $(\text{psi})^2/\text{Hz}$.

When one tone has a frequency twice as large as another, we say the first tone is *one octave* higher than the second. It is common to employ a logarithmic scale for frequencies, which somewhat complicates the computation process. For example, the center frequency of the band between f_l and f_u is

$$f_c = (f_l f_u)^{1/2} \qquad (18.21)$$

Quite often a sound spectrum is presented as a set of values at the center frequencies of one-third octave bands. This also implies dividing the bands in the logarithmic scale (see Prob. 18-9). The relation between mean-square forcing pressure and mean-square displacement response of a simple oscillator is obtained with the aid of Eq. 18.11:

$$\overline{u^2} = \mu_r^2 \overline{u_r^2} = \frac{A_0^2}{k^2}\mu_r^2 \overline{p_r^2} \qquad (18.22)$$

where A_0 is the area on which the pressure acts, k is the spring constant of the oscillator, μ_r is the magnification factor given by Eq. 18.13, and $\overline{p_r^2}$ is the mean-square acoustic pressure applied with frequency Ω_r. Equation 18.22 defines the response due to forcing with a pure tone. When there is a continuous-spectrum sound involved, the total oscillator response is obtained by integration. For a uniform sound spectrum called white noise (constant G), the result of that integration is

$$\overline{u^2(t)} = \frac{A_0^2}{k^2}\frac{\pi f_n G}{4\zeta} \qquad (18.23)$$

Suppressing the factor A_0^2/k^2, we again obtain Eq. 18.17.

18G. Random Base Motion. When the base of a structure is subjected to a prescribed displacement, we can use some of the results obtained in Chapter 5 to calculate the response parameters. For the amplitude of relative displacement $\bar{u} = u - u_b$ in Fig. 18.8 denoted by $\bar{A}$ and the amplitude of base deflection by u_0, we have the following transmissibility μ'' from

Eq. 5.8:

$$(\mu'')^2 = \frac{\overline{A}^2}{u_0^2} = \frac{(\Omega/\omega)^4}{(1 - \Omega^2/\omega^2)^2 + (2\zeta\Omega/\omega)^2} \quad (18.24)$$

(Note that in this chapter a bar over a symbol is almost exclusively used to designate the mean value of the quantity involved. If the bar is used to designate a motion relative to a base, as in this paragraph, it is always clearly stated.) If the base displacement $u_b(t)$ is a random function with the spectral density $S_u(f)$, Eqs. 18.15 and 18.16 remain valid, except that μ is replaced by μ''. There is, however, an important difference between the two sets of results. When the forcing is white noise so that S_u can be taken out of the integral sign in Eq. 18.16, the integral will not converge when μ is replaced by μ''. This indicates that the mean-square response is, in this case, infinite, and the upper frequency limit Ω_u must be imposed. If this is done, the value of the integral in Eq. 18.16 (and thereby the mean-square relative displacement) can be computed by numerical methods.

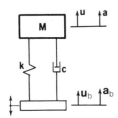

Figure 18.8 Excitation (u_b or a_b) and response (u or a) of oscillator with moving base.

Suppose now that the spectral density S_b of the base acceleration a_b in Fig. 18.8 is given and we want to find the absolute acceleration $a(t)$ of the mass. The acceleration transmissibility is

$$(\mu')^2 = \frac{a_{max}^2}{a_{b\,max}^2} = \frac{1 + (2\zeta\Omega/\omega)^2}{(1 - \Omega^2/\omega^2)^2 + (2\zeta\Omega/\omega)^2} \quad (18.25)$$

The mean-square response $\overline{a^2(t)}$ can again be determined from Eqs. 18.15 or 18.16 after replacing μ by μ'. When the spectral density S_b is constant over the entire range of frequencies, we have

$$\overline{a^2(t)} = \frac{\omega S_b}{8\zeta}(1 + 4\zeta^2) = \frac{\pi f_n S_b}{4\zeta}(1 + 4\zeta^2) \quad (18.26)$$

18H. Lightly Damped Oscillator Under Wide-Band Forcing. Figure 18.9 shows the spectrum of the magnification factor (squared) μ^2 for a lightly damped oscillator having the natural frequency ω. The other curve in the picture is the forcing spectrum $S(\Omega)$ extending over a sizable range of frequencies and nearly constant in the vicinity of ω. The term *wide-band* is used with regard to the latter spectrum $S(\Omega)$, to indicate the large number of forcing frequencies it contains. Conversely, the term *narrow-band* is applied to the former spectrum of $\mu^2(\Omega)$ plotted on the basis of Eq. 18.13, provided ζ is a small number.

The two functions in Fig. 18.9 can be multiplied by each other and plotted as a function of Ω to represent what is called the *response spectrum*. The calculations of the mean-square response in Eq. 18.16 can then be viewed as an integration of the response spectrum over the range of forcing frequencies. To get an approximate value of the integral of $\mu^2 S$, we can replace $S(\Omega)$ by a constant value $S(\omega)$ and take it outside the integration symbol. This is made possible by μ^2 forming a sharp peak at ω and S being relatively flat in the vicinity of ω.

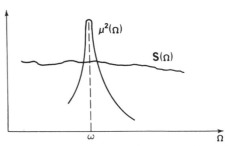

Figure 18.9 Forcing spectrum S and spectrum of magnification factor (squared) μ^2.

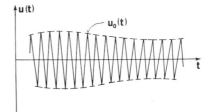

Figure 18.10 Random sine wave.

It is now obvious that when a lightly damped oscillator is forced by a wide band, flat spectrum, it responds by vibrating at predominantly its own natural frequency ω. It can be shown that the response may be described by

$$u(t) = u_0(t)\sin[\omega t - \theta(t)] \quad (18.27)$$

This is quite similar in appearance to a steady-state response, except that $u(t)$, $u_0(t)$, and $\theta(t)$ are random functions. The *apparent frequency* of response ω is identical with the natural frequency of the system. An example of this type of response, called the *random sine wave*, is given in Fig. 18.10.

The same type of response can be observed in lightly damped MDOF systems whenever the effect of a single vibratory mode predominates.

18J. Rayleigh Distribution of Probability. When investigating a plot of a random response versus time, one notes a number of peaks and valleys at various levels. The peaks are of particular significance, since they determine the local maxima of response. In the plot of the random sine wave (Fig. 18.10) the number of peaks is the same as the number of pseudo-cycles of motion. The probability of a peak being in a specified range, say, between $u_0 = 1.5\sigma$ to 1.7σ may be found from the record by calculating the number of peaks in this range and dividing it by the total number of peaks in the record. The law governing the probability density of peaks in the random sine wave is called the *Rayleigh distribution*, which is defined by

$$\mathfrak{p}(u_0) = \frac{u_0}{\sigma^2} \exp\left(-\frac{u_0^2}{2\sigma^2}\right) \qquad (18.28)$$

and illustrated in Fig. 18.11. The probability of u_0 being in a specific range of values may be found by integration of this probability density, similarly to what was done in Sect. 18C. A more convenient procedure is to approximately calculate the probability as

$$P(u_{01}, u_{02}) \approx \mathfrak{p} \cdot (u_{02} - u_{01}) \qquad (18.29)$$

where $\mathfrak{p}$ is taken midway between u_{01} and u_{02}, provided $\mathfrak{p}$ is nearly a straight line over the segment of interest.

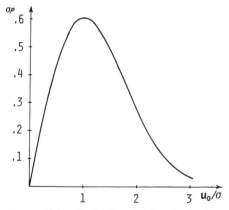

Figure 18.11 Rayleigh probability distribution.

18K. Strength Considerations When Random Loading Is Applied. The longer the exposure to random loading, the larger the instantaneous response may be experienced, even if the RMS of response remains constant. Although this statement is correct from the theoretical veiwpoint, the requirements of design demand and practical experience suggest placing a limit on what can actually be attained. In aerospace practice, for example, 3σ has traditionally been used as the largest vibratory response amplitude. Although exceeding three times the standard deviation should happen less than 0.3% of time according to the Gaussian distribution and is therefore quite likely to happen even in short-duration testing, experience has shown the "3σ criterion" to be safe.

Checking on fatigue strength is the closing part of random response analysis of structures. Everything that was said about the subject in Chapter 7 applies here as well, and only the number and amplitude of anticipated vibratory cycles remain to be discussed. Focusing our attention on the response in the form of the random sine wave in Fig. 18.10, let us think about a fatigue test that is to last t_m seconds. If the natural frequency of a system is f_n (Hz), the number of pseudo-cycles experienced during the test will be $f_n t_m$. The distribution of amplitudes is defined by the Rayleigh curve, Eq. 18.28, which specifically means that if there is a probability P of the stress amplitude being between, say, 3000 and 4000 psi, we can anticipate

$$n = Pf_n t_m$$

cycles of alternating stress not to exceed 4000 psi, but larger than 3000 psi. Once we have calculated how many cycles in each designated amplitude range will occur, we are ready to use the $\tilde{n}$–σ curve to assess the damage ratio. A common analytical expression of that curve is Eq. 7.3:

$$\tilde{n}(\sigma_a')^b = r$$

(As mentioned before, a small number of cycles theoretically exceeds the 3σ level. In practice we assume the amplitudes to be limited to that level.)

A somewhat different approach would be to treat the number of applied stress cycles as a continuous function of amplitude. For the Rayleigh distribution the damage ratio is then

$$D = \frac{f_n t_m}{r}\left(\sqrt{2}\,\sigma\right)^b \Gamma\left(1 + \frac{b}{2}\right) \qquad (18.30)$$

as demonstrated in Ref. 10. In Eq. 18.30 σ is the standard deviation of applied stress and Γ is the

Euler gamma function. To be quite precise, we may add that D is a statistical quantity itself and Eq. 18.30 gives its mean value.

18L. MDOF Systems. It is not practical to attempt a full, manual random-response analysis when there is more than one DOF involved. Some realistic estimates, however, may often be obtained for such systems when enough is known about them. When it is likely that the effect of a single mode will predominate, we can use an equivalent oscillator in the manner described in Chapter 11. When the harmonic response of a structure is known, the random response may be accurately computed as outlined below.

Suppose that an acceleration amplitude at some point of the structure is a_n, due to a harmonic force $P = P_0 \sin(2\pi f t)$ applied at some other point. Let us introduce a ratio of response to excitation amplitude $\tilde{\mu}$,

$$\tilde{\mu}^2 = \frac{a_n^2}{P_0^2}$$

which we might call a *generalized magnification factor* and which differs from such factors used elsewhere in this text only by not being a dimensionless ratio. If $S(f)$ is a spectrum of a force applied in the same direction as $P_0 \sin(2\pi f t)$, the mean-square response acceleration is

$$\overline{a^2} = \int_0^\infty \tilde{\mu}^2 S \, df \qquad (18.31)$$

where $\tilde{\mu}^2$ is also a known function of f. The statement above was merely an example, since forcing may be differently prescribed and/or a different response parameter may have to be computed. However, the right-hand side of Eq. 18.31 always retains this form.

Quite often, the mean-square response of MDOF systems may be approximated by the sum of mean-square modal contributions. This approach is presented in Prob. 18-16.

SOLVED PROBLEMS

18-1 Find the mean value $\overline{P}$, the mean-square $\overline{P^2}$, the variance σ^2, and the RMS (P) of the deterministic forcing function P in the figure.

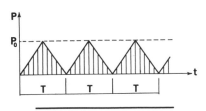

We can take $t_m = T$ as an integration range, since the function is periodic. We have

$$\overline{P} = \frac{1}{T} \int_0^T P \, dt = \frac{P_0}{2}$$

and noting that

$$P = \frac{2P_0}{T} t \qquad \text{for} \quad 0 \leqslant t \leqslant \frac{T}{2}$$

we obtain

$$\overline{P^2} = \frac{2}{T} \int_0^{T/2} \left(\frac{2P_0}{T} \right)^2 t^2 \, dt = \frac{P_0^2}{3}$$

The variance can be found from Eq. 18.4:

$$\sigma^2 = \frac{P_0^2}{3} - \left(\frac{P_0}{2} \right)^2 = \frac{P_0^2}{12}$$

Finally,

$$\text{RMS}(P) = \left(\frac{P_0^2}{3} \right)^{1/2} = \frac{P_0}{\sqrt{3}}$$

18-2 Calculate and plot a portion of the discrete spectrum of the forcing function described in Prob. 18-1. Note that the resolution into harmonic components is given (Prob. 2-28) as follows:

$$P(t) = \frac{P_0}{2} - \frac{4P_0}{\pi^2} \left(\cos \Omega_0 t \right.$$
$$\left. + \frac{1}{3^2} \cos 3\Omega_0 t + \frac{1}{5^2} \cos 5\Omega_0 t + \cdots \right)$$

in which $\Omega_0 = 2\pi/T$.

The amplitude of the zero-frequency component is $P_0/2$, while for the remaining harmonics we have

$$P_{0r} = \frac{4P_0}{\pi^2 (2n-1)^2} \qquad n = 1, 2, 3, \dots .$$

The mean-square values of these components are

$$\overline{P_r^2} = \frac{1}{2} P_{0r}^2 = \frac{8P_0^2}{\pi^4 (2n-1)^4} \qquad n = 1, 2, 3, \dots$$

which gives us $82.13 \times 10^{-3} P_0^2, 1.014 \times 10^{-3} P_0^2, 0.131 \times 10^{-3} P_0^2$, $\dots$, for successive values of n. To plot such vastly differing quantities, a log scale must be used. Notice that for the constant component $P_0/2$ (not shown in the figure) the mean-square value is $P_0^2/4$.

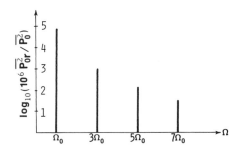

18-3 The *overall* mean-square acceleration associated with a particular spectrum is the area under the S–f curve. Compute this value for the given plot using Eq. 18.10b (with S_u replaced by S_a). Determine the overall level if we construct a corresponding plot using rad/s as units of frequency and employing a modified Eq. 18.10a.

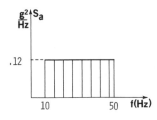

For the given plot the overall level is:

$$0.12 \left(\frac{g^2}{\text{Hz}} \right) 40\,\text{Hz} = 4.8 g^2 = 715{,}181 \text{ in.}^2/\text{s}^4$$

When Ω is used instead of f, we employ the same numerical value for S_a,

$$S_a = 0.12 \frac{g^2}{(\text{rad}/\text{s})} \qquad \Delta\Omega = 40 \times 2\pi$$

$$\overline{a^2} = 0.12 \frac{g^2}{(\text{rad}/\text{s})} \times \frac{40 \times 2\pi}{2\pi} \text{ rad}/\text{s} = 4.8 g^2$$

The overall level, which is the measure of the total energy supplied by a forcing system, is the same no matter which frequency units we happen to use.

18-4 The PSD of force P is 0.5 Lb^2/(rad/s). Treating the beam as weightless, derive the RMS displacement response of mass M. Use the following data: $L = 40$ in., $M = 0.056$ Lb-s^2/in., $\zeta = 0.354$, $E = 29 \times 10^6$ psi, and $I = 0.0005$ in.^{4}

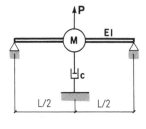

The beam stiffness is

$$k = \frac{48EI}{L^3} = \frac{48 \times 29 \times 10^6 \times 0.0005}{40^3} = 10.875 \text{ Lb}/\text{in.}$$

When computing the natural frequency we must include the effect of damping, which is quite sizable. As determined in Sect. 2D,

$$\omega_d^2 = \frac{k}{M}(1 - \zeta^2) = \frac{10.875}{0.056}(1 - 0.354^2)$$

$$\therefore \ \omega_d = 13.03 \text{ rad}/\text{s}$$

The response is determined by Eq. 18.17, except that now the force density S_P rather than displacement density S_u is prescribed. Employing Eq. 18.18 we find

$$\overline{u^2(t)} = \frac{\omega_d S_P}{8\zeta k^2} = \frac{13.03 \times 0.5}{8 \times 0.354 \times 10.875^2} = 0.01945 \text{ in.}^2$$

or RMS$(u) = 0.139$ in.

18-5 The mass of an oscillator is subject to two forcing accelerations, $a_{01}\sin\Omega_1 t$ and $a_{02}\sin\Omega_2 t$. Compute the mean-square forcing and response accelerations for the following data: $a_{01} = 2.1g$, $a_{02} = 6.5g$, $\Omega_1/\omega = 0.8$, $\Omega_2/\omega = 1.1$, and $\zeta = 0.1$.

The mean-square forcing acceleration is

$$\tfrac{1}{2}a_{01}^2 + \tfrac{1}{2}a_{02}^2 = 2.205g^2 + 21.125g^2 = 23.33g^2$$

From Sect. 18E and Eq. 2.27,

$$\overline{a^2} = \mu_1^2 \left(\frac{\Omega_1}{\omega} \right)^4 \frac{a_{01}^2}{2} + \mu_2^2 \left(\frac{\Omega_2}{\omega} \right)^4 \frac{a_{02}^2}{2}$$

Replacing Ω/ω by x, we have,

$$\mu_1^2 \left(\frac{\Omega_1}{\omega} \right)^4 = \frac{x_1^4}{\left(1 - x_1^2\right)^2 + (2\zeta x_1)^2} = 2.639 \qquad \text{for} \quad x_1 = 0.8$$

and $\mu_2^2 \left(\dfrac{\Omega_2}{\omega} \right)^4 = 15.828$ for $x_2 = 1.1$

The mean-square response is

$$\overline{a^2} = 2.639 \times 2.205g^2 + 15.828 \times 21.125g^2 = 340.19g^2$$

18-6 An SDOF system with the natural frequency $\omega = 100$ rad/s and damping ratio $\zeta = 0.16$ is subjected to random forcing with a constant spectral density S_u applied to the mass. Calculate the mean-square displacement response by replacing the continuous spectrum in (a) by three discrete amplitudes: one at $\Omega_2 = \omega$, representing the bandwidth $2\omega\zeta$ and one on each side of the bandwidth. Repeat the calculation using two amplitudes on each side of the bandwidth, a total of five amplitudes. Compare the results.

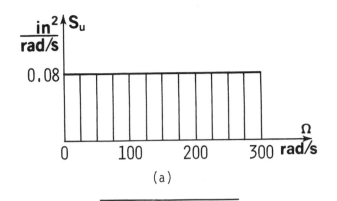

(a)

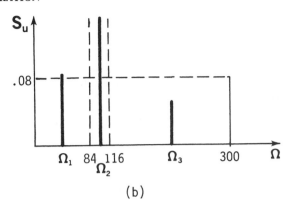

(b)

The bandwidth is $2\omega\zeta = 2\times 100\times 0.16 = 32$ rad/s, and (b) shows how the frequency range is subdivided. Using the center frequencies of the segments, we have

$$\Omega_1 = 42, \qquad \Delta\Omega_1 = 84 \text{ rad/s}$$

$$\Omega_2 = 100, \qquad \Delta\Omega_2 = 32 \text{ rad/s}$$

$$\Omega_3 = 208, \qquad \Delta\Omega_3 = 184 \text{ rad/s}$$

The magnification factors are computed from Eq. 18.13:

$$\frac{1}{\mu_1^2} = \left(1 - \frac{42^2}{100^2}\right)^2 + \left(2\times 0.16\times \frac{42}{100}\right)^2$$

$$\mu_1^2 = 1.4360 \qquad \mu_2^2 = 9.7656 \qquad \mu_3^2 = 0.0869$$

We can now apply Eq. 18.15 with $S_u = S_0 =$ constants.

$$\overline{u^2(t)} = \frac{S_0}{2\pi}\left[\mu_1^2(\Delta\Omega_1) + \mu_2^2(\Delta\Omega_2) + \mu_3^2(\Delta\Omega_3)\right]$$

Substituting the calculated figures and $S_0 = 0.08$ in.^{2}/rad, we have

$$\overline{u^2(t)} = 5.7183 \text{ in.}^2$$

The second part of the problem is solved similarly, except that the two frequency bands outside the bandwidth are halved.

r	Ω_r (rad/s)	$\Delta\Omega_r$ (rad/s)	μ_r^2
1	21	42	1.0890
2	63	42	2.4730
3	100	32	9.7656
4	162	92	0.3439
5	254	92	0.0329

This time the result is

$$\overline{u^2(t)} = \frac{0.08}{2\pi}\times 496.77 = 6.3251 \text{ in.}^2$$

Increasing the number of frequency bands brings us closer to an exact solution.

18-7 The isolator in (a) has a random acceleration applied to its base. The PSD plot in (b) consists of straight lines when drawn in log-log scale and is defined by the following points:

f (Hz):	20	300	1200	2000
S_b (g^2/Hz):	0.017	0.25	0.25	0.09

$$S_b = 0 \text{ outside } 20\text{–}2000 \text{ Hz range.}$$

Determine the RMS value of the absolute acceleration response a of mass M if the natural frequency of the system (when the base is fixed) is $f_n = 54$ Hz. The damping ratio is $\zeta = 0.02$

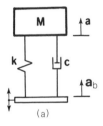

(a)

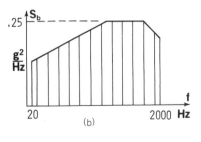

(b)

The acceleration analog of Eq. 18.10b is

$$\Delta \overline{a_b^2} = S_b(\Delta f)$$

in which S_b is the base forcing PSD and $\Delta \overline{a_b^2}$ is the component of the mean-square forcing coming from the frequency band Δf.

Using Eq. 18.25, we get the absolute response as follows:

$$\overline{a^2(t)} = (\mu'_1)^2 S_b(f_1) \cdot (\Delta f)_1 + \cdots + (\mu'_n)^2 S_b(f_n) \cdot (\Delta f)_n$$

The form of this expression is similar to Eq. 18.15, but the factor of 2π does not appear, since the frequency is expressed in hertz. (The ratio Ω/ω is the same as f/f_n.) The flow of computation is presented in the table below. The intermediate values of S_b for the selected frequencies can be read off the curve plotted on graph paper or algebraically interpolated.

f (Hz)	f_c (Hz)	S_b (g^2/Hz)	$(\mu')^2$	Δf (Hz)	$\Delta \overline{a^2}$ (g^2)
20					
	28.5	0.024	1.9205	17	0.7836
37					
	45.0	0.0378	10.597	16	6.4091
53					
	54.0	0.045	626	2	56.34
55					
	65.0	0.0544	4.9174	20	5.3501
75					
	87.5	0.073	0.3794	25	0.6924
100					
	200.0	0.167	0.0063	200	0.2104
300					
	650.0	0.25	0.00006	700	0.0105
1000					
	1100.0	0.25	0.00001	200	0.0005
1200					
	1600.0	0.14	0	800	0
2000					
					$\Sigma = 69.80$

In the table f_c is the average frequency in the segment under consideration; S_b and μ' are both calculated at f_c. We conclude that the RMS value of the absolute acceleration of M is

$$\text{RMS}(a) = \left[\sum (\Delta \overline{a^2}) \right]^{1/2} = (69.80)^{1/2} = 8.355 g$$

18-8 A simple torsional oscillator has a shaft with length $L = 28$ in., diameter $d = 4$ in. and shear modulus $G = 11 \times 10^6$ psi. The disc has $J = 20$ Lb-s^2-in. and the damping ratio is $\zeta = 0.03$. The random torque applied to the disc has zero mean value and the constant spectral density $S = 10^7$ (Lb-in.)^2/(rad/s). Treating the shaft as weightless, determine its RMS stress response.

The shaft stiffness is

$$K = \frac{G}{L} C = \frac{G}{L} \frac{\pi d^4}{32} = 9.8736 \times 10^6 \text{ Lb-in./rad}$$

and the frequency is:

$$\omega = \left(\frac{K}{J} \right)^{1/2} = 702.62 \text{ rad/s}$$

The mean-square rotation is found from Eqs. 18.17 and 18.18,

$$\overline{\alpha^2(t)} = \frac{\omega S}{8\zeta K^2} = \frac{702.62 \times 10^7}{8 \times 0.03 \times 9.8736^2 \times 10^{12}} = 0.3003 \times 10^{-3}$$

$$\therefore \text{ RMS}(\alpha) = 0.01733 \text{ rad}$$

When the end rotation is known, the shear stress due to twisting is found from

$$\tau = \frac{16 M_t}{\pi d^3} = \frac{16 K \alpha}{\pi d^3}$$

or

$$\text{RMS}(\tau) = \frac{16 K}{\pi d^3} \text{RMS}(\alpha) = 13,616 \text{ psi}$$

18-9 A single-octave band extends from f_l to f_u Hz. Determine the upper, the lower and the center frequency of each of the one-third-octave subbands. Calculate for $f_l = 71$ and $f_u = 142$ Hz.

The figure shows the location of the limiting frequencies in the proportional as well as in the logarithmic scale. Our only initial information is $f_u/f_l = 2$. (All logarithms used here are to the base 10.) $\log f_1$ will fall at one-third of the segment $\log f_u - \log f_l$, hence

$$\log f_1 = \frac{1}{3} (\log f_u - \log f_l) + \log f_l$$

or

$$\log \frac{f_1}{f_l} = \log \left(\frac{f_u}{f_l} \right)^{1/3} = \log 2^{1/3}$$

$$\therefore f_1 = 1.26 f_l = 89.46 \text{ Hz}$$

Similarly,

$$\log f_2 = \frac{2}{3} (\log f_u - \log f_l) + \log f_l$$

$$f_2 = 1.5874 f_l = 112.71 \text{ Hz}$$

The center frequency f_c' of the first band is found from

$$\log f_c' = \frac{1}{2} (\log f_l + \log f_1) = \log (f_l f_1)^{1/2}$$

$$f_c' = (f_l f_1)^{1/2} = (71 \times 89.46)^{1/2} = 79.7 \text{ Hz}$$

Similarly

$$f_c'' = (89.46 \times 112.71)^{1/2} = 100.41 \text{ Hz}$$

$$f_c''' = (112.71 \times 142)^{1/2} = 126.51 \text{ Hz}$$

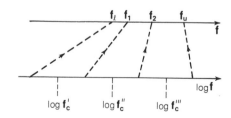

18-10 The fundamental frequency of a square, simply supported plate is

$$\omega = \frac{5.7}{a^2} \left[\frac{Et^2}{\rho(1-\nu^2)} \right]^{1/2}$$

When uniform pressure p is applied to the surface, the lateral deflection u and the maximum bending stress σ_b at the center of the plate are, respectively,

$$u = 0.0444 \frac{pa^4}{Et^3} \quad \text{and} \quad \sigma_b = 0.2874 \frac{pa^2}{t^2}$$

Calculate the RMS bending stress from the fundamental mode when the plate is subjected to a white noise with spectral density $G = 12 \times 10^{-6}$ psi^2/Hz. Put $a = 12$ in., $t = 0.04$ in., $E = 29 \times 10^6$ psi, $\nu = 0.3$, $\gamma = 0.282$ Lb/in.^{3}, and $\zeta = 0.025$.

The natural frequency is:

$$f_n = \frac{5.7}{2\pi \times 12^2} \left[\frac{29 \times 10^6 \times 0.04^2}{(0.282/386)(1-0.3^2)} \right]^{1/2} = 52.63 \text{ Hz}$$

The plate stiffness, relative to the applied pressure, is:

$$k = \frac{p}{u} = \frac{Et^3}{0.0444 a^4} = 2.016 \text{ psi/in.}$$

We can employ Eq. 18.23 now, with A_0 suppressed, since the static deflection is p/k, not $A_0 p/k$,

$$\overline{u^2} = \frac{\pi f_n G}{4k^2 \zeta} = \frac{\pi \times 52.63 \times 12 \times 10^{-6}}{4 \times 2.016^2 \times 0.025} \quad \therefore \text{RMS}(u) = 0.06987 \text{ in.}$$

The equivalent static pressure that would induce such displacement is $p = ku = 2.016 \times 0.06987 = 0.1409$ psi, and the RMS value of bending stress is:

$$\text{RMS}(\sigma_b) = 0.2874 \times \frac{0.1409 \times 12^2}{0.04^2} = 3643 \text{ psi}$$

18-11 A nodal point of a structure vibrates in a random sine-wave pattern. If we consider 1000 successive pseudo-cycles, how will the deflection amplitudes u_0 be distributed? Draw a diagram showing how many cycles out of 1000 there are in each 0.25σ increment of u_0 in the range $u_0 = 0$ to 3σ (σ is the standard deviation of u_0).

The distribution of amplitudes u_0 is governed by the Rayleigh law, Eq. 18.28. Suppose that we want to find out how many cycles there is likely to be between $u_0 = 1.25\sigma$ and 1.50σ. Using Eq. 18.28 with

the center amplitude of 1.375σ we obtain

$$\mathrm{p}(1.375\sigma) = \frac{1.375}{\sigma} \exp\left(-\frac{1}{2} \times 1.375^2\right) = \frac{0.53427}{\sigma}$$

From the approximate Eq. 18.29:

$$P(1.25\sigma, 1.50\sigma) \approx \left(\frac{0.53427}{\sigma} \right) \cdot (0.25\sigma) = 0.13357$$

When 1000 cycles are involved, there will be

$$1000 \times 0.13357 \approx 134$$

within the limits of interest. Proceeding analogously with the remaining segments, we obtain the plot shown. The total number of cycles in the figure is 992. This means that 8 cycles will have a magnitude larger than 3σ (provided the net round-off error is zero). If 3σ is used as an upper bound of response, the last segment will have 20, not 12 cycles.

To use a more tangible description, let us say that the mean deflection is zero and the RMS value is 2 in. Out of 1000 cycles, there will be 134 with the amplitude between 2.5 and 3.0 in.

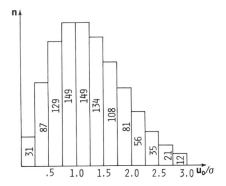

18-12 A lightly damped oscillator is subjected to a wide-band random forcing that produces a stress response characterized by a zero mean value and the standard deviation $\sigma = 11,000$ psi. The natural frequency of the oscillator is $f_n = 85$ Hz and the time of exposure to forcing is $t_m = 12,600$ s. The material fatigue strength σ_a' and the corresponding number of cycles $\tilde{n}$ are connected by the exponential law

$$\tilde{n}(\sigma_a')^b = r$$

The material is an aluminum alloy, for which $b = 6.09$ and $r = 192 \times 10^{30}$. Calculate the damage ratio D.

The total number of cycles experienced is

$$f_n t_m = 85 \times 12,600 = 1.071 \times 10^6$$

Equation 18.30 gives us

$$D = \frac{1.071 \times 10^6}{192 \times 10^{30}} (11,000\sqrt{2})^{6.09} \Gamma(4.045)$$

The gamma function is reduced to the tabulated region between 1.0 and 2.0:

$$\Gamma(4.045) = 3.045 \times 2.045 \times 1.045\Gamma(1.045) = 6.3507$$

where $\Gamma(1.045) = 0.97595$ was read from math tables. Substituting, $D = 1.1968$.

18-13 Two samples of material are tested, each in a different way. One is subjected to random sine-wave oscillations with zero mean stress and the RMS stress equal to σ_1. The second is under a sinusoidal stress of constant amplitude σ_e. How does one select the value of σ_e so that after an equal number of cycles both samples experience the same damage ratio D? Perform the calculations for two materials using the exponential fatigue law,

1. Aluminum alloy, $b = 6.09$ and $r = 192 \times 10^{30}$.
2. Alloy steel, $b = 5.6276$ and $r = 36.6 \times 10^{30}$.

For the sinusoidal stress, the damage ratio is

$$D = \frac{n}{\tilde{n}}$$

where n is the number of experienced cycles and $\tilde{n}$ is the number of cycles the sample can withstand at the level σ_e. From the exponential law

$$\tilde{n} = \frac{r}{\sigma_e^b} \quad \text{and} \quad D = \frac{n}{r}\sigma_e^b$$

Equation 18.30 for this case is

$$D = \frac{n}{r}\left(\sqrt{2}\,\sigma_1\right)^b \Gamma\left(1 + \frac{b}{2}\right)$$

Equating both expressions we get the equivalent constant amplitude

$$\sigma_e = \sqrt{2}\,\sigma_1 \Gamma^{1/b}$$

The function $\Gamma \equiv \Gamma(1 + b/2) = 6.3507$ was determined in Prob. 18-12 for the aluminum alloy. For the other material

$$\Gamma\left(1 + \frac{5.6276}{2}\right) = \Gamma(3.8138) = 4.7725$$

The equivalent sinusoidal stress levels are

1. $\sigma_e = \sqrt{2} \times 6.3507^{1/6.09}\sigma_1 = 1.9158\sigma_1$
2. $\sigma_e = \sqrt{2} \times 4.7725^{1/5.6276}\sigma_1 = 1.8669\sigma_1$

18-14 The beam described in Prob. 9-26 is acted on by a random force having spectral density $S = 0.64$ Lb^2/Hz in the range 0.2–0.8 Hz and zero outside this range. The force is applied at node 3, while the displacement of interest is at node 1. When a harmonic response test was conducted, a force with a constant amplitude $P_0 = 1000$ Lb applied at node 3 gave the amplitudes of response at node 1, tabulated below in row u_1. What is the RMS displacement response at node 1 due to the prescribed random forcing?

f (Hz):	0.20	0.25	0.30	0.35	0.40	0.45	0.50
u_1 (in.):	0.4868	0.5515	0.6534	0.8198	1.0936	1.4114	1.2914

	0.55	0.60	0.65	0.70	0.75	0.80
	0.8992	0.6311	0.4711	0.3705	0.3030	0.2552

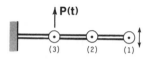

Referring to Sect. 18L, our generalized magnification factor in this case is

$$\bar{\mu}^2 = \frac{u_1^2}{P^2}$$

The PSD of forcing is constant and may be taken out of the summation sign. We can tell by inspection that there are no sharp peaks in the $u_1(f)$ curve and Δf may be taken as a constant, $\Delta f = 0.05$ Hz. The RMS displacement response at node 1 is, by (modified) Eq. 18.31:

$$\overline{u^2} = \sum \bar{\mu}^2 S_p \Delta f = \frac{1}{P^2}S_p \Delta f \sum u_1^2$$

$$= \frac{1}{1000^2} \times 0.64 \times 0.05 \times (0.4868^2 + \cdots + 0.303^2)$$

$$= 0.2609 \times 10^{-6} \quad \therefore \text{RMS}(u) = 0.5108 \times 10^{-3} \text{ in.}$$

A more accurate procedure employing a computer program gave the answer RMS$(u) = 0.5083 \times 10^{-3}$ in.

18-15 The model of the three-story building analyzed in Prob. 9-9 is subjected to base movement with a constant spectral density of acceleration $S_b = 50$ (in.^{2}/s^4)/(rad/s). Calculate the relative displacement response using the equivalent oscillator developed for this structure in Prob. 11-8, for which $\omega = 15.989$ rad/s. The deflected shape is assumed to be described by

$$\vec{X}^T = [1.0 \quad 2.1429 \quad 3.7429]$$

Take the damping ratio to be $\zeta = 0.005$.

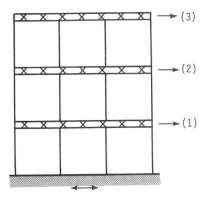

Equation 18.24 tells us the relative deflection amplitude $\overline{A}$ if the displacement amplitude of base forcing is u_0. Since the equation refers to a harmonic motion, we know that the *acceleration* amplitude of base forcing is $a_{b\,max} = \Omega^2 u_0$, and

$$\tilde{\mu}^2 = \frac{\overline{A}^2}{a_{b\,max}^2} = \frac{1}{\left(1 - \Omega^2/\omega^2\right)^2 + \left(2\zeta\Omega/\omega\right)^2} \cdot \frac{1}{\omega^4}$$

is our counterpart of Eq. 18.24. By analogy with Eqs. 18.16 and 18.17, we can write

$$\overline{u^2(t)} = \frac{1}{2\pi}\int_0^\infty \tilde{\mu}^2 S_b \, d\Omega = \frac{S_b}{8\omega^3\zeta}$$

for the mean-square relative displacement. As we know from Chapter 12, there is a difference between the effect of ground movement on an actual SDOF oscillator and on an equivalent oscillator representing a larger system. Reviewing Prob. 12-11 we find that the base forcing has to be multiplied by the factor of 1.2787 to obtain the proper response level. Since we are dealing here with *squared* forcing and response, our result is

$$\overline{u^2} = \frac{50\times 1.2787^2}{8\times 15.989^3 \times 0.005} = 0.500 \text{ in.}^2$$

This corresponds to the top level. Using the definition of the deflected shape, we have the RMS values

$$u_1 = 0.1889 \text{ in.} \qquad u_2 = 0.4048 \text{ in.} \qquad u_3 = 0.7071 \text{ in.}$$

The solution for a model of a building with nearly the same stiffness and inertia parameters is given on page 408 of Ref. 12. After adjusting for a difference in forcing levels we find our solution to give RMS(u_3) nearly 7% less. The main reason is the differences in the first-mode shape of both structures. Although Ref. 12 uses three modes of vibrations, the contribution of higher modes to RMS(u_3) is insignificant.

18-16 A structure is subjected to a set of random joint forces,

$$P_1 \qquad P_2 \qquad P_3 \qquad \text{etc.}$$

Although these forces vary in time, their ratios remain constant, which means that it is enough to define the spectrum $S_P(\Omega)$ of $P_1 \equiv P_0$, for example. Develop an expression for the mean-square displacement response, assuming that it may be approximated by the sum of mean-square modal contributions. All modal properties of the system are presumed to be known. The spectrum is nearly flat in the vicinity of each natural frequency and the modal damping ratios ζ_r are small.

Let us first consider the system to be under the action of a set of harmonic forces, as described in Sect. 10F. A generalized force with amplitude

$$Q_{0r} = B_r P_0$$

may then be formed, since any joint force amplitude is a multiple of some selected reference value P_0. The constant B_r will be, in general, different for each mode. The amplitude in the rth generalized coordinate is

$$s_r = \mu_r \frac{Q_{0r}}{k_r} = \mu_r \frac{B_r P_0}{\omega_r^2 \overline{M}_r}$$

The essential difference between this harmonic loading and the random load applied to our structure is that instead of the function $\sin\Omega t$ in the former we have a random variable characterized by the spectrum $S_P(\Omega)$. A contribution of the incremental force, acting with frequency Ω, to the mean-square response is

$$\Delta \overline{s_r^2} = \left(\frac{B_r}{\omega_r^2 \overline{M}_r}\right)^2 \frac{\mu_r^2}{2\pi} S_P \cdot \Delta\Omega$$

Following Sect. 18E, the increment is replaced by a differential and integrated with respect to Ω to obtain the total response of this rth mode. Since the spectrum is said to be nearly flat in the vicinity of the natural frequencies, we can put

$$\frac{1}{2\pi}\int_0^\infty \mu_r^2 S \, d\Omega \approx \frac{S_r \omega_r}{8\zeta_r}$$

where S_r is the spectrum value at ω_r, according to Sect. 18H. The total response becomes

$$\overline{s_r^2} = \left(\frac{B_r}{\omega_r^2 \overline{M}_r}\right)^2 \frac{S_r \omega_r}{8\zeta_r}$$

To make our development more comprehensive, let us assume we have a three-DOF system and that we intend to calculate the displacement u_2. If the modal displacements are available, we can write, following Eq. 9.14a,

$$u_2 = x_{21}s_1 + x_{22}s_2 + x_{23}s_3$$

where

$$[x_{21} \quad x_{22} \quad x_{23}]$$

is the second row of the modal matrix. The problem statement allows us to use the summation of the modal contributions (squared) in calculation of the mean-square response, thus we put

$$\overline{u_2^2} = x_{21}^2 \overline{s_1^2} + x_{22}^2 \overline{s_2^2} + x_{23}^2 \overline{s_3^2}$$

Generalizing this equation to some nth physical direction, we find

$$\overline{u_n^2} = \sum_r x_{nr}^2 \overline{s_r^2} = \sum_r \left(\frac{x_{nr} B_r}{\omega_r^2 \overline{M_r}} \right)^2 \left(\frac{S_r \omega_r}{8 \zeta_r} \right) \qquad (*)$$

where $\begin{bmatrix} x_{n1} & x_{n2} & x_{n3} \cdots \end{bmatrix}$

is the nth row of the modal matrix of the system.

EXERCISES

18-17 Suppose that the acceleration spectrum presented in Prob. 18-3 pertains only to the vibratory component and that this vibration is superposed on the mean acceleration value of $1.5g$. What is the actual mean-square value of forcing?

18-18 An oscillator is excited by the force

$$P(t) = P_{01} \sin \Omega_1 t + P_{02} \sin(\Omega_2 t - \alpha_2)$$
$$+ P_{03} \sin(\Omega_3 t - \alpha_3)$$

Calculate the mean-square response, assuming the integration time t_m to be many times larger than the longest of the three periods involved.

18-19 When the spectral density S_u of the forcing displacement u_0 is given, the mean-square value $\overline{u^2(t)}$ can be calculated from Eq. 18.16. Develop the corresponding equations for velocity $\overline{\dot{u}^2(t)}$ and acceleration $\overline{\ddot{u}^2(t)}$ as functions of $S_u(\Omega)$.

18-20 The spectral density of forcing acceleration applied to the mass of a basic oscillator is a known function $S_a(\Omega)$. Develop the formulas for the mean-square acceleration response $\overline{a^2(t)}$ analogous to Eqs. 18.15 and 18.16.

18-21 Equation 18.17 gives the mean-square displacement response to a constant-density forcing spectrum extending over the entire frequency range. Find an approximate expression for $\overline{u^2}$ when the range of frequencies is limited, as shown, in the figure. Assume $\Omega_l \ll \omega$ and $\Omega_u \gg \omega$. *Hint.* Ignore damping outside (Ω_l, Ω_u) segment. Use two-term approximations for the hyperbolic functions obtained from integration.

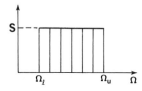

18-22 Find an approximate answer to Prob. 18-7 by assuming the PSD of forcing acceleration to be con-

stant, equal to the value of S_b for $f_n = 54$ Hz, and extending from $f = 0$ to $f = \infty$.

18-23 An oscillator having natural frequency $f_n = 1.55$ Hz and damping $\zeta = 0.05$ is subjected to a random acceleration spectrum applied to the base. The PSD is constant, $S_b = 1.0 g^2/\text{Hz}$ from $f_l = 0.1$ Hz to $f_u = 10$ Hz and zero outside this range. Find the RMS of the absolute acceleration response.

18-24 What is the RMS response in Prob. 18-23 if the forcing frequency band is narrowed to $f_l = 0.75$ Hz and $f_u = 3.0$ Hz?

18-25 An instrument package is placed on a base that may be subjected to random vibration with $S_b = 0.18 \ g^2/\text{Hz}$. If the RMS acceleration response of the package is not to exceed $4.0g$ for design reasons, what condition must the natural frequency fulfill? Use $\zeta = 0.08$.

18-26 The pressure at equilibrium is 14 psi. What is the maximum sound pressure level possible, in decibels?

18-27 An oscillator is subjected to random pressure load applied to the area $A_0 = 16$ in.^{2} The spectral density $G = 15 \times 10^{-3}$ psi$^2/\text{Hz}$ is constant over the frequency range. Find the RMS spring force if $k = 750$ Lb/in., $M = 0.18$ Lb-s^2/in., and $\zeta = 0.04$.

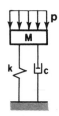

18-28 The fundamental frequency of a semicircular plate with fixed edges is $f_n = 60$ Hz. When the uniform pressure p is applied to the surface, the maximum bending stress occurs at point C, $\sigma_b \approx 0.42 \, pr^2/t^2$, while the maximum deflection is

$$u \approx 0.014 \frac{pr^4}{Et^3}$$

Calculate the RMS stress induced by the fundamental mode at point C when the plate is subjected to white noise. The pressure level at 63 Hz is 127 dB ($p_{\text{ref}} = 2.9 \times 10^{-9}$ psi, while 63 Hz is the center frequency of one-third-octave band). The spectral density G should be calculated at $f_c = 63$ Hz. Solve for $r = 36.7$ in., $t = 0.66$ in., $E = 62,155$ psi, $\gamma = 0.0027$ Lb/in.^{3}, and $\zeta = 0.03$.

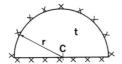

18-29 The simply supported beam whose random response was determined in Prob. 18-4 has the depth $2c = 2.0$ in. and is made of steel with fatigue properties defined by $b = 5.6276$ and $r = 36.6 \times 10^{30}$. How long can random forcing be applied, if a fatigue failure is anticipated when the damage ratio $D = 0.6$?

18-30 The plate in Prob. 18-10 is subjected to a random acoustic pressure that induces the RMS bending stress of 3643 psi. The pressure is to be applied for 120 hr. The natural frequency of the plate is 52.63 Hz. The material is the alloy steel whose fatigue properties are defined in Prob. 18-13 and whose ultimate strength is 180,000 psi. What constant bending stress may be applied so that the damage ratio does not exceed 0.6?

18-31 A shaft can experience 2700 pseudo-cycles of random-sine stress with the RMS value of 13,444 psi. The material is aluminum alloy with the fatigue properties as described in Prob. 18-13. The damage ratio at which the failure occurs is 0.5. When the latter figure is divided by the damage ratio actually experienced by the shaft, the factor of safety in fatigue is obtained. Calculate this factor using the energy-of-distortion theory to convert the tensile to the shear stress quantities.

18-32 A structure is subjected to a random base shaking characterized by the acceleration forcing spectrum $S_a(\Omega)$. Develop an expression for the mean-square relative displacement, approximating the total response by the sum of mean-square modal contributions. All modal properties of the system are presumed to be known. The spectrum is flat in the vicinity of each natural frequency and the modal damping ratios ζ_r are small. *Hint*: The rth normal coordinate of relative displacement is

$$s_r = \Gamma_r \mu_r \frac{a_0}{\omega_r^2}$$

where Γ_r is defined by Eq. 12.2.

18-33 A random deflection pattern has a zero mean and the mean-square value of 6.25 in.^{2} What is the probability that $2 < u < 2.5$ at any instant? What is $P(|u| > 7.5)$ and $P(|u| > 10)$? Use the tables of Gaussian distribution in, for example, Ref. 9.

Generation of Dynamic Models

The previous 18 chapters of this book dealt with the determination of dynamic properties and the responses of simple structures, or as we may prefer to see it, with simplified models of real structures. From the point of view of modern engineers equipped with powerful computer programs, the objective of those preceding chapters was to help in calculating the order of magnitude of the results so that they can rationally design models for dynamic analysis. Such models cannot be too simple, because something important might be missed; nor can they be too involved, since that could make the analysis overly costly. The question of how sophisticated a model to use is most often related to another one, namely, how dense should the network of joints be? That question could be resolved in a simple experiment by, for example, doubling the number of joints and repeating the calculation to see whether there would be any meaningful difference in results. In most real situations, however, we cannot afford the time to carry out such an experiment, and we attempt to create a reasonable model from the outset.

To facilitate this task, several criteria leading to proper spacing of joints are presented for beams, plates, and the second-order elements (the latter group is defined in Chapter 15). Those criteria are spelled out with reference to how the loading pattern changes relative to time and space variables. To appreciate the problem, one may note that doubling the number of dynamically active joints may easily double the cost of computer work, to say nothing of the increased engineering effort in dealing with a larger model. Among the marginal problems associated with accuracy of modeling, the shear deformation of beams and plates is discussed in detail.

19A. Overview of Dynamic Modeling Problems. An ideal model should closely resemble the original structure with respect to all elastic and inertial features. This means that the action of static forces should have the same effect and that mass distribution should also be closely related. Unfortunately, because of time and cost limitations, the practical model usually does not come close to meeting the ideal objectives.

In any real structure at least some of the mass is continuously distributed along the physical dimensions. In contrast to this, a discrete (finite-element) model typically employs lumped masses connected with weightless elements. This gives rise to a *discretization error*, which may show itself in various ways. A model (or a fragment of it) that is designed to preserve the natural frequency may give inaccurate results in response computation. To come up with a good model and to be able to interpret and/or adjust the results, an analyst must be aware of those inherent inaccuracies.

The process of analyzing a structure often involves two major phases. In the first, a static model is established and the responses due to static load cases are determined. The second phase involves setting up a dynamic model and obtaining the natural frequencies and mode shapes, as well as a dynamic response. If a structure is simple, the transition from the first model to the second may be accomplished by merely requesting a generation of mass properties by the program. Most often, however, the computer program involved distinguishes between the static and the dynamic DOFs. If this is the case, the user must decide which coordinates are to be employed as the dynamic DOFs. Occasionally, a dynamic model does not even resemble its static counterpart because of the need to limit the size of a dynamic problem.

There are many good reasons to use relatively few dynamic DOFs in a model. Not only is the flow of computation more involved in dynamics than in statics, but many more data must be stored as well. Although it is possible to solve problems involving thousands of dynamic DOFs, this does not mean that it is often practical to do so. Generating too large models has another disadvantage in the form of excessive size of output. Although it is possible to selectively request some portions of the output in most programs, this option often creates some additional complications.

As in the rest of the book, many conclusions in this chapter are derived from the use of the normal-mode method in response calculations. Although many programs employ a step-by-step integration of the equations of motion to determine transient response, it is expected that the conclusions pertaining to the required density of joint network will remain valid regardless of the method used.

When designing a dynamic model, it is important that all possible flexibilities be included. Consider as an example a beam on two supports, loaded at the center. The most obvious deflection component is flexure, but others (e.g., shear, deformation of the cross section under the load and over the supports as well as deflection of supports themselves) may be equally important. Not all these factors are included in static stress calculations, because they are often unimportant. A disregard for those "hidden flexibilities" leads to overestimation of natural frequencies as compared with tested values, which is a commonplace occurrence. A crude estimate of those marginal deflection components is much better than disregarding them altogether.

The manner in which loads are applied often dictates the model properties. Loads may be classified according to how they are distributed in space and time. Two major types of spatial distribution can be distinguished: continuously applied and concentrated. Either of the two may be uniform or nonuniform and applied directly or through the base. Since the discrete models typically allow only concentrated loads, often a careful thought must be given to how to discretize the actual loading, which is almost always distributed.

With regard to time properties, the forcing functions may be classified as follows:

1. Periodic: (*a*) harmonic; (*b*) nonharmonic.
2. Nonperiodic.
3. Random prescribed by response spectrum.
4. Random prescribed by spectral density of forcing.

As we remember from Chapter 2, a nonharmonic, periodic function may be represented as a sum of its harmonic components. Some of the loads in category 2 may also be represented in this manner. A single-shock load of finite duration, for instance, may be treated as a series of identical shocks equally spaced in time, which makes it a periodic function. Representation of such loading via its harmonic components is not valid, of course, after the second shock begins.

Categories 3 and 4 may refer to the same dynamic event, an earthquake, for example. Yet, depending which category we choose for load definition, there will be decisive differences in the response calculations. The response spectrum method reduces the problem to a harmonic response procedure as described in Chapter 12. When spectral density of forcing is specified, the statistical approach outlined in Chapter 18 is used.

19B. Selection of Dynamic DOFs and Mass Matrix Generation. Designing a discrete model of a structure amounts to replacing the actual configuration by a series of joints (or nodes) connected with elastic elements. At certain places, such as a change of cross section or a point of application of a concentrated load, the joints can be placed without hesitation, but more often than not the analyst must decide on how many joints to use over which area.

Let us say that we have selected some preliminary network of joints and wish to associate a few dynamic DOFs with each of them. Any general-purpose computer program will then be capable of lumping a continuously distributed mass into those joints in the manner described in Sect. 9G (at least for the translational DOFs). The mass matrix obtained from this procedure (called a *lumped mass matrix*) will have a diagonal form, and the terms on the main diagonal will be joint masses and mass moments of inertia, if needed. It is not necessary, however, to place masses at every joint. Usually, the analyst will have some control over the lumping process and does not have to accept all static DOFs as being associated with inertia.

A *consistent mass matrix* is generated by a more sophisticated approach. Starting with a 4×4 beam stiffness matrix (e.g., Eq. 8.13), we can calculate the kinetic energy of the beam element, assuming the dynamic and static deflected patterns to be the same as long as the end displacements are equal. The entries of the mass matrix are then computed so that the true kinetic energy of the deformed pattern is preserved. This method gives us a full 4×4 mass matrix with all off-diagonal coefficients present, even though only the translational inertia is taken into account. Introducing the rotatory inertia yields another 4×4 matrix, and the total mass matrix is the sum of the two. (As in the lumped-mass approach, the rotatory terms are used only when they are significant.) The use of consistent mass approach gives us slightly higher natural frequencies, closer to the exact theoretical results. The additional calculating effort as well as the need for larger memory capacity, however, makes it hard to justify the use of this method.

Much of the effort in this chapter is expended on discretizing a continuous mass distribution. More specifically, the objective is to create a lumped inertia model that is adequate for the task. The examples of computer-generated solutions were obtained using an established program that employs a lumped inertia formulation of the mass matrix.

19C. Natural Modes of Vibration for Systems with Distributed Inertia.

Consider an axial bar with both ends fixed (Fig. 19.1). The solution of the equation of motion (15.4) with $q=0$ may be represented as

$$X_n = \sin \frac{\pi x}{\Lambda_n} \qquad (19.1a)$$

$$\Lambda_n = \frac{L}{n} = \frac{\pi c}{\omega_n} = \frac{\pi}{\omega_n} \left(\frac{EA}{m} \right)^{1/2} \qquad (19.1b)$$

for the nth vibratory mode (see Prob. 19-3). The dashed line in Fig. 19.1 is the plot axial deflection corresponding to $n=4$. This pattern is a harmonic function of time and, upon a proper selection of the

reference time point, may always be written as

$$u_n(x,t) = B \sin \frac{\pi x}{\Lambda_n} \sin \omega_n t$$

where B is an amplitude of deflection. The length of the bar has therefore been divided into four segments, vibrating independently from one another, each behaving like a bar with fixed ends. The deflected pattern, when projected across the bar, looks like four half-waves. The length of the half-wave is associated with the natural frequency by Eq. 19.1b.

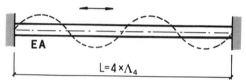

Figure 19.1 Lateral projection of vibration of axial bar.

The vibrating pattern in Fig. 19.1 is exceptional in the sense that the length is divided in a round number of full half-waves. For any other end conditions the picture is different, but at a location far enough from the ends, Eq. 19.1b is a good approximation as the mode number n becomes large. If the end fixity is thought of as the "perfect" boundary conditions for an axial bar, some other type of end constraint will disturb the ideal wave pattern, but its effect on the vibratory shape will manifest itself only over some distance from the ends.

What was said about vibrations of an axial bar with a distributed mass is also valid for a shaft, a cable, or a shear beam, provided changes in notation are made (see Prob. 19-5). For a flexural beam, the "perfect" boundary conditions occur when the ends are simply supported, since in the nth vibratory mode the length is then divided into n half-waves,

$$X_n = \sin \frac{\pi x}{\Lambda_n} \qquad (19.2a)$$

$$\Lambda_n^2 = \left(\frac{L}{n} \right)^2 = \frac{\pi^2}{\omega_n} \left(\frac{EI}{m} \right)^{1/2} \qquad (19.2b)$$

Each half-wave may be treated separately as a simply supported beam with length Λ_n. Some other end conditions will make the wave pattern deviate from what is given by Eq. 19.2a, but at a certain distance from the ends and for a sufficiently large n, this equation will again be a good approximation.

A beam in which not only flexural but also shear deformations are considered and where rotatory inertia due to beam depth is included is referred to as a *Timoshenko beam*. When the ends are simply supported, the mode shape expression is again given by Eq. 19.2a (See Prob. 19-8), but the relation between the frequency and the half-wave length becomes more involved, namely,

$$\frac{1}{\omega} = \frac{1}{\omega_b}\left[1 + \frac{\pi^2}{\Lambda_n^2}\left(\frac{EI}{GA_s} + \frac{I}{A}\right)\right]^{1/2} \quad (19.3)$$

where ω_b is the same as ω_n in Eq. 19.2b,

$$\omega_b = \left(\frac{\pi}{\Lambda_n}\right)^2\left(\frac{EI}{m}\right)^{1/2}$$

In most technical applications the use of a flexural beam concept is sufficient, because the influence of shear and rotatory inertia is minor. (Of these two factors, the first is several times more important than the second.) When the ratio of Λ to the depth of the beam is as low as 4, ω is about 10% less than ω_b. (This is for a solid section. For other cross-sectional types the influence of shear may be larger—see Sect. 14B.)

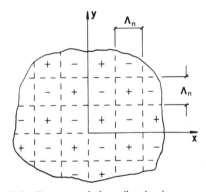

Figure 19.2 Fragment of plate vibrating in square pattern.

Two-dimensional bodies offer a greater variety of modal shapes. As far as plates are concerned, our attention will focus on a square pattern shown in Fig. 19.2. It is defined by

$$X_n(x, y) = \sin\frac{\pi x}{\Lambda_n}\sin\frac{\pi y}{\Lambda_n} \quad (19.4a)$$

$$\Lambda_n^2 = \frac{2\pi^2}{\omega_n}\left(\frac{D}{m}\right)^{1/2} \qquad D = \frac{Eh^3}{12(1-\nu^2)} \quad (19.4b)$$

for a plate of thickness h and mass per unit surface $m = h\rho$. The dashed lines in Fig. 19.2 go through the points having zero deflection. A plus sign in a square designates the surface convex toward the viewer at this selected instant. This pattern can appear in its pure form only in rectangular, simply supported plates that have the ratio of side lengths equal to a ratio of two integers. (A square plate is obviously included in this category.) For other boundary conditions, the square pattern can be a limit of a modal shape, when n is increasing and in a region away from the edges. Although for plates of some contours the square pattern may not be attained, it is still an important reference shape.

A *thick plate* is an extension of a flexural plate concept discussed thus far to include the effects of shear distortion and rotatory inertia. Since the second of the two influences is relatively minor, we will be concerned only with the additional effect of shear. One may show by a Dunkerley-type formula that the approximate natural frequency of a thick plate ω is given by

$$\frac{1}{\omega} = \frac{1}{\omega_b}\left[1 + \frac{\pi^2}{6(1-\nu^2)}\frac{1}{\Lambda_n^2}\frac{Eh_b^3}{Gh_s}\right]^{1/2} \quad (19.5)$$

in which h_b and h_s are calculated in such a manner that the terms Eh_b^3 and Gh_s are proportional to the bending and the shear stiffness, respectively. (For a solid, homogeneous plate with thickness h, $h_b = h$, and $h_s = h/1.2$). The ω_b is the same as ω_n in Eq. 19.4b,

$$\omega_b = \frac{2\pi^2}{\Lambda_n^2}\left(\frac{D}{m}\right)^{1/2}$$

The shear term in Eq. 19.5 remains small as long as the ratio Λ/h is large. When Λ/h is as low as 5, ω is about 10% less than ω_b. This figure may be used as a guide for solid, homogeneous plates only, because in a sandwich design, for example, the effect of shear is much more pronounced (see Prob. 19-11).

The concept of a half-wave length of a vibratory pattern has thus far been exclusively used with reference to sinusoidal deflected shapes. This limitation is not necessary, however. Any structure built of identical segments will, when properly excited, exhibit a deformed pattern repeated in space. If $X(x)$ is the deflection, the wavelength 2Λ is defined by

$$X(x + 2\Lambda) = X(x) \quad\text{and}\quad X(x + \Lambda) = -X(x)$$

$$(19.6)$$

That is, the change of x by Λ shifts us to a place where the sign of deflection is opposite, and if we increase x by another Λ, we reach a location where X is again the same. All discrete-element models can be regarded as segmented structures.

19D. Discrete Models of Continuous Systems. A flexural beam in Fig. 19.3a vibrating in a natural mode is represented by a discrete model with two lumped masses per half-wave (Fig. 19.3b) or one with three masses (Fig. 19.3c). In the first case a lumped mass is $M_0' = m\Lambda/2$, in the second $M_0 = m\Lambda/3$. It can be shown that in both instances the natural frequency computed for the discrete model (each half-wave treated as a simply supported beam) is less than 1% below the true value obtained for the continuous beam.

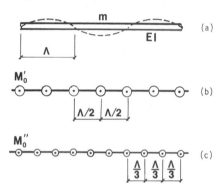

Figure 19.3   Flexural beam and its discrete models.

The comparison of dynamic response of both systems is not a very straightforward process. If some distributed, time-dependent load is applied to a continuous system, we must first convert it to a set of concentrated loads applied to the joints. (This can be done by the method outlined in Sect. 11F.) The bending moment and the shear force induced in the model can then be compared to the corresponding quantities in the original structure. Although the results are strongly influenced by the type of forcing, such comparisons indicate that the lumped models are not very accurate, especially if there is only one vibrating mass per half-wave, as in Fig. 19.3b. Analyzing an axial bar (or any other of the second-order elements described) we reach similar conclusions. This is when we place $M_0 = m\Lambda/2$ at each joint of a model with two segments per half-wave. Increasing the number of segments obviously improves the accuracy.

The plate vibrating as in Fig. 19.2 may be most simply modeled by the structure in Fig. 19.4, which

has only a single moving joint per square (encircled). When the concept of a tributary length introduced in Sect. 9G is extended to a *tributary area* (i.e., the plate area assigned to a joint on purely geometric basis), we find the mass at each joint to be $M_0 = m\Lambda^2/4$. The natural frequency of such a model is about 6% less than that of a plate with continuously distributed mass m. The response, however, is grossly underestimated (see Prob. 19-15).

The model in Fig. 19.4 will be referred to as having *two segments per half-wave*. The actual discrete model for computer analysis would have four plate elements per square (elements are not marked in Fig. 19.4) and an average of four masses per half-wave. (This is counted as follows: since a stationary line divides a nodal mass in half, one half-wave square has four quarters, four halves, and one whole mass assigned to it.) The next, more detailed representation has three segments per half-wave, which corresponds to nine discrete plate elements per square between the stationary lines.

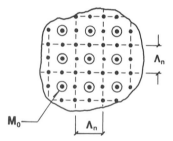

Figure 19.4   Modeling of plate vibrating in square pattern.

Returning to Fig. 19.3b, one may observe that the highest frequency of vibrations in the model corresponds to a shape in which the adjacent masses move in the opposite directions (Fig. 19.5). Based on symmetry, one may conclude that each segment Λ' acts as a simply supported beam. The mass lumped at a center is $M_0' = \frac{1}{2}m\Lambda = m\Lambda'$, where Λ is the half-wave length from Fig. 19.3a. A simple calculation shows that only $0.4928m\Lambda'$ should be placed at the center of the span to achieve the natural frequency equal to that of a continuous beam with a corresponding deflected pattern. The maximum model frequency is therefore underestimated by about 30%.

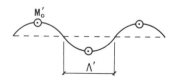

Figure 19.5   Highest mode of discrete model of a flexural beam.

This particular case illustrates a certain general property of discrete models. Only a portion of all possible modes of vibration can be related to those of the actual structure, with frequencies of both systems closely matching. The vibratory mode in Fig. 19.3*a* will still make sense for a beam in Fig. 19.3*b*, but any mode associated with a higher frequency will not be reliable for that beam. To generalize this let us introduce the terms *highest significant frequency* and *highest significant mode* with reference to a vibratory shape, which is still representative of the original structure. If we accept this as a guide, we find that, for a beam, about half of all frequencies in a discrete model should be rejected. For a plate, the highest significant frequency is associated with even a smaller number of vibrating masses (Fig. 19.4), and consequently a smaller portion of frequencies can be regarded as significant.

From the viewpoint of response calculations, we know that typically the higher the mode number, the lesser its contribution to overall response. The group of modes above the highest significant frequency contributes so little that it need not be included in a superposition.

When discussing a deformed shape we have shown, in Fig. 19.3 as well as in other places, a set of joints symmetrically placed with respect to the deflected line. Such nice symmetry will almost never take place in an actual response problem—the joints will not coincide with peaks and valleys, as a rule. First, because more than one mode shape is usually excited, and second because "nonperfect" boundary conditions will disturb the pattern.

Reviewing the libraries of finite elements employed by computer programs, we find that shear deformation is a standard feature available with most beam elements, but not with plates. When there is reason to believe that the shear component in the plate elements used is important, some adjustments in the geometry or material properties of the plate may have to be made.

19E. Models Designed for Frequency Range.
Suppose a harmonic force with frequency $f = 100$ Hz is applied to the structure under consideration. While it is difficult to formulate a general recipe for designing a computer analysis model, one thing is certain—the parts of the structure that are capable of vibrating at 100 Hz must be allowed to do so. By investigating what vibratory shapes the structure may assume and by using the formulas given earlier in this chapter, we place a sufficient number of lumped masses or

dynamic DOFs to make the response at 100 Hz possible. This is usually enough to assure a good representation of the mode shapes associated with frequencies below 100 Hz. When some other forcing frequencies, say 72 and 15 Hz are present, a model designed for 100 Hz is most likely to accurately respond to those lower frequencies.

This simple approach will often produce a satisfactory model. It may happen, however, that when high forcing frequencies are involved (say, 10 or more times the fundamental frequency of a structure), the network of joints may be denser than we can afford to analyze. To determine whether all those joints with masses are really needed, we must do some more preparatory work; namely, we must evaluate the relative importance of the modal contributions.

19F. Response Level as Criterion for Model Design.
When the maximum forcing frequency is not specified because of the nature of loading, we have to employ other criteria for joint spacing. Our usual interest is to have the maximum calculated response close to that experienced by the original structure. A simple example of how the density of the joint pattern and the accuracy of computation are related is provided by an axial bar fixed at one end and suddenly loaded at the other by force P_0 (Prob. 19-22). The maximum end displacement can be expressed as

$$u_{max} = B_1 \left(\frac{1}{1^2} + \frac{1}{3^2} + \frac{1}{5^2} + \cdots \right) \quad (19.7)$$

in which B_1 is some multiple of static deflection under P_0. The successive terms in parentheses are the contributions of the successive natural modes and the sum of the series is $\pi^2/8$. It is easy to verify that if we take, for example, the first five modes into account, we obtain about 96% of the true answer. When a discrete model is constructed, which is capable of closely predicting those modes, the same accuracy will be obtained. Using Eq. 19.1b with $n = 5$ gives us the half-wavelength of the highest mode, which should be accurately reflected by the model. (We may also use the half-wave expression in Prob. 19-3, based on the actual boundary conditions.)

The equation for the maximum axial force at the fixed end of the bar is

$$N_{max} = B_2 \left(\frac{1}{1} - \frac{1}{3} + \frac{1}{5} - \frac{1}{7} + \cdots \right) \quad (19.8)$$

in which B_2 is some multiple of P_0 and the sum of the series is $\pi/4$. Comparing this with Eq. 19.7, we find the convergence of the series to be slower, which means that if the same number of modes is taken into account, the answer will be less accurate than previously. This happens because the force is proportional to the first derivative of displacement and differentiation of a modal expansion always slows the convergence. For this reason a discrete model with a fixed number of contributing modes will be more accurate in terms of displacements than of a force or stress.

So much for axial bars. The conclusions achieved thus far may be applied, upon an appropriate change of terminology, to any of the three remaining, second-order elements, as defined in Chapter 15 (shaft, shear beam, and stretched cable). Let us now consider a flexural beam, supported at both ends and suddenly loaded at the center with a force P_0. The response in terms of modal contributions, at the end of the first half-period, is given by the following:

$$u_{max} = B_3 \left(\frac{1}{1^4} + \frac{1}{3^4} + \frac{1}{5^4} + \cdots \right) \quad \text{(deflection)} \quad (19.9)$$

$$u'_{max} = B_4 \left(\frac{1}{1^3} - \frac{1}{3^3} + \frac{1}{5^3} - \cdots \right) \quad \text{(slope)} \quad (19.10)$$

$$\mathcal{M}_{max} = B_5 \left(\frac{1}{1^2} + \frac{1}{3^2} + \frac{1}{5^2} + \cdots \right) \text{(moment)} \quad (19.11)$$

$$V_{max} = B_6 \left(\frac{1}{1} - \frac{1}{3} + \frac{1}{5} - \cdots \right) \quad \text{(shear)} \quad (19.12)$$

(This example differs from the previous one in that the antisymmetric modes are not excited. Adding up the first three terms of the series, for example, is equivalent to including the contributions of the first five modes.) The formulas clearly show that the best accuracy for a particular model will be obtained for a deflection and the worst for a shear force (or shear stress).

The convergence criteria for plates could be derived in a similar manner, but their application would not be easy because of the second spatial dimension that is added to the picture. A rectangular plate vibrating in a natural mode will have, in general, a different number of half-waves along the length and width. Since there are strong similarities between beams and plates, we can use the beam criteria for joint spacing along both directions of the plate surface.

As mentioned before, modeling is affected by distribution of load in space and time. The most favorable spatial distribution is in the shape of a natural mode—the structure will then respond in that mode only (see Prob. 15-11, e.g.). A concentrated force, on the other hand, is most unfavorable, because a superposition of many mode shapes is needed to approximate it. Somewhere in between there is a linearly distributed force; although the response is expressed by an infinite series, the convergence is better than that of a concentrated force. In case of a bar under a uniformly distributed loading, the series in Eq. 19.7, for instance, will have cubes instead of squares of integers in the denominators (see Ref. 2, p. 386). This allows us to use fewer modes to obtain a certain accuracy, which means fewer dynamic DOFs in a model. (A discrete model accepts only concentrated joint forces. To minimize the error of the lumping operation when the original load is distributed, we should use an adjusted Eq. 11.18, choosing the deflected pattern to be that of the predominating mode.)

We have thus far been focusing our attention on forces directly applied to a structure. When excitation takes place as a result of the base motion (that base often being another structure itself), the forces are imparted to our system by the attachment points. This clearly suggests an excitation by concentrated loads, but there is another way of looking at the problem, according to Chapter 12. If we limit our interest to the displacements relative to the moving base (and this displacement component dictates the dynamic stress), we can replace the base movement by a kinetically equivalent inertia load. This approach reduces the effect of the base motion on a distributed loading, uniform or nonuniform depending on mass distribution and on whether translation or rotation of the base is involved.

A response spectrum is an indirect specification of external loading. It is most often employed when the base of a structure performs a complicated, irregular motion. A spectrum allows us to compute the individual mode responses, but it does not say how to combine them—this must be separately specified to make the load definition complete. Of the two basic summation methods, the absolute-value sum and the RSS as described in Sect. 12C, the second offers a considerably faster convergence. For example, a simply supported beam subjected to a lateral, constant-value response acceleration spectrum has the maximum bending moment expressed by

$$\mathcal{M}_{max} = B_7 \left(\frac{1}{1^3} + \frac{1}{3^3} + \frac{1}{5^3} + \cdots \right) \quad (19.13)$$

when absolute-value summation is employed and

$$\mathfrak{M}_{max} = B_7 \left(\frac{1}{1^6} + \frac{1}{3^6} + \frac{1}{5^6} + \cdots \right)^{1/2} \quad (19.14)$$

when RSS is used. This means that a model will have fewer dynamic DOFs if the second method is involved. (If the model is already established, there will be fewer modes to be computed and employed in the response calculation.)

A random load prescribed by the spectral density of forcing is a more involved topic, but the RMS response may again be presented as a series of modal contributions.

One of the features that distinguishes the step load from other loading patterns is the infinitely large slope of the force-time curve. For any real shock event, this slope must have a finite value. The reason for using the step load so often in this section is the mathematical convenience; moreover, it is a conservative way to approach model design. Taking the slope of the force-time curve into account may lead to a smaller number of the dynamic DOFs in the model.

All the response formulas in this section were made in reference to some point of a structure, which experiences the maximum response under a particular load pattern. In most real structures, however, there are stronger and weaker points, and the *largest* response is not always the *most critical* response. The convergence criteria in other locations may be different, and so may be the conclusions regarding the number of modes, which ought to be involved in the calculation. Consider, for example, a simply supported beam, loaded antisymmetrically by concentrated forces. Instead of $1, 3, 5, \ldots,$ in Eq. 19.11, we would have $2, 4, 6, \ldots,$ in the denominators and more modes would have to be involved to obtain the same portion of the true maximum bending.

For most types of loading considered here the first few modes were decisive as far as magnitude of response was concerned. It can be clearly seen that including more and more mode shapes to get closer and closer to the true response is an effort with quickly diminishing returns. A common-sense procedure suggested here is to use only a few modes and multiply the response by some uncertainty factor, larger than unity.

When analyzing structure response to the step load, the attention is usually focused at the time point $t = \tau_1/2$ (i.e, the end of the first half-period of the

fundamental mode). This instant (or the vicinity of it) is characterized not only by the maximum response in the first mode, but also by a relatively orderly deformation pattern. It takes many more mode shapes to analyze the response at, say, $t = \tau_1/20$. When such a calculation is attempted, we refer to it as the analysis of *strongly transient response*. Included in this class may also be other force-time functions associated with sudden imposition or removal of the applied load. (For example, what happens *just after* the harmonic loading is applied?)

When designing a model based on the response criterion, we usually do not concern ourselves with the discretization error, already mentioned. The chief reason is that this error is meaningful only in the higher modes, which contribute little to the magnitude of the response.

19G. Special Requirements for Strongly Transient Response. Sometimes it is important to learn about a deformed pattern and a stress distribution in the vicinity of a suddenly applied force, just after the force was applied. Before attempting to find an answer by means of computer-aided analysis, we must have some idea of how fast the deformation spreads across the structure. Although the subject of *elastic wave propagation* is outside the scope this book, it will be helpful to quote some results. Consider, for example, the bar in Fig. 19.6. When a axial force $P(t)$ is applied in the form of a step function, the deformation after time t is spread over a segment whose length is ct. The force in this segment is constant, equal to the applied force P_0. The static approach may be employed in calculating the end displacement,

$$u_m(t) = \frac{P_0 ct}{EA} \quad (19.15)$$

and differentiation with respect to time gives us the velocity of the end

$$v_m = \frac{P_0 c}{EA} = \frac{P_0}{A(E\rho)^{1/2}} \quad (19.16)$$

The moving left end of the deformed segment is called the *wave front* of an elastic wave, spreading with velocity c (refer to Sect. 15A). The action of the applied load amounts to stretching a bar whose length changes with time. The period of time needed for the wave to traverse the length of bar, $t_L = L/c$ is the same as one-quarter of the fundamental period of vibration. Our interest in this section is limited to the interval of time well before the elastic wave has spread over the entire length L. (More information on this subject may be found in Ref. 3, Sect. 142.)

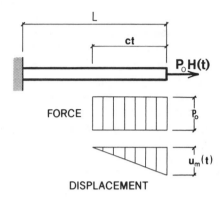

Figure 19.6 Transient deformation pattern of axial bar.

Everything said about the axial bar applies equally well to a shear beam. Figure 19.7 presents two such beams, one semi-infinite and the other infinite. (Since we said we were not interested in a time long enough to allow the deformation to reach the boundary, we could as well move the boundaries to infinity.) The lateral displacements are

$$u_{s1} = \frac{P_0 c_3 t}{GA_s} \quad \text{or} \quad u_{s2} = \frac{P_0 c_3 t}{2GA_s} \quad (19.17)$$

for Figs. 19.7a and 19.7b, respectively. The constant defined by $c_3^2 = GA_s/m$ is the propagation velocity. The shear force in the beam remains constant, provided the load is not changing and the wave front does not reach a boundary.

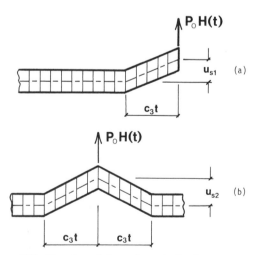

Figure 19.7 Transient deformation of shearbeam: (a) semi-infinite; (b) infinite.

In a purely flexural beam, the situation is more complicated. Instead of one elastic wave, there are many, each with a different speed of propagation. Problem 15-13 serves as a good example of the type of deflected line to expect if a step load is applied. A detailed investigation shows (Ref. 16, p. 1585) that we can distinguish a *prime flexural wave* or a portion of a beam significantly deformed by bending at any time point following the impact. The length l_b of this wave, depicted in Fig. 19.8, is

$$l_{b1} = \left(\frac{4EI}{m} \right)^{1/4} t^{1/2}$$

or

$$l_{b2} = 1.96 \left(\frac{EI}{m} \right)^{1/4} t^{1/2} \quad (19.18)$$

for a semi-infinite and an infinite beam, respectively. Since l_b is not proportional to time, the propagation mechanism is obviously different from that of a shear beam. The lateral deflection is computed assuming the segment deformed by the flexural wave to have its end(s) fixed;

$$u_{b1} = \frac{P_0 l_{b1}^3}{3EI} = \frac{0.9428 P_0 t^{3/2}}{(EI)^{1/4} m^{3/4}} \quad (19.19a)$$

$$u_{b2} = \frac{P_0 l_{b2}^3}{24EI} = \frac{0.3137 P_0 t^{3/2}}{(EI)^{1/4} m^{3/4}} \quad (19.19b)$$

for Figs. 19.8a and 19.8b, respectively.

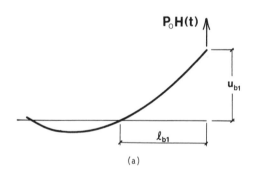

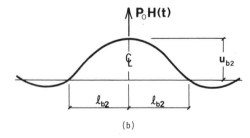

Figure 19.8 Prime flexural wave in a beam: (a) semi-infinite; (b) infinite.

A real beam exhibits both flexural and shear flexibility. A reasonable approximation to the actual displacement response is to superpose both components after separately calculating them.

SOLVED PROBLEMS

19-1 In (a) there is a fragment of a prismatic beam with the second area moment equal to I. In (b) there is a discrete model of the same beam, which has translational mass mL and rotatory inertia ρIL at each node. Determine how accurately this discrete model represents the inertia properties of the beam fragment as far as a motion in the xz-plane is concerned.

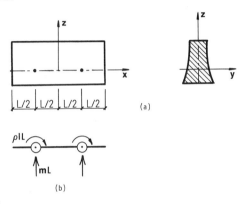

Since the joint spacing is L, the mass per joint is $mL = \rho AL$, which shows that the translational inertia is properly reflected by the model. The rotatory inertia about the center of fragment is

$$J = J_1 + J_2$$

where J_1 is the moment of inertia component relative to the xy-plane and J_2 is the same with respect to the yz-plane. From standard formulas of mechanics we find that

$$J_1 = 2\rho IL \qquad J_2 = \frac{1}{12}(2mL)(2L)^2 = \frac{2}{3}mL^3$$

In the model beam the first component is the same, while the second is

$$J_2' = 2(mL)\left(\frac{L}{2}\right)^2 = \frac{1}{2}mL^3$$

This component of angular inertia is therefore underestimated by one-fourth by the model. This, however, is not the only way of looking at the problem. We may also think of the fragment as having a full joint at the center and one-half a joint at each end. Nothing will change with regard to the inertia except J_2. In this new approach we have a lumped mass of $mL/2$ at either end, therefore

$$J_2' = 2\left(\frac{mL}{2}\right)L^2 = mL^3$$

This time the inertia component is overestimated by one-third. The average of these two methods give us a result quite close to the actual J_2, which we deem satisfactory.

19-2 The two discs on a continuous shaft vibrate with the same amplitude and in (1) the same or (2) in the opposite direction. Assume that the rotation changes linearly between the discs, and calculate what portion of the distributed inertia to assign to each disc to preserve the kinetic energy of the system. If we decide to place a node at the midpoint of L, how would the inertia be reassigned?

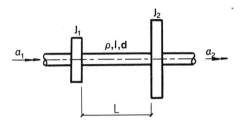

Assuming that the motion of discs is a harmonic function of time will not make our reasoning less general. We can therefore make this assumption and use Eq. 11.21 to calculate the kinetic energy.

$$T = \frac{\omega^2}{2}\rho I_0 \int \alpha^2(x)\, dx$$

where ρI_0 is the angular inertia per unit length and $\alpha(x)$ is the displacement function assumed to be linear with respect to x. When both discs rotate in the same direction, we have $\alpha_1 = \alpha_2 = \vartheta$ =constant, and the entire shaft segment is moving like a rigid body. We can therefore assign

$$\Delta J = \tfrac{1}{2}\rho I_0 L$$

to each of the discs. When the discs are out of phase, the displacement function is

$$\alpha = \vartheta\left(1 - \frac{2x}{L}\right)$$

It is easy to find that

$$\int_0^L \alpha^2\, dx = \tfrac{1}{3}L\vartheta^2$$

and since ϑ may be taken as unity,

$$T = \frac{\omega^2}{2} \times \frac{1}{3}\rho I_0 L = \frac{\omega^2}{2}(2\Delta J)$$

where ΔJ is the angular inertia

$$\Delta J = \tfrac{1}{6}\rho I_0 L$$

added to each disc to preserve the kinetic energy. If we place an extra joint midway between the disc, half the shaft inertia will be assigned to it in the symmetric motion ($\alpha_1 = \alpha_2 = \vartheta$) and a negligible amount will be assigned in the antisymmetric case. If nothing is known about the actual displacement function, we can place there an average value of the two cases:

$$\Delta J \approx \tfrac{1}{4}\rho I_0 L$$

The inertia assigned to the discs from this shaft segment must then be reduced.

19-3 Consider an axial bar in Prob. 15-1 vibrating in one of the natural modes defined by

$$X_n = \sin \frac{n\pi x}{2L} \qquad \omega_n = \frac{n\pi c}{2L} \qquad \text{for} \quad n = 1, 3, 5, \dots .$$

How can the half-wavelength Λ_n, defined in Sect. 19C, be expressed by the associated natural frequency ω_n? What is the limiting value of Λ_n when n is increasing?

The expression for the natural mode may be rewritten:

$$X_n = \sin \frac{\omega_n x}{c}$$

According to the definition of a half-wave, the increase in x by $2\Lambda_n$ should change the argument of X_n by one period, which is 2π in this case:

$$\frac{\omega_n}{c}(x + 2\Lambda_n) - \frac{\omega_n}{c}x = 2\pi \qquad \therefore \; \Lambda_n = \frac{\pi c}{\omega_n}$$

To obtain the answer to the second question, we substitute ω_n from the original definition into the last expression,

$$\Lambda_n = \frac{\pi c}{n\pi c/(2L)} = \frac{2L}{n} \qquad n = 1, 3, 5, \dots$$

To get a better picture we put $n = 2r - 1$,

$$\Lambda_n = \frac{L}{r - \tfrac{1}{2}} \qquad r = 1, 2, 3, \dots$$

and conclude that for a large r, Λ_n will be only slightly larger than L/r. We can easily find that for some other boundary conditions Λ_n, may be less than L/r, while for others the condition

$$\Lambda_r = \frac{L}{r}$$

holds exactly true, which means that in each mode there is a round number of half-waves along the length. This happens, for example, when both ends of the bar are fixed.

19-4 An axial bar with stiffness EA and distributed mass m is vibrating in a natural mode which has half-wave length Λ and natural frequency

$$\omega = \frac{\pi c}{\Lambda} = \frac{\pi}{\Lambda}\left(\frac{E}{\rho}\right)^{1/2}$$

In a discrete model, the mass is to be lumped so that

there are two segments per Λ. How big should each lumped mass be so that the natural frequency is properly reflected? If $M_0 = m\Lambda/2$ is placed at a joint, what is the error in the modal frequency?

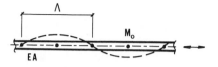

Each half-wave may be treated as a bar with fixed ends and our reasoning can be limited to only one bar with length Λ. To construct a discrete model, we have to place mass M_0 at the center of weightless bar. Since the bar is symmetric about the center and the motion is antisymmetric, there is no interaction at the center and the bar (including M_0) may be cut in half, each half independent from the other. We find

$$k = \frac{2EA}{\Lambda} \qquad M = \frac{M_0}{2} \qquad \therefore \; \omega^2 = \frac{4EA}{M_0\Lambda}$$

Equating this with the actual frequency determines M_0,

$$M_0 = \frac{4}{\pi^2}\rho A \Lambda = \frac{4}{\pi^2} m\Lambda \approx 0.4053 m\Lambda$$

When we use $M_0 = 0.5 m\Lambda$ instead, we obtain the frequency, which is 0.9 of the true value, which means the error in frequency is then 10%.

19-5 Calculate the natural frequencies of several prismatic elements with distributed mass: (1) axial bar, (2) shaft, (3) stretched cable, and (4) shear beam, all of which have both ends fixed. Discuss the length Λ_n of the modal wave.

From Sects. 15A and 15B as well as from Prob. 15-15 we know that the differential equations of free vibrations for all four types of element are the same except for the constants:

1. Axial bar,

$$c^2 = \frac{E}{\rho}$$

2. Shaft

$$c_1^2 = \frac{GC}{\rho I_0}$$

3. Stretched cable,

$$c_2^2 = \frac{N}{\rho A}$$

4. Shear beam,

$$c_3^2 = \frac{GA_s}{\rho A}$$

Fixing the ends has the same significance for all four types, namely, that the displacements in the direction of interest are excluded. It is therefore sufficient to determine the vibratory properties of one of them, for example, a shaft with fixed ends, and the rest will follow automatically.

When the left end at $x=0$ is fixed, the characteristic equation is (Prob. 15-3)

$$X(x) = D_2 \sin \frac{\omega x}{c_1}$$

and for $X(L) = 0$, we have

$$\sin \frac{\omega L}{c_1} = 0 \quad \text{or} \quad \omega_n = \frac{n\pi c_1}{L}$$

while the mode shapes are given by

$$X_n = \sin \frac{n\pi x}{L} \quad \text{for} \quad n = 1, 2, 3, \dots$$

The relation between the frequency and the half-wavelength in a particular mode is similar to that in Prob. 19-3, while (unlike Prob. 19-3) the length is divided into a round number of half-waves,

$$\Lambda_n = \frac{\pi c_1}{\omega_n} \quad \text{and} \quad \Lambda_n = \frac{L}{n}$$

The results for the remaining three elements are obtained by replacing c_1 with c or c_2 or c_3.

19-6 In Prob. 7-12 the natural frequencies of three weightless beams, each with a lumped mass M_0 at the center, were determined. The boundary conditions were: simply supported (S-S), fixed at both ends (F-F), and one end fixed, one supported (F-S). The corresponding fundamental frequencies of beams with a distributed mass are determined by

$$\omega^2 = \frac{\alpha^4}{L^4} \frac{EI}{m}$$

where
$$\alpha = \pi \quad \text{(S-S)}$$

$$\alpha = 4.73 \quad \text{(F-F)}$$

$$\alpha = 3.9266 \quad \text{(F-S)}$$

What portion of the total mass mL should M_0 be in each case to preserve the fundamental frequency? How do the end conditions influence mass lumping in a discrete model?

The formulas for beams with lumped masses may be presented as

$$\omega^2 = \frac{\beta EI}{M_0 L^3}$$

Equating both expressions, we have

$$\frac{\alpha^4}{mL} = \frac{\beta}{M_0} \quad \text{or} \quad M_0 = \frac{\beta}{\alpha^4} mL$$

Our three sets of end conditions give us

$$\beta = 48 \qquad M_0 = 0.4928 mL \quad \text{(S-S)}$$

$$\beta = 192 \qquad M_0 = 0.3836 mL \quad \text{(F-F)}$$

$$\beta = 109.71 \qquad M_0 = 0.4615 mL \quad \text{(F-S)}$$

Since a simply supported beam in its first mode represents a half-wave of length Λ, placing a lumped mass $0.5m\Lambda$ at the center gives us a frequency that is

$$\left(\frac{0.4928}{0.5} \right)^{1/2} = 0.9928$$

of the true value. This error is quite small and the lumping is acceptable for these end conditions. A beam that is fixed at both ends may have its first mode approximately described by

$$X(x) = 1 - \cos \frac{2\pi x}{L}$$

(see Prob. 16-11), which would take three or four segments over the length to adequately model it. Using only two segments with $0.3836 mL$ at the center would give a correct natural frequency, but the error in response would be quite large. We therefore see that the consequence of having the end conditions different from simple supports is the need for more joints in the model.

19-7 A simply supported, flexural beam with the following properties

$$E = 29 \times 10^6 \text{ psi} \qquad I = 4009 \text{ in.}^4$$

$$L = 242.5 \text{ in.} \qquad m = 0.02811 \text{ Lb-s}^2/\text{in.}^2$$

was analyzed by a computer program and the following results were noted:

$f_1 = 54.320$ Hz   (first frequency)

$f_{25} = 33,704$ Hz   (highest significant frequency, per Sect. 19D as determined by inspection of deflected shapes)

$f_{49} = 95,140$ Hz   (maximum model frequency)

The model of beam had 50 segments. How do these results compare with the expectations based on Sects. 19C and 19D with regard to the fundamental, the significant and the maximum frequency of the model?

By Eq. 19.2b, the first frequency of the actual beam is

$$f_1' = \frac{\pi^2}{2\pi L^2} \left(\frac{EI}{m} \right)^{1/2} = 54.323 \text{ Hz}$$

The theoretical frequency corresponding to the half-wavelength $\Lambda = L/25$ would be

$$25^2 \times 54.323 = 33,952 \text{ Hz}$$

According to Prob. 19-6, the lumped-mass model should give us

$$f'_{25} = 0.9928 \times 33,952 = 33,708 \text{ Hz}$$

as the highest significant frequency. The maximum model frequency (Fig. 19.5) uses the same mass as f'_{25}, but only one-half of the length. Noting that for a beam with a lumped mass the stiffness is inversely proportional to the cube of length, we get

$$f'_{49} = 2^{3/2} \times 33,708 = 95,341 \text{ Hz}$$

This is the only of the three frequencies involved for which the error of manual prediction is noticeable (0.2%) and the end conditions must be blamed for the discrepancy. To have f_{49} and f'_{49} closer to each other, the segments adjacent to the beam ends would have to be only half as long as the remaining ones. (This is to allow the shape depicted in Fig. 19.5.) A review of the mode shape associated with f_{49} shows that it closely follows Fig. 19.5 near the center, but the similarity diminishes toward the beam ends.

19-8 The equation of free vibrations of a simply supported Timoshenko beam with distributed inertia is satisfied by the function

$$u_n(x,t) = (B_{1n}\cos\omega_n t + B_{2n}\sin\omega_n t)\sin\frac{n\pi x}{L}$$

for $n = 1, 2, 3, \ldots$ (see Ref. 2, Sect. 5.12).

One may check by reviewing Sect. 15C and Prob. 15-5 that the solution for a purely flexural beam has the same form, which indicates that the only difference may lie in the magnitude of constants, especially ω_n. Using the approximate formula from Prob. 14-6, derive an equation for the natural frequency of a Timoshenko beam vibrating in the nth mode.

_____

The approximate frequency is given by

$$\frac{1}{\omega^2} = \frac{1}{\omega_{11}^2} + \frac{1}{\omega_{22}^2} + \frac{1}{\omega_{33}^2}$$

The fictitious component frequencies were defined in Prob. 14-6. For the flexural beam

$$\omega_{11}^2 = \left(\frac{n\pi}{L}\right)^4 \frac{EI}{A\rho}$$

from Prob. 15-5. For a shear beam we refer to Prob. 19-5,

$$\omega_{22}^2 = \left(\frac{n\pi}{L}\right)^2 \frac{GA_s}{\rho A}$$

In Prob. 14-7 a flexural beam with rotatory inertia was analyzed. When the mass moment of inertia per unit length, $j = \rho I$ is in-

volved, we have

$$\omega^2 = \frac{\pi^2 E}{L^2 \rho}$$

at the first mode and at the nth,

$$\omega_{33}^2 = \left(\frac{n\pi}{L}\right)^2 \frac{E}{\rho}$$

Substituting all the fictitious frequencies into initial formula and introducing the half-wave length $\Lambda_n = L/n$, we get

$$\frac{1}{\omega^2} = \left(\frac{\Lambda_n}{\pi}\right)^4 \left(\frac{A\rho}{EI}\right)\left[1 + \frac{\pi^2}{\Lambda_n^2}\left(\frac{EI}{GA_s} + \frac{I}{A}\right)\right]$$

When this equation is written the same as Eq. 19.3, we realize that the content of the square brackets may be looked on as a correction term with respect to the basic flexural frequency.

19-9 The fundamental frequencies of a circular plate with radius R are

$$\omega = \frac{4.977}{R^2}\left(\frac{D}{m}\right)^{1/2} \quad \text{and} \quad \omega = \frac{10.22}{R^2}\left(\frac{D}{m}\right)^{1/2}$$

for the simply supported and clamped-edge conditions respectively. Suppose we want to create a simplified model, in which the plate is a weightless, elastic element with a lumped mass at the center. What fraction of the total plate mass $\pi R^2 m$ should be placed at the center in each case? (Note that only the flexural deformations are considered.) Refer to Prob. 7-14 and use $\nu = 0.3$.

_____

Equating the lumped-mass frequencies found in Prob. 7-14 with those given above, the center mass M_s for simply supported edges is

$$\frac{16\pi(1+0.3)D}{(3+0.3)R^2 M_s} = \frac{4.977^2}{R^4}\frac{D}{m}$$

$$\therefore M_s = 0.2545(\pi R^2 m)$$

Similarly, the center mass M_c for the clamped-edge plate is found from

$$\frac{16\pi D}{R^2 M_c} = \frac{10.22^2}{R^4}\frac{D}{m}$$

$$\therefore M_c = 0.1532(\pi R^2 m)$$

9-10 Using the results of Prob. 14-2, calculate an approximate natural frequency of a circular shear plate. Next, by comparing the formulas for frequencies of circular and square flexural plates (both shapes simply supported), develop an equation for a square shear plate.

The reference problem gives us the static deflection at the center of a shear plate under a uniform surface load q. If the gravity loading $q=mg$ is applied, we have

$$u_{st} = \frac{mgR^2}{4Gh_s}$$

The natural frequency may be found by the static deflection method

$$\omega^2 \approx \frac{g}{u_{st}} = \frac{4Gh_s}{R^2 m} \quad \text{or} \quad \omega = \frac{2}{R}\left(\frac{Gh_s}{m}\right)^{1/2}$$

From Prob. 19-9 and Eq. 19.4b, the flexural frequencies are

$$\omega = \frac{4.977}{R^2}\left(\frac{D}{m}\right)^{1/2} \quad \text{and} \quad \omega = \frac{2\pi^2}{L^2}\left(\frac{D}{m}\right)^{1/2}$$

for the circular plate of radius R and the square plate with side length L, respectively. To relate these equations, we put $L=2R$ in the second one. The square plate expression then becomes

$$\omega = \frac{4.935}{R^2}\left(\frac{D}{m}\right)^{1/2}$$

This result is interpreted as follows. The natural frequency of a square flexural plate is only slightly (1%) lower than that of a round plate inscribed in it. Extrapolating this result to shear plates, we obtain the frequency of a square plate ($R=L/2$) as

$$\omega = \frac{4}{L}\left(\frac{Gh_s}{m}\right)^{1/2}$$

One may note that both shear plate formulas developed here are based on the static deflection method, which underestimates the natural frequency.

19-11 Determine the influence of shear deformation on natural frequency of a sandwich plate. In particular, calculate the ratio ω/ω_b (the natural frequency to the frequency based on flexural properties alone) for the following proportions:

$$\frac{\text{half-wave length}}{\text{total plate thickness}} = \frac{\Lambda}{h} = 1, 2, 3, 4, 5, 10, \text{ and } 20$$

Use the same properties of facing sheets and sandwich core as in Prob. 14-8 and put $\nu=0.325$.

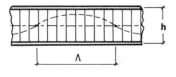

Symbol h_b in Eq. 19.5 stands for the thickness of a homogeneous plate, which is effective in bending. The second area moment of such a plate would be $h_b^3/12$ per unit width. Equating this with the analogous quantity found in the reference problem, we have

$$I = 0.03308 = \frac{h_b^3}{12} \quad \therefore\ h_b = 0.7349 \text{ in.}$$

or $\qquad h_b = 0.6124h.$

The shear thickness is

$$h_s = 1.1 \text{ in.} = 0.9167h$$

The purpose in writing both thicknesses in this manner was to isolate the ratio Λ/h in Eq. 19.5, to make it an independent variable. Using this equation we keep in mind that E refers to the material of the facing sheets, while G is used for the core:

$$\frac{\omega_b}{\omega} = \left[1 + \frac{\pi^2}{6(1-0.325^2)}\frac{h^2}{\Lambda^2}\frac{10.6\times10^6\times0.6124^3}{68,000\times0.9167}\right]^{1/2}$$

$$= \left(1 + \frac{71.83h^2}{\Lambda^2}\right)^{1/2}$$

The following table shows the results of computation.

$\frac{\Lambda}{h}$:	1	2	3	4	5	10	20
$\frac{\omega}{\omega_b}$:	0.1172	0.2297	0.3337	0.4268	0.5081	0.7629	0.9207

This table illustrates the strength of the influence of shear distortion on this type of plate. For short waves, say $\Lambda/h<5$, the bending deformation has less influence on frequency than shear, which is a reversal of what we are accustomed to think of as a normal situation.

19-12 A sinusoidal loading, in (a)

$$q = q_0 \sin\frac{\pi x}{\Lambda}$$

is applied to a continuous beam. Two finite-element models of the beam are considered, one with three and one with two segments per half-wave, Figs. 19.3c and 19.3b, respectively. What equivalent concentrated loads, must be applied to the respective models, so that the work performed is the same in every case?

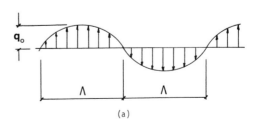

(a)

Based on symmetry alone, we can determine there is zero displacement and bending moment at the ends of half-waves of load, which means that the deflected shape is as depicted in Fig. 19.3a. Each half-wave Λ may therefore be treated as a beam on simple supports. Since this is a static problem, we put $\ddot{u}=0$ in Eq. 15.15 and obtain

$$EIu''''=q=q_0\sin\frac{\pi x}{\Lambda}$$

as a differential equation of the deflected line. Integrating four times, we get the deflected shape as

$$u=\left(\frac{\Lambda}{\pi}\right)^4\frac{q_0}{EI}\sin\frac{\pi x}{\Lambda}$$

The work performed by the load is

$$U=\frac{1}{2}\int_0^\Lambda q(x)u\,dx=\left(\frac{\Lambda}{\pi}\right)^4\frac{q_0^2}{2EI}\int_0^\Lambda\sin^2\frac{\pi x}{\Lambda}dx=\left(\frac{\Lambda}{\pi}\right)^4\frac{\Lambda q_0^2}{4EI}$$

The equivalent forces are shown in (b). The joint deflections are, respectively,

$$u_2=\frac{5P_2\Lambda^3}{162EI}\quad\text{and}\quad u_1=\frac{P_1\Lambda^3}{48EI}$$

and the work expressions,

$$U_2=2\times\tfrac{1}{2}P_2u_2\quad\text{and}\quad U_1=\tfrac{1}{2}P_1u_1$$

Equating these with the formula for U, we obtain

$$P_2=0.2884q_0\Lambda\quad\text{and}\quad P_1=0.4964q_0\Lambda$$

These forces must be applied to the discrete models to perform the same work as the original load performs on the continuous structure.

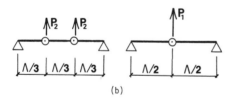

(b)

19-13 Two beam models were considered in Prob. 19-12 and the concentrated forces, equivalent to the actually applied, distributed load were determined. How accurately will those models predict the maximum bending and shear? To what extent are the conclusions valid for a beam vibrating in the natural mode?

Once the deflection formula is known, we can determine bending $\mathfrak{M}$ and shear V from Eqs. 15.16 and 15.17:

$$\mathfrak{M}=EIu''=-\left(\frac{\Lambda}{\pi}\right)^2q_0\sin\frac{\pi x}{\Lambda}$$

$$V=EIu'''=-\left(\frac{\Lambda}{\pi}\right)q_0\cos\frac{\pi x}{\Lambda}$$

The extreme values are, respectively,

$$\mathfrak{M}_m=\frac{q_0\Lambda^2}{\pi^2}\quad\text{and}\quad V_m=\frac{q_0\Lambda}{\pi}$$

The corresponding quantities for the discrete models are found from (b) in the reference problem. For three segments per half-wave:

$$\mathfrak{M}_m=\frac{P_2\Lambda}{3}=\frac{(0.2884q_0\Lambda)\Lambda}{3}=0.09613q_0\Lambda^2$$

$$V_m=P_2=0.2884q_0\Lambda$$

For two segments we have:

$$\mathfrak{M}_m=\frac{P_1\Lambda}{4}=\frac{(0.4964q_0\Lambda)\Lambda}{4}=0.1241q_0\Lambda^2$$

$$V_m=\frac{P_1}{2}=0.2482q_0\Lambda$$

The following are the errors of discrete models as compared with the original structure.

$$\Delta\mathfrak{M}=-5.1\%\qquad\Delta V=-9.2\%\qquad\text{(three segments)}$$

$$\Delta\mathfrak{M}=22.5\%\qquad\Delta V=-22\%\qquad\text{(two segments)}$$

The curve in Fig. 19-12a may be thought of as a vibrating beam on equally spaced supports. (Problem 15-5 would then relate to one span.) The amplitude of that curve is determined by a time function, so that we have

$$u(x,t)=f(t)\sin\frac{n\pi x}{L}$$

The inertia load, $-m\ddot{u}(x,t)$, is therefore also a sinusoid, differing from the deflected line only by a constant factor. This indicates the validity of static reasoning in Prob. 19-12, as well as here, for vibratory motion in the natural mode.

19-14 The surface of a square, simply supported plate is subjected to a uniform pressure q_0 varying harmonically in time with frequency $\Omega=\omega_1$, where ω_1 is the fundamental frequency of the plate. The magnification factor (which depends on damping) is μ_1. Find the expressions for the maximum deflection, the bending moment in the plate, and the lateral force applied to the plate in the fundamental mode of vibration.

The first mode is expressed by (Eq. 19.4)

$$X_1(x,y)=\sin\frac{\pi x}{a}\sin\frac{\pi y}{a}$$

The component of the total, uniform load q_0 affecting this mode may be found on the basis of Sect. 15E, when extended to plates:

$$\int_0^a\int_0^a X_1(x,y)q\,dx\,dy=q_1\int_0^a\int_0^a X_1^2(x,y)\,dx\,dy$$

where $q = q_0 \sin \omega_1 t$. Integration gives us

$$\left(\frac{2a}{\pi}\right)\left(\frac{2a}{\pi}\right) q_0 = q_1 \left(\frac{a}{2}\right)\left(\frac{a}{2}\right)$$

or

$$q_1 = \frac{16}{\pi^2} q_0 \sin \omega_1 t$$

Let us consider a static problem of a sinusoidal load with center magnitude $16 q_0/\pi^2$ applied to the plate. The solution is found in Ref. 15, Sect. 27, upon some minor adjustments in notation. This also gives the answer to our dynamic problem when we multiply that solution by the magnification factor μ_1:

$$u_{\max} = \mu_1 \left(\frac{16}{\pi^2}\right) \frac{q_0 a^4}{4\pi^4 D} = \mu_1 \frac{4 q_0 a^4}{\pi^6 D} \quad \text{(center deflection)}$$

$$\mathfrak{M}_{\max} = \mu_1 \left(\frac{16}{\pi^2}\right) \frac{(1+\nu) q_0 a^2}{4\pi^2} = \mu_1 \frac{4(1+\nu) q_0 a^2}{\pi^4}$$

(center bending moment per unit width)

The resultant lateral force is

$$Q_{\max} = \mu_1 \left(\frac{16}{\pi^2}\right) \frac{4 q_0 a^2}{\pi^2} = \mu_1 \frac{64 q_0 a^2}{\pi^4}$$

19-15 The base of a square, simply supported plate is subjected to lateral acceleration $a_0 \sin \Omega t$, where $\Omega = \omega_1$ is the fundamental frequency. The plate is weightless and has a lumped mass M_0 at the center. Determine the amplitudes of center deflection and the lateral force applied to the plate if the magnification factor is μ_1. Referring to the results of Prob. 19-14, how accurately can this lumped-mass model represent the actual plate if $M_0 = 0.25 m a^2$?

When the base of the plate moves with an acceleration a_0, the center mass is subjected to the inertial force of magnitude $M_0 a_0$. According to Ref. 7 (p. 225), static deflection under force P applied at the center is

$$u = 0.1391 (1-\nu^2) \frac{P a^2}{E h^3} \approx 0.0116 \frac{P a^2}{D}$$

Introducing the magnification factor and substituting $P = M_0 a_0$ yields

$$u_{\max} = 0.0116 \mu_1 \frac{M_0 a_0 a^2}{D}$$

The lateral force is simply

$$Q_{\max} = \mu_1 M_0 a_0$$

When the base of a plate with a continuously distributed mass m is subjected to the lateral acceleration a_0, the magnitude of distributed load is $m a_0$. The results from Prob. 19-14 for a plate vibrating in the natural mode may be used for this dynamic case if q_0 is

replaced by $m a_0$. The deflection and the lateral force are, respectively,

$$u_{\max} = \mu_1 \frac{4 m a_0 a^4}{\pi^6 D} \quad \text{and} \quad Q_{\max} = \mu_1 \frac{64 m a_0 a^2}{\pi^4}$$

Taking the ratio of maximum displacement in a discrete model to that of the original plate, we obtain

$$0.697 \quad \text{or} \quad -30.3\% \text{ error}$$

For the lateral force:

$$0.3805 \quad \text{or} \quad -61.9\% \text{ error}$$

The discrete model with a single lumped mass at the center is therefore not an accurate representation of a simply supported plate.

19-16 A portal frame consists of two columns ($A = 14.4$ in.^{2}, $I = 272.9$ in.^{4} each) and the horizontal beam ($A = 11.77$ in.^{2}, $I = 515$ in.^{4}). There is an additional distributed dead weight $mg = 32$ Lb/in. on the horizontal beam. Construct a discrete model capable of responding to excitation frequencies up to 90 Hz. Place three segments per half-wave of the highest mode. If shear areas are needed, use $A_s/A = 0.4$. The material is steel, $E = 29 \times 10^6$ psi, $G = 11 \times 10^6$ psi, $\gamma = 0.284$ Lb-s^2/in.^{4} Consider only the dynamic DOFs that are perpendicular to the frame axis.

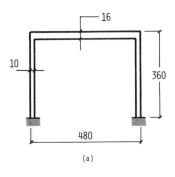

(a)

The distributed masses for a column and for a beam, respectively, are:

$$m_c = \frac{A \gamma}{g} = 14.4 \times \frac{0.284}{386} = 0.01059$$

$$m_b = \frac{32 + 11.77 \times 0.284}{386} = 0.09156 \text{ Lb-s}^2/\text{in.}^2$$

The half-waves obtained from Eq. 19.2b are, respectively,

$$\Lambda_c^2 = \frac{\pi^2}{2\pi \times 90} \left(\frac{29 \times 10^6 \times 272.9}{0.01059}\right)^{1/2} \quad \therefore \Lambda_c = 122.8 \text{ in.}$$

$$\Lambda_b^2 = \frac{\pi^2}{2\pi \times 90} \left(\frac{29 \times 10^6 \times 515}{0.09156}\right)^{1/2} \quad \therefore \Lambda_b = 83.96 \text{ in.}$$

Using three joints per half-wave yields joint spacing as follows:

$$\frac{122.8}{3} = 40.93 \text{ in.} \quad \text{(on column)}$$

$$\frac{83.96}{3} = 27.99 \text{ in.} \quad \text{(on beam)}$$

The numbers of elements, respectively, are:

$$\frac{360 - 16/2}{40.93} = 8.6 \approx 9$$

$$\frac{480 - 10}{27.99} = 16.79 \approx 17$$

Before making a final decision on the number of elements to use, let us calculate the ratio of half-wave length to the depth of the beam;

$$\frac{\Lambda_b}{h} = \frac{83.96}{16} = 5.25$$

This ratio is too large for shear distortion to have an appreciable influence on frequency, and this effect will be even less for the column. We may, however, decrease the beam joint spacing slightly by using 18 instead of 17 segments. The model is presented in (*b*). According to the problem statement, our dynamic DOFs are associated with the motion normal to the frame axis. Only the corner joints 9 and 27, which belong to columns and beams at the same time, have two mutually perpendicular, dynamic DOFs (coinciding with dots in (*b*)) assigned to each of them. The problem of axial vibrations of members must be separately considered.

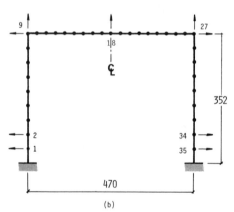

(b)

19-17 Establish the number of joints needed in the portal frame in Prob. 19-16 for the axial response of the members. Use about three segments per half-wave, unless the joint spacing established in the reference problem imposes a denser pattern.

The half-wave length of axial vibrations at 90 Hz is found from Eq. 19.1b. For the column and beam, respectively,

$$\Lambda_c = \frac{\pi}{2\pi \times 90} \left(\frac{29 \times 10^6 \times 14.4}{0.01059} \right)^{1/2} = 1103.2 \text{ in.}$$

$$\Lambda_b = \frac{\pi}{2\pi \times 90} \left(\frac{29 \times 10^6 \times 11.77}{0.09156} \right)^{1/2} = 339.2 \text{ in.}$$

Using three joints per half-wave yields joint spacing as follows:

$$\frac{1103.2}{3} = 367.7 \text{ in.} \quad \text{(on column)}$$

$$\frac{339.2}{3} = 113.1 \text{ in.} \quad \text{(on beam)}$$

Since the effective column length is only 352 in., it is obvious that the natural frequency of the column is much higher than 90 Hz. The forcing is thus not able to give any significant response, and the vertical DOFs at corners 9 and 27 of Fig. 19-16*b* are sufficient to represent column inertias. The joint spacing along the beam is at $470/18 = 26.11$ in., as dictated by the flexural properties. Placing a longitudinal mass at every fourth joint gives us spacing of $4 \times 26.11 = 104.44$ in., which is denser than 113.1 in. required. The figure shows all joints with axial, dynamic DOFs.

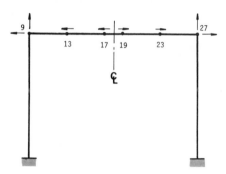

19-18 The base of Timoshenko beam is subjected to harmonic shaking with frequency $f = 50$ Hz. We have to develop its discrete model with masses evenly spaced at a distance *l*. There are to be at least three segments per half-wave. Determine the largest possible mass spacing *l* if the cross section is of the wide-flange type and the following are given:

$$A = 38.21 \text{ in.}^2 \qquad I = 4009 \text{ in.}^4$$

$$h = 24.25 \text{ in.} \qquad A_s = 13.701 \text{ in.}^2$$

$$L = 300 \text{ in.} \qquad m = 0.3 \text{ Lb-s}^2/\text{in.}^2$$

$$E = 29 \times 10^6 \text{ psi} \qquad G = 11 \times 10^6 \text{ psi}$$

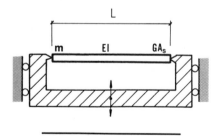

The forcing frequency is

$$\Omega = 2\pi f = 314.16 \text{ rad/s}$$

and it will strongly excite the modes that have ω close enough to the value above. Limiting ourselves at first to flexural deformation, we obtain from Eq. 19.2b,

$$\Lambda_n^2 = \frac{\pi^2}{\omega_n}\left(\frac{EI}{m}\right)^{1/2} = \frac{\pi^2}{314.16}\left(\frac{29\times10^6\times4009}{0.3}\right)^{1/2}$$

$$\therefore \Lambda_n = 139.8 \text{ in.}$$

The nearest higher frequency, which will be associated with half-wave fitting a round number of times into L, is one corresponding to $\Lambda_3 = 100$ in. Since this is only about four times the beam depth h, we are going to use Eq. 19.3 for the natural frequency of a Timoshenko beam:

$$\frac{1}{\omega_3^2} = 2.649\times10^{-6}\left[1+0.987\times10^{-3}(771.42+104.92)\right]$$

$$\therefore \omega_3 = 449.91 \text{ Hz}$$

This natural frequency has a higher value than Ω. Thus

$$l = \frac{\Lambda_3}{3} = 33.3 \text{ in.}$$

that is, dividing the beam length into nine segments will give us a satisfactory model.

19-19 A rectangular plate with thickness $h=0.09$ in. is 45 in. long and 18 in. wide. The long sides are simply supported, the short sides clamped. The material is aluminum, $E=10.6\times10^6$ psi, $\nu=0.33$, and $\gamma=0.1$ Lb/in.^{3} The model of the plate is to respond to all frequencies up to 2000 Hz. Design a discrete model with no fewer than two segments per half-wave.

From Eq. 19.4b:

$$\Lambda^2 = \frac{2\pi^2}{2\pi\times2000}\left[\frac{10.6\times10^6\times0.09^2\times386}{12\times0.1(1-0.33^2)}\right]^{1/2} \quad \therefore \Lambda=2.957 \text{ in.}$$

Using two segments per half-wave would give us the joint spacing

$$\frac{2.957}{2} = 1.479 \text{ in.}$$

Choose the number of segments in the long and short directions, respectively, as follows:

$$\frac{45}{1.479} = 30.4 \quad \text{use 32 segments}$$

$$\frac{18}{1.479} = 12.17 \quad \text{use 13 segments}$$

The plate will therefore have 416 elastic elements and

$$(32-1)(13-1)=372$$

dynamic DOFs. No reference has thus far been made in the solution to the fact that the short sides are not simply supported, but clamped. To have the desired accuracy, one could add a segment or two in the long direction of the plate. This, in fact, has been done, by rounding up the number of required segments.

19-20 A large containment building is made of pre-stressed concrete and has a constant wall thickness $h=42$ in. (only the median surface is shown in (a)). The section area of the ring at the upper edge is $A_t=24,640$ in.^{2} Construct a beam model capable of responding to frequencies up to 40 Hz. There should be more than two elements per half-wave. Forcing is applied in the horizontal direction only. Material data: $E=5.3\times10^6$ psi, $\nu=0.15$, and $\gamma=0.0868$ Lb/in.^{3} (This example is from Ref. 16, p. 1242.)

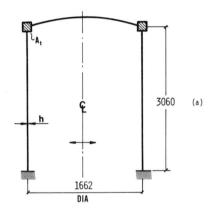

The cylindrical shell structure will be replaced by a beam coinciding with the axis of the shell. The cross section is a thin annulus with a mean radius $R=831$ in. and

$$A=2\pi Rh=219,296 \text{ in.}^2$$

$$I= R^3h=75.72\times10^9 \text{ in.}^4$$

$$A_s=0.5A=109,648 \text{ in.}^2$$

The shear modulus is found from

$$G=\frac{E}{2(1+\nu)}=2.304\times10^6 \text{ psi}$$

Also, $m=A\gamma/g=49.313$ Lb-s^2/in.^{2}
Let us first try the flexural beam formula,

$$\Lambda^2 = \frac{\pi^2}{2\pi\times40}\left(\frac{5.3\times10^6\times75.72\times10^9}{49.313}\right)^{1/2} \quad \therefore \Lambda=1882 \text{ in.}$$

The ratio $2R/\Lambda=1662/1882=0.88$ indicates that there will be a strong influence of shear deformation. Equation 19.3, representing a Timoshenko beam, becomes

$$\frac{1}{\omega}=\left(\frac{\Lambda}{29,839}\right)^2\left[1+\left(\frac{4368.8}{\Lambda}\right)^2\right]^{1/2}$$

After a few trials, we find $\Lambda\approx799$ in. for $\omega=2\pi\times40=251.33$ rad/s. The joint spacing should therefore be $799/2=399.5$ in. This gives the number of beam elements in the model as

$$\frac{3060}{399.5}=7.65 \quad \text{use 8 beam elements}$$

The beam model is shown in (b). The lumped mass and the angular inertia at each intermediate joint are, respectively,

$$M_0 = ml = 49.313 \times 382.5 = 18,862 \text{ Lb-s}^2\text{-in.}$$

$$J_0 = \rho Il = \frac{\gamma}{g} Il = 6.513 \times 10^9 \text{ Lb-s}^2\text{-in.}$$

At the top joint there is an additional weight of the roof and the edge ring,

$$W_1 \approx \gamma \pi R^2 h = 7.909 \times 10^6 \text{ Lb}$$

$$W_2 = 2\pi R A_t \gamma = 11.167 \times 10^6 \text{ Lb}$$

The lumped mass at the top is:

$$M_0' = \frac{M_0}{2} + \frac{W_1 + W_2}{g} = 58,851 \text{ Lb-s}^2/\text{in.}$$

To evaluate the angular inertia associated with the edge ring, put $A_t = L_t h$, where h is the shell thickness and L_t is some equivalent length measured along the shell axis. The diametral mass moment of inertia is then

$$J_t = \rho L_t I = \rho L_t (\pi R^3 h) = \rho \pi R^3 A_t = 9.989 \times 10^9$$

The lumped angular inertia at the top is

$$J_0' = \frac{J_0}{2} + J_t = 13.246 \times 10^9 \text{ Lb-s}^2\text{-in.}$$

The beam models of shells are not very accurate in that they do not reflect some three-dimensional deformation patterns that shells can undergo. Nevertheless, they are quite often used because of their simplicity.

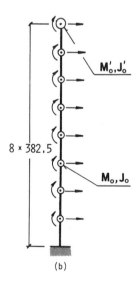

8 × 382.5

M_0', J_0'

M_0, J_0

(b)

19-21 In Prob. 19-20 a beam model of a cylindrical building was designed with lateral motion in mind. Construct a vertical model for the same range of forcing frequencies. Treat the roof of the building as a circular plate clamped at the boundary. The vibra-

tory modes of the plate, beyond the first one, need not be considered.

The frequency equation for a round, fixed-edge plate from Prob. 19-9 may be written as

$$\omega = \frac{10.22}{R^2} \left(\frac{Eh^2 g}{12\gamma(1-\nu^2)} \right)^{1/2} \qquad \therefore \omega = 27.86 \text{ rad/s} = 4.434 \text{ Hz}$$

According to the same problem, we lump 0.1532 of the plate mass at its center and the remainder at the edge. After Prob. 19-20 we have

$$W_c = 0.1532 \times 7.909 \times 10^6 = 1.212 \times 10^6 \text{ Lb}$$

$$W_e = 0.8468 \times 7.909 \times 10^6 = 6.697 \times 10^6 \text{ Lb}$$

The spacing of joints along the axis of building is determined by the frequency of 40 Hz or 251.33 rad/s. From Eq. 19.1b we have

$$\Lambda = \frac{\pi}{\omega} \left(\frac{Eg}{\gamma} \right)^{1/2} = 1919 \text{ in.}$$

Using three segments per half-wave would correspond to joint spacing of

$$l = \frac{1919}{3} = 639.7 \text{ in.}$$

The number of segments is therefore

$$\frac{3060}{639.7} = 4.8 \approx 5$$

and the joint spacing is

$$l = \frac{3060}{5} = 612 \text{ in.}$$

Reviewing the lateral vibration model, we see the joints spaced 382.5 in. apart. It would be convenient to have both joint networks at least partially coincide. If we use every other joint from the lateral model, the axial model will have joints at 765 in., as shown in the figure. After Prob. 19-20 we find the mass at any intermediate joint to be

$$M_0 = 2 \times 18,862 = 37,724 \text{ Lb}$$

The mass at the very top of the model represents the vibrating inertia of the plate,

$$M_c = \frac{W_c}{g} = 3140 \text{ Lb-in./s}^2$$

The new mass M_0' at the top of cylindrical part is

$$M_0' = 58,851 - 3140 + \frac{18,862}{2} = 65,142 \text{ Lb-in./s}^2$$

The model so designed is reasonably accurate as far as the cylindrical part of the structure is concerned. It will, however, underestimate the response of the plate, even in the first mode. The higher modes, which will undoubtedly be excited in the real structure, were not reflected in the model of the roof.

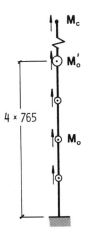

4 × 765

The displacement solution is

$$u(x,t) = \frac{q_0 L^2}{\pi^2 EA} \sin\frac{\pi x}{L}(1-\cos\omega_1 t)$$

where ω_1 is the fundamental frequency. Construct a discrete model with a single mass at the center, apply to it an equivalent concentrated force and compare the peak dynamic displacement and axial force in both systems.

Noting that the deflected shape is a half-wave of a sinusoid, $X(x) = \sin(\pi x/L)$, we calculate the equivalent load P_0 from Eq. 11.18:

$$P_0\sin\frac{\pi}{2} = \int_0^L q_0\sin^2\frac{\pi x}{L}dx = q_0\frac{L}{2} \qquad \therefore P_0 = \frac{1}{2}q_0 L$$

The natural frequency will be preserved if $M_0 = 0.4053 mL$ is placed at the center, per Prob. 19-4. Our loaded model is shown in the figure. When P_0 is suddenly applied, we can obtain the peak dynamic response by multiplying the static values by 2.0. According to what was said about symmetry of this model in Prob. 19-4, we have

$$u_{max} = 2\frac{(P_0/2)(L/2)}{EA} = \frac{1}{2}\frac{P_0 L}{EA} = \frac{1}{4}\frac{q_0 L^2}{EA}$$

$$N_{max} = 2\frac{P_0}{2} = \frac{1}{2}q_0 L$$

From the formula in the problem statement, we find

$$u_{max} = \frac{2}{\pi^2}\frac{q_0 L^2}{EA}$$

$$N_{max} = (EAu')_{max} = \frac{q_0 L^2}{\pi^2}\frac{\pi}{L}\times 2 = \frac{2}{\pi}q_0 L$$

Our discrete model overestimates the maximum displacement by 23.3% and underestimates the axial force by 21.5%.

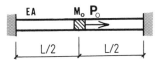

19-22 When the bar is under the action of a step load at the end, the normal-mode method gives the maximum end deflection as

$$u_{max} = \frac{16 P_0 L}{\pi^2 EA}\left(\frac{1}{1^2} + \frac{1}{3^2} + \frac{1}{5^2} + \cdots\right)$$

(Ref. 2, p. 377). How many masses must a discrete model have to give 90% of the true value of maximum deflection? How many to give 95%? (Assign twice as many masses as the nominal number of modes required.) What result can we expect if there are 16 masses in the model?

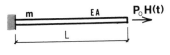

We may first note that the sum of the series is $\pi^2/8$, which gives us $u_{max} = 2PL/(EA)$. (This is twice the value induced by the static action of P.) Taking only the first two terms (or two modes, see Prob. 19-3) gives us slightly over 90% of the answer. To satisfy the requirements of the problem, we place four masses along the length. When the first four terms of the series are added, we get nearly 95% of the true value. To attain this, we need eight masses along the length.

When 16 masses are placed along the bar, we can count on the first eight modes and their contributions to be accurately reflected, which gives us 97.47% of the true answer. Doubling this number of masses (and corresponding modal terms, up to $1/31^2$) brings us to 98.73% of the true answer.

19-23 In Prob. 15-11 an axial bar of length L and distributed mass m, with both ends fixed, was subjected to a suddenly applied load, distributed according to

$$q(x) = q_0\sin\frac{\pi x}{L}$$

19-24 In Prob. 15-13 the displacement response of a simply supported beam, loaded at the center with $P = P_0 H(t)$ was derived. Find the corresponding formulas for the maximum slope, bending moment and shear force at $t = \tau_1/2$ (i.e., at the end of the first half-period). Discuss the accuracy of computation that we can expect for these components of response once a discrete model of the beam has been selected.

From the reference problem we have

$$u\left(x,\frac{\tau_1}{2}\right) = \sum_{n=1,3,5,\ldots}^{\infty} \sin\frac{n\pi x}{L} f_n\left(\frac{\tau_1}{2}\right)$$

where

$$f_n\left(\frac{\tau_1}{2}\right) = \frac{192}{\pi^4 n^4} u_{st}(-1)^{(n-1)/2}$$

when n is odd. After successive differentiations and setting $x = L/2$ or $x = 0$ (to get the extreme value), we have

$$u\left(\frac{L}{2},\frac{\tau_1}{2}\right) = \frac{192}{\pi^4} u_{st}\left(1 + \frac{1}{3^4} + \frac{1}{5^4} + \cdots\right)$$

$$u'\left(0,\frac{\tau_1}{2}\right) = \frac{192}{\pi^3}\frac{u_{st}}{L}\left(1 - \frac{1}{3^3} + \frac{1}{5^3} - \cdots\right)$$

$$\mathfrak{M} = EIu''\left(\frac{L}{2},\frac{\tau_1}{2}\right) = -EI\frac{192}{\pi^2}\frac{u_{st}}{L^2}\left(1 + \frac{1}{3^2} + \frac{1}{5^2} + \cdots\right)$$

$$V = EIu'''\left(0,\frac{\tau_1}{2}\right) = -EI\frac{192}{\pi}\frac{u_{st}}{L^3}\left(1 - \frac{1}{3} + \frac{1}{5} - \cdots\right)$$

Writing the answer in the form of an infinite series, as above, allows us to clearly visualize how the modal contributions add up. The sums of the four series are, respectively,

$$\frac{\pi^4}{96} \qquad \frac{\pi^3}{32} \qquad \frac{\pi^2}{8} \qquad \frac{\pi}{4}$$

Suppose, now, that we have a model that accurately combines the effects of the first five modes and the remaining modes, if any, are ignored. (This will be equivalent to adding up of the first three terms, since the antisymmetric modes are not excited here.) The following are the fractions of the true answers for our quantities of interest.

$$0.9993 \qquad \text{for} \quad u$$

$$1.0021 \qquad \text{for} \quad u'$$

$$0.9331 \qquad \text{for} \quad \mathfrak{M}$$

$$1.1035 \qquad \text{for} \quad V$$

It is seen that the higher the derivative of u with respect to x, the less accurate is the answer for a given number of contributing modes. It is clearly seen that more modes are needed to achieve, say, 95% of the true bending moment than to achieve 95% of deflection.

19-25 The base of the system in Fig. 19-18 is subjected to a motion whose acceleration response spectrum Z_a is constant. Using the results from Sect. 12C, we can determine the relative displacement in each mode. We would like to generate a discrete model with at least two segments per half-wave. What joint spacing would make the model fine enough to give us no more than 1% error in the estimation of maximum bending? (The true value would be the limit obtained

by including an infinite number of modes.) Assume that the successive contributions of modal components have alternating signs.

A discrete model of this beam for the nth vibratory mode was analyzed in Prob. 14-12, where we learned that its participation factor

$$\Gamma_n = \frac{1}{n} \qquad \text{for} \quad n = 1,3,5,\ldots$$

and zero for all even-numbered modes. The relative displacement response of each vibrating mass, according to Eq. 12.3, is

$$|u_n| = \frac{1}{\omega_n^2} Z_a \Gamma_n$$

(For any vibrating mass, there is either $+1$ or -1 in the modal vector.) Substituting $\omega_n^2 = k_n/M_0$, with both parameters referring to a half-wave with a mass lumped at the center, we have

$$|u_n| = \frac{M_0}{k_n} Z_a \frac{1}{n}$$

while the bending moment is

$$\mathfrak{M}_n = \frac{1}{4} P_0\left(\frac{L}{n}\right) = \frac{1}{4}(k_n|u_n|)\frac{L}{n} = \frac{Z_a L}{4n^2} M_0$$

Substitution ($M_0 = mL/2n$) gives us the final expression for the amplitude of the bending moment in the nth mode:

$$\mathfrak{M}_n = \frac{Z_a m L^2}{8n^3}$$

Adding the modal amplitudes with alternating signs we have

$$\sum_{n=1,3,5,\ldots} \mathfrak{M}_n = \frac{1}{8} Z_a m L^2\left(1 - \frac{1}{3^3} + \frac{1}{5^3} - \cdots\right)$$

It is known from mathematics that instead of trying to determine the actual sum, we may take only the first few terms of the alternating series, commiting an error not larger then the first discarded term. In our case the permissible error is 1% and the total in parentheses is approximately unity, hence

$$\frac{1}{100} \approx \frac{1}{p^3} \qquad \therefore \; p \approx 4.6$$

The first rejected term is $n = 5$. The approximate sum of the series is

$$1 - \frac{1}{3^3} = 0.963$$

which means that we include all modes up to $n = 3$. The node spacing is therefore

$$l = \frac{L}{2n} = \frac{L}{6}$$

In solving Prob. 14-12, from which we quoted some results, we were concerned with a model responding to a single excitation frequency. A different model would be constructed for each such frequency. In this problem there is a single model having each natural frequency associated with a certain level of response. This, however, should be of no concern, because if we generate a model responding reasonably well to, say, $n=3$, the number of elements is sufficiently large to make it quite accurate for $n=1$ and 2. (This is not to say that $\Gamma_n = 1/n$ holds exactly true in this problem, but the approximation is sufficiently accurate for our purpose.) Note that the manner in which the modal contributions were combined was not typical for the response spectrum calculation.

19-26 The acceleration response spectrum presented in the figure is the envelope of what was used in Prob. 12-8. The lower level of $0.5g$ holds for all natural frequencies higher than 35 Hz. Suppose that a discrete model of the beam similar to one in Prob. 19-18 is subjected to this spectrum. The fundamental frequency is 2.4 Hz. Assuming that the modal components will be combined in their absolute values, determine the ratio of the calculated maximum bending moment to the actual value if (1) three symmetric modes are included in the calculation, and (2) six modes are included.

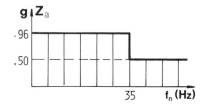

Let us first establish the natural frequencies of interest. From Eq. 19.2b,

$$\omega_n = n^2 \frac{\pi^2}{L^2} \left(\frac{EI}{m} \right)^{1/2}$$

which means that frequencies are proportional to the square of the mode number. Using only symmetric modes, we have

$$n=1 \qquad f_1 = 2.4 \text{ Hz} \qquad \text{(problem statement)}$$

$$n=3 \qquad f_2 = 21.6 \text{ Hz}$$

$$n=5 \qquad f_3 = 60.0 \text{ Hz}$$

Thus we find that all the mode numbers beginning with $n=5$ pertain to frequencies above 35 Hz. If our spectrum had a constant value, the maximum bending moment would be proportional to

$$1 + \frac{1}{3^3} + \frac{1}{5^3} + \cdots = C$$

as we can determine by reviewing Prob. 19-25 and changing signs to suit the postulated method of summation. The value of C is not given in Ref. 9, but we can estimate it as the average value of two similar series, one with $1/n^2$ and the other with $1/n^4$ in the denominator, thus obtaining $C \approx 1.124$. Since our spectrum is not

of constant value, the first two terms are actually

$$\frac{0.96}{0.50} \left(1 + \frac{1}{3^3} \right) = 1.991$$

and the sum changes from C to $C' = 2.078$. The first three modes give us

$$1.991 + \frac{1}{5^3} = 1.999$$

which is 96.2% of the estimated true value of the bending moment. The first six modes (up to $n=11$) yield 2.004 or 96.4% of the true value. The reason for such a small difference is that the effect of the first two modes is strongly predominating.

19-27 When a bar, as in Fig. 19-22, is subjected to a harmonic forcing $P = P_0 \sin \Omega t$, the displacement of the end is given by

$$u = \frac{2P_0}{\rho A L} \sum_{r=1,3,5,\ldots} \frac{1}{\omega_r^2} \frac{1}{1 - \Omega^2/\omega_r^2}$$

where $\qquad \omega_r = \frac{r\pi c}{2L}$

What will be the RMS response to the random force defined by the spectrum S_P shown in the figure when there is a damping ratio ζ, the same for every mode? Evaluate the answer by taking (1) a single contributing mode, (2) the first three modes, and (3) the first six modes.

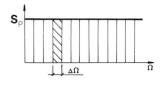

The deflection response to the harmonic forcing may be represented by the modal components as

$$u = u_1 + u_2 + u_3 + \cdots$$

where $\qquad u_r = \frac{2P_0}{\rho A L} \frac{1}{\omega_r^2} \frac{1}{1 - \Omega^2/\omega_r^2} = \frac{8P_0 L}{\pi^2 EA} \frac{1}{r^2} \frac{1}{1 - \Omega^2/\omega_r^2}$

This is exactly true when the system is undamped and is also a good approximation when Ω is not too close to ω_r. Otherwise, the last factor in the equation for u_r must be replaced by the damped magnification factor (Sect. 2G). Our problem now is to calculate the modal response to an increment of a harmonic force ΔP_0 given by

$$(\Delta P_0)^2 = S_P \cdot \frac{\Delta \Omega}{2\pi}$$

in accordance with Eq. 18.14. The mean-square response to one such increment is

$$\Delta \overline{u_r^2} = \left(\frac{8L}{\pi^2 EA} \frac{1}{r^2} \right)^2 \frac{1}{2\pi} \mu^2 S_P \cdot \Delta \Omega$$

Summing the increments and proceeding as in Sect. 18E, we obtain the mean-square response,

$$\overline{u_r^2} = \left(\frac{8L}{\pi^2 EA} \frac{1}{r^2} \right)^2 \frac{S_P}{2\pi} \int_0^\infty \mu^2 \, d\Omega$$

or

$$\overline{u_r^2} = \left(\frac{8L}{\pi^2 EA} \frac{1}{r^2} \right)^2 \frac{\omega_r S_P}{8\zeta} = B \frac{1}{r^3}$$

The constant B is the same for every mode. Assuming the total mean-square to be the sum of the corresponding modal quantities, we have

$$\overline{u^2} = B \left(1 + \frac{1}{3^3} + \frac{1}{5^3} + \cdots \right)$$

The sum of the series is about 1.124, as evaluated in Prob. 19-26. Taking successively one, three, and six terms, we obtain

$$1.0 \qquad 1.045 \qquad 1.0501$$

for those partial sums. To find the RMS instead of mean-square values, we simply calculate the square roots of the latter. This makes it possible to evaluate the percentage of the true answer,

94.3%	for	one mode included
96.4%	for	three modes included
96.7%	for	six modes included

19-28 The shear beam in Fig. 14-9 is subjected to a random base acceleration characterized by the constant forcing spectrum $S_a(\Omega)$. Using Probs. 14-10 and 18-32 as a reference, express the relative RMS displacement at the center in terms of mode numbers. What is the spacing of masses needed to get at least 95% of the true RMS displacement? Request three masses per half-wave and ignore the error in the individual modes due to discretization. Damping is the same in all modes.

Assuming that the successive mode shapes are correctly represented by the respective discrete models, we put $\Gamma_r = 1/r$ for the odd-numbered modes. (See Prob. 14-10. This is a reasonable approximation here, just as in Prob. 19-25.) The largest absolute value of any entry in a modal vector need not be larger than unity, so we can put $x_{nr}^2 = 1$ in the solution of Prob. 18-32. (This will make the convergence appear to be slightly worse than it is.) That equation becomes now

$$\overline{u_c^2} = \frac{S_a}{8\zeta} \sum_{n=1,3,5,\ldots} \frac{1}{r^2} \frac{1}{\omega_r^3}$$

Problem 14-10 tells us that ω_r is proportional to r, so finally we have

$$\overline{u_c^2} = B \left(\frac{1}{1^5} + \frac{1}{3^5} + \frac{1}{5^5} + \cdots \right)$$

where B is a constant. The sum of the series may be estimated as 1.0081. As we find by taking the square root, the first term of the series gives us 99.6% of the RMS(u_c). Since three masses per half-wave are requested, the joint spacing will be

$$l = \frac{L}{3}$$

19-29 Step load $P_0 H(t)$ is applied laterally to an infinite beam. Calculate the shear-deformed length $c_3 t$, the length of prime flexural wave l_b, and the deflection at the loaded point for $t = 0.0002$, 0.0005, and 0.0015 s. Draw conclusions regarding the relative influence of shear and bending for these time points. The beam data are:

$$A_s = 27.1 \text{ in.}^2 \qquad\qquad I = 15{,}375 \text{ in.}^4$$

$$E = 30 \times 10^6 \text{ psi} \qquad\qquad G = 11.54 \times 10^6 \text{ psi}$$

$$m = 0.06684 \text{ Lb-s}^2/\text{in.}^2 \quad h = 36.48 \text{ in. (height)}$$

The force $P_0 = 22 \times 10^6$ Lb.

The length deformed by shear is

$$c_3 t = \left(\frac{GA_s}{m} \right)^{1/2} t = 68{,}400 t$$

The length of the flexural wave, Eq. 19.18 is:

$$l_{b2} = 1.96 \left(\frac{EI}{m} \right)^{1/4} t^{1/2} = 3176.7 t^{1/2}$$

The deflections at the loaded points, Eq. 19.17 and 19.19b:

$$u_{s2} = \frac{P_0 c_3 t}{2GA_s} = 2405.9 t$$

$$u_{b2} = \frac{P_0 l_{b2}^3}{24EI} = 63{,}709 t^{3/2}$$

Assuming shear and bending deflections to be independently induced, we have $u = u_{s2} + u_{b2}$ in the table below.

t(s):	0.0002	0.0005	0.0015
$c_3 t$ (in.):	13.68	34.2	102.6
l_{b2} (in.):	44.93	71.03	123.03
u_{s2} (in.):	0.481	1.203	3.609
u_{b2} (in.):	0.180	0.712	3.701
u (in.):	0.661	1.915	7.310

At the initial phase after the impact, the shear deformation clearly predominates, but later the bending component becomes more significant. After a sufficiently long time ($t=0.0015$ s in this case), their relative contributions are about equal. The reader may note that at $t=0.0002$ s the shear-deformed length is only 0.375 of the section height, while at $t=0.0015$ s this ratio is 2.81.

19-30   An early stage of the pipe whip phenomenon, which is a subject of consideration in nuclear plant design, is represented in (*a*). Following a sudden circumferential rupture above the elbow, the end of the pipe in (*a*) is separated from the rest of the system and subjected to the reaction of the escaping fluid, $P_0 H(t)$. In (*b*) the same pipe is simplified to a cantilever plus a restraint, not depicted in (*a*). Assuming that we are interested only in what happens before the pipe hits the restraint, establish joint spacing for computer-aided analysis. At least six segments must be present over the portion of pipe subjected to significant deformation. Use the following data:

$A_s = 35.06$ in.^{2} $I = 4650$ in.^{4}

$E = 25.7 \times 10^6$ psi $G = 9.885 \times 10^6$ psi

$m = 0.0617$ Lb-s^2/in.^{2} $D = 24$ in. (diameter)

$L = 900$ in. $b = 2.0$ in. $P_0 = 108,000$ Lb

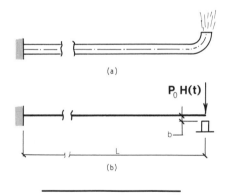

(a)

(b)

Let us first determine the lateral deflection at the tip using Eqs. 19.17 and 19.19a.

$$c_3 = \left(\frac{GA_s}{m} \right)^{1/2} = 74{,}947$$

$$u_{s1} = \frac{108{,}000 \times 74{,}947}{9.885 \times 10^6 \times 35.06} t = 23.356t$$

$$u_{b1} = \frac{0.9428 \times 108{,}000 \times t^{3/2}}{(25.7 \times 10^6 \times 4650)^{1/4} 0.0617^{3/4}} = 1398.9t^{3/2}$$

After putting $u = u_{s1} + u_{b1}$, the time needed to reach the stop is

found from $u = b$:

$$23.356t + 1398.9t^{3/2} = 2.0 \qquad \therefore t = 0.0115 \text{ s}$$

The deformed segments are $c_3 t$ and l_{b1}, independent from each other,

$$c_3 t = 861.9 \text{ in.}$$

$$l_{b1} = \left(\frac{4EI}{m} \right)^{1/4} t^{1/2} = 178.9 \text{ in.}$$

The shear-deformed length is so large that it nearly reaches the boundary. Yet the shear deflection is quite small, only about 0.27 in. out of 2.0 in. total. The significant part of the deformation comes from bending, and for this reason we may decide to space the joints as shown in (*c*). The first six segments span l_{b1}, while the other six cover the remaining portion of the beam.

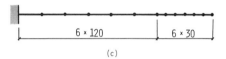

(c)

19-31   A large structure is made of sandwich panels having the cross section shown. It is desirable, for reasons of economy, to create a dynamic model consisting of homogeneous plate elements. Establish thickness and material constants of such elements so that the elastic and inertial properties of the original panel are preserved. Once the computer analysis has been completed, how shall we interpret the results, especially those relating to stress response?

$E_1 = 5.4 \times 10^6$ psi   (isotropic)

$G_1 = 2.25 \times 10^6$ psi   (isotropic)

$G_2 = 3200$ psi   (out of plane, no in-plane resistance)

$mg = 1.778 \times 10^{-3}$ Lb/in.^{2} (entire plate)

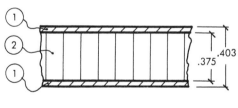

The key stiffness components are bending and shear, and our objective is to preserve them both. The bending stiffness parameter is

$$E_1 \left(h_1^3 - h_2^3 \right) = 5.4 \times 10^6 (0.403^3 - 0.375^3) = 68{,}670$$

while for the shear stiffness

$$G_2 h_2 = 3200 \times 0.375 = 1200$$

Let us denote by E and G the elastic constants of the equivalent homogeneous plate material and by $\bar{h}$ the thickness of this plate. The stiffness parameters will be preserved when

$$E\bar{h}^3 = 68{,}670 \quad \text{and} \quad G\frac{\bar{h}}{1.2} = 1200$$

Dividing the sides of these equations, we get

$$\frac{E}{G}\bar{h}^2 = 47.688$$

For an elastic material, $E/G = 2(1+\nu)$. Setting $\nu = 0.25$ we get

$$\frac{E}{G} = 2.5 \quad \text{and} \quad \bar{h} = 4.3675 \text{ in.}$$

From previous expressions,

$$E = 824.3 \text{ psi} \quad G = 329.7 \text{ psi}$$

To preserve the inertia of the original plate, we put

$$mg = 1.778 \times 10^{-3} = \bar{\gamma}\bar{h} \quad \therefore \bar{\gamma} = 0.4071 \times 10^{-3} \text{ Lb/in.}^3$$

The equivalent plate appears to be quite thick and flexible. This excessive thickness should be of no concern, however, because it is the stiffness and inertia, rather than thickness alone, that determine the response of the model.

The results of analysis of such a model will be correct as far as displacements and internal forces are concerned, but the stress components will require some interpretation. Suppose that at some place we have the bending moment $\mathfrak{M}$ (Lb-in./in.). The computer printout will then show the bending stress

$$\sigma_b = \frac{6\mathfrak{M}}{\bar{h}^2} = 0.3145\,\mathfrak{M}$$

while the true value is

$$\sigma_b' = \frac{12\mathfrak{M}}{h_1^3 - h_2^3}\frac{h_1}{2} = 190.15\,\mathfrak{M}$$

The printout values should therefore be multiplied by the factor

$$\frac{190.15}{0.3145} = 604.6$$

to obtain the true bending stress. This illustrates one of many situations in which the results of computer analysis must be carefully interpreted.

EXERCISES

19-32 The lateral, free vibrations of a simply supported flexural beam are defined in Prob. 15-5:

$$X_n = \sin\frac{n\pi x}{L} \quad \omega_n^2 = \left(\frac{n\pi}{L}\right)^4 \frac{EI}{m}$$

$$\text{for} \quad n = 1, 2, 3, \ldots$$

Express the half-wave length Λ_n as a function of

natural frequency ω_n. Also discuss briefly Λ_n for other boundary conditions.

19-33 Determine the first natural frequency of a Timoshenko beam for $\Lambda/h = 2$, 5, and 10 using Eq. 19.3. The properties are as follows

$$A = 38.21 \text{ in.}^2 \qquad I = 4009 \text{ in.}^4$$

$$h = 24.25 \text{ in.} \qquad A_s = 13.701 \text{ in.}^2$$

$$E = 29 \times 10^6 \text{ psi} \qquad G = 11 \times 10^6 \text{ psi}$$

$$m = 0.02811 \text{ Lb-s}^2/\text{in.}^2$$

Calculate the errors given by Eq. 19.3, assuming the true values to be:

$\dfrac{\Lambda}{h}$	ω
2	4091.0
5	1091.2
10	318.94

(These are calculated by a computer program for a discrete model having 50 segments and all necessary features of a Timoshenko beam. When a purely flexural model is used, the program underestimates the frequency by less than 0.01%.)

19-34 Using Eq. 19.3, determine the influence of shear deformation and rotatory inertia on natural frequencies of a thin-wall pipe. In particular, calculate the ratio ω/ω_b (the natural frequency to the frequency based on flexural properties alone) for the following proportions:

$$\frac{\text{half-wave length}}{\text{pipe diameter}} = \frac{\Lambda}{2R} = 1, 2, 3, 4, 5, \text{ and } 10$$

Use $E/G = 29/11$ and $A_s/A = 0.5$.

19-35 The deflected surface of a rectangular, simply supported plate, vibrating in the natural mode, is given by

$$u = f(t)\sin\frac{i\pi x}{a}\sin\frac{j\pi y}{b}$$

in which $f(t)$ is a time function, while i and j are arbitrary integers. The natural frequency is given by

$$\omega_{ij}^2 = \pi^4 \frac{D}{m}\left(\frac{i^2}{a^2} + \frac{j^2}{b^2}\right)^2$$

in which

$$D = \frac{Eh^3}{12(1-\nu^2)}$$

where h is the plate thickness and m is the mass per unit surface. For a certain proportion of sides a and b, the vibrating surface (shown in plan view in the figure) can be subdivided into squares by the stationary lines (also called the nodal lines). A plus sign in a square indicates that when $f(t)$ is positive, that square deflects out of the paper. Find the relationship between the half-wave length and the natural frequency associated with a described mode shape.

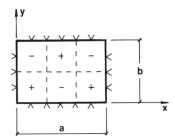

19-36 Let us introduce the following four natural frequencies of a flexural plate: ω_1 is the frequency associated with a mode shape in Fig. 19.2 with half-wave length Λ, while ω_2 refers to a similar shape, except that the length is $\Lambda/2$; the remaining two frequencies will apply to the lumped-mass model in Fig. 19.4. Also, ω_1' is for the model with the stationary lines as in Fig. 19.4 and half-wave length Λ (the highest significant mode), while ω_2' is the maximum natural frequency of the discrete model. (The latter is associated with $\Lambda/2$ and the masses along a horizontal line vibrating opposite each other.) If $\omega_1 = 100$ Hz, what are the values of the remaining three frequencies? The lumped mass is $M_0 = 0.25m\Lambda^2$. Refer to Prob. 15-14.

19-37 The fundamental frequency of a shear beam with length L (ends restrained) is known from Prob. 19-5,

$$\omega_b = \frac{\pi}{L}\left(\frac{GA_s}{m}\right)^{1/2}$$

The approximate frequency ω_b' of the same beam based on the static deflection method may easily be established. Using the same approximation, the natural frequency ω_p' of a square shear plate was found in Prob. 19-10,

$$\omega_p' = \frac{4}{L}\left(\frac{Gh_s}{m}\right)^{1/2}$$

Assuming an analogy between both types of element, estimate the actual frequency ω_p of a square shear plate.

19-38 Determine the influence of shear deformation on the natural frequency of a solid aluminum plate with the following material constants: $E = 10.6 \times 10^6$ psi, $G = 4 \times 10^6$ psi, and $\nu = 0.325$. The method of approaching the problem and the selected ratios Λ/h are to be the same as in Prob. 19-11.

19-39 A simply supported square plate with the following properties:

$$E = 29 \times 10^6 \text{ psi}$$

$$h = 0.25 \text{ in.}$$

$$L = 36 \text{ in. (side length)}$$

$$\gamma = 0.284 \text{ Lb/in.}^3$$

$$\nu = 0.3182$$

was analyzed by computer. The model had six segments per side, a total of 36 identical elements. The first natural frequency was $f_1 = 36.897$ Hz, the highest significant frequency (determined by reviewing the deflected shapes) was $f_{11} = 338.39$ Hz and the maximum model frequency $f_{25} = 626.58$ Hz. Calculate the first frequency of the continuous model f_1' and evaluate the remaining two for the lumped model. Compare with the computer-generated figures.

19-40 A platform, mounted on several parallel columns, supports rotating machinery, which has the working speed of 6000 rpm. Each column has length $L = 600$ in. and is made of steel, $E = 29 \times 10^6$ psi and $\gamma = 0.284$ Lb/in.^{3} Determine the joint spacing suitable for axial response analysis of the system assuming the ends to be fixed. Use about three segments per half-wave.

19-41 The square plate in Prob. 18-10 is subjected to an acoustic loading with high-frequency components. Design two discrete models of the plate, one responding to all frequencies up to 2000 Hz and one up to 8000 Hz. Each model must have at least two segments per half-wave.

19-42 Lacking better criteria, a certain company uses the following procedure to establish the side length l of a square plate element, which is used to build a discrete model. When the highest frequency to which the model should respond is ω_m, l is determined by the side length of a plate with fixed edges and having the fundamental frequency equal to ω_m. Find the highest significant frequency ω_1 of the model with joints spaced at l and the maximum frequency that the discrete model will exhibit, if $\omega_m = 1000$ Hz. The natural frequency of a plate with all edges fixed is

$$\omega = \frac{36}{l^2} \left(\frac{D}{m} \right)^{1/2}$$

According to Sect. 19D there must be at least two segments per half-wave if the natural frequency is to be called significant.

19-43 The displacement solution for the bar in Prob. 19-22, at $t_0 = 2L/c$, is

$$u = \frac{16P_0 L}{\pi^2 EA} \sum_{n=1,3,5,\ldots} \frac{(-1)^{(n-1)/2}}{n^2} \sin \frac{n\pi x}{2L}$$

where t_0 is equal to one-half the fundamental period and for $t = t_0$ the displacement reaches its maximum (see Ref. 2, p. 376). Derive the expression for the axial force at the fixed end of the bar and find the answer that is obtained by summing the contributions of (1) the first two modes and (2) the first 10 modes.

19-44 Rework Prob. 19-25, assuming that the true bending moment is obtained by RMS summation.

19-45 The base of the beam in Prob. 19-18 is subjected to harmonic acceleration $a = a_0 \sin \Omega t$, where $\Omega = \omega_1$ is the fundamental frequency. The maximum bending moment and a shear force are available from Prob. 15-42. Calculate the corresponding quantities for a discrete model of this beam having one-half of the mass lumped at the center. Compare the moment and shear response if the magnification factor μ_1 is the same for both systems.

19-46 The flexural beam, like one in Fig. 14-11, when subjected to a harmonic base acceleration $a_b = a_0 \sin \Omega t$, has the rth mode bending moment

$$\mathfrak{M}_r = \mu \frac{mL^2 a_0}{8r^3}$$

(see Prob. 14-12). Using the same method as in Prob. 18-32, calculate the mean-square bending moment at the center of the beam when the acceleration forcing

is a random function of time with a spectrum $S_a(\Omega)$. Present the result as a series involving the mode numbers. What portion of the true answer would be provided by the first mode? The rth mode frequency is given by

$$\omega_r = \frac{r^2}{L^2} \left(\frac{96EI}{m} \right)^{1/2}$$

19-47 A simply supported, square plate consisting of 36 square elements is subjected to a step load at the center. The calculated displacement and the stress response depend on how many vibratory modes were included in the computation:

Numbers of Modes	u_c (in.)	σ_c (psi)
1	0.6238	22,965
11	0.6791	29,238
20	0.6875	29,609

In this table u_c is the center displacement and σ_c is the bending stress in an element adjacent to the center, both being peak values with respect to time. The eleventh mode corresponds to the deformed pattern illustrated in Fig. 19.2. Why did u_c change by only 8.9% while σ_c went up by 27.3% when the number of contributing modes was changed from 1 to 11? Why did the results change so little when the number of modes was further increased to 20?

19-48 An axial bar with $E = 10.6 \times 10^6$ psi is suddenly loaded at the free end with stress $\sigma_0 = 25,000$ psi. As a result, a local thickening of the bar diameter will impact a rigid stop. If our interest is limited to what happens prior to that impact, which part of the bar will require many joints in modeling, and which will not? Ignore the effect of the local discontinuity of the cross section and use the data in the figure.

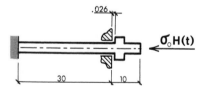

Answers to Exercise Problems

Chapter 1

1-17 $\tau = 11.63 \times 10^{-3}$ s

1-18 $\omega = \left(\dfrac{73}{33} \dfrac{EAg}{WL} \right)^{1/2}$

1-19 (a) $\dfrac{1}{k^*} = \dfrac{L^3}{3EI} + \dfrac{1}{k}$; (b) $k^* = \dfrac{3EI}{L^3} + 2k$

$\omega = \left(\dfrac{k^* g}{W} \right)^{1/2}$

1-20 $\omega^2 = \dfrac{24EIg}{Wh^3}$, $f = 7.1$ Hz

1-21 $\dfrac{1}{\omega^2} = M \left(\dfrac{L^3}{3EI} + \dfrac{L^2}{K} \right)$

1-22 $f = 1.337$ Hz

1-23 $\omega^2 = \dfrac{K}{(M_1 + M_2)h^2 + J}$

1-24 (1), (2) $\omega^2 = \dfrac{2}{M} \left(k_1 + \sqrt{2} \, \dfrac{N_0}{a} \right)$;

(3) $\omega^2 = \dfrac{4\sqrt{2}}{M} \dfrac{N_0}{a}$

1-25 (1) $\omega^2 = \dfrac{3}{M} \left(k_1 + \dfrac{N_0}{a} \right)$; (2) $\omega^2 = \dfrac{6N_0}{Ma}$

1-26 $f = 297.8$ Hz

1-27 For both cases $\omega^2 = K^* \left(\dfrac{1}{J_1} + \dfrac{1}{J_2} \right)$.

In (a), $K^* = \dfrac{GC}{L}$; $C = \dfrac{\pi d^4}{32}$.

In (b), $\dfrac{1}{K^*} = \dfrac{L_1}{GC_1} + \dfrac{L_2}{GC_2}$; $C_1 = \dfrac{\pi d_1^4}{32}$; $C_2 = \dfrac{\pi d_2^4}{32}$.

1-28 $\omega^2 = \dfrac{k}{M + 2J/r^2}$

1-29 $\omega^2 = \dfrac{K/L^2 + 2k}{M}$

1-30 $\omega^2 = \dfrac{kL_2^2}{J_2 + Me^2 + J_1 L_1^2/r^2}$

Chapter 2

2-19 $u_0 = \dfrac{v_{max}}{\omega}$

2-20 Define the angle α by $\cos \alpha = u_0/A$ and $\sin \alpha = v_0/(\omega A)$ (refer to Prob. 2-1). Note that α can be anywhere in the range $(-180°, 0°, +180°)$ depending on the initial conditions.

2-21 $\Pi = \frac{1}{2}kA^2$, $T = \frac{1}{2}MA^2\Omega^2$,

$U_p = \dfrac{1}{2}kA^2 \left(1 - \dfrac{\Omega^2}{\omega^2} \right)$

2-22 $\dfrac{u}{u_{st}} \approx -\dfrac{\Omega}{2\Delta} \cos \Omega t \sin \Delta t$

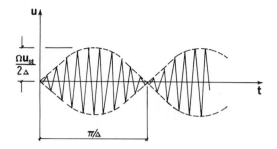

2-23 (1) $\Omega \ll \omega$ $\theta \approx 0$, $u_p = \dfrac{P_0}{k} \sin \Omega t$

(2) $\Omega = \omega$ (resonance), $\theta = \dfrac{\pi}{2}$,

$$u_p = \frac{P_0}{k}\frac{1}{2\zeta}\sin\left(\Omega t - \frac{\pi}{2}\right)$$

(3) $\Omega \gg \omega$ $\theta \approx \pi$, $u_p = -\dfrac{P_0}{M\Omega^2}\sin\Omega t$

2-24 $u_{max} = 5.2183 P_0/k$ when all three maxima happen at the same time.

2-25 $P_1 = \dfrac{\mu_0}{\mu_1}P_0 = 0.8267\,P_0$

2-26 Displacement at $\Omega_c = 95.9$ Hz,
velocity at $\Omega_c' = 100$ Hz,
acceleration at $\Omega_c'' = 104.3$ Hz

2-27 $N_{max} = 155$ Lb

2-28 $P(t) = \dfrac{P_0}{2} - \dfrac{4P_0}{\pi^2}\left(\cos\Omega t + \dfrac{\cos(3\Omega t)}{3^2} + \dfrac{\cos(5\Omega t)}{5^2} + \cdots\right)$

2-29

$$P(t) = \frac{8P_0}{\pi^2}\left(\sin\Omega t - \frac{\sin(3\Omega t)}{3^2} + \frac{\sin(5\Omega t)}{5^2} - \cdots\right)$$

2-30 $u = u_0 - (u_1 + u_3 + u_5 + \cdots)$

where $u_0 = \dfrac{P_0}{2k}$

$$u_1 = \frac{4P_0}{\pi^2 k}\frac{\cos\Omega t}{1 - \Omega^2/\omega^2}$$

$$u_3 = \frac{4P_0}{(3\pi)^2 k}\frac{\cos(3\Omega t)}{1 - (3\Omega/\omega)^2}; \qquad \text{etc.}$$

2-31 $u = u_{st}\left(\dfrac{t}{t_0} - \dfrac{\sin\omega t}{\omega t_0}\right)$ where $u_{st} = \dfrac{P_0}{k}$

2-32 $u = u_{st}\left[1 - \dfrac{\tau}{\pi t_0}\sin\dfrac{\pi t_0}{\tau}\cos\omega\left(t - \dfrac{t_0}{2}\right)\right]$

The amplitude of the second term in brackets is of the form $\sin\alpha/\alpha$, where $\alpha = \pi t_0/\tau$. Only for $\alpha = 0$, which is the case of a step function with magnitude P_0, can the content of brackets be as large as 2.0. The shortest t_0 after which oscillations do not occur is $t_0 = \tau$.

2-33 $u = -\dfrac{2(\Omega/\omega)u_{st}}{1 - \Omega^2/\omega^2}\sin\omega\left(t - \dfrac{\pi}{2\Omega}\right)\cos\dfrac{\pi\omega}{2\Omega}$

for $t > t_0$

2-34 The dynamic twisting moment amplitude is 354.2 Lb-in.

2-35 $u(t) = \dfrac{P_0}{M\omega}[\sin\omega t - \sin\omega(t - t_0)]$

for $t_0 = \dfrac{\pi}{\omega}$

2-36 $\alpha = \dfrac{M_{t0}}{J\omega^2}[(1 - \cos\omega t) + H(t - t_0)\times$
$(1 - \cos\omega(t - t_0)) - 2H(t - 2t_0)(1 - \cos\omega(t - 2t_0))]$

2-37 $u_{max} = \dfrac{2W}{k}$,   with rebound to the initial position.

$$v_{max}^2 = \frac{gW}{k}$$

2-38 $u = \dfrac{P_0}{M\omega_d}H(t - t_0)e^{-\omega\zeta(t - t_0)}\sin\omega_d(t - t_0)$
due to $P = P_0\delta(t - t_0)$

2-39 $u = \dfrac{P_0}{M\omega^2}\left(\dfrac{t}{t_0} - \dfrac{\sin\omega t}{\omega t_0}\right)$

2-40 $\alpha = \dfrac{M_{t0}}{J\omega^2}\left\{\dfrac{1}{\omega t_1}[\sin\omega t - \sin\omega(t - t_1)] - \cos\omega t\right\}$

$\mu_{max} = 0.3107$

2-41 $u(t) = \dfrac{P_0}{M}\dfrac{1}{a^2 + \omega^2}\left(e^{-at} - \cos\omega t + \dfrac{a}{\omega}\sin\omega t\right)$

Chapter 3

3-15 $u_m = \left[\left(\dfrac{R_0}{k} \right)^2 + \left(\dfrac{v_0}{\omega} \right)^2 \right]^{1/2} - \dfrac{R_0}{k}$

$R_m = R_0 + k u_m = k \left[\left(\dfrac{R_0}{k} \right)^2 + \left(\dfrac{v_0}{\omega} \right)^2 \right]^{1/2}$

3-16 $\tau = \dfrac{\pi}{\omega_1} + \dfrac{2}{\omega_1} \arcsin\left(\dfrac{b\omega_1}{v_0} \right) + \dfrac{2}{\omega_2} \arctan\left(\dfrac{k_2 \dot{u}_1}{\omega_2 k_1 b} \right)$

$\omega_1^2 = \dfrac{k_1}{M}; \quad \omega_2^2 = \dfrac{k_2}{M}; \quad \dot{u}_1 = v_0 \left[1 - \left(\dfrac{b\omega_1}{v_0} \right)^2 \right]^{1/2}$

3-17 $\omega = \dfrac{2\pi}{\tau}; \quad \tau = \pi\left(\sqrt{\dfrac{M}{k_1}} + \sqrt{\dfrac{M}{k_2}} \right)$

3-18 $\omega = 15.571$ rad/s

3-19 $\omega = 15.574$ rad/s

3-20 $c = 0.96$ Lb-s/in., $A = 1.2$ in. (A would be 1.5625 in. if the system were treated as linear).

Chapter 4

4-12 The force transmitted to the foundation is proportional to μ. At $\Omega/\omega = 1.2$ increasing damping tenfold reduces μ from 2.272 to 2.193. At $\Omega/\omega = 1.6$ this reduction is even less. *Conclusion*. Increasing damping is not an effective way of reducing the dynamic response in this particular case.

4-13 $\zeta = \dfrac{\Omega_2 - \Omega_1}{\Omega_2 + \Omega_1}$

4-14 $\zeta_1 = 0.0443, \quad \zeta_2 = 0.0747, \quad \zeta_3 = 0.1325$

4-15 $\psi = 0.1076$

4-16 $N_{max} = 150.6$ Lb

Chapter 5

5-14 Acceleration is 205.1 in./s^2; shear 41.02 Lb, and bending moment 820.4 Lb-in.

5-15 Maximum spring force is 150 Lb. Maximum possible damper force is 95.3 Lb.

5-16 At $\Omega = \omega$ or $v = 3.18$ mph $n = 0.777$; at $\Omega = 2\omega$ or $v = 6.36$ mph $n = 0.774$. Note that n does not include the acceleration of gravity.

5-17 $F = 1288$ Lb

5-18 $\left(\dfrac{\overline{A}}{u_0} \right)_{max} = 1.7761$ at $\dfrac{\Omega}{\omega} = 1.21$

Chapter 6

6-9 The impact force will increase by a factor of 1.587. *Hint*. Note that the radius of the ball also increases.

6-10 (1) $R_m'' = 5.278 R_m', \quad t_1'' = 0.758 t_1';$

(2) $R_m'' = 0.5274 R_m', \quad t_1'' = 1.896 t_1'.$

6-11 $t_1 = 0.1275 \times 10^{-3}$ s

6-12 $R_m = 149{,}500$ Lb $(k^* = 1.9527 \times 10^6$ Lb/in.)

6-13 From the work-energy balance $u_m = 0.2888$ in. and from the characteristic, $R_m = 45.76$ Lb.

6-14 $t_1 = 0.1273 \times 10^{-3}$ s

6-15 $t_1 = 0.1288 \times 10^{-3}$ s, 1% error

6-16 (a) $\mu' = 0.618, \quad \mu'' = 0.628;$
(b) $\mu' = 0.3963, \quad \mu'' = 0.4$

6-17 Rectangular impulse: $\mu' = 1.4142, \mu'' = 1.481,$ $e = 4.7\%$; triangular impulse: $\mu' = 0.7458, \mu'' = 0.7405,$ $e = 0.7\%$

6-18 $\sigma_{max} = 7.25$ Lb/in.^{2}

6-19 $\sigma = 28{,}142$ psi

6-20 (1) $h = 4.586$ in. ($n = 23.76$), (2) $n = 102$.

6-21 $n = 5.878; \quad u_{max} = 0.4538$ in.

6-22 $n = 4.939; \quad u_{max} = 0.3813$ in.

6-23 $R_m = 22{,}859$ Lb

6-24 The impact force will be 19,454 Lb, which is less than 22,859 Lb calculated in the reference problem. The gap is undesirable.

6-25 $R_m = 11{,}200$ Lb

6-26 $W_1 = 8159$ Lb

Chapter 7

7-12 Both ends supported, $\omega^2 = \dfrac{48EI}{M_0 L^3}$;

both ends fixed, $\omega^2 = \dfrac{192EI}{M_0 L^3}$

One end fixed, one supported, $\omega^2 = \dfrac{109.71EI}{M_0 L^3}$

7-13 $\omega^2 = \dfrac{7ka^2}{J}$

7-14 $\omega^2 = \dfrac{16\pi(1+\nu)D}{(3+\nu)R^2 M}$ (simply supported)

$\omega^2 = \dfrac{16\pi D}{R^2 M}$ (clamped)

The ratio of the clamped-plate frequency to that of the simply supported plate is 1.5933.

7-15 $u = |A| e^{i(\Omega t - \theta)}$ or

$u = |A|[\cos(\Omega t - \theta) + i\sin(\Omega t - \theta)]$

Note that Prob. 7-4 is a special case of this solution. Figure 7-4 gives the answer for the cosine forcing if a projection is made on the horizontal axis.

7-16 $\sigma_{\max} = 134{,}892$ psi, $N_{\max} = 2250$ Lb

7-17 Combine the step function (Prob. 2-12) with the exponential decay function (Prob. 2-41), getting

$$P = P_0(1 - e^{-at})$$

Displacement response:

$$\frac{u}{u_{st}} = \frac{\omega^2}{a^2 + \omega^2}\left(\frac{a^2}{\omega^2} + 1 - e^{-at} - \frac{a}{\omega}\sin\omega t - \frac{a^2}{\omega^2}\cos\omega t\right)$$

7-18 The forcing function may be reduced to a single vector by summing the three components by means of vector addition. Afterward, the reference problem is followed.

7-19 Minimizing the impact force is the same as minimizing the peak acceleration. A set of trial values inserted in place of ζ in Eq. 6.13 gives us a minimum acceleration of

$$a = 0.7833 v_0 \left(\frac{k}{M}\right)^{1/2} \quad \text{at} \quad \zeta \approx 0.31$$

7-20 After four cycles.

7-21 $n = 2.78$

7-22 $n = 7.19$

7-23 $D = 0.6224$

Chapter 8

8-19
$$\mathbf{k} = \begin{bmatrix} 3 & -1 & 0 & 0 \\ -1 & 2.5 & -1 & 0 \\ 0 & -1 & 3 & -2 \\ 0 & 0 & -2 & 5 \end{bmatrix} \frac{EA}{L}$$

8-20
$$\mathbf{k} = \frac{3EI}{L^3}\begin{bmatrix} 1 + \dfrac{kL^3}{3EI} & -1 \\ -1 & 1 \end{bmatrix}$$

8-21
$$\mathbf{k} = \begin{bmatrix} \dfrac{12}{L^2} & \dfrac{-6}{L} \\ \dfrac{-6}{L} & 4 \end{bmatrix} \frac{EI}{L}$$

8-22
$$\mathbf{a} = \begin{bmatrix} 0.33333 & 0.17284 & 0.04938 \\ & 0.09877 & 0.03086 \\ \text{symmetric} & & 0.01235 \end{bmatrix} \frac{L^3}{EI}$$

8-23 Only nonzero coefficients of the lower diagonal matrix are specified:

$$k_{11} = K_1, \qquad k_{21} = -K_1, \qquad k_{22} = K_1 + K_2\left(\frac{r_2}{r_{12}}\right)^2,$$

$$k_{23} = -K_2 \frac{r_2}{r_{12}},$$

$$k_{33} = K_2 + K_3\left(\frac{r_3}{r_{13}}\right)^2 + K_4\left(\frac{r_3}{r_{13}}\right)^2,$$

$$k_{34} = -K_3\frac{r_3}{r_{13}}, \qquad k_{35} = -K_4\frac{r_3}{r_{13}}, \qquad k_{44} = K_3,$$

$$k_{55} = K_4 + K_5 \left(\frac{r_5}{r_{15}} \right)^2, \qquad k_{56} = -K_5 \frac{r_5}{r_{15}}$$

$$k_{66} = K_5 + K_6 \left(\frac{r_6}{r_{16}} \right)^2, \qquad k_{67} = -K_6 \frac{r_6}{r_{16}},$$

$$k_{77} = K_6 + K_7, \qquad k_{78} = -K_7,$$

$$k_{88} = K_7 + K_8 \left(\frac{r_8}{r_{18}} \right)^2,$$

$$k_{89} = -K_8 \frac{r_8}{r_{18}}, \qquad k_{99} = K_8$$

8-24 $\quad \mathbf{k}_q = (A_1 + A_2) \dfrac{E}{L}$

8-26

$$\mathbf{k}_q = \begin{bmatrix} \dfrac{12}{L^2} I_2 & -\dfrac{6}{L} I_2 & -\dfrac{12}{L^2} I_2 & -\dfrac{6}{L} I_2 \\[2mm] -\dfrac{6}{L} I_2 & 4 I_2 & \dfrac{6}{L} I_2 & 2 I_2 \\[2mm] -\dfrac{12}{L^2} I_2 & \dfrac{6}{L} I_2 & \dfrac{12}{L^2}(I_1 + I_2) & \dfrac{6}{L}(I_2 - I_1) \\[2mm] -\dfrac{6}{L} I_2 & 2 I_2 & \dfrac{6}{L}(I_2 - I_1) & 4(I_2 + I_1) \end{bmatrix} \dfrac{E}{L}$$

8-28

$$\mathbf{k} = \begin{bmatrix} 2 & -1 & 0 \\ -1 & 2 & 1 \\ 0 & 1 & 2 \end{bmatrix} \dfrac{EA}{L} r^2$$

8-30

$$\mathbf{a} = \begin{bmatrix} 2.6114 & & & \\ 1.7148 & 1.2117 & & \text{symmetric} \\ 0.7898 & 0.6144 & 0.3953 & \\ 0.208 & 0.208 & 0.208 & 0.208 \end{bmatrix} 10^{-6}$$

8-31

$$\mathbf{k} = \begin{bmatrix} 2.9494 & 3.9809 & 0 \\ 3.9809 & 7.9618 & 0 \\ 0 & 0 & 0.07161 \end{bmatrix} 10^3$$

Chapter 9

9-20

$$\mathbf{k} = \begin{bmatrix} 3k & -2k & 0 \\ -2k & 6k & -4k \\ 0 & -4k & 4k \end{bmatrix}$$

$$\boldsymbol{\Phi} = \begin{bmatrix} 1.0 & 1.0 & 1.0 \\ 1.4404 & 0.185 & -0.6254 \\ 1.5317 & -0.5874 & 0.5557 \end{bmatrix}$$

$$\omega_1^2 = \frac{0.11927k}{M} \quad \omega_2^2 = \frac{2.63k}{M} \quad \omega_3^2 = \frac{4.251k}{M}$$

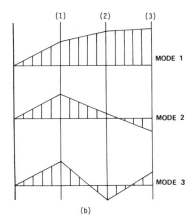

(b)

9-21

$$\boldsymbol{\Phi} = \begin{bmatrix} 1.0 & 1.0 \\ 0.8776 & -0.4558 \end{bmatrix}$$

$$\omega_1^2 = 0.3737 \times 10^6 \ (\text{rad/s})^2,$$
$$\omega_2^2 = 7.1499 \times 10^6 \ (\text{rad/s})^2$$

9-22

$$\mathbf{k} = \begin{bmatrix} 337{,}500 & -150{,}000 \\ -150{,}000 & 150{,}000 \end{bmatrix}$$

$$\boldsymbol{\Phi} = \begin{bmatrix} 1.0 & 1.0 \\ 1.7711 & -0.6211 \end{bmatrix}$$

$$\omega_1 = 7.1 \ \text{rad/s}, \qquad \omega_2 = 17.385 \ \text{rad/s}$$

9-23 $\quad \omega_1 = 558.0 \ \text{rad/s}, \qquad \omega_2 = 851.3 \ \text{rad/s}$

$$\boldsymbol{\Phi} = \begin{bmatrix} 1.0 & 1.0 \\ 1.1427 & -1.108 \end{bmatrix}$$

9-24

$$\mathbf{M} = \begin{bmatrix} 0.3120 & 0 \\ 0 & 4.9288 \end{bmatrix}$$

$$\omega_1 = 880.75 \ \text{rad/s}, \qquad \omega_2 = 5923.3 \ \text{rad/s}$$

9-25 Symmetric, $\omega^2 = \dfrac{162}{5}\dfrac{EI}{ML^3}$;

antisymmetric, $\omega^2 = 486\dfrac{EI}{ML^3}$

9-26
$$\mathbf{k} = \begin{bmatrix} 43.512 & -99.350 & 74.277 \\ -99.350 & 273.02 & -284.97 \\ 74.277 & -284.97 & 496.07 \end{bmatrix}\dfrac{EI}{L^3}$$

$$\boldsymbol{\Phi} = \begin{bmatrix} 1.0 & 1.0 & 1.0 \\ 0.5401 & -0.7068 & -1.5919 \\ 0.1618 & -0.7307 & 2.2242 \end{bmatrix}$$

$\omega_1 = 2.982$ rad/s, $\omega_2 = 16.831$ rad/s,
$\omega_3 = 41.912$ rad/s

9-27 $\omega_1 = 77.79$ rad/s, $\omega_2 = 88.62$ rad/s,
$\omega_3 = 145.7$ rad/s

$$\boldsymbol{\Phi} = \begin{bmatrix} 1.0 & 0 & 1.0 \\ 0 & 1.0 & 0 \\ -0.005 & 0 & 0.0371 \end{bmatrix}$$

9-28 Noting that the initial displacement is proportional to the second mode shape, we have

$$\begin{Bmatrix} u_1 \\ u_2 \\ u_3 \end{Bmatrix} = \begin{Bmatrix} 1.0 \\ 0.185 \\ -0.5874 \end{Bmatrix} 0.1\cos\omega_2 t$$

9-29
$-u_1 = 0.4556\sin\omega_1 t + 0.0292\sin\omega_2 t + 0.0091\sin\omega_3 t$
$-u_2 = 0.6562\sin\omega_1 t + 0.0054\sin\omega_2 t - 0.0057\sin\omega_3 t$
$-u_3 = 0.6978\sin\omega_1 t - 0.0172\sin\omega_2 t + 0.0051\sin\omega_3 t$

9-30 $\omega_1^2 = 0$ $\omega_2^2 = k\left(\dfrac{1}{M_1} + \dfrac{1}{M_2}\right)$

$$\boldsymbol{\Phi} = \begin{bmatrix} 1 & 1 \\ 1 & \dfrac{-M_1}{M_2} \end{bmatrix}$$

9-31
$$\boldsymbol{\Phi} = \begin{bmatrix} 1.0 & 1.0 & 1.0 \\ 2.197 & 2.0024 & -2.0586 \\ -4.4602 & 1.1305 & -0.0420 \end{bmatrix}$$

$\omega_1 = 0$ rad/s, $\omega_2 = 44.71$ rad/s,
$\omega_3 = 120.19$ rad/s

9-32 $M_{t12} = (0.8638\cos\omega_2 t + 0.1362\cos\omega_3 t)M_{t0}$

$M_{t23} = (0.4852\cos\omega_2 t - 0.4852\cos\omega_3 t)M_{t0}$

$\omega_2^2 = \dfrac{1.2192K}{J}$, $\omega_3^2 = \dfrac{3.2808K}{J}$

9-33 Six rigid-body modes and six elastic modes. The deflected patterns are as in Prob. 9-19, except that the xz-plane is involved as well as the xy-plane. The only new item is the twisting mode.

Chapter 10

10-18 $\zeta_r = \frac{1}{2}b\omega_r + \dfrac{\beta}{2\omega_r}$

10-19 $\omega_{1d} = 0$, $\omega_{2d}^2 = \omega_2^2(1 - \zeta_2^2)$

with $\zeta_2 = \dfrac{c}{2(kM^*)^{1/2}}$

This is the same as the damping ratio of a basic oscillator with a spring constant k, damping constant c, and mass M^*.

10-20 $\begin{Bmatrix} u_1 \\ u_2 \\ u_3 \end{Bmatrix} = \begin{Bmatrix} -3.1145 \\ -0.5762 \\ 1.8295 \end{Bmatrix} 10^{-3}P_0\sin(\Omega t - \theta_2)$

$N_{01} = 3.1145P_0$, $N_{12} = 5.0766P_0$,
$N_{23} = 9.6228P_0$

10-21 $s_1 = \dfrac{P_0}{6746.2}\cos(\Omega t + 1.8001)$

$s_2 = \dfrac{P_0}{164.9}\cos(\Omega t - 2.0775)$

$s_3 = \dfrac{P_0}{2139.6}\cos(\Omega t + 0.9767)$

10-22 $\Omega_0 = \left(\dfrac{k_{11}}{M_{11}}\right)^{1/2} = \left(\dfrac{3k}{M}\right)^{1/2}$

10-23
$$\boldsymbol{\Phi}^T\mathbf{c}\boldsymbol{\Phi} = \begin{bmatrix} 64.105 & -23.2 \\ -23.2 & 8.396 \end{bmatrix}$$

$\zeta_1 = 0.0465$, $\zeta_2 = 0.0289$

10-24 The joint forces from matrix multiplication are $P_1 = 1957$, $P_2 = -2994$, $P_3 = -2405$ Lb. These are the same as the forces in Fig. 10-14, within the accuracy of computation.

10-25 $\alpha_1 = 0.3199$ rad, $\alpha_2 = 0.3707$ rad,
$\alpha_3 = -0.3199$ rad

10-26 BC, 220.9; CD, 132.6; DF, -162.0 Lb;
$(N_0)_{min} = 220.9$ Lb, $(M_{t0})_{max} = 2240.8$ Lb-in.

10-27 Adding the absolute values of the symmetric and antisymmetric models, each affected by P_1 and P_2, we get $ub(\mathfrak{M}) = 2898$ Lb-in.

10-28 The largest static bending is $\mathfrak{M}_{st} = -(\frac{5}{9})P_0 L$. The upper bound of dynamic bending is $ub(\mathfrak{M}_d) = 0.2542 P_0 L$ $ub(\mu) = 0.4576$.

10-29 Denoting by Q_{0i} the multiplier of $H(t)$ in the ith generalized force, we have

$$ub\begin{Bmatrix} u_1 \\ u_2 \end{Bmatrix} = \begin{bmatrix} |x_{11}| & |x_{12}| \\ |x_{21}| & |x_{22}| \end{bmatrix} \begin{Bmatrix} \dfrac{2|Q_{01}|}{\bar{k}_1} \\ \dfrac{2|Q_{02}|}{\bar{k}_2} \end{Bmatrix}$$

for a two-DOF system. A generalization to n degrees is obvious.

10-30
$$ub\begin{Bmatrix} u_1 \\ u_2 \\ u_3 \end{Bmatrix} = \begin{Bmatrix} 1.4964 \times 10^{-3} P_0 \\ 1.2491 \times 10^{-3} P_0 \\ 1.5831 \times 10^{-3} P_0 \end{Bmatrix}$$

Displacement is in inches if P_0 is in pounds.

10-31 The required forces are: $P_1 = -9100$ Lb, $P_2 = 28,983$ Lb, and $P_3 = -40,518$ Lb.

10-32 $N_{12} = N_1'' + N_1' = (0.75 - 0.5\cos\omega_1 t - 0.25\cos\omega_2 t)P_0$
$N_{23} = N_1'' - N_1' = (0.25 - 0.5\cos\omega_1 t + 0.25\cos\omega_2 t)P_0$

Chapter 11

11-13 $f = 2.528$ Hz (Compare with Prob. 9-20.)

11-14 $\omega = 3.494\left(\dfrac{EI}{mL^4}\right)^{1/2}$

11-15 $\omega = 1.1147\left(\dfrac{k}{M}\right)^{1/2}$

and $\vec{\Phi}^T = [1.0 \quad 0.9167]$
(Compare with Prob. 9-3.)

11-16 $\omega = 464.54$ rad/s   or $f = 73.93$ Hz
(A computer-aided analysis gives $f = 72.73$ Hz.)

11-17 $\omega = \dfrac{4.635}{L^2}\left(\dfrac{EI_1}{\rho A_1}\right)^{1/2}$

11-18 $f = 2.373$ Hz

11-19 $f = 78.58$ Hz

11-20 $\omega = 0.3536\left(\dfrac{K}{J}\right)^{1/2}$

11-21 $f = 20.22$ Hz

11-22 $f = 91.632$ Hz

11-23 $f = 71.85$ Hz

11-24 $\omega = \dfrac{1.4627}{R^2}\left(\dfrac{EI}{m}\right)^{1/2}$

11-25 $\omega = \dfrac{3.113}{L^2}\left(\dfrac{EI_1}{\rho A_1}\right)^{1/2}$

11-26 $\omega_1 = 11.042$ rad/s $\omega_2 = 18.11$ rad/s

11-27 $\omega = 3.012$ rad/s $\vec{\Phi} = \begin{Bmatrix} 1.000 \\ 0.433 \\ -0.096 \\ -0.394 \end{Bmatrix}$

11-28 The amplitudes are $u_1 = 1.0835 P_0/k$ and $u_2 = 0.9932 P_0/k$. The upper bounds from Prob. 10-3 are, respectively, $1.189 P_0/k$ and $1.006 P_0/k$.

11-29 $\omega = 69.78$ rad/s

11-30 New $\omega_1 \approx 66.86$ rad/s

Chapter 12

12-13

$$\begin{Bmatrix} \ddot{u}_1 \\ \ddot{u}_2 \end{Bmatrix} = \begin{Bmatrix} -2 + \cos\omega_1 t + \cos\omega_2 t \\ -0.5 + 0.8792\cos\omega_1 t - 0.3792\cos\omega_2 t \end{Bmatrix} a_0$$

12-14 The forcing frequency at which M_1 is motionless is defined by $\Omega^2 = \Omega_0^2 = k_d/M_2$.

12-15 Joint forces, mode 1, $1.193W$ and $1.920W$; joint forces, mode 2, $0.567W$ and $-0.320W$; RSS resultants, $1.321W$ and $1.946W$

12-16 The differences in computed values are well below 1%.

12-17 $V = 110,595$ Lb, $\mathfrak{M} = 9.573 \times 10^6$ Lb-in., $\tau = 751$ psi, $\sigma = 3682$ psi. *Note.* The shear area was assumed to be one-half of the total area.

12-18 $\vec{j}^T = [0 \quad 0 \quad 1 \quad 0 \quad 0 \quad 0 \quad 1 \quad 1]$

12-19 $P_t = 2700$, $P_1 = 99,500$, $P_2 = 133,300$, $P_3 = 298,500$ Lb, $V_c = 133,500$ Lb, $\mathfrak{M}_c = 12.015 \times 10^6$ Lb-in.

12-20 Required thickness at the base is 0.7663 in. (due to seismic bending and gravity load).

12-21 The axial load in the legs induced by overall bending is $\sqrt{2}$ times larger for the diagonal direction. There is no difference in the local bending caused by the base shear.

Chapter 13

13-15 The velocities will be interchanged, that is, $V_1 = v_2$ and $V_2 = v_1$.

13-16 $\dfrac{M_2}{M_1} = 0.5405$

13-17 $\alpha = 81.7°$

13-18 A second collision takes place after rebound from the barrier; $R_m = 97,416$ Lb.

13-19 Prescribed formula, $t_1 = 84.91 \times 10^{-6}$ s; Sect. 13C, 85.84×10^{-6} s

13-20 $\sigma_m = v(E\rho)^{1/2} = 43,667$ psi

13-21 (1) 12,775 Lb, (2) 9590 Lb

13-22 $R_m = v_r \left(\dfrac{M^* k^*}{1 - \zeta^2} \right)^{1/2} \exp\left[\left(2\zeta - \dfrac{\pi}{2} \right) \zeta \right]$

where $\dfrac{1}{M^*} = \dfrac{1}{M_1} + \dfrac{1}{M_2}$

$$\frac{1}{k^*} = \frac{1}{k_1} + \frac{1}{k_2}$$

$$\zeta = \frac{c^*}{2(k^* M^*)^{1/2}}$$

$$\frac{1}{c^*} = \frac{1}{c_1} + \frac{1}{c_2}$$

13-23 The upper bound of u_2 is 0.0211 in.

13-24 The new overload factor of the lighter car is $n_2 = 5.167$.

13-25 $u_m = 0.04522 \times 0.9694 = 0.04384$ in.

13-26 $\tan\beta = \dfrac{1}{\kappa}\tan\alpha$, $V = (\kappa^2 \cos^2\alpha + \sin^2\alpha)^{1/2} v$

13-27 By equating the work of force of gravity with strain energy, we get $\delta_m = 2.3695$ in. and $R_m = 36.44 \times 10^6$ Lb.

13-28 $\kappa = 0.675$

13-29 $c = 2L/3$. The point at this location is called the center of impact.

13-30 $d = \dfrac{a}{4}$.

13-31 $R_m = 5946$ Lb and $\mathfrak{M} = 356,730$ Lb-in.

13-32 $\mathfrak{M}_{max} = 28.78 \times 10^6$ Lb-in.

13-33

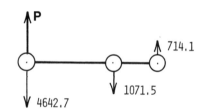

(b)

13-34

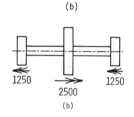

(b)

13-35 600 Lb acting upward on the beam.

13-36 10,000 Lb acting downward on the beam.

13-37 $\sigma_{max} = \sigma_a + \sigma_b = 2136 + 11,560 = 13,697$ psi, $\sigma_{max} < 36,000$ psi, yielding not initiated

Chapter 14

14-15 $\mathbf{a} = \mathbf{a_b} + \mathbf{a_s} = \begin{bmatrix} 3.2603 & 1.1395 \\ 1.1395 & 0.6489 \end{bmatrix} 10^{-6}$

14-16 $f_z = 3.383$ Hz, $f_{yz1} = 2.763$ Hz, $f_{yz2} = 5.442$ Hz

14-17 (b) $\omega^2 = 48 \dfrac{EI}{ML^3}$;   (c) $\omega^2 = 192 \dfrac{EI}{ML^3}$

14-18 Displacement, 2.292×10^{-3} in., rotation, 0.0653×10^{-3} rad,   shear force, 7701 Lb

14-19 $u_{z\,\text{max}} = 2.241 \times 10^{-3} + 31.45 \times 10^{-6} \times 105 + 51.09 \times 10^{-6} \times 72 = 9.222 \times 10^{-3}$ in.

14-20 Write the equations for a two-DOF system, set $M_2 = 0$, and obtain the following:

$$M_1 \ddot{u}_1 + 3\left(\dfrac{EI}{L^3}\right) u_1 = 0.3125P$$

14-21 $\alpha = 21.435 \times 10^{-6}\left(t - \dfrac{\sin \omega t}{1075.5}\right)$ (rad)

14-22 When the average of two bending flexibilities from Prob. 7-14 is used and when the shear flexibility is included, we get $R_m = 41{,}590$ Lb.

Chapter 15

15-17 $f_0 = 0$, $f_1 = 498.1$ Hz, $f_2 = 996.2$ Hz

15-18 $EAu' = k_1 u + M_1 \ddot{u}$ at $x = 0$;
$EAu' = -k_2 u - M_2 \ddot{u}$ at $x = L$

15-19 $\tan z = \dfrac{3z}{2z^2 - 1}$ with $z = \dfrac{\omega L}{c}$

Frequency $\omega_1 = 1.1362(k/M)^{1/2}$. The coefficient is 1.2247 in the reference problem.

15-20 $\bar{J} = 1.3511J$

15-21 Frequencies to be calculated from $\sin z = 0$ and $\tan z = -4z \dfrac{EA}{2Lk}$ where $z = \omega L/c$. Mode shape:

$$X_n = \sin \dfrac{n\pi x}{L} \qquad n = 1, 2, 3, \ldots$$

15-22 $\cos pL \cdot \cosh pL = -1$, $\omega_1 = \dfrac{1.875^2}{L^2}\left(\dfrac{EI}{m}\right)^{1/2}$

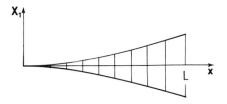

15-23 Frequency equation, $\cos pL \cdot \cosh pL = 1$, first vibratory frequency:

$$p_1 L = 4.73 \qquad \text{or} \qquad \omega_1 = \dfrac{4.73^2}{L^2}\left(\dfrac{EI}{A\rho}\right)^{1/2}$$

15-24 $EIu'' = K_1 u'$ and $EIu''' = -ku - M\ddot{u}$ at $x = 0$; $EIu'' = -K_2 u'$ and $EIu''' = 0$ at $x = L$

15-25 $\cos pL \cdot \cosh pL = 1$

15-26 $\dfrac{1}{\alpha}(\tan pL - \tanh pL) = 1 + \dfrac{1}{\cos pL \cdot \cosh pL}$

with $\dfrac{1}{\alpha} = \left(\dfrac{M}{mL}\right) pL$

15-27 (1) $p_1 L = 1.4200$, $\omega_1 = \dfrac{2.0164}{L^2}\left(\dfrac{EI}{m}\right)^{1/2}$

(2) $p_1 L = 1.1464$, $\omega_1 = \dfrac{1.3142}{L^2}\left(\dfrac{EI}{m}\right)^{1/2}$

15-28 $M = 0.2427mL$

15-29 $u(x, t) = \displaystyle\sum_{n=1,2,3,\ldots,}^{\infty} X_n f_n$, $X_n = \sin \dfrac{n\pi x}{L}$

$\omega_n = \dfrac{n\pi c_2}{L}$, $f_n(t) = \dfrac{32}{3}\left(\dfrac{\delta}{n^2 \pi^2}\right) \sin \dfrac{n\pi}{4} \cos \dfrac{n\pi c_2}{L} t$

15-30 $u(x, t) = \displaystyle\sum_{n=1,3,5,\ldots}^{\infty} X_n f_n$, $X_n = \sin \dfrac{n\pi x}{2L}$

$f_n = \dfrac{q_0}{m} \dfrac{32 L^2}{c^2 \pi^4 n^4} (1 - \cos \omega_n t)(-1)^{(n-1)/2}$

15-31 Maximum dynamic deflection at $t = 2L/c$:

$$u(L) = \dfrac{2}{3} \dfrac{q_0 L^2}{mc^2}, \qquad u_{\text{st}} = \dfrac{q_0 L^2}{3EA} \qquad \therefore \mu = 2.0$$

15-32 $u(x, t) = \dfrac{P_0}{mL} \times$

$$\left[\dfrac{t^3}{6t_0} + \sum_{n=1}^{\infty} \dfrac{2}{\omega_n^2} \cos \dfrac{n\pi}{2}\left(\dfrac{t}{t_0} - \dfrac{\sin \omega_n t}{\omega_n t_0}\right) \cos \dfrac{n\pi x}{L}\right]$$

15-33 $u(x,t)=X_2 f_2 = \dfrac{q_0}{m\omega_2^2}\sin\dfrac{2\pi x}{L}(1-\cos\omega_2 t)$

(There is only a single modal load.)

15-34 $u(x,t)=u(x)\sin\Omega t$

$u(x)=\dfrac{q_0}{m\Omega^2}\left[\dfrac{1-\cos(\Omega L/c_2)}{\sin(\Omega L/c_2)}\sin\dfrac{\Omega x}{c_2}+\cos\dfrac{\Omega x}{c_2}-1\right]$

15-35 $u(x,t)=\dfrac{4q_0\sin\Omega t}{\pi m}\times$

$$\sum_{n=1,3,5,\ldots}\dfrac{1}{n}\dfrac{1}{\omega_n^2-\Omega^2}\sin\dfrac{n\pi x}{L}$$

15-36 $u(x)=\dfrac{P_0 c}{EA\Omega}\dfrac{\sin(\Omega x/c)}{\cos(\Omega L/c)},$

$u(x,t)=u(x)\cos\Omega t$

15-37

$u(x,t)=\dfrac{2P_0}{mL}e^{i\Omega t}\sum_{n=1}^{\infty}\dfrac{1}{\omega_n^2-\Omega^2}\sin\dfrac{n\pi a}{L}\sin\dfrac{n\pi x}{L}$

15-38 $\mathfrak{M}=-\dfrac{2P_0 L}{\pi^2}e^{i\Omega t}\sum_{n=1,3,5,\ldots}\dfrac{1}{n^2}\dfrac{1}{1-\Omega^2/\omega_n^2}$

Notice that for $\Omega=0$, $\mathfrak{M}=-P_0 L/4$.

15-39 By analogy with an axial bar, Prob. 15-36,

$u(x)=\left(\dfrac{P_0 c_3}{GA_s\Omega}\right)\dfrac{\sin(\Omega x/c_3)}{\cos(\Omega L/c_3)}\cos\Omega t,$

$c_3^2=\dfrac{GA_s}{m},$

$m=\dfrac{M}{L},\qquad A_s=\dfrac{kL}{G}$

15-40 (1) $\omega_1=\dfrac{2.0098}{L^2}\left(\dfrac{EI}{m}\right)^{1/2},$

(2) $\omega_2=\dfrac{1.3120}{L^2}\left(\dfrac{EI}{m}\right)^{1/2}$

Both answers are less than 0.5% below those in Prob. 15-27.

15-41 $f=19.62$ Hz

15-42 $\mathfrak{M}=\dfrac{4a_0 mL^2}{\pi^3}\mu_1,\qquad V=\dfrac{4a_0 mL}{\pi^2}\mu_1$

Chapter 16

16-8 (1) Tension, $\tilde{f}=111.8$ Hz; (2) compression, $\tilde{f}=86.6$ Hz. A load of the same magnitude brings about a bigger relative change when applied in compression.

16-9 The vector of the external forces associated with the given displacement pattern is

$$\vec{\mathbf{N}}^T=[N\theta \quad 0 \quad -N\theta \quad 0]$$

which means we need to apply a force of magnitude $N\theta$ at each end, normal to the beam. The same was found before, in a much simpler way, for an axial bar.

16-10 $\tilde{\omega}=5.7966\left(\dfrac{EI}{ML^3}\right)^{1/2}$

16-11 Using the Ritz-Galerkin method gives us $\tilde{\omega}=0.8367\omega$ and $N_{cr}=4N_e$.

16-12 $N_0=12EI/L^2$ is the critical force obtained from Eq. 16.15.

16-13 The errors in the determination of the buckling force are (a) 1.3%, (b) 48.6%, and (c) 0.8%. Poor accuracy is the rule rather than the exception when approximate stiffening formulas are used to determine the buckling load. It is preferable to estimate the latter using a static calculation.

16-14 Not only the sign of the axial force changes, but also its magnitude grows to $\frac{5}{8}\Omega^2 mL^2$ at the center. The new expression is

$$\tilde{\omega}=\omega\left(1+\dfrac{5}{8}\dfrac{\Omega^2 mL^4}{\pi^2 EI}\right)^{1/2}$$

The natural frequency increases 1.271 times as a result of switching of supports.

16-15 Factor of safety $=1.265$

16-16

$u(x,t)=\dfrac{2P_0}{mL}\sin\Omega t\sum_{n=1}^{\infty}\dfrac{1}{\tilde{\omega}_n^2-\Omega^2}\sin\dfrac{n\pi}{2}\sin\dfrac{n\pi x}{L}$

The same equation is obtained from the solution of Prob. 15-37 if the latter is modified somewhat.

16-17 $\alpha_{max} \approx \dfrac{\alpha_{st}}{1-\Omega^2/\tilde{\omega}^2} = 5.982 \times 10^{-3}$ rad

16-18 $\tilde{\omega}_n^2 = \left(\dfrac{n\pi}{L}\right)^4 \dfrac{EI}{m} + \left(\dfrac{n\pi}{L}\right)^2 \dfrac{N}{m}$

Notice there is no difference between a cable that is simply supported and one that is fixed at both ends.

Chapter 17

17-10 When the disc is in the position shown in Fig. 17-10 and $l<0.866D$, the gyroscopic moment acts clockwise. This moment acts counterclockwise for $l>0.866D$. For $l=0.866D$ the moment is equal to zero.

17-11 $ec = \dfrac{2}{M}(J-J')\phi$ (in general)

$ec = \dfrac{1}{8}D^2\phi$ (for a thin disc)

17-12 $\rho_0 = \dfrac{\pm e}{1-(\Omega^2/\omega^2)}$ (with plus sign for $\Omega<\omega$)

17-13 Equations 17.7. Synchronous whirl, $\lambda=\Omega$.

17-14 $r_0 = 0.008333$ in., $\rho_0 = 0.00857$ in.
at $\Omega=\omega=481.25$ rad/s

17-15 $P(t)=\Omega^2 Me\cos\Omega t$ to obtain

$x = \pm \dfrac{\Omega^2 Me\cos\Omega t}{k(1-\Omega^2/\omega^2)}$ (plus sign for $\Omega<\omega$)

17-16 $P_B = \dfrac{M\Omega^2 e\left[1+(2\zeta\Omega/\omega)^2\right]^{1/2}}{\left[(1-\Omega^2/\omega^2)^2+(2\zeta\Omega/\omega)^2\right]^{1/2}}$

17-17 $r_0 = \dfrac{1}{1-\Omega^2/\omega^2}\dfrac{\Omega^2 M}{k}(\delta\pm e)$
(plus sign when $\Omega<\omega$)

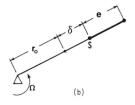
(b)

17-18 $\alpha = \dfrac{(J'-J)\Omega^2\phi}{\pm(1-\Omega^2/\omega_c^2)K}$ with $K=\dfrac{1}{2}k_1L^2$

and $\omega_c^2 = \dfrac{K}{J'-J}$

The plus sign in the denominator is for subcritical speeds, $\Omega<\omega_c$.

17-19 $\alpha = \dfrac{\phi}{1+\omega_c^2/\Omega^2}$, $\omega_c^2 = \dfrac{K}{J-J'}$, $K=\dfrac{1}{2}k_1L^2$

$\Omega_1 = \omega_c$ (α has always opposite sign to ϕ).

17-20 $x = \dfrac{1}{1-(\Omega/\omega_x)^2}\dfrac{\Omega^2 Me}{2k_x}\cos\Omega t$, $\omega_x^2 = \dfrac{2k_x}{M}$

$y = \dfrac{1}{1-(\Omega/\omega_y)^2}\dfrac{\Omega^2 Me}{2k_y}\sin\Omega t$, $\omega_y^2 = \dfrac{2k_y}{M}$

17-21 $x = \dfrac{1}{1-(\Omega/\omega_x)^2}\dfrac{\Omega^2 Me}{k_x^*}\cos\Omega t$

where $\dfrac{1}{k_x^*} = \dfrac{1}{k} + \dfrac{1}{2k_1}$ $\omega_x^2 = \dfrac{k_x^*}{M}$

Similarly for y.

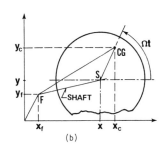
(b)

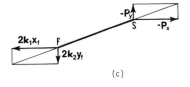
(c)

17-22 Assuming that $f=91.632$ Hz is the true first frequency, the synchronous whirl speed is 577.36 rad/s or 91.89 Hz.

17-23 $A_1 = 0.00515$ in., $A_2 = 0.00455$ in.

17-24 Left bearing, 1915.5 Lb; right bearing, 3039.5 Lb

17-25 A_2 is never zero. $A_1 = 0$ for $\Omega_0 = 1640.2$ rad/s. Disc 2 is then a dynamic absorber for disc 1.

Chapter 18

18-17 $\sigma^2 = 4.8g^2$, $\overline{a}^2 = (1.5g)^2$; $\therefore \overline{a^2} = 7.05g^2$

18-18 $\overline{u^2} = \frac{1}{2}(\mu_1^2 u_{01}^2 + \mu_2^2 u_{02}^2 + \mu_3^2 u_{03}^2)$

in which $u_{0r} = P_{0r}/k$.

18-19 $\overline{\dot{u}^2(t)} = \frac{1}{2\pi}\int_0^\infty \mu^2\Omega^2 S_u\, d\Omega$ and

$$\overline{\ddot{u}^2(t)} = \frac{1}{2\pi}\int_0^\infty \mu^2\Omega^4 S_u\, d\Omega$$

18-20 $\overline{a^2(t)} = \frac{1}{2\pi}\int_0^\infty \frac{\Omega^4}{\omega^4}\mu^2 S_a\, d\Omega$

and similarly the other formula.

18-21 $\overline{u^2(t)}\Big|_{\Omega_l}^{\Omega_u} = \frac{\omega S_u}{8\zeta} - \frac{\omega S_u}{4\pi}\left[x_l\left(\frac{1}{1-x_l^2}+1+\frac{x_l^2}{3}\right)\right.$

$\left. + \frac{1}{x_u}\left(\frac{x_u^2}{x_u^2-1}+1+\frac{1}{3x_u^2}\right)\right]$

in which $x_l = \Omega_l/\omega$ and $x_u = \Omega_u/\omega$.

18-22 RMS$(a) = 9.78g$

18-23 RMS$(a) = 4.924g$

18-24 RMS$(a) = 4.766g$

18-25 $f_n \leqslant 8.828$ Hz by Eq. 18.26.

18-26 193.7 dB

18-27 RMS$(u) = 0.0371$ in., RMS$(N) = 27.83$ Lb

18-28 RMS$(\sigma_b) = 87.45$psi, $G = 2.887 \times 10^{-6}$ psi^2/Hz

18-29 39,955 cycles or 19,267 s

18-30 $\overline{\sigma}_b = 46,850$ psi (mean stress)

18-31 Factor of safety $= 1.721$

18-32 $\overline{u_n^2} = \sum_r \left(\frac{x_{nr}\Gamma_r}{\omega_r^2}\right)^2\left(\frac{S_r\omega_r}{8\zeta_r}\right)$

The notation is the same as in Prob. 18-16.

18-33 $\sigma = 2.5$, $P(2.0, 2.5) = 0.0532$; $P(|u| > 7.5) = 0.0027$, $P(|u| > 10) = 0.00006$

Chapter 19

19-32 Equation 19.2b provides the answer. Consulting Ref. 13 (p. 203), we find that the frequency equation quoted here is the limit as n increases, regardless of how the ends are constrained.

19-33 The following table shows the frequency values computed from Eq. 19.3 and their errors as compared with the numbers in the problem statement.

$\dfrac{\Lambda}{h}$	Λ (in.)	ω (rad/s)	Δ (%)
2	48.5	3945.7	−3.6
5	121.25	1083.3	−0.7
10	242.5	318.69	−0.1

19-34

$\dfrac{\Lambda}{2R}$:	1	2	3	4	5	10
$\dfrac{\omega}{\omega_b}$:	0.3383	0.5837	0.7333	0.8210	0.8739	0.9634

19-35 Equation 19.4b

19-36 Actual plate, $\omega_1 = 100$ Hz, $\omega_2 = 400$ Hz. Corresponding values for the model are $\omega_1' = 94.1$ Hz and $\omega_2' = 188.2$ Hz.

19-37 $\omega_b' = \frac{\sqrt{8}}{L}\left(\frac{GA_s}{m}\right)^{1/2}$

$\therefore \omega_p = \frac{4.443}{L}\left(\frac{Gh_s}{m}\right)^{1/2}$

19-38

$\dfrac{\Lambda}{h}$:	1	2	3
$\dfrac{\omega}{\omega_b}$:	0.3821	0.6373	0.7785

4	5	10	20
0.8558	0.9002	0.9720	0.9928

19-39 $f_1' = 36.636$ Hz (computer program 0.7% off)

$$f_{11}' = 310.24 \text{ Hz} \qquad (9.1\% \text{ off})$$

$$f_{25}' = 620.47 \text{ Hz} \qquad (1.0\% \text{ off})$$

19-40 The lowest frequency $\omega_1 = 1040$ rad/s. A single joint at the center with one-half the column mass in it is sufficient.

19-41 2000 Hz model: 13 plate elements per side, 144 dynamic DOF; 8000 Hz model: 25 plate elements per side, 576 dynamic DOFs.

19-42 The highest significant frequency is 129 Hz and the maximum frequency exhibited by the model is twice that, or 258 Hz. (Refer to Prob. 19-36.) The criterion is grossly inaccurate.

19-43 $N = \dfrac{8}{\pi} P_0 \left(\dfrac{1}{1} - \dfrac{1}{3} + \dfrac{1}{5} - \dfrac{1}{7} + \cdots \right).$

the sum in parentheses is equal to $\pi/4$. (1) With 2 modes, 84.9% of true answer; (2) with 10 modes, 96.8%.

19-44
$$\text{RMS}(\mathfrak{M}) = \frac{1}{8} Z_a mL^2 \left(1 + \frac{1}{3^6} + \frac{1}{5^6} + \cdots \right)^{1/2}$$

The first mode is sufficient.

19-45 The following are the lumped-model results and their deviations from the true (first-mode) quantities of Prob. 15-42:

$$\mathfrak{M} = \mu_1 \frac{a_0 mL^2}{8}, \qquad \Delta = -3.1\%;$$

$$V = \mu_1 \frac{a_0 mL}{4}, \qquad \Delta = -38.3\%$$

19-46

$$\overline{\mathfrak{M}_c^2} = \frac{S}{8\zeta} \left(\frac{mL}{8} \right)^2 \left(\frac{96EI}{m} \right)^{1/2} \left(\frac{1}{1^4} + \frac{1}{3^4} + \frac{1}{5^4} + \cdots \right)$$

Sum of the series is $\pi^4/96$. The first mode gives 98.6% of the answer.

19-47 The bending stress in a plate is proportional to the second derivative of deflection; therefore it takes a superposition of more modes to obtain the same accuracy in stress as it does in deflection. In this example one could expect a reasonable single-mode approximation in u_c, but not in σ_c. Mode 11 apparently corresponds to the highest significant frequency for this model. The modes associated with still larger frequencies will not contribute much to the response.

19-48 The distance $l = 21.0$ in. from the free end is deformed prior to impacting of the stop. This part requires a much denser grid of joints than the remaining, undeformed part.

APPENDIX

I

Resolution of Arbitrary Periodic Force into Harmonic Components

Fig. A1.1 shows an arbitrary periodic force with period T and frequency $\Omega = 2\pi/T$. This force may be represented by an infinite sum of its harmonic components, which is called the Fourier series:

$$P(t) = a_0 + a_1 \cos \Omega t + a_2 \cos 2\Omega t + a_3 \cos 3\Omega t + \cdots$$
$$+ b_1 \sin \Omega t + b_2 \sin 2\Omega t + b_3 \sin 3\Omega t + \cdots$$

or

$$P(t) = a_0 + \sum_{n=1}^{\infty} (a_n \cos n\Omega t + b_n \sin n\Omega t) \quad (A1.1)$$

The unknown coefficients a_n and b_n are found from the following formulas:

$$a_0 = \frac{1}{T} \int_0^T P(t)\, dt \quad (A1.2a)$$

$$a_n = \frac{2}{T} \int_0^T P(t) \cos n\Omega t\, dt \quad (A1.2b)$$

$$b_n = \frac{2}{T} \int_0^T P(t) \sin n\Omega t\, dt \quad (A1.2c)$$

where a_0 is the average value of the forcing function. To make the calculation of the two remaining integrals easier, note that integration is extended over n

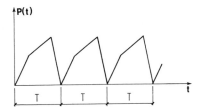

Figure A.1.1 Arbitrary function with period T.

periods of $\cos n\Omega t$ or $\sin n\Omega t$. A single period of these functions is

$$T_n = \frac{1}{n} \frac{2\pi}{\Omega}$$

Some of the integrals encountered quite often are:

$$\int_0^{T_n} \cos n\Omega t\, dt = \int_0^{T_n} \sin n\Omega t\, dt = 0 \quad (A1.3)$$

$$\int_{-T_n/4}^{T_n/4} \cos n\Omega t\, dt = \int_0^{T_n/2} \sin n\Omega t\, dt = \frac{2}{\pi} \frac{T_n}{2} \quad (A1.3a)$$

$$\int_0^{T_n} \cos^2 n\Omega t\, dt = \int_0^{T_n} \sin^2 n\Omega t\, dt = \frac{T_n}{2} \quad (A1.4)$$

$$\int_0^{T_n/2} \cos^2 n\Omega t\, dt = \int_0^{T_n/2} \sin^2 n\Omega t\, dt = \frac{T_n}{4} \quad (A1.4a)$$

$$\int_0^T \cos n\Omega t \sin m\Omega t \, dt = 0 \qquad \text{if} \quad n \neq m \qquad \text{(A1.5)}$$

$$\int_0^{T/2} \cos n\Omega t \sin m\Omega t \, dt = 0,$$

when both m and n are either even or odd

$$= \frac{T}{2\pi}\left(\frac{1}{m+n} + \frac{1}{m-n}\right)$$

$$\text{(A1.5a)}$$

when m is even and n is odd, or vice versa.

$$\int_0^{T/2} \cos m\Omega t \cos n\Omega t \, dt = \int_0^{T/2} \sin m\Omega t \sin n\Omega t \, dt$$

$$= 0 \qquad \text{for} \quad m \neq n \qquad \text{(A1.6)}$$

Any of these equalities valid over one period of the integrand are valid over a multiple of that period. Some of the results may be easily memorized by presenting them as follows.

$$\int_0^a f(t)\, dt = \chi a f_m$$

where a is the length of segment of integration and f_m is the maximum value of the function integrated. The term χ is what might be called a *filling factor*, which shows what portion of the rectangle $a f_m$ is occupied by the area under the curve $f(t)$. (See Fig. A1.2.) It is useful to remember that for a half-wave of the sine curve (integration over half the period) we have

$$\chi = \frac{2}{\pi}$$

per Eq. A1.3a, while for a $\sin^2$ or $\cos^2$

$$\chi = \tfrac{1}{2}$$

according to Eq. A1.4a. These factors are also true for a quarter of the full period because of symmetry. They are also valid regardless whether the time variable t or the length variable x is the argument.

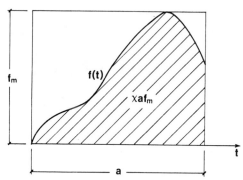

Figure A.1.2   Area under curve $f(t)$:

Although the calculation of coefficients as defined by Eqs. A1.2 involves integration over T, it may sometimes be convenient (e.g., in the case of symmetry) to perform integration over $T/2$ and adjust the result.

APPENDIX
II

Matrix Algebra

A.2.1. Definitions. A matrix of order $m \times n$ is a rectangular array of symbols consisting of m rows and n columns. For example, a 3×5 matrix may be represented by

$$\mathbf{A} = [a_{ij}] = \begin{bmatrix} a_{11} & a_{12} & a_{13} & a_{14} & a_{15} \\ a_{21} & a_{22} & a_{23} & a_{24} & a_{25} \\ a_{31} & a_{32} & a_{33} & a_{34} & a_{35} \end{bmatrix}_{3 \times 5}$$

A row matrix is a matrix consisting of a single row. The 3×5 array shown above may be thought of as composed of three separate *row matrices*, each of order 1×5. Alternatively, it may be treated as five *column matrices*, each of order 3×1. The term *vector* is very frequently used instead of "column matrix." Because of the special role this type of matrix plays in structural analysis, it has a specific designation:

$$\vec{\mathbf{A}} = \{a_{ij}\} = \begin{Bmatrix} a_{11} \\ a_{21} \\ a_{m1} \end{Bmatrix}$$

(Note that the arrow is not commonly used in texts dealing with matrices, but is believed to be useful by this writer.) Since a vector is a sequence of a certain number of quantities, the second index, as shown above, is often not necessary.

The symbols appearing in a matrix are called *entries* or *elements*. When all entries are equal to zero,

the array is called *zero matrix*, $\mathbf{0}$. When the number of rows is the same as the number of columns, $m = n$, the array is called a *square matrix*.

The *main diagonal* of a matrix is the diagonal from the upper left to the lower right corner of the array. A square matrix in which each entry not on the main diagonal is zero is called *diagonal matrix*. Two alternative ways of writing a diagonal matrix are shown below:

$$\mathbf{A} = \begin{bmatrix} a_{11} & 0 & 0 & 0 \\ 0 & a_{22} & 0 & 0 \\ 0 & 0 & a_{33} & 0 \\ 0 & 0 & 0 & a_{44} \end{bmatrix} = \lceil a_{11} \quad a_{22} \quad a_{33} \quad a_{44} \rfloor$$

Very often blank spaces are used instead of zeros. A *unit matrix* is one that has unit entries on the main diagonal and zeros elsewhere. A unit matrix of order 4×4 is denoted by $\mathbf{I}_4$ and is written as

$$\mathbf{I}_4 = \lceil 1 \quad 1 \quad 1 \quad 1 \rfloor$$

When all entries on one side of the main diagonal are zeros, the array is called a *triangular matrix*. When the elements of a square matrix fulfill the condition $a_{ij} = a_{ji}$, the matrix is said to be *symmetric*. When $a_{ij} = -a_{ji}$ and the elements on the main diagonal are zero, the matrix is *skew-symmetric* or *antisymmetric*.

Matrix **C** below is symmetric, while **D** is antisymmetric.

$$C=\begin{bmatrix} 2 & 3 & 4 \\ 3 & 5 & 6 \\ 4 & 6 & 7 \end{bmatrix} \qquad D=\begin{bmatrix} 0 & 1 & 3 \\ -1 & 0 & -5 \\ -3 & 5 & 0 \end{bmatrix}$$

The operation of dividing a larger matrix into smaller arrays is called *partitioning*. When the matrix **E** below is partitioned, as the broken lines indicate, one obtains a set of *submatrices*, E_{11}, E_{12}, E_{21}, and E_{22}.

$$E=\begin{bmatrix} e_{11} & e_{12} & e_{13} & e_{14} \\ e_{21} & e_{22} & e_{23} & e_{24} \\ e_{31} & e_{32} & e_{33} & e_{34} \\ e_{41} & e_{42} & e_{43} & e_{44} \end{bmatrix}=\begin{bmatrix} E_{11} & E_{12} \\ E_{21} & E_{22} \end{bmatrix}$$

A.2.2. Elementary Operations

Equality of matrices: Matrices **A** and **B** are equal if their corresponding entries are equal; that is, $A=B$ if $a_{ij}=b_{ij}$ for every i and j. Of course, the number of rows and columns in both matrices must be the same. The concept of zero matrix is a natural result of that equality, because we can write

$$A-B=0$$

where **0** is the zero matrix of the same order as **A** and **B**.

Addition of matrices is defined as the addition of their respective entries. For example:

$$\begin{bmatrix} a & a & 0 \\ -a & 0 & 2a \end{bmatrix}+\begin{bmatrix} 0 & a & -a \\ -a & 0 & -a \end{bmatrix}$$
$$=\begin{bmatrix} a & 2a & -a \\ -2a & 0 & a \end{bmatrix}$$

Again, the matrices must be of the same order to make the addition possible. Subtraction of matrices amounts to subtracting their respective entries. The sequence of addition and subtraction is immaterial:

$$(A+B)-(C+D)=(A-C)+(B-D)$$

Multiplication of a matrix by a constant results in a matrix whose every entry is multiplied by that constant.

Multiplication of matrices is best explained using an example. Suppose that we want to multiply two matrices in the following order:

$$\underset{2\times 3}{A} \times \underset{3\times 2}{B} = C$$

The number of columns in matrix **A** agrees with the number of rows in **B**. If a pair of matrices satisfies this condition, we say they are *conformable*, that is, they can be multiplied. After checking on conformability, we first write **A** and then write **B** to the right and above **A**:

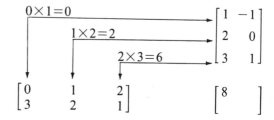

The entries of matrix **C** are located at the intersections of the appropriate rows of **A** and columns of **B**. Instead of circles, which merely indicate the location of elements, the numerical values will now be used. The second scheme shows how to calculate the entry of **C** at the first row and first column:

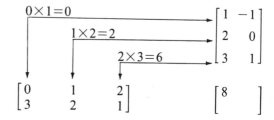

The products of the elements joined by arrows are added, giving the first entry:

$$c_{11}=0+2+6=8$$

Similarly, c_{12} is obtained by combining the first row and second column:

$$c_{12}=0\times(-1)+1\times 0+2\times 1=2$$

Repeating this operation for the remaining two entries, we get

$$C=\begin{bmatrix} 8 & 2 \\ 10 & -2 \end{bmatrix}$$

The most striking difference between the matrix multiplication and that of ordinary algebra is that the commutative property does not in general apply:

$$A \times B \neq B \times A$$

Some of the exceptions involve the unit matrix I and the zero matrix 0. When they are of the same order as A, we have

$$A \times I = I \times A = A$$

$$A \times 0 = 0 \times A = 0$$

Because the sequence of multiplication is so important, we sometimes refer to the product AB as A *postmultiplied* by B or B *premultiplied* by A.

Another interesting feature is that the product of two nonzero matrices may be a zero matrix. In other words, $AB = 0$ does not in general imply that either $A = 0$ or $B = 0$.

Now that the process of multiplication is described, we can write the opening equation of this section more fully:

$$\underset{2\times3}{A} \times \underset{3\times2}{B} = \underset{2\times2}{C}$$

This was written to visualize how to find the order of a product of two matrices—simply by omitting the indices that are joined by arrows. The same procedure applied to a chain multiplication:

$$\underset{1\times3}{A} \times \underset{3\times5}{B} \times \underset{5\times7}{C} = \underset{1\times7}{D}$$

Matrices are in many respects handled like ordinary algebraic quantities. One of the exceptions is that the order of multiplication must not be changed. For example:

$$ABC = (AB)C = A(BC)$$

$$(A+B)C = AC + BC$$

Transposition is performed by interchanging rows with columns. Example:

$$A = \begin{bmatrix} 1 & 2 \\ 3 & 4 \\ 5 & 6 \end{bmatrix} \qquad A^T = \begin{bmatrix} 1 & 3 & 5 \\ 2 & 4 & 6 \end{bmatrix}$$

A^T is called the transpose of A. It is seen that in the geometrical sense transposing a matrix means rotating it by 180° about the main diagonal. When transposition is performed twice, the original matrix is recovered:

$$(A^T)^T = A$$

The transpose of a product of two matrices is equal to the product of the transposed matrices taken in the reverse order:

$$(AB)^T = B^T A^T$$

The latter may be generalized to a product of any number of conformable matrices:

$$(ABC)^T = C^T B^T A^T$$

From the definitions of symmetric or antisymmetric matrices given before, one can deduce that

$$A^T = A \qquad \text{if} \quad A \text{ is symmetric}$$

$$A^T = -A \qquad \text{if} \quad A \text{ is antisymmetric}$$

Vectors. It was stated previously that "vector" is just another term for "column matrix." Equivalently, we may say that a vector is a sequence of symbols or numbers in particular. The idea is similar to what is used in geometry, except that n-dimensional, rather than three-dimensional quantities are involved. The scalar product of two vectors $\vec{X}$ and $\vec{Y}$

$$\vec{X} = \begin{Bmatrix} x_1 \\ x_2 \\ x_3 \\ x_4 \end{Bmatrix} \qquad \vec{Y} = \begin{Bmatrix} y_1 \\ y_2 \\ y_3 \\ y_4 \end{Bmatrix}$$

is defined as $x_1 y_1 + x_2 y_2 + x_3 y_3 + x_4 y_4$. Using matrix notation, we can write the scalar product as

$$\vec{X}^T \vec{Y} = \begin{bmatrix} x_1 & x_2 & x_3 & x_4 \end{bmatrix} \begin{Bmatrix} y_1 \\ y_2 \\ y_3 \\ y_4 \end{Bmatrix}$$

We say that two vectors are perpendicular, or *orthogonal* if their scalar product is zero:

$$\vec{X}^T \vec{Y} = 0$$

Using the concept of a scalar product, a fairly brief definition of a product of two rectangular matrices may be given: if $A \times B = C$, then the entry c_{ij} is the scalar product of the ith row in A and the jth column in B.

A.2.3. Determinants. The *nth order determinant* is a number (or a symbol) that results from certain operations performed on an $n \times n$ matrix. A determinant of a matrix $\mathbf{A}$ is denoted by $|\mathbf{A}|$. If matrix $\mathbf{A}$ is 2×2:

$$\mathbf{A} = \begin{bmatrix} a_{11} & a_{12} \\ a_{21} & a_{22} \end{bmatrix}$$

then

$$|\mathbf{A}| = \begin{vmatrix} a_{11} & a_{12} \\ a_{21} & a_{22} \end{vmatrix} = a_{11}a_{22} - a_{12}a_{21} \quad (A2.1)$$

For a 3×3 matrix, the scheme of calculating a determinant is shown below:

$$|\mathbf{A}| = \begin{array}{c} \\ \begin{vmatrix} a_{11} & a_{12} & a_{13} \\ a_{21} & a_{22} & a_{23} \\ a_{31} & a_{32} & a_{33} \end{vmatrix} \begin{matrix} a_{11} & a_{12} \\ a_{21} & a_{22} \\ a_{31} & a_{32} \end{matrix} \end{array}$$

$$(A2.2)$$

The procedure amounts to repeating the first two columns, forming six products as the diagonal lines indicate and adding the signed products. Example:

$$|\mathbf{A}| = \begin{vmatrix} 0 & 1 & 2 \\ 1 & 1 & 0 \\ 2 & -1 & 1 \end{vmatrix}$$

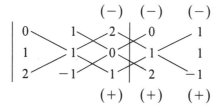

$$|\mathbf{A}| = 0 \times 1 \times 1 + 1 \times 0 \times 2 + 2 \times 1 \times (-1)$$

$$- [2 \times 1 \times 2 + (-1) \times 0 \times 0 + 1 \times 1 \times 1] = -7$$

Equation A2.1 is good for the second-order determinants and Eq. A2.2 for the third. For the fourth and higher orders we must use a more general method. Before this can be done, the concept of a *cofactor* will be developed. The cofactor of an entry a_{ij} of $|\mathbf{A}|$ is a determinant obtained by striking out the row and the column in which a_{ij} is located, multiplied by $(-1)^{i+j}$. (For practical purposes one can restate the latter by saying the multiplier of -1 will appear in the cofactors where $i+j$ is odd.) To calculate a determinant we choose a certain row (or column) and calculate the cofactors of that row (or column). The determinant is the sum of products of entries in that row with their respective cofactors. This example will illustrate the procedure:

$$|\mathbf{A}| = \begin{vmatrix} 0 & 1 & 2 & 3 \\ 3 & 1 & 1 & 0 \\ 1 & 1 & 1 & 1 \\ 1 & 2 & 2 & 1 \end{vmatrix}$$

The row with the largest number of zeros is the natural candidate to choose. The first and the second are equal in that respect. Suppose we select the second row. The multiplier of the first term is -1, because $i+j = 2+1 = 3$. We have

$$|\mathbf{A}| = (-1)(3)\begin{vmatrix} 1 & 2 & 3 \\ 1 & 1 & 1 \\ 2 & 2 & 1 \end{vmatrix} + (+1)(1)\begin{vmatrix} 0 & 2 & 3 \\ 1 & 1 & 1 \\ 1 & 2 & 1 \end{vmatrix}$$

$$+ (-1)(1)\begin{vmatrix} 0 & 1 & 3 \\ 1 & 1 & 1 \\ 1 & 2 & 1 \end{vmatrix} + (+1)(0)\begin{vmatrix} 0 & 1 & 2 \\ 1 & 1 & 1 \\ 1 & 2 & 2 \end{vmatrix}$$

The values of the first three determinants calculated by Eq. A2.2 are, respectively 1, 3, and 3. The result is

$$|\mathbf{A}| = -3$$

This method of calculation brings out the following properties of determinants.

1. If all the elements in any row (or column) are zero, the value of the determinant is zero.
2. If each entry in one row (or column) is multiplied by c, the value of $|\mathbf{A}|$ becomes $c|\mathbf{A}|$.
3. The determinants of a matrix and its transpose are equal.

The other, easily derivable properties are:

4. Interchanging any two rows (or columns) changes the sign of a determinant.
5. If any two rows (or columns) are proportional or identical, the value of the determinant is zero.
6. The value of the determinant remains unchanged if we add a constant multiple of the entires in one row (or column) to the corresponding entries in another row (or column).

This last property is useful in modifying the determinant to obtain more zero entries, which makes the rest of computation much easier.

A.2.4. Matrix Inversion. Division by a matrix is not defined. Instead there is a multiplication by an *inverse* of a matrix. The inverse of **A** is denoted by $\mathbf{A}^{-1}$ and it satisfies the following equalities:

$$\mathbf{A}\mathbf{A}^{-1} = \mathbf{I} = \mathbf{A}^{-1}\mathbf{A}$$

in which **I** is the unit matrix of the same order as **A**. A procedure for calculating the inverse of a given matrix **A** is shown below. First, we construct the *cofactor matrix* of **A**, cof **A**, by replacing the entries with their respective cofactors:

$$\text{if } \mathbf{A} = \begin{bmatrix} 0 & 1 & 2 \\ 1 & 1 & 0 \\ 2 & -1 & 1 \end{bmatrix}$$

$$\text{cof } \mathbf{A} = \begin{bmatrix} \begin{vmatrix} 1 & 0 \\ -1 & 1 \end{vmatrix} & -\begin{vmatrix} 1 & 0 \\ 2 & 1 \end{vmatrix} & \begin{vmatrix} 1 & 1 \\ 2 & -1 \end{vmatrix} \\ -\begin{vmatrix} 1 & 2 \\ -1 & 1 \end{vmatrix} & \begin{vmatrix} 0 & 2 \\ 2 & 1 \end{vmatrix} & -\begin{vmatrix} 0 & 1 \\ 2 & -1 \end{vmatrix} \\ \begin{vmatrix} 1 & 2 \\ 1 & 0 \end{vmatrix} & -\begin{vmatrix} 0 & 2 \\ 1 & 0 \end{vmatrix} & \begin{vmatrix} 0 & 1 \\ 1 & 1 \end{vmatrix} \end{bmatrix}$$

or

$$\text{cof } \mathbf{A} = \begin{bmatrix} 1 & -1 & -3 \\ -3 & -4 & 2 \\ -2 & 2 & -1 \end{bmatrix}$$

Next, the *adjoint* of **A** is formed by transposing the cofactor matrix:

$$\text{adj } \mathbf{A} = (\text{cof } \mathbf{A})^T$$

The inverse of **A** is calculated from the equation

$$\mathbf{A}^{-1} = \frac{\text{adj } \mathbf{A}}{|\mathbf{A}|} \qquad (A2.3)$$

The determinant of this matrix was calculated in the preceding section as -7, so we can write:

$$\mathbf{A}^{-1} = \frac{1}{7}\begin{bmatrix} -1 & 3 & 2 \\ 1 & 4 & -2 \\ 3 & -2 & 1 \end{bmatrix}$$

Multiplying **A** by $\mathbf{A}^{-1}$ we obtain a unit matrix that verifies the correctness of computation. If Eq. A2.3 is applied to the second-order matrix

$$\mathbf{A} = \begin{bmatrix} a_{11} & a_{12} \\ a_{21} & a_{22} \end{bmatrix}$$

we find

$$\mathbf{A}^{-1} = \frac{1}{|\mathbf{A}|}\begin{bmatrix} a_{22} & -a_{12} \\ -a_{21} & a_{11} \end{bmatrix} \qquad (A2.4)$$

The inverse of a diagonal matrix is very easy to remember:

$$\mathbf{A} = \begin{bmatrix} a_{11} & a_{22} & a_{33} & a_{44} \end{bmatrix}$$

$$\mathbf{A}^{-1} = \begin{bmatrix} \dfrac{1}{a_{11}} & \dfrac{1}{a_{22}} & \dfrac{1}{a_{33}} & \dfrac{1}{a_{44}} \end{bmatrix} \qquad (A2.5)$$

Only a square matrix may have an inverse. In addition, the determinant of a matrix must be different from zero. Whenever the determinant is equal to zero, the matrix is called *singular* and no inverse exists. When the inversion process is performed twice, the original matrix is obtained:

$$(\mathbf{A}^{-1})^{-1} = \mathbf{A}$$

The inverse of a product of two matrices is equal to the product of the inverted matrices taken in the reverse order:

$$(\mathbf{A}\mathbf{B})^{-1} = \mathbf{B}^{-1}\mathbf{A}^{-1}$$

The latter applies to a product consisting of any number of nonsingular matrices.

As stated before, the equality $\mathbf{A}\mathbf{B} = \mathbf{0}$ does not in general imply that either **A** or **B** is equal to **0**. Only if **A** is not singular, we have $\mathbf{B} = \mathbf{0}$. Similarly, in the expression $\mathbf{A}\mathbf{B} = \mathbf{A}\mathbf{C}$ we can "cross out" **A** and say $\mathbf{B} = \mathbf{C}$ only if $|\mathbf{A}| \neq 0$.

A.2.5. Simultaneous Equations. Consider a system of linear simultaneous equations:

$$a_{11}x_1 + a_{12}x_2 + a_{13}x_3 = b_1$$
$$a_{21}x_1 + a_{22}x_2 + a_{23}x_3 = b_2 \qquad (A2.6)$$
$$a_{31}x_1 + a_{32}x_2 + a_{33}x_3 = b_3$$

Using matrix notation, it can be presented as:

$$\begin{Bmatrix} a_{11} \\ a_{21} \\ a_{31} \end{Bmatrix}x_1 + \begin{Bmatrix} a_{12} \\ a_{22} \\ a_{32} \end{Bmatrix}x_2 + \begin{Bmatrix} a_{13} \\ a_{23} \\ a_{33} \end{Bmatrix}x_3 = \begin{Bmatrix} b_1 \\ b_2 \\ b_3 \end{Bmatrix}$$

or

$$\begin{bmatrix} a_{11} & a_{12} & a_{13} \\ a_{21} & a_{22} & a_{23} \\ a_{31} & a_{32} & a_{33} \end{bmatrix}\begin{Bmatrix} x_1 \\ x_2 \\ x_3 \end{Bmatrix} = \begin{Bmatrix} b_1 \\ b_2 \\ b_3 \end{Bmatrix}$$

The symbolic form of Eq. A2.6 is

$$A\vec{X} = \vec{B} \tag{A2.7}$$

If the *coefficient matrix* **A** is nonsingular, the *solution vector* $\vec{X}$ is found by premultiplying both sides of Eq. A2.7 by A^{-1}:

$$\vec{X} = A^{-1}\vec{B} \tag{A2.8}$$

Finding the unknowns is reduced to the process of matrix multiplication. This method of approach is used not necessarily to save time, but to achieve greater generality of solution. Equation A2.6 is what we often see in engineering applications. All terms on the right side must not be equal to zero at the same time. If they are, we have a homogeneous instead of a nonhomogeneous system and two possibilities exist:

1. If $|A| \neq 0$, then $x_1 = x_2 = x_3 = 0$.
2. If $|A| = 0$, then a solution exists for which not all x's are zero.

A.2.6. Eigenvalues and Eigenvectors.

In a matrix equation

$$k\vec{x} = \lambda M\vec{x} \tag{A2.9a}$$

k and **M** are known, square matrices, and λ is an undetermined coefficient. The unknown, which has to be calculated, is the vector $\vec{x}$. To put it differently, we want to find some peculiar values of vector $\vec{x}$, which, when premultiplied by **k**, give the same result obtained when $\vec{x}$ is premultiplied by λM. A more suitable form of Eq. A2.9a is

$$(k - \lambda M)\vec{x} = \vec{0} \tag{A2.9b}$$

There is a nonzero solution for $\vec{x}$ only when the condition

$$|k - \lambda M| = 0 \tag{A2.10}$$

is satisfied. The procedure is better explained using an example. Put

$$k = \begin{bmatrix} 4 & 6 & 6 \\ 1 & 3 & 2 \\ -1 & -4 & -3 \end{bmatrix} \qquad M = \begin{bmatrix} 1 & 0 & 0 \\ 0 & 1 & 0 \\ 0 & 0 & 1 \end{bmatrix}$$

Equation A2.10 gives

$$\begin{vmatrix} 4-\lambda & 6 & 6 \\ 1 & 3-\lambda & 2 \\ -1 & -4 & -3-\lambda \end{vmatrix} \begin{vmatrix} 4-\lambda & 6 \\ 1 & 3-\lambda \\ -1 & -4 \end{vmatrix} =$$

$$-\lambda^3 + 4\lambda^2 + \lambda - 4 = 0$$

The three solutions are:

$$\lambda_1 = -1 \qquad \lambda_2 = 1 \qquad \lambda_3 = 4$$

Equation A2.10 is called the *characteristic equation* of the system described by Eq. A2.9a. The values of λ that satisfy the characteristic equation are known as *characteristic values* or *eigenvalues*. If the system is nth order, there are n eigenvalues, although some of them may repeat. Only when λ is equal to one of the eigenvalues can we determine an unknown vector $\vec{x}$. To achieve this, we return to Eq. A2.9b and set $\lambda = \lambda_1 = -1$, obtaining the following system of simultaneous equations:

$$\begin{bmatrix} 5 & 6 & 6 \\ 1 & 4 & 2 \\ -1 & -4 & -2 \end{bmatrix} \begin{Bmatrix} x_1 \\ x_2 \\ x_3 \end{Bmatrix} = \begin{Bmatrix} 0 \\ 0 \\ 0 \end{Bmatrix}$$

We knew before that the matrix of coefficients of these equations would be singular. Now the reason is quite clear, since the second and the third equations are identical except for sign. (However, note that sometimes a matrix may be singular for a less obvious reason.) This means we have only two independent equations, and therefore only two variables can be calculated. One of the three unknowns must be assumed, which we do by putting $x_1 = 1$. The first two equations may now be written as

$$6x_2 + 6x_3 = -5$$

$$4x_2 + 2x_3 = -1$$

from which we find $x_2 = \frac{1}{3}$ and $x_3 = -\frac{7}{6}$. This means our solution vector for $\lambda = \lambda_1$ is determined. For $\lambda = \lambda_2 = 1$ we get the set of equations in which the first and second have proportional coefficients. The first one is discarded and the remaining two are

$$\begin{bmatrix} 1 & 2 & 2 \\ -1 & -4 & -4 \end{bmatrix} \begin{Bmatrix} x_1 \\ x_2 \\ x_3 \end{Bmatrix} = \begin{Bmatrix} 0 \\ 0 \\ 0 \end{Bmatrix}$$

An attempt to set $x_1 = 1$ leads to a contradiction, which indicates that x_1 must have some particular value and may not be assumed. Trying again with $x_3 = 1$ gives $x_1 = 0$ and $x_2 = -1$. Finally, for λ_3 we have $x_1 = 1$, $x_2 = \frac{1}{3}$ and $x_3 = -\frac{1}{3}$.

A solution vector $\vec{x}$ satisfying Eq. A2.9b is also called a *characteristic vector* or an *eigenvector* of the system of equations symbolically represented by Eq. A2.9. When all eigenvectors are arranged in a square matrix, that array is called a *modal matrix* of the system and is denoted by $\mathbf{\Phi}$,

$$\mathbf{\Phi} = \begin{bmatrix} 1 & 0 & 1 \\ \frac{1}{3} & -1 & \frac{1}{3} \\ -\frac{7}{6} & 1 & -\frac{1}{3} \end{bmatrix}$$

$$\quad (1) \quad\quad (2) \quad\quad (3)$$

Note that a solution vector satisfying Eq. A2.9 will still satisfy that equation if its entries are multiplied by a constant. This is because we cannot determine the absolute, but merely the relative values of those entries.

A very useful property of the modal matrix is the ability to turn $\mathbf{k}$ and $\mathbf{M}$ into diagonal matrices. In our example the third-order system was employed and we can get:

$$\mathbf{\Phi}^T \mathbf{k} \mathbf{\Phi} = \begin{bmatrix} \bar{k}_1 & \bar{k}_2 & \bar{k}_3 \end{bmatrix}$$

$$\mathbf{\Phi}^T \mathbf{M} \mathbf{\Phi} = \begin{bmatrix} \bar{M}_1 & \bar{M}_2 & \bar{M}_3 \end{bmatrix} \quad (A2.11)$$

If we now perform a change of variables in Eq. A2.9b by putting $\vec{x} = \mathbf{\Phi}\vec{s}$ and then premultiply the equation by $\mathbf{\Phi}^T$, we obtain a system of three independent equations. The condition that the new variables s_1, s_2, and s_3 are not equal to zero, gives us

$$\bar{k}_1 - \lambda_1 \bar{M}_1 = 0$$

$$\bar{k}_2 - \lambda_2 \bar{M}_2 = 0$$

$$\bar{k}_3 - \lambda_3 \bar{M}_3 = 0$$

APPENDIX

III

Laplace Transforms

A.3.1. Some Special Functions. One of the useful functions often mentioned in connection with the main topic of this appendix is the *unit step function*, defined as

$$H(t) = 0 \quad \text{for} \quad t \leqslant 0$$

$$H(t) = 1 \quad \text{for} \quad t > 0$$

Figure A3.1 shows $H(t)$ as well as $H(t-t_0)$, which is the same as $H(t)$, but shifted to the right of the t-axis by the distance t_0.

With the aid of the step functions some expressions are written more compactly. Take as an example the function shown in Fig. A3.2. A conventional way of defining it would be

$$f(t) = \begin{cases} 0 & \text{for} \quad t < 1 \\ 2t & \text{for} \quad 1 < t < 3 \\ 0 & \text{for} \quad t > 3 \end{cases}$$

Alternatively, we may form the expression

$$H(t-1) - H(t-3)$$

which when multiplied by any function preserves only the portion of it that is between $x = 1$ and $x = 3$. We may thus write

$$f(t) = [H(t-1) - H(t-3)](2t)$$

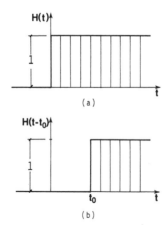

(a)

(b)

Figure A.3.1 Unit step function.

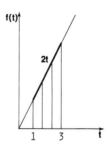

Figure A.3.2 Function defined over a finite segment.

292

Figure A3.3 shows a rectangle with the center at $t=t_0$, and the area equal to 1. If we begin to reduce the length a, the base shrinks and the height grows, while the area remains unchanged. At the limit of this process we obtain what is called a *unit delta function* $\delta(t-t_0)$, which is more formally defined as

$$\delta(t-t_0)=\begin{cases} 0 & \text{for} \quad t<t_0 \\ \infty & \text{for} \quad t=t_0 \\ 0 & \text{for} \quad t>t_0 \end{cases}$$

and $\quad \int_{-\infty}^{+\infty} \delta(t-t_0)\,dt=1$

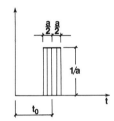

Figure A.3.3 Derivation of delta function.

Although the concept of a rectangle with an infinitely short base and infinitely large height may seem somewhat abstract, some of the quantities used in mechanics have a similar appearance. One example is the impulse that may be defined as a product of a force and the time of its duration. For the sake of convenience we often treat an impulse of a finite magnitude as an infinitely large force acting for an infinitely short time. Another example is obtained when we change the variable of time to that of length. The function $\delta(x-x_0)$ will then represent a unit concentrated force at $x=x_0$, which is a limiting case of a distributed load in the vicinity of that point. An interesting property of a unit delta function is given by the following:

$$\int_0^\infty \delta(t-t_0)\cdot f(t)\,dt=f(t_0) \qquad \text{(A3.1a)}$$

and the particular case when the delta function is at the origin, $t_0=0$,

$$\int_0^\infty \delta(t)\cdot f(t)\,dt=f(0) \qquad \text{(A3.1b)}$$

A.3.2. Definitions and Terminology. Let $f(t)$ be a function defined for $t>0$. The expression

$$\mathcal{L}[f(t)]=\int_0^\infty e^{-st}f(t)\,dt=\phi(s) \qquad \text{(A3.2)}$$

is called the *Laplace transform* of $f(t_0)$. To calculate the transform of a function from this definition, we multiply it by e^{-st}, integrate from $t=0$ to $t=b$ and then find the limit for b going to infinity. The t is a variable of integration that vanishes once the integration has been performed, and we are left with a function of a new variable s. It is usual to briefly call $\phi(s)$ a *transform* and to write the transformation symbolically as

$$f(t)\rightarrow\phi(s)$$

The first nine entries in the table at the end of this appendix show the transforms of some basic functions of t. All those functions could have been shown multiplied by $H(t)$, since they are to be defined only for $t>0$.

When using Laplace transformation in dynamics, an alternative notation is often used, namely

$$f(t)\rightarrow\bar{f}(s)$$

so that fewer symbols are needed for a particular set of operations.

To return from the domain of the variable s to that of t we use an inverse transformation, which we write as

$$\phi(s)\rightarrow f(t) \qquad \text{or} \qquad \phi(s)=\mathcal{L}^{-1}[f(t)]$$

In the great majority of practical cases the Laplace transform and its inverse are unique. This means that going from $f(t)$ to $\phi(s)$ and performing the inverse transformation gives the same original function $f(t)$. Exceptions are possible when a function has zero values at its discontinuity points. Unless a restriction is made, we always assume that an inverse transformation is possible.

Let there be two functions such that

$$f_1(t)\rightarrow\phi_1(s) \qquad \text{and} \qquad f_2(t)\rightarrow\phi_2(s)$$

The definition (Eq. A3.2), clearly shows the linearity property of the Laplace transformation,

$$c_1 f_1(t)+c_2 f_2(t)\rightarrow c_1\phi_1(s)+c_2\phi_2(s)$$

A.3.3. Shifting Properties. First we write

$$e^{-at}f(t)\rightarrow\phi(s+a) \qquad \text{(A3.3)}$$

This theorem is interpreted as follows. If $f(t)$ transforms into $\phi(s)$, then to obtain the transform of

$e^{-at}f(t)$, we must merely replace s by $(s+a)$ in the transform. Example:

$$t \to \frac{1}{s^2} \quad \text{then} \quad e^{-at}t \to \frac{1}{(s+a)^2}$$

The counterpart of Eq. A3.3 when the roles of the variables are interchanged is

$$H(t-a) \cdot f(t-a) \to e^{-as}\phi(s) \qquad (A3.4)$$

The original function on the left side of the transformation has the same shape as $f(t)$, but is translated to the right by a along the t-axis and cut off for $t \leqslant a$. The theorem says that to find the transform of such a function, one determines $\phi(s)$ corresponding to $f(t)$ and then multiplies $\phi(s)$ by e^{-as}. Example: if

$$f_1(t) = \begin{cases} t-3 & \text{for} \quad t>3 \\ 0 & \text{for} \quad t<3 \end{cases}$$

we can write it as $f_1(t)=H(t-3)\cdot(t-3)$. By Eq. A3.4:

$$H(t-3)\cdot(t-3) \to e^{-3s}\frac{1}{s^2} \quad \text{because} \quad t \to \frac{1}{s^2}$$

The following equation, which pertains to a function that is treated as cut off, but not shifted, is also found useful:

$$H(t-a)\cdot f(t) \to e^{-as}\mathcal{L}[f(t+a)] \qquad (A3.5)$$

To obtain the transform, we first calculate $f(t+a)$ and transform it into a function of s. The result is multiplied by e^{-as}. As an example consider the same $f_1(t)$ defined before,

$$f_1(t+a)=(t+3)-3=t$$

this means

$$\mathcal{L}[f_1(t+a)]=\frac{1}{s^2} \quad \text{and} \quad H(t-3)\cdot f_1(t) \to e^{-3s}\frac{1}{s^2}$$

Equation A3.5 is more general than Eq. A3.4.

A.3.4. Multiplication, Division, Differentiation, and Integration.

The formulas given below show how the operations named in the title are related when a transformation is performed. If

$$f(t) \to \phi(s)$$

then

$$tf(t) \to -\phi'(s) \qquad (A3.6)$$

$$f'(t) \to s\phi(s)-f(0^+) \qquad (A3.7)$$

$$\frac{f(t)}{t} \to \int_s^\infty \phi(s)\,ds \qquad (A3.8)$$

$$\int_0^t f(t)\,dt \to \frac{1}{s}\phi(s) \qquad (A3.9)$$

The symbol $f(0^+)$ in Eq. A3.7 means, in most cases, the value of f at $t=0$. If there is a discontinuity at

Table of Selected Laplace Transforms

No.	$f(t)$	$\phi(s)$
1	$H(t)$	$\dfrac{1}{s}$
2	t	$\dfrac{1}{s^2}$
3	t^n	$\dfrac{n!}{s^{(n+1)}}$
4	$\sin at$	$\dfrac{a}{s^2+a^2}$
5	$\cos at$	$\dfrac{s}{s^2+a^2}$
6	$\sinh at$	$\dfrac{a}{s^2-a^2}$
7	$\cosh at$	$\dfrac{s}{s^2-a^2}$
8	e^{at}	$\dfrac{1}{s-a}$
9	$\delta(t)$	1
10	$e^{-at}f(t)$	$\phi(s+a)$
11	$H(t-a)\cdot f(t-a)$	$e^{-as}\phi(s)$
12	$H(t-a)\cdot f(t)$	$e^{-as}\mathcal{L}[f(t+a)]$
13	$tf(t)$	$-\phi'(s)$
14	$f'(t)$	$s\phi(s)-f(0^+)$
15	$\dfrac{f(t)}{t}$	$\displaystyle\int_s^\infty \phi(s)\,ds$
16	$\displaystyle\int_0^t f(t)\,dt$	$\dfrac{1}{s}\phi(s)$
17	$f''(t)$	$s^2\phi(s)-sf(0^+)-f'(0^+)$
18	$\dfrac{\cos at-\cos bt}{b^2-a^2}$	$\dfrac{s}{(s^2+a^2)(s^2+b^2)}$
19	$\dfrac{(1/a)\sin at-(1/b)\sin bt}{b^2-a^2}$	$\dfrac{1}{(s^2+a^2)(s^2+b^2)}$

$t=0$, but the limit of f exists, as t tends to zero on the positive side, $f(0^+)$, stands for the value of that limit. In Eq. A3.8 the limit of $f(t)/t$ must exist as t tends to zero. Examples:

$$e^{at} \to \frac{1}{s-a} \qquad \therefore\, te^{at} \to \frac{1}{(s-a)^2} \qquad \text{(for Eq. A3.6)}$$

$$t \to \frac{1}{s^2} \qquad \therefore\, 1 \to \frac{1}{s} - 0 \qquad \text{(for Eq. A3.7)}$$

This result has the same meaning as the first transform in the table. It is of course understood that we are not including the negative side of the t-axis.

$$t^2 \to \frac{2}{s^3} \qquad \therefore\, t = \frac{t^2}{t} \to \int_s^\infty \frac{2}{s^3}\,ds = \frac{1}{s^2}$$

$$\text{(for Eq. A3.8)}$$

$$\cos t \to \frac{s}{s^2+1} \qquad \therefore\, \int_0^t \cos t\,dt \to \frac{1}{s}\frac{s}{s^2+1} = \frac{1}{s^2+1}$$

$$\text{(for Eq. A3.9)}$$

The examples used were rather trivial and their purpose was only to familiarize the reader with handling of that group of transformations.

References

(1) C. M. Harris and C. E. Crede. *Shock and Vibration Handbook*. 2nd ed., McGraw-Hill, 1976.

(2) S. P. Timoshenko. *Vibration Problems in Engineering*. Wiley, 1974.

(3) S. Timoshenko and J. N. Goodier. *Theory of Elasticity*. 2nd ed., McGraw-Hill, 1951.

(4) M. R. Spiegel. *Theory and Problems of Laplace Transforms*. Schaum Publishing Co., 1965.

(5) R. B. Green. "Gyroscopic Effects on the Critical Speeds of Flexible Rotors," Pap No. 48-SA-3, *American Society of Mechanical Engineers Journal of Applied Mechanics*, December 1948.

(6) J. S. Przemieniecki. *Theory of Matrix Structural Analysis*. McGraw-Hill, 1968.

(7) R. J. Roark. *Formulas for Stress and Strain*. 4th ed., McGraw-Hill, 1965.

(8) R. J. Roark and W. C. Young. *Formulas for Stress and Strain*. 5th ed., McGraw-Hill, 1975.

(9) M. R. Spiegel. *Mathematical Handbook*. Schaum Publishing Co., 1968.

(10) S. H. Crandall. *Random Vibrations in Mechanical Systems*. Academic Press, 1963.

(11) W. Nowacki. *Dynamika Budowli*. Arkady, Warsaw, 1961.

(12) Uniform Building Code. Conference of Building Officials, Pasadena, California, 1970.

(13) W. C. Hurty and M. F. Rubinstein. *Dynamics of Structures*. Prentice-Hall, 1964.

(14) N. Willems and W. M. Lucas. *Structural Analysis for Engineers*. McGraw-Hill, 1978.

(15) S. Timoshenko. *Theory of Plates and Shells*. 2nd ed., McGraw-Hill, 1959.

(16) Second ASCE Specialty Conference on Structural Design of Nuclear Plant Facilities. American Society of Civil Engineers, New York, 1975.

(17) R. D. Blevins. *Formulas for Natural Frequency and Mode Shape*. Van Nostrand Reinhold, 1979.

(18) F. E. Richart, Jr., et al. *Vibrations of Soils and Foundations*. Prentice-Hall, 1970.

(19) R. W. Clough and J. Penzien. *Dynamics of Structures*. McGraw-Hill, 1975.